CONCISE HUMAN PHYSIOLOGY

Concise Human Physiology

EDITED BY

M.Y. SUKKAR
MB, BS (Khartoum), PhD (Edinburgh)
Professor of Physiology
Faculty of Medicine and Allied Sciences
King Abdulaziz University
Jeddah, Saudi Arabia

H.A. EL-MUNSHID
MB, BS (Khartoum), PhD (Newcastle), MD (Lund)
Professor of Physiology
College of Medicine and Medical Sciences
King Faisal University
Dammam, Saudi Arabia

M.S.M. ARDAWI
BMedSc, MA, DPhil (Oxford), FACB (USA)
Professor of Clinical Biochemistry
Department of Clinical Biochemistry
Faculty of Medicine and Allied Sciences
King Abdulaziz University
Jeddah, Saudi Arabia

OXFORD

BLACKWELL SCIENTIFIC PUBLICATIONS

LONDON EDINBURGH BOSTON

MELBOURNE PARIS BERLIN VIENNA

© 1993 by
Blackwell Scientific Publications
Editorial Offices:
Osney Mead, Oxford OX2 0EL
25 John Street, London WC1N 2BL
23 Ainslie Place, Edinburgh EH3 6AJ
238 Main Street, Cambridge
    Massachusetts 02142, USA
54 University Street, Carlton
    Victoria 3053, Australia

Other Editorial Offices:
Librairie Arnette SA
2, rue Casimir-Delavigne
75006 Paris
France

Blackwell Wissenschafts-Verlag
Meinekestrasse 4
D-1000 Berlin 15
Germany

Blackwell MZV
Feldgasse 13
A-1238 Wien
Austria

First published 1993

Set by Setrite Typesetters Ltd, Hong Kong
Printed and bound in Great Britain
at The University Press, Cambridge

DISTRIBUTORS

Marston Book Services Ltd
PO Box 87
Oxford OX2 0DT
(Orders: Tel. 0865 791155
        Fax: 0865 791927
        Telex: 837515)

USA
  Blackwell Scientific Publications, Inc.
  238 Main Street
  Cambridge, MA 02142
  (Orders: Tel: 800 759−6102
          617 876−7000)

Canada
  Times Mirror Professional Publishing, Ltd
  130 Flaska Drive
  Markam, Ontario L6G 1B8
  (Orders: Tel: 800 268−4178
          416 470−6739)

Australia
  Blackwell Scientific Publications Pty Ltd
  54 University Street
  Carlton, Victoria 3053
  (Orders: Tel: 03 347−5552)

A catalogue record for this book is
available from the British Library.

ISBN 0−632−03383−5

Library of Congress
Cataloging in Publication Data

Concise human physiology/edited by
M.Y. Sukkar, H.A. E.-Munshid, M.S.M. Ardawi.
    p.   cm.
  Includes bibliographical references and index.
  ISBN 0−632−03383−5
  1. Human physiology.
  2. Human physiology — Study and teaching —
  Developing countries.
  I. Sukkar. Mohamed Yousif.   II. El-Munshid, H.A.
  III. Ardawi, M.S.M.
  [DNLM: 1. Physiology.
  QT 104 C744]
  QP34.5.C66   1992
  612 − dc20
  DNLM/DLC

Contents

Contributors

A.M. ABDUL GADER MB, BS (Khartoum), PhD (Dundee). *Professor of Physiology, Faculty of Medicine, King Saud University, Riyadh, Kingdom of Saudi Arabia. (Formerly Associate Professor, Faculty of Medicine, University of Khartoum, Khartoum, Sudan.)*

T.S. AHMED MB, BS (Khartoum), PhD (Bristol). *Associate Professor, Department of Physiology, Faculty of Medicine, University of Khartoum, Khartoum, Sudan, at present on secondment to King Saud University, Riyadh, Kingdom of Saudi Arabia.*

A.G. ALZUBEIR MB, BS (Khartoum), DTPH (London), DTM&H (London), MRCP (Ireland). *Assistant Professor in Community Medicine, King Faisal University, Dammam, Kingdom of Saudi Arabia.*

M.S.M. ARDAWI BMedSc (Oxford), MA (Oxford), DPhil (Oxford), FACB (USA). *Professor of Clinical Biochemistry, Department of Clinical Biochemistry, Faculty of Medicine and Allied Sciences, King Abdulaziz University, Jeddah, Kingdom of Saudi Arabia.*

M.A. BALLAL MB, BS (Khartoum), PhD (Nottingham). *Associate Professor, Department of Physiology, Faculty of Medicine, University of Khartoum, Khartoum, Sudan, at present on secondment to King Saud University, Riyadh, Kingdom of Saudi Arabia.*

H.A. EL-MUNSHID MB, BS (Khartoum), PhD (Newcastle), MD (Lund). *Professor of Physiology, College of Medicine and Medical Sciences, King Faisal University, Dammam, Kingdom of Saudi Arabia. (Formerly Associate Professor, Faculty of Medicine, University of Khartoum, Khartoum, Sudan.)*

H.A. HAMAD-ELNEIL MB, BS (Khartoum), PhD (Cambridge). *Director, Emergency Relief Operations, WHO, Geneva. (Formerly Professor of Physiology, Dar es Salaam University, Dar es Salaam, Tanzania, and Associate Professor, Faculty of Medicine, University of Khartoum, Khartoum, Sudan.)*

M.O. HASSAN MB, BS (Khartoum), PhD (Glasgow). *Professor of Physiology, College of Medicine, Sultan Qaboos University, Muscat, P.O. Box 32485, AL-Khod, Sultanale of Oman. (Formerly Chairman, Department of Physiology, Faculty of Medicine, King Abdulaziz University, Jeddah, Kingdom of Saudi Arabia.)*

M.M.T. KORDY MB, ChB (Cairo), PhD (London). *Professor of Physiology, Faculty of Medicine, King Saud University, Riyadh, Kingdom of Saudi Arabia.*

N.A. MAHMOUD MB, BS (Khartoum), PhD (Edinburgh). *Professor of Physiology, Faculty of Medicine, University of Khartoum, Khartoum, Sudan. (Formerly Professor of Physiology, Faculty of Medicine, University of Kuwait, Kuwait.)*

H.A. NASRAT MB, BCh (Cairo), MRCOG (London), FRCS (Edinburgh). *Consultant, King Abdulaziz University Hospital, and Assistant Professor, Department of Obstetrics and Gynaecology, Faculty of Medicine, King Abdulaziz University, Jeddah, Kingdom of Saudi Arabia.*

Y. OWNALLA YOUNIS MB, BS (Khartoum), DTPH, MRCPsych (London). *Psychiatrist, AlGimi Hospital, AlAin, United Arab Emirates. (Formerly Associate Professor, Faculty of Medicine, University of Khartoum, Khartoum, Sudan.)*

M.Y. SUKKAR MB, BS (Khartoum), PhD (Edinburgh). *Professor of Physiology, Department of Physiology, Faculty of Medicine and Allied Sciences, King Abdulaziz University, Jeddah, Kingdom of Saudi Arabia. (Formerly Professor of Physiology, Faculty of Medicine, University of Khartoum, Khartoum, Sudan.)*

M. ZAIN UL ABEDIN MB, ChB (Punjab), MPhil (Karachi), PhD (Manchester). *Professor of Physiology, Department of Physiology, College of Medicine and Medical Sciences, Arab Gulf University, Manama, Bahrain.*

A.M. ZIADA MB, BS (Khartoum), PhD (Birmingham). *Assistant Professor, Department of Physiology, Faculty of Medicine, King Abdulaziz University, Jeddah, Kingdom of Saudi Arabia.*

Preface

During the last 15 years more colleges of medicine and allied health sciences have been established in the developing countries than ever before. The special needs of students in these institutions are inadequately met by available textbooks, which have been written primarily for medical students in Europe and the United States — more often than not for a wide readership which includes science and postgraduate students. This textbook aims to address the special needs of students in developing countries, in both content and presentation. For this purpose, there are four main features which make this book different from all others.

The first of these features is the language. Most of the colleges in developing countries use the English language as the medium of instruction. Most countries have started to use their own languages as the medium of instruction in pre-university education. For this reason, most students find it difficult to use existing textbooks of basic medical sciences.

This textbook sets out to use as simple language as possible through the use of simple vocabulary and sentence structure. It is anticipated that the student will be helped in three ways: (i) comprehension of the basic concepts of human physiology; (ii) economic utilization of the learner's time; and (iii) a smooth introduction to the use of the scientific textbook by avoiding the frustration that goes with the absence of the above two.

The second feature addresses the selection of content. During the last two decades the medical curriculum has witnessed two apparently contradictory phenomena. On the one hand, there has been a tremendous growth in the amount of knowledge in basic medical sciences. This has resulted in overcrowding of the curricula and the growth in size of almost all known textbooks. On the other hand, recent developments in curriculum design have brought into focus the need for integration of basic sciences with the clinical disciplines. One way to achieve this is the introduction of clinical sciences in the early years of the programme. Thus, pathology, microbiology and pharmacology have crept back into the medical curriculum. Consequently, the time allocated for learning basic medical sciences has decreased.

Under these conditions selection of content presents a challenge to both the student and the teacher. This textbook aims to present the basic principles and concepts of human physiology required for the study and practice of medicine and to present these within a reasonable length. Two ways have been adopted to solve this problem: (i) the contents have been selected to give adequate scope and depth relevant to the study of medicine; and (ii) experimental evidence and descriptions of animal experiments have been cut down to a minimum. Except where such experiments help to reinforce the scientific method, such content has been totally eliminated. It is realized, however, that some students will find the time to consult other reference works. A bibliography drawn upon in the selection of contents is provided. Some special titles are also recommended for further reading at the end of each chapter.

The third feature of this book is its orientation to the tropical environment with its characteristic health problems. The populations to be served by the graduates largely consist of women and children. The health problems of pregnant women and small children present the majority of cases seen in primary health care. Thus, separate chapters on nutrition and on pregnancy and perinatal physiology have been included. Most of the graduates will practise medicine as general practitioners in a tropical and largely rural environment. The special physiological mechanisms invoked in the adaptation of the individual and the community to the tropical environment have been given special attention. Nutrition is presented as a basic science which calls upon the biochemical and physiological mechanisms concerned with the utilization of nutrients. The orientation of nutritional information focuses on the use of locally available foods and the prevention of malnutrition. This also serves as a unique opportunity to introduce basic community medicine concepts at an early stage of the curriculum.

The fourth feature is the selection of authors, all of whom have had significant experience in teaching medicine and allied health science students in developing countries. Each has contributed to areas in which they have special interest and published research. The focus of editing was to harmonize the treatment and style of presentation in the various chapters.

We would like to express our appreciation of the co-operation of all the contributors without whose support this work would not have been possible. We would like to thank all those who have helped in the making of this textbook, especially the members of the Department of Physiology, Faculty of Medicine, King Abdulaziz University, for their continuous support and advice during the preparation of the manuscript. Special thanks go to Mrs Kay Gari for excellent secretarial help and for the preparation of the final manuscript. The credit for preparation of the illustrative material goes to Mr Mannan Azmi of the King Fahad Medical Research Centre, King Abdulaziz University. Last, but not least, we would like to thank our families for their patience in bearing with us during the lengthy period of the preparation of this textbook.

We hope the product will be found useful by students and teachers. The editors will be delighted to receive comments and suggestions from experienced teachers and from students for the purpose of improving future editions.

M.Y.S.
H.A.E.
M.S.M.A.

1: An Introduction to Human Physiology

Objectives

On completion of the study of this chapter, the student should be able to:

1 Appreciate the relationship between man and his environment so as to identify the place of human physiology in the study of medicine as a health and social science.

2 Comprehend the functional organization of the organ systems of the human body in order to develop a holistic approach to the study of medicine.

3 Understand the concepts of the internal environment and its constancy as an important condition for normal health.

4 Acquire basic information on body composition so as to interpret changes which occur in disease processes, such as dehydration, oedema, hyper- and hypo-osmolality and acid−base disturbances.

5 Describe the major components and properties of control systems so as to explain the physiological mechanisms encountered in the study of the organ systems.

Man and the environment

The human individual can be viewed as an integral part of the community and the environment in which he lives. Human physiology is a science which analyses the body functions and their interaction with the environment. In so doing, we seek to understand the mechanisms whereby these interactions take place. The study of human physiology is an essential prerequisite to the practice of preventive and curative medicine.

Man, as one of the living organisms, is influenced by the environment in which he lives. This external environment has its physical, chemical and biological as well as its psychosocial components. The human individual interacts with the climate of his environment, thrives on the air, water and foods available in it, and becomes victim to the pathogenic organisms and other hazards prevalent in his surroundings. Man is also influenced by the customs, beliefs and values which shape the psychosocial structure of his community. The major environmental factors with which the human organism interacts can be summarized as oxygen, water, food, physical environment, social environment, micro-organisms and parasites (Fig. 1.1). These components are the main life support. The human body uses or adjusts to these components throughout life. Most of the concerns of medicine attempt to maintain the balance of this interaction in favour of man. Social and preventive medicine attempt to maintain the equilibrium of the individual, the family and the community with their environment. Curative medicine helps to readjust the equilibrium of the human body when it is disturbed by disease.

The human physiologist is concerned with the following aspects of this system:

1 The way in which the human body is functionally equipped to interact with its environment.

2 The mechanisms by which the individual adjusts to the environment.

3 The maintenance of internal balance, normal function and, therefore, the health of the body and mind.

1

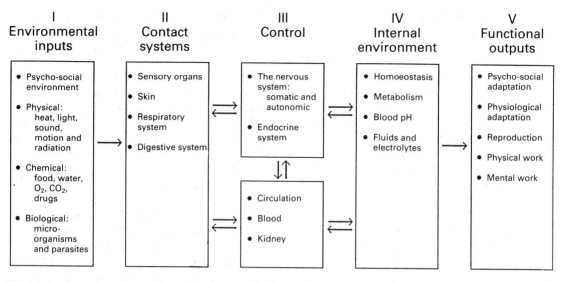

Fig. 1.1 A schematic representation of the functional relationships of body systems.

4 The functions of physical and mental work and productivity.

Through the following discussion, each of the above statements will be analysed and explained.

In the first place, the human body interacts with its external environment through the systems listed in the second column in Fig. 1.1. These have been labelled the contact systems because they come into direct contact with the external environment. The respiratory system is in contact with the atmospheric air. The digestive system is in contact with food and drink. The skin comes into close contact with and senses the objects, materials and temperature of the environment. The sensory organs help us to smell, taste, hear or see the various components of our surroundings. The contact systems differ from all the other functional divisions of the human body in this respect, i.e. direct contact with the outside world. It is not surprising, therefore, that many of the defence and protection mechanisms are found in these systems. It is also not surprising that hypersensitivity (allergy) occurs mainly in the skin (allergic skin conditions), respiratory system (asthma and allergic rhinitis) and the gastrointestinal tract (allergic diarrhoea and colitis).

The internal environment and homoeostasis

To adjust to the changes which take place in the external environment, many detectors and sensory mechanisms are involved. These send messages to the systems responsible for control of body functions, i.e. the nervous system and the endocrine glands. These organs will initiate and control the processes by which the body adjusts to changes in the external environment. They also respond to changes which take place inside the body such as the ingestion and absorption of food and water, the change in composition and pH of the body fluids and the accumulation of products of metabolism. These two systems can be called the organs of control (Fig. 1.1). Together or singly, they provide the mechanisms of communication and adjustment of the internal environment. This is the second concern of human physiology. These adjustments in body functions are required to meet external variations such as heat and cold and internal changes due to various activities such as muscular exercise or absorption of nutrients from the gastrointestinal tract. The nervous system and the endocrine glands provide a means of communication between the various parts of the body. The nervous system works by means of electrical signals which travel in the nerves, and the endocrine system by means of hormones which are transported in the blood to the target organs.

THE INTERNAL ENVIRONMENT

The human body, like other complex living organisms, consists of organs and tissues which, in turn, are made up of cells. The smallest unit of life in this complex structure is the cell. The cells of the body live in a fluid environment. It is this tissue fluid (i.e. interstitial fluid) which represents the internal environment. The function of all the organ systems of the body is to ensure that the physical and chemical characteristics of the tissue fluid remain constant within narrow limits. Thus, the temperature, the pH and the concentration of chemical components of the internal environment are precisely regulated by physiological processes collectively described as *homoeostatic mechanisms*. This is the third and major concern of human physiologists. The concept of 'internal environment' was introduced by the French physiologist Claude Bernard in 1857. He stated that all the life processes have but one goal, that of maintaining the constancy of the internal milieu (environment), and that the fixity of this internal milieu was the necessary condition for free existence.

THE BODY FLUIDS

The body of a 65 kg man contains about 40 litres of water. The distribution of this water can be measured by special methods which are discussed later in this chapter. It has been found that about 25 litres are found inside the cells (intracellular) and 15 litres are outside the cells (extracellular) (Fig. 1.2).

The extracellular fluid can be further subdivided into a part which is found within the vascular system of the circulation (intravascular) and one which lies between the cells (interstitial). The interstitial fluid volume includes lymph, which cannot be measured separately.

There is a third component of extracellular fluid, which is referred to as transcellular fluid. This is fluid secreted by epithelial linings, e.g. cerebrospinal fluid (CSF), fluid in transit in the gastrointestinal tract and urinary system, and fluid in potential spaces such as the joint cavities and in the pleural and peritoneal spaces. The total amount of transcellular fluid is normally small. However, large volumes can be formed in disease states, such as pleural effusion and collection of fluid in the peritoneal cavity (ascites). Large losses of gastrointestinal secretions occur in diarrhoea and vomiting, which lead to dehydration and electrolyte disturbances.

HOMOEOSTASIS

The constancy of the internal environment (homoeostasis) is achieved in spite of changes in the surroundings. It is also possible to maintain the steady state during changes of body function, e.g. increased muscular activity, which generates heat, CO_2 and other metabolic products. Homoeostatic mechanisms, therefore, deal with

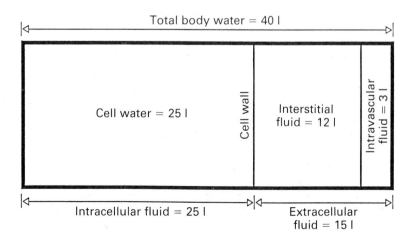

Fig. 1.2 The volumes and distribution of body fluids.

all the automatic reactions which take place in the body in response to changes in the external or internal environments and which tend to correct all deviations from normal.

The term 'homoeostasis' was coined by the American physiologist, Walter B. Cannon, in 1929. He emphasized that independent organisms are open systems which can be influenced by changes in their surroundings. These organisms require some automatic adjustments which keep the variation in the system within narrow limits. These regulatory adjustments were called 'homoeostasis.' Examples of homoeostatic control will be given later on in this section.

The internal equilibrium is assisted by the blood and circulatory systems. Efficient distribution of heat and nourishment is achieved through the bloodstream and lymphatic drainage. These systems also remove the waste products of metabolism and normal wear and tear of tissues. These are carried to the kidneys, where waste products such as urea and acids (H^+) are excreted in the urine, and to the lungs, where CO_2 is removed. The kidneys have many important functions which maintain the chemical composition of the internal environment. It should be noted, however, that this maintenance is achieved under the close control of the nervous system and the endocrine glands.

Energy supplies, protein, minerals and vitamins maintain normal nutrition and balanced metabolism. Equilibrium exists among intake, storage and metabolism. Metabolic processes are also homoeostatically controlled by the endocrine system (see column IV, Fig. 1.1).

It appears from the above discussion that the organ systems listed in column III of Fig. 1.1 serve the all-important tasks of maintaining internal equilibrium. This is essential both for normal physiological functions for manual and mental work and for healthy psychosocial interactions in the community. Once these adjustments are made, then the person can lead a productive life. The body can then be said to be healthy, i.e. in a state of physical and mental well-being. The United Nations World Health Organization (WHO) gives an extended definition of health to include the environment in which people live. Health is defined as 'a state of complete physical, social and mental well-being' and not merely as the absence of disease or infirmity.

The physiologist is also concerned with the functions required for physical and mental productivity (column V, Fig. 1.1). Skeletal muscles undergo physical and chemical changes in order to do work. They use chemical energy and are therefore able to convert it into mechanical work.

The functions of the reproductive system ensure the continuity of the species, and in man they also provide one of the most cherished joys of life. The physiologist is concerned with the neurological and endocrine regulatory mechanisms of reproduction. One of the most important physiological outputs is growth and development. On this process, which spans childhood and adolescence, every other accomplishment in adult life will depend. Growth is concerned with attainment of the physical dimensions of the skeletal and organ systems. Development, however, is more concerned with functional attainments, particularly related to neurological and mental capacities.

It goes without saying that the maintenance of adequate supplies and a stable internal environment allows the higher functions of the brain (such as learning and creative thinking) to work. For these functions to flourish, a conducive socio-economic environment is required, which further emphasizes the wisdom of the above definition of health.

Adaptation

The functional outputs (Fig. 1.1, column V) also refer to the adaptation of the individual to the physical and social environments. Unlike homoeostasis, adaptation is concerned with the long-term adjustments required for a 'normal' life. Adaptation to the environment can be considered successful when the relationship of the human individual or community is in a state of balance and productivity. The practice of community medicine (social and preventive medicine) aims to enhance this process through environmental health, health education and other promotional and supportive health measures. Thus, immunization of children against prevalent diseases makes them more adapted to survive in health in a particular locality.

Examples of physical adaptation are seen in the body shape, body composition and skin colour of people who live under different climatic conditions. Cold climates favour deposition of fat, a short stature and white skin (e.g. the Eskimos of North Canada). Hot climates favour a lean and tall body frame as well as the formation of protective pigment in the skin (e.g. the Dinka of southern Sudan and the Masai of Kenya). The density of skin pigment varies directly with the amount of sunshine and protects the body from a harmful excess of ultraviolet rays. The body shape tends to be related to the ability to lose and conserve heat. A more or less spherical shape has a small surface area proportional to mass and favours heat conservation. On the other hand, a long cylindrical body shape has a greater surface area proportional to mass and therefore favours heat loss.

There are also physiological adaptations which maintain body functions under extremes of environmental conditions. For instance, people living at high altitudes make more red blood cells and haemoglobin so as to be able to live and work at a lower oxygen pressure. Similarly, people living in cold environments increase their thyroid function so as to generate more heat in the body through a higher rate of metabolism.

Psychosocial adaptations take place under a variety of conditions. People who migrate to different countries will acquire behaviour and habits more acceptable to the new society. Individuals who make new partnerships in business or in friendship have to adjust to the new partner by understanding, developing new concepts or giving up old ones. Marriage is an excellent example where each of the partners adjusts to the attitudes and behaviour of the other so as to build a home and to raise a family. The discipline of social and preventive medicine considers the individual as a member of a family and the family as a unit of the community. These subdivisions have profound effects in the determination of health and the patterns of prevailing disease.

Measures taken by the physician, whether preventive or curative, will have to take into account the family and the social environment as well as the various physical, chemical and biological factors referred to in column I of Fig. 1.1.

In conclusion, it seems that the doctor is dealing with a complex situation. Disease is basically an expression of the failure of body mechanisms to maintain homoeostasis in the face of challenge, which may take many forms, e.g. hereditary defects, infection, injury, neoplasms and degenerative conditions. In simple terms, the doctor tries to help the body maintain the internal environment through curative medicine practices. He also tries to help the process of physical and social adaptation in his practice of community medicine and primary health care activities, which endeavour to take health to the community rather than wait for the individual to become sick and to seek treatment.

The composition of the human body

BODY WATER

The human body is made mainly of water, which constitutes about 60% of body-weight in the adult. However, the amount of body water varies with age. The new-born have about 82% of body-weight as water at birth and the elderly about 52% (Table 1.1). Loss of body water is a common cause of death in children who suffer from dehydration in developing countries.

The amount of water in the body is affected by the quantity of body fat. Lean subjects have a higher percentage of water than those with more body fat. This is because adipose tissue contains far less water than muscle, skin and other soft tissues (Table 1.2). Women have more fat than men; therefore, as expected, they have a lower percentage of body water than men (Table 1.1).

Table 1.1 Variation of total body water with age and sex expressed as % of body-weight

	Male	Female
At birth	82%	82%
Children and adolescents	70%	70%
18–20 years	59%	57%
20–40 years	61%	51%
40–60 years	55%	47%
Over 60 years	52%	46%

Table 1.2 Water content of plasma and various tissues

Plasma	90–92%
Muscle	72–78%
Bone	45%
Adipose tissue	13%
Enamel	5%

PROTEIN

Protein is the second largest component in the human body. Protein is found in the structure of all tissues, but by far the largest amount is found in skeletal muscle.

FAT

The next largest component in lean individuals is fat. This is found in the form of triacylglycerol in adipose tissue, which constitutes the main energy store of the body (Table 1.3). These stores are found around abdominal viscera and in subcutaneous tissues. Normally, men have less fat than women. In women, fat distribution represents one of the secondary sex characteristics. It gives the female body its typical feminine configuration. Fats, in the form of phospholipids, are found in the structure of cell membranes. An appreciable part of the structure of the central nervous system is fat.

MINERALS

The minerals are present in the human body in relatively small quantities with the exception of calcium. The total body calcium is about 1.2 kg in a young adult, the bulk of which is found in the bones. Minute, but functionally significant, amounts of calcium are found in body fluids.

In adults the total amount of iron is about 3 to 4 g, 70% of which is the haem parts of haemoglobin and myoglobin. These proteins are associated with carriage of oxygen. Iron is also found in iron stores, in transport forms and in iron-containing enzymes (e.g. cytochrome oxidase).

Other minerals and electrolytes are found in the body fluids in minute concentrations which are closely regulated to maintain the composition of the internal environment. Their concentrations in the intracellular fluid are different from those in extracellular fluid (Table 1.4). This difference depends on the permeability of the cell membrane and on active transport at the living cell membrane. Passive equilibrium is also attained by means of physical factors such as concentration gradients and electrical charge distribution (see Chapter 4).

The body fluids

MEASUREMENT OF BODY FLUIDS

The volumes of the various compartments of the body fluids can be measured using well-known techniques employing the *dilution principle*. If a known quantity of a substance is dissolved in an unknown volume of fluid, we can calculate the volume from the following relationship:

Table 1.3 Body composition of a 70 kg man

	Weight (kg)	% Body-weight	% Lean mass
Water	42.0	60%	75%
Protein	12.6	18%	20%
Fat	12.6	18%	—
Minerals	2.8	4%	5%
Total	70.0	100%	100%

Table 1.4 The concentrations of major cations and anions in body fluids (mmol/l)

	Extracellular fluid		Intracellular fluid
	Plasma	Interstitial	
Cations (mmol/l)			
Na^+	145	140	10
K^+	4	4	145
Ca^{2+}	5	5	1
Mg^{2+}	2	2	40
Total (+)	156	151	196
Anions (mmol/l)			
Cl^-	105	110	3
HCO_3^-	28	31	10
Protein	17	4	45
HPO_4^{2-}	6	6	138
Total (−)	156	151	196

$$V = \frac{Q}{C}$$

where V = volume; Q = quantity of substance; and C = final concentration of substance after even mixing in the fluid to be measured.

The substance chosen is injected into the bloodstream. The choice of substance for measurement of a certain compartment of body fluid, e.g. extracellular fluid (ECF), should take into consideration that the substance should distribute itself only in the ECF and should not enter the cells. Other prerequisites are that the substance should distribute itself uniformly in all parts of the fluid to be measured and that it should not be rapidly excreted or metabolized. The method for measurement of its concentrations should not provide practical difficulties. Obviously, if a substance is to be administered into a patient, it should not be toxic.

If a small amount of the substance is metabolized or excreted during the period of mixing, this amount must be subtracted from the total amount administered before dividing by the final concentration.

Total body water

The total body water is measured by injection of a substance which will diffuse freely in all the fluid compartments. Examples of suitable substances are deuterium oxide (D_2O), tritiated water (^3H_2O), antipyrine and amino antipyrine. A sample of plasma is obtained for measurement of the concentration (C) of the substance used. The total body water can be obtained by calculating Q/C. Slightly different results are obtained depending on the characteristics of the substance used.

Extracellular fluid

Two types of substances are used for measurement of extracellular fluid: (i) saccharides, e.g. inulin, sucrose, raffinose and mannitol (an alcohol derived from mannose); and (ii) diffusible ions, e.g. sulphate, thiosulphate, thiocyanate, chloride, bromide and sodium. The saccharides give lower volumes because they enter some compartments rather slowly (especially CSF). They are also excreted in the urine. The ions, however, tend to give higher results because small amounts penetrate the intracellular compartment. Recently, radio-isotopes, especially (^{35}S) sulphate, have been used with good results. Because of errors due to using different substances, it is advisable to qualify the result, e.g. sulphate space, thiocyanate space, etc.

The volume of intracellular fluid can be determined by subtraction of ECF from total body water.

Plasma volume

In the case of plasma volume the dilution method uses a dye which binds itself to plasma albumin, such as Evans' blue. Radioactively labelled albumin using iodine (^{131}I) has also been used. These substrates give a measurement of plasma volume. The blood volume can be calculated from the measured volume of plasma. First, the haematocrit or packed-cell volume (PCV) is measured by centrifugation of a sample of blood. The *haematocrit* is the proportion of blood volume occupied by red blood cells expressed as a percentage. The blood volume is given by the formula:

$$\text{Blood volume} = \text{Plasma volume} \times \frac{100}{100 - \text{Haematocrit}}$$

Thus, if the haematocrit is 45%, the proportion of plasma would be 55% of the blood volume:

$$\text{Blood volume} = \text{Plasma volume} \times \frac{100}{55}$$

If red blood cells labelled with radioactive chromium (^{51}Cr) are used, then a direct value of the blood volume is obtained.

When plasma volume is subtracted from the volume of ECF, the volume of interstitial fluid including lymph is obtained.

MECHANISMS OF FLUID EXCHANGE

The intravascular fluid consists of the blood plasma. The interstitial fluid is similar to plasma in composition except for its low content of protein. These two parts of the extracellular fluid are in constant exchange. The circulation renews the tissue fluid and maintains its constancy by supplying materials needed by the cells and removing those produced by cellular metabolism. Exchange of fluid between the blood capillaries and the interstitial compartment takes place continuously. It is estimated that about 150 litres of fluid diffuse out of the blood capillaries and re-enter again every hour.

The forces which govern this exchange of fluid between the plasma and the interstitial fluid are: (i) *hydrostatic pressure* due to fluid tension within the circulation (the blood pressure); and (ii) the *colloid osmotic pressure* of the plasma proteins, also known as oncotic pressure (see Fig. 1.3).

The hydrostatic pressure is much higher in the capillaries than in the tissue spaces and tends to drive fluid out of the capillaries by filtration. The colloid osmotic pressure is much higher in the blood plasma than in the interstitial fluid because the plasma proteins are retained inside the capillaries. This tends to draw water into the capillaries by osmosis. These two forces act in opposite directions. While the colloid osmotic pressure is uniform throughout the capillary length, the hydrostatic pressure falls from the arteriolar to the venular end. At the arteriolar end of the capillary the hydrostatic pressure is greater than the colloid osmotic pressure and therefore fluid tends to pass to the outside of the capillaries. At the venous end of the capillaries, however, the hydrostatic pressure is less than the colloid osmotic pressure and therefore water is reabsorbed into the capillaries at this end (Fig. 1.3). It is important to note that, in the lungs, capillary hydrostatic pressure is uniformly less than the colloid osmotic pressure and therefore there is no filtration. The alveoli are kept dry. In the glomerular capillaries of the kidney, the hydrostatic pressure is much higher than the colloid osmotic pressure. Therefore, filtration takes place without reabsorption.

Starling's hypothesis states that there exists an equilibrium between the colloid osmotic pressure and the hydrostatic pressure so that about 90% of the fluid which passes into the interstitial space goes back into the circulation at the venular end of the capillaries. The remaining 10% is drained via the lymph vessels which start blindly in the interstitial space.

The colloid osmotic pressure in the interstitial fluid is about 5 mmHg. The interstitial fluid hydrostatic pressure is usually of the order of a few mmHg negative or positive and varies in the different body tissues. These two forces tend to cancel each other out and therefore play a minor role under normal conditions. Small amounts of albumin normally pass into the interstitial spaces but these are constantly removed via the lymphatics and returned to the bloodstream together with some of the interstitial fluid formed.

APPLIED PHYSIOLOGY OF TISSUE FLUID EXCHANGE

Oedema
Oedema is defined as the abnormal collection of fluid in the interstitial spaces. There are three main causes which occur frequently in clinical practice:
1 *Increased capillary hydrostatic pressure.* This is usually due to obstruction of blood in the venous system. Failure of the right ventricle (as in congestive heart failure) results in a rise in the central venous pressure in the great veins near the heart and consequently a rise in capillary hydrostatic pressure. If this becomes close to or exceeds the value of the colloid osmotic pressure (normally about 25 mmHg), then the return of fluid by osmosis will be opposed and accumulation of fluid in the interstitial space results.
2 *A decrease in the plasma colloid osmotic pressure.* This is either due to excessive loss of

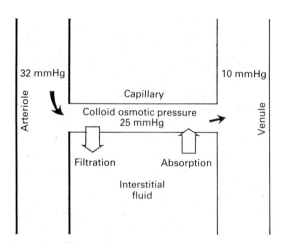

Fig. 1.3 Diagrammatic representation of the forces responsible for exchange of fluids across the blood capillary membrane. The hydrostatic pressure and the colloid osmotic pressure of the interstitial fluid have been omitted for simplification.

albumin in the urine (as in kidney disease) or deficiency of albumin production (as in chronic liver disease or malnutrition). The fall in plasma albumin results in failure of reabsorption of water and therefore accumulation of fluid in the interstitial space.

3 *Obstruction of lymph vessels.* This results in accumulation of albumin in the interstitial space, which is normally cleared away by the lymphatics. Thus, a significant rise in the colloid osmotic pressure of the interstitial fluid results in oedema.

The forces which govern fluid exchange across the capillary wall are delicately balanced. This allows a continuous flow of fluid through the tissue spaces so that equilibrium with the blood plasma is ensured and transport of nutrients and removal of waste products are achieved.

Control mechanisms

The functions of the different systems of the body are closely monitored and adjusted to meet the demands of the body, which change with various activities, e.g. intake of food and water, running or doing heavy manual work. The adjustments also respond to changes in the external environment to which the body is exposed at any given time, e.g. sitting in a hot room or getting out into the cold night air.

CONTROL THEORY

When a system is described as controlled, this means that its variables are kept within a prescribed range. Control theory deals with the analysis of the whole system so that the relationships between inputs and outputs are defined. In engineering science, these relationships are expressed in mathematical terms. In biological systems, mathematical models have been used in analysing and studying biological phenomena.

Example:

$$\text{Arterial blood-pressure} = \frac{\text{Cardiac output (flow)}}{\text{Resistance to flow}}$$

In the control of arterial blood-pressure, variables such as the cardiac output and arteriolar resistance are manipulated so as to maintain arterial blood-pressure within the normal range. The purpose of

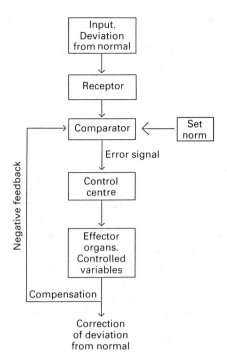

Fig. 1.4 The major components of a negative feedback control system in simple form.

control is to maintain the variables of a system within the range of reference or standard values.

CONTROL SYSTEMS

A control system can be analysed into various components for the purpose of understanding the relationships between them. The model in Fig. 1.4 shows a simplified control system divided into four major components:

1 *A receptor*, or sensing device, sensitive to the changes which take place in the internal environment. These receptors are usually specialized and will respond to change in one variable, e.g. blood-pressure, oxygen, body temperature, osmotic pressure, etc.

2 *A controller.* The controller in biological systems is usually in the central nervous system, where receptor signals are received, analysed and compared with reference values or standards. After comparison, an error signal is sent out. The appropriate response is given by the controlled system.

3 *The controlled system.* This represents one or

more organ systems which are responsible for the functions involved. In the above example the heart and blood-vessels represent the controlled systems responsible for the function of regulation of blood-pressure.

4 *Regulated variable.* The blood-pressure is the regulated variable in this case. However, it can be further analysed into subsidiary variables which may affect the blood-pressure such as the heart rate and the strength of cardiac muscle contraction (together these two determine the output of the heart). Another subsidiary variable is the diameter of blood-vessels, which determines the resistance to blood flow to the tissues. The regulated variable is monitored by the sensing device. Therefore, a *feedback mechanism* exists between the output and the input.

Examples of homoeostatic mechanisms
The homoeostatic mechanisms described below

are everyday life examples selected to illustrate the control systems outlined above:

1 *Core body temperature* in man is maintained within a fraction of a degree on both sides of 37 °C. If this temperature starts to rise, even by a small fraction of a degree, homoeostatic regulation is immediately called for and the error in body temperature is immediately corrected.

The temperature rise is measured by means of sensitive receptors situated in the skin and at the base of the brain (in the hypothalamus). These receptors send signals to the heat-regulatory centres in the hypothalamus. Here the receptor signals are compared with the set normal temperature (standard). The output of the control centres, e.g. those controlling skin blood-vessels and the sweat glands is adjusted accordingly. The responses promote heat loss, for example: (i) more blood flows to the skin so that heat can be lost from the surface of the body by radiation; and

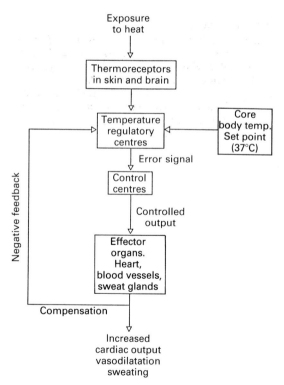

Fig. 1.5 The control of body temperature used to illustrate the interrelations of the components of a control system.

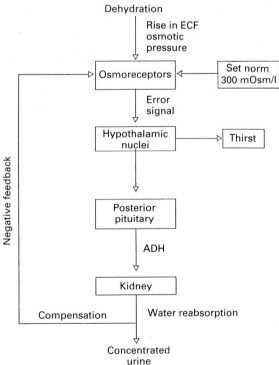

Fig. 1.6 The control of water excretion through a control system involving the osmotic pressure of extracellular fluid (ECF).

(ii) production of sweat, which evaporates from the surface of the skin producing a cooling effect (Fig. 1.5).

The temperature receptors continue to monitor the compensation of body temperature and keep the hypothalamus informed through a feedback mechanism so that excessive heat loss does not occur, thus completing the control system of receptor, comparator, control centre, controlled variables and feedback.

2 *Control of water intake and water loss.* The body water is maintained within narrow limits. Changes in the content of body water occur when we go without drinking as in fasting or being lost in the desert. Changes can also occur when we become dehydrated through loss of water in sweating, vomiting and diarrhoea. Conversely, a large intake of fluid by mouth or through intravenous infusion can produce changes in body water in the opposite direction.

These physiological and pathological changes call for immediate responses. In the case of dehydration, there is loss of water resulting in a rise of the osmotic pressure of the extracellular fluid (ECF). The osmotic pressure of plasma is 290 mosm/litre. Any change is detected by receptors in the hypothalamus of the brain. These receptors are called *osmoreceptors*. Signals are sent in the form of electric impulses to other parts of the hypothalamus (thirst centre) and we feel thirsty and therefore seek water to drink (Fig. 1.6). At the same time signals are sent to the posterior pituitary gland just below the hypothalamus. A hormone is released in the circulation, which is called antidiuretic hormone (ADH) because it acts on the kidney to produce a low volume of more concentrated urine. The ultimate effect is to retain body water by decreasing water loss in the urine. Thus, the osmotic pressure of body water can be corrected both by drinking and by decreasing the volume of urine. Conversely, if excessive water is taken, the osmotic pressure of extracellular fluid falls and secretion of antidiuretic hormone is stopped. Large volumes of dilute urine are passed and body water is adjusted.

In conclusion, this approach to the analysis of homoeostatic control provides a model suitable for studying and describing all physiological regulatory mechanisms. A large number of such mechanisms will be encountered in the study of body functions in the following chapters.

Further reading

1 Emslie-Smith, D., Paterson, C.R., Scratcherd, T. & Read, N.W. (eds) (1988) The body fluids. In *BDS Textbook of Human Physiology*, 11th edn, pp. 16–22. Churchill Livingstone, Edinburgh.
2 Guyton, A.C. (1991) Functional organisation of the human body and control of the internal environment. In *Textbook of Medical Physiology*, 8th edn, pp. 1–8. Saunders, Philadelphia.

2: The Social Environment

Introduction

The environment includes all the surroundings in which man lives: earth, air, water and people. These environmental factors have long been known to influence the state of health of man. But, while the influence of physical factors, for example, climate, housing and water pollution, have been extensively studied, the enquiry into the role of the social environment is relatively new, perhaps because many of these social influences are not easily measured.

The social environment encompasses systems of relationships, education, occupations, customs and beliefs. These factors contribute towards shaping people's concept of health and their attitudes towards disease and may influence the management of patients. Some of them are desirable and should be encouraged for promotion of good health, while others are undesirable and must be discouraged and replaced with good ones. In some rural communities in developing countries, children below 2 years of age are not given eggs, though available, because it is believed that this will delay speech development. Thus, they are denied a good source of protein. Also, they may be given supplementary feeding only in the form of sweetened fluids, again in the belief that these fluids are nutritious for the child. As a

| **Objectives** |
| On completion of the study of this chapter, the student should be able to: |
| **1** Appreciate the concept of the social environment as one of the contributing factors to health and disease. |
| **2** Describe the organization of a social environment and the interactions which occur within a given social system so as to recognize the relation of these interactions to physiological processes. |
| **3** Recognize the interactions between social and economic activities and the physical environment so as to develop an awareness of the effects of these interactions on community health. |
| **4** Understand the basics of human behaviour and its development so as to appreciate its relevance to psychosomatic disease and mental illness. |

result of these feeding practices, there is a high prevalence of protein–calorie malnutrition.

Society is a system of relationships between individuals regulated by law and customs. Taking society as a living organism and using the medical model in discussion, we may draw an analogy between society and the human body. We may talk about the anatomy of society, its physiology and its behaviour in the same manner as we talk about the anatomy of the human body, its physiology and its behaviour. Thus, we consider the social organization, e.g. family, class and tribes, as the anatomy of society, the social institutions, e.g. education, economy and religion, as its physiology and social dynamics, e.g. migration, procreation and recreation, as its behaviour.

The organizational system in a society, its type of institutions and the dynamics operating in it very much influence the state of health of its people. We may now look at each of these in some detail.

Social organization

People are dependent on each other for their survival and progress. Life is spent largely in interaction. So the milieu (environment) in which one lives, family, household or tribe, very much influences one's style of life.

Every person is born in a family, whatever the concept of a family one has. The family is the smallest social unit. It consists of a man and a woman and their children. This is what is called the nuclear family. In some places this nuclear family may live together with grandparents and other close blood relatives, in the same house or in a complex of adjacent houses, forming what is called an extended family. Family grouping is affected by the degree of urbanization of the society. Life conditions in urban areas force people to live in small households and so nuclear family living arrangements are favoured. In rural areas, conditions permit and favour the extended family pattern of living. In many developing countries, where the major part of the population lives in rural areas, the extended family is the commoner pattern of family grouping. There is a great interrelationship between the health of an individual and that of his family. The family functions as a source of nutrition, protection and recreation, providing the individual with the necessities for growth and maturation. The loving atmosphere within the family is equally important for the individual's mental well-being.

It is within the family that the person develops his first habits of hygiene and forms his attitudes towards health issues through, for example, toilet training and treatment-seeking behaviour. At the same time, the individual's behaviour and health conditions affect the general state of health of the family. For example, when the head of a nuclear family with a modest income falls ill and is confined to bed for some time, the limited resources of the family will be strained. The wife will have to meet the continuous demands of children, nurse her husband and keep the house tidy. She will be under stress. The relations within the family will also be stressed. Physical and emotional states of family members are then affected. On the other hand, in an extended family this state of affairs is unlikely to happen because the pooled resources of the family assure every member of the family of a continuous material and emotional support in health and in sickness. Also, within the extended family the deleterious effects of weakening of the nuclear family ties,

through death of a parent, divorce or separation, are greatly modified.

SOCIAL INSTITUTIONS

Every society has its set of rules and codes of behaviour which regulates the life of the individuals. Conformity to these regulations by individuals is important for the smooth running of social life. Deviation beyond the limit or pattern of behaviour accepted by the community leads to ill effects in both the deviant individual and the social group to which he belongs. The situation is very similar to the physiological functioning of the human body. Every functional unit in the body must be working normally and in harmony with other units to give the individual his state of well-being. Deviation of the functional unit from the normal pattern results in a state of ill health.

The system of education, economy and religion and the political situation are the institutions and systems through which society functions. They give the society its shape. Their form and content very much influence people's behaviour and life style including their health condition.

It is well known that illiteracy, poverty and disease go in a circle; each one increases the chances of the others. People's concern with their health and the proper use of health facilities available for them are directly associated with their educational standard. Education usually increases the chance of being in a good social position, which, in turn, provides one with the opportunity to enjoy a fair standard of health through better nutrition and better living arrangements. Conversely, illiteracy usually reduces the chances of attaining a high social position and consequently limits the chances of getting adequate nutrition and accommodation, and hence one is deprived of a good state of health.

Religion, as a social institution, influences people's attitudes and behaviour in various life situations including health. Islam, for example, prohibits the consumption of alcohol and deplores suicide. At the same time, it calls for supporting fellow Muslims, especially during crisis and personal tragedies. Thus, while creating an unfavourable public attitude towards such practices as the drinking of alcohol, it provides a protective mechanism during periods of weakness that push one to indulge in alcohol or to commit suicide. Perhaps this is one important reason why the incidence of alcoholism and suicide in Muslim societies is not as high as in other places. This leads some practising doctors to advocate the use of religion in health education. It is argued that it could be very effective in deterring people from wrong health practices and encourage them to adopt correct ones. Surely in Christianity, as in Islam, there are teachings pertaining to health practices which can be used for health promotion.

Social dynamics

There are physical characteristics and psychological traits which are inherited. Some of these characteristics predispose to disease, and the degree of blood relation between parents determines the probability with which the offspring may or may not get them. Their manifestation may be enhanced by marriage between close relatives and masked by marrying outside the family. For example, phenylketonuria is a condition which results from the absence of a specific enzyme (L-phenylalanine dehydroxylase). Consequently, phenylalanine is not converted into tyrosine and, together with other related compounds, it accumulates in the blood, causing mental retardation. The condition is recessively inherited—that is to say, there is a chance of one in four that the offspring will get it if both parents have the gene responsible for it. In random marriages, the likelihood of rare identical genes appearing together in a child is small, and it is increased when blood relatives marry each other since a proportion of their genes are necessarily identical. Thus, the system of marriage preferred in a certain community influences the health endowment of the family.

Population movement within a country and between countries is influenced by various economical, social and environmental factors. Development of small industries in cities brings an influx into the city from rural areas; inadequate rainfall and drought in some areas push people into more prosperous places. The harvest season brings in migrant labourers from other regions. This influx will create a strain on resources and facilities. Hospitals will have to serve more people; schools must increase their capacity; public transport and the market-place will have an extra load; housing, water supply and electricity will have to be shared by more people than they were originally intended for. This state of affairs may lead to lowering of the standard of services or they may become completely inadequate. As a result, there may be insanitary conditions that favour the spread of infection and disease. Also, the newcomers may introduce diseases which were not there before.

At the same time, the newcomers will meet difficulties of adaptation to their new environment and circumstances of living. These movements disrupt family bonds and social relations. Great expectations may not come true and the new environment may be hostile. Frustration then occurs and leads to psychological disturbances among the vulnerable migrants.

On the other hand, as occurred in some historical events, movement of a certain group into a new land may enrich the indigenous culture and the new blend may be for the better.

TRADITIONS AND CUSTOMS

Every society has its traditions and customs which influence behaviour. Traditions and customs are not just a peculiarity of developing societies. For example, infant feeding is a universal practice. Every mother has to feed her baby one way or another. The way the baby is fed is influenced by the particular culture. A rural African mother would feed her baby from her breast any time the baby needs it and will not be ashamed to do that in public. On the other hand, her urban European counterpart may feed her baby from the bottle rather than give him the breast and she would probably be ashamed to breast-feed in public. These are different customs influencing behaviour. However, modern medical practice supports the rural African mother in her customs, for it considers breast-feeding on demand as healthier to both baby and mother.

There are among the traditional practices and customs of societies some which directly or indirectly affect the health of the people. And in these societies there are men and women whose advice is sought in matters pertaining to health and who are entrusted with the treatment of sick people. The methods of these traditional healers are based on the traditional concepts of health and disease in these particular societies. While some of these practices have a negative effect on health, others are of proved value and are in line with modern scientific medical practices. Learned groups, such as the World Health Organization, have shown great interest in the study of these traditional practices affecting the health of people, with the aim of ferreting out the good ones to promote them and perhaps integrate them with the existing modern medical service whenever possible.

The effect of customs and traditions on physical health may be illustrated by various examples of feeding habits and food restrictions during pregnancy, confinement, infancy and convalescence or rituals done at various periods of one's life as in circumcision, marriage and childbirth. What may perhaps be overlooked is their effect on mental health.

It is known that the personality of an individual is affected by experiences in early childhood — that is to say, that child-rearing practices influence the development of personality and affect future mental well-being. Heredity, no doubt, has its effect on personality. But, given that this is the same in a group of people, the influence of the social environment becomes manifest. Studies comparing children reared within their own families with others of the same attributes who were reared in institutions away from their biological families have shown that children deprived of their natural homes are more prone to develop maladjusted behaviour and to have more psychological disturbances. Love and care given by the mother are found to be important for one to be able to develop meaningful emotional bonds with others. The practice of carrying the baby on the back or at the side of the mother, as in some African and Middle Eastern countries, and having the baby sleep beside the mother in the same bed is therefore not to be forsaken for nurseries or otherwise disturbed.

Interaction between the social and the physical environments

The discussion of the social and physical environments separately is arbitrary. They are two sides of the same coin. They are directly affected by each other and the state of one closely follows that of the other. For example, water is a main constituent of man's physical environment. An adequate, safe water supply is a prerequisite for life. Its quantity and quality determine how people live and interact. If it is rare and difficult to get, people will be scattered and in constant search for it. The work they do, the type of houses they build, their play, their assets and their whole style of life will be determined by this kind of water supply. They will live as nomads, their life full of change and continuous adaptation. On the other hand, if safe water is available in abundance, people will be settled. With settlement, industries grow, agriculture develops, more people come together and a whole new complex system of social relationships is created.

Man's relation with his physical environment is a series of adaptations and he is not always in harmony with it. Sometimes his attempts to make his life easy and pleasurable bring disastrous results, e.g. when new irrigation schemes are constructed without the necessary precautions to prevent water-borne diseases. Notable examples are the irrigation schemes in Africa. Thousands of miles of unprotected canals traverse the land to irrigate the plantations. People are in frequent contact with the canal. They drink from it, bathe in it, wash their clothes in it and, of course, wade through it when irrigating their fields. The canal water is a suitable habitat for the freshwater snails which act as intermediate hosts for bilharzia. It is enough for a few persons infected with bilharzia to pollute the canals with urine or faeces and the cycle of bilharzia is completed. This is why irrigation schemes are endemically infested with bilharzia. Studies done on work efficiency of infected persons in the Gezira (Sudan) and in sugar plantations in Tanzania have shown that the work capacity of farmers is greatly reduced by the disease. Thus, the main purpose of the agricultural schemes, which is to augment the national economy, is defeated.

Brain, behaviour and environment

The effects of the external environment (social and/or physical) on human behaviour are regulated by the brain. On receiving stimuli from the environment, the brain processes, modifies and organizes them into relevant behaviour. Biological factors such as heredity and somatic diseases, as well as psychosocial events such as past experience and current life events, determine how the brain regulates behaviour. The brain is a very complex communication centre. A healthy mental state depends on the proper physiochemical functioning of its 10 billion neurones. The cells concerned mainly with behaviour are organized into three main organizational systems. Innermost is the reticular formation system, surrounded by the limbic system, and outermost is the cerebral cortex. The crude, more nonspecific components of behaviour are dealt with at the reticular formation level and the limbic system exercises a higher emotional control. The cerebral cortex is where the fine and precise activity is carried out. It is the highest level of integration where multiple stimuli are transformed into highly complex responses.

As a great deal of social behaviour is learned and as learning requires that the individual should be alert, attentive and able to retain information, then proper functioning of these brain processes is essential for normal behaviour.

Normal behaviour depends on normal inputs from the environment into the brain and on normal functioning of the brain systems. Abnormal inputs or stimuli will be expected to result in abnormal output or behaviour. Likewise, malfunctioning of the brain systems will result in distorted behaviour even if the environmental input is normal. Normality in behaviour is relative, depending on the particular social concepts and local beliefs. By normal behaviour we mean behaviour which is in keeping with one's educational, maturation and social background.

The processes that work in the brain to enable one to respond with appropriate behaviour to changes in the environment are collectively called cognitive processes, namely perception, learning and remembering. The work of these processes is interrelated and any discussion of them separately is arbitrary. Also, individuals differ one from another in terms of their behavioural responses to environmental stimuli. Among the factors that influence this difference are one's intelligence and one's personality.

By way of illustration we shall briefly discuss some of these factors, namely memory, intelligence and personality, and refer to some practical aspects.

MEMORY

Memory involves perception of information, registration, retention and then recall later on. It is a function of certain areas of the brain. Hence, remembering depends on the integrity of these areas and forgetfulness occurs when they are damaged. For example, destructive lesions of the mammillary bodies, as in Wernicke's encephalopathy or temporal lobectomy (especially bilateral), lead to memory disturbances.

The emotional state of the individual and the environmental conditions at the time influence, to a greater or lesser extent, the process of remembering. For example, when one is emotionally involved in a certain task and distracting stimuli in the surrounding environment are kept to a minimum, there is a greater chance for attention, concentration and remembering.

In old age there is the likelihood for memory to fail. This failure of memory may be related to one of its components, namely registration. An old person may fail to register new information because of inattention or because of defective perception. Also, in old age dementia may occur. This is a condition characterized by intellectual deficit, manifested first by loss of memory for recent events.

INTELLIGENCE

The concept of 'intelligence' is difficult to define; one way is to follow the school of thought which considers intelligence as the ability to solve problems and modify behaviour in the light of experience. Intellectual capacity develops during childhood and declines in old age and the degree of intelligence can be measured quantitatively using standardized intelligence tests based on both verbal and non-verbal responses. The measurement is expressed as the ratio of mental age (as established by the intelligence test) and the chronological age. It is expressed as a percentage and is called the intelligence quotient (IQ). For example, a child of 12 years with a mental age of 6 has an IQ of 50 (i.e. $6/12 \times 100$).

Intelligence is normally distributed in the population, with the majority of individuals falling within the middle of an average block of IQ scores. A small percentage will be found on either side of the normal distribution: those with very high scores, who may be called superior or geniuses, and those with very low scores, who may be called mentally retarded or mentally subnormal.

In mentally subnormal individuals the process of development and maturation of the brain has been arrested at an early stage. As a result, the cognitive functions of those individuals is impaired. When the insult to the brain is severe enough, this will result in 'severe' subnormality of intelligence and in gross abnormality of behaviour.

The causes of the arrested or incomplete development of the brain leading to severe mental subnormality cover a wide range of factors. These include genetic and chromosomal abnormalities, birth injuries, infections and other diseases. However, the cause is often untraceable. Milder degrees of mental subnormality may not be the result of biological factors alone, but may be due to an intellectually non-stimulating environment. Individuals with subnormal intelligence have limited insight, lack imagination and are unable to adapt to changing circumstances.

PERSONALITY

People behave differently in life and each one has his or her own style of responses in various situations which is characteristic of him or her. Hence, it is common in everyday experience to describe people on the basis of their responses as having a weak personality, forceful personality, inadequate personality, etc. This sum total of responses to thinking, feeling and behaving which makes one unique among others is called personality.

What could be considered as a 'normal' personality depends on the particular culture in which one lives. 'Normal' is the average person who conforms to customs, traditions and values of the particular society and is not in gross conflict with the accepted social system. This concept of normal is statistical and is not a value judgement. Thus 'normal' is not equivalent to 'good.'

Various theories have been put forward to explain deviations in the development of personality: (i) to some, the disorders of the development of personality are constitutionally or genetically determined; (ii) psychodynamic schools of thought see the development of personality as a result of instinctive unconscious tendencies together with acquired habits and reactions with the environment; and (iii) a third group holds an intermediate position and sees the disorders in the development of personality as an outcome of multiple factors including heredity, upbringing during early childhood and interaction with the environment.

Social stress and disease

Each individual inevitably comes in contact with other people in his or her daily activities. It is not always a smooth interaction. One's wishes are not always fulfilled and obstacles sometimes have to be overcome. These obstacles in life can be considered a stress. The physical environment presents such obstacles as very rough, impassable roads and very hot, dry weather which will not allow one to reach places where one wants to be. Likewise, the social environment presents obstacles in the form of restrictions imposed on the individual by customs of social living. These thwarting circumstances may lead to immediate reactions on the part of an individual. The immediate consequences of frustrating, stressful situations may include restlessness and tension associated with many actions indicating unhappiness, aggression and hostility which may lead to destructiveness, apathy, withdrawal and detachment, or the individual may become absorbed in fantasy and escape into a dream world.

More enduring mechanisms for coping with stressful situations occur frequently to protect the individual's self-esteem and protect him against excessive anxiety. These are called defence mechanisms. By analogy with the physiological mechanisms for maintaining equilibrium within the body, we can think of these defence mechanisms as a process of psychological homoeostasis. Among these defence mechanisms, for example, are rationalization, which is a justification of conduct according to personally desirable motives, and projection, in which the individual attributes his undesirable feelings to others. One may be aggressive towards other people but, instead of admitting one's aggression, one would say that others are aggressive and hence one's own aggression is justifiable. There are other unconscious ways of adjustment to stressful situations that diminish the individual's self-esteem.

An individual facing any social stress will either be able to adjust and cope successfully or fail to cope, succumb under stress and become ill. Illness may then manifest itself physically in the form of a peptic ulcer, hypertension or diabetes or psychologically in the form of anxiety or depression, or both may overlap in the same individual.

The social concept of health

Social behaviour, e.g. in the form of occupational activities, social roles and life styles, has multiple connections with the health of the population. For example, becoming ill is not purely an abnormal physiological state but has more to it than that. A young farmer in a village may have chronic complaints of abdominal pain. During the harvest he may continue to carry out his duties in the field in spite of the pain while at other times he may take to bed and absent himself from work. Thus, circumstances such as the social role and obligations influence declaration of one's illness and treatment-seeking behaviour.

If we turn to mental illness we also find the influence of social factors in its genesis, form and course. Mental illness is found in all societies, poor or rich, developed or underdeveloped, but there are variations in these societies as to what is included among mental illnesses. There are differences in the degree of tolerance of abnormal behaviour in each society. In less developed so-cieties, for example, where social organization is not complicated, abnormal mental behaviour of various degrees of severity is more or less tolerated, while in societies with a complicated organization similar abnormal behaviour may readily lead to the isolation of the abnormally behaving individual from society.

As far back as 1946 the World Health Organization defined health as 'a state of complete physical, social and mental well-being' and not merely the absence of disease or infirmity. This indicates that social well-being is an integral part of what is considered as a healthy state. To attain such a balanced state of health needs multidisciplinary action which involves medical, social and policy-making agencies.

Community medicine is a branch of medicine concerned with the protection and promotion of the health of the population. It studies the distribution and determinants of health and disease and investigates in depth the role of social factors in these. It uses epidemiological techniques in its investigation. In light of the information obtained, it applies medical knowledge and mobilizes society resources to resolve the health problems in the community. It is thus one of the medical disciplines highly involved in the collaborative effort to achieve physical, mental and social well-being in individuals and social groups.

Further reading

1 Armstrong, D. (1983) Introduction: health as a social concept. In *An Outline of Sociology as Applied to Medicine*, 2nd edn, pp. 1–3. Wright, Bristol.
2 Pritchard, M.J. (1986) Psychosocial aspects of disease. In *Medicine and the Behavioural Sciences*, pp. 297–335. Arnold, London.
3 Weinman, J, (1987) *An Outline of Psychology as Applied to Medicine*. Wright, Bristol.

3: The Blood

Objectives

On completion of the study of this chapter, the student should be able to:

1 Understand the processes involved in blood cell formation and their requirements so as to explain the causes of defects in haemopoiesis.

2 Describe the morphology and functions of different types of cells found in the blood so as to recognize abnormalities such as anaemias and leukaemias.

3 Explain the basic mechanisms of the immune system.

4 Enumerate the types of plasma proteins and describe their origins, concentrations and functions so as to recognize disturbances of their kinetics.

5 Describe the basis of blood group classification and its application in blood transfusion.

6 Understand the mechanisms of haemostasis so as to explain the causes of excessive bleeding and intravascular clotting.

Introduction

Blood is classified as a connective tissue with an excessive and complex liquid intercellular material, the plasma, in which the blood cells or formed elements are suspended. The proportion of plasma to cells can easily be determined by spinning a sample of blood in a centrifuge tube. The percentage volume of whole blood occupied by packed red cells, i.e. packed-cell volume or the haemotocrit, is 37–47% in males and slightly less in females. Blood may be obtained easily from the superficial veins of the body and its analysis may give a good idea of the status of body functions.

FUNCTIONS OF THE BLOOD

Blood is the main transportation vehicle of the body. It carries oxygen and nutrients to tissues and waste products of metabolism, e.g. carbon dioxide and urea, to the lungs and kidneys. Most of the hormones are also carried from the endocrine glands to target organs. Blood also has important homoeostatic functions exemplified in the following: blood circulation helps to distribute heat around the body from metabolically active and warmer organs, e.g. liver and gut, to peripheral organs, thereby helping to maintain an even body temperature. Buffers in the blood like haemoglobin, plasma proteins, bicarbonate and others help to keep the hydrogen concentration of the extracellular fluid constant at a pH of 7.4. Blood plays a vital protective function against infection by virtue of its leucocytes and antibodies (immunoglobulins) in the plasma. Furthermore, injury to blood-vessels is followed by blood clotting, which stops further loss of this vital fluid.

COMPOSITION OF THE BLOOD

Blood consists of:
1 Formed elements (blood cells); of which there are three types:
  (a) Red cells — erythrocytes.
  (b) White cells — leucocytes.
  (c) Platelets — thrombocytes.
2 Plasma.

Formation of blood (haemopoiesis)

DEFINITIONS

Erythropoiesis: formation of erythrocytes.
Leucopoiesis: formation of leucocytes.
Thrombopoiesis: formation of thrombocytes (platelets).

SITES OF BLOOD FORMATION

1 *In the foetus*:
  (a) *The yolk sac* is the only site of blood formation in the first 3 months of life. Thereafter, haemopoiesis starts in the liver and spleen and continues until the 7th month of intrauterine life.
  (b) *Bone marrow*. Blood formation in the bone marrow commences in the 4th month of intrauterine life and from the 7th month of intrauterine life until birth it becomes the only haemopoietic organ.
2 *After delivery and throughout life*. Bone marrow (red marrow) fills all the bone cavities throughout childhood but it is gradually replaced by fatty yellow marrow. This is particularly noticed in long bones, where red marrow shrinks towards the ends (epiphyses), leaving the shaft cavity (diaphysis) to be filled by yellow marrow. In addition and throughout life, red marrow continues to occupy the cavities of flat bones, e.g. skull, iliac bones, sternum, ribs and scapula, as well as vertebrae and small bones of the hands and feet. Twenty-five per cent of the cells in red marrow are erythrocyte precursors while the remaining 75% are leucocyte precursors. This is perhaps due to the longer survival time of erythrocytes — 120 days as compared with only a few days for leucocytes; therefore, the need to produce red cells is less pressing than producing the short-lived leucocytes. In certain blood diseases, e.g. chronic anaemias in children, spleen and liver resume their foetal haemopoietic function and start producing blood cells. This is referred to as extramedullary haemopoiesis.

Red blood cells (erythrocytes)

DEVELOPMENT

The production of erythrocytes starts from the primitive haemopoietic stem cell, which becomes successively a proerythroblast; early, intermediate and late normoblast; reticulocyte; and finally an erythrocyte or mature red cell. In the early stages of this production line there is mitotic multiplication of developing cells, followed later by maturation, i.e. the development of both functional and structural cell characters, and finally release into the bloodstream. The process of maturation involves reduction in size, shrinkage and disappearance of the nucleus and the acquisition of the vital red pigment haemoglobin. The appearance of reticulocytes in peripheral blood indicates rapid production of erythrocytes, which is a sign of good response to treatment of anaemia.

STRUCTURE

The erythrocyte is a non-nucleated cell shaped like a flat biconcave disc. Its dimensions are shown in Fig. 3.1. It is made of a framework of protein (stromatin) containing a concentrated solution of haemoglobin. The red cell is flexible and can easily be distorted during its passage in capillaries.

Red cell membrane

The red cell membrane is 75 Å (7.5 nm) thick and is composed of phospholipid and protein. It is semipermeable, allowing controlled movement of water but maintaining a potassium/sodium ratio as found in the intracellular/extracellular fluid compartments.

CHEMICAL COMPOSITION

Red blood cells consist of 60% water and 40% solids (90% haemoglobin and 10% stromatin). The cytoplasm contains adenosine triphosphate (ATP), which is generated mainly by the process of glycolysis.

LIFESPAN

Red cells live for about 120 days, i.e. 1/120 circulating erythrocytes is renewed daily and the total number of erythrocytes is renewed once every three to four months.

Red blood cell count in peripheral blood:
Males   4.8−5.8 million cells/mm^3
Females   4.2−5.2 million cells/mm^3

It is estimated that about 3 million red blood cells are normally broken down every second and, since in a normal healthy person the total number of red cells remains within normal limits, about 3 million red cells must be produced per second. When this balance is disturbed, i.e. production is decreased or destruction increased, the red cell count and haemoglobin level drop, resulting in anaemia.

Destruction and removal of red cells occur in the reticuloendothelial system. Ageing cells become progressively deficient in the enzymes needed for deriving energy from glucose and therefore it becomes difficult for the cell to maintain its integrity. The cells then burst or succumb to osmotic lysis, fragmentation and erythrophagocytosis by reticuloendothelial cells. The components of the destroyed cells are dealt with in the following manner: iron is reused in haemoglobin synthesis, globin is degraded and its amino acids are returned to the amino acid pool of the body, and the pigment, haem, is ultimately converted to bilirubin, which is excreted in bile.

FUNCTIONS

By virtue of their content of haemoglobin, the function of red cells is the transport of oxygen from lungs to tissues and carbon dioxide from tissues to the lungs for excretion. The biconcave discoidal shape of mature erythrocytes is well adapted to its function of gas transport; the surface area of the red blood cell is larger than the minimal area needed to enclose its volume which would be provided if the cell were a simple sphere. Furthermore, in the narrow channels of the microcirculation where O_2 is given up, more red cells can be accommodated in a given volume of blood

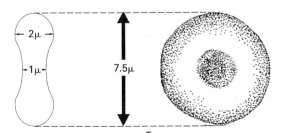

Fig. 3.1 Outline and dimensions of a typical red blood cell.

than would be possible if red cells were spherical in shape. Lastly, the red cell membrane, being elastic and distortable, facilitates the movement of these cells through the 4 μm wide terminal capillaries where oxygen is delivered and carbon dioxide removed.

METABOLISM

The erythrocyte is metabolically active and gets its energy from glucose, both aerobically and anaerobically. Knowledge of red cell metabolism is essential to the clear understanding of methods used in red cell preservation in blood banks, where blood is collected in an anticoagulant solution containing glucose (dextrose) as a source of energy. Blood is then stored in the cold (temperatures of 4 °C) to reduce the metabolic rate of erythrocytes; consumption of energy (i.e. ATP) at 4 °C is 30 times slower than at 37 °C.

Nutritional requirements for red cell production

Most of the substances required for erythropoiesis are available from the metabolic pool of the body. However, the following substances are described as 'essential' for normal erythropoiesis since deficiency of any of them results in detectable changes in the blood:

1 *Amino acids.* The supply of proteins in the diet provides the amino acids needed for the synthesis of the globin of haemoglobin. It is known that haemoglobin, being a vital pigment, has a high priority for the available protein and therefore, in humans, protein deficiency must be very severe before haemoglobin synthesis is impaired. In kwashiorkor a hypoplastic anaemia develops which disappears spontaneously on feeding. This anaemia is characterized by a reduction in the cellularity of the bone marrow (hence the name hypoplastic).

2 *Iron.* This is dealt with in detail at a later stage in this chapter. Iron deficiency causes microcytic hypochromic anaemia, i.e. smaller than normal red cells with reduction in the haemoglobin content.

3 *Vitamins.* There are several vitamins essential for normal haemopoiesis:

(a) Vitamin B_{12} (cyanocobalamin) and folic acid play an important role in the synthesis of nucleoproteins. The deficiency of either leads to a disturbance of deoxyribonucleic acid (DNA) metabolism in developing nucleated red cell precursors. The result is megaloblastic (mega = large) erythropoiesis and megaloblastic (macrocytic) anaemia (pernicious anaemia).

(b) As a reducing agent, vitamin C (ascorbic acid) facilitates iron absorption in the gut by helping to keep iron in the reduced (ferrous, Fe^{2+}) state. Ascorbic acid is also involved in the normal metabolism of cyanocobalamin and folic acid. Diets deficient in vitamin C are likely to be deficient in folic acid. Both vitamins are present in vegetable foods and both are heat-labile and therefore deficiency of both vitamins is common. The result can be normocytic, macrocytic or microcytic anaemia with normochromic or hypochromic red cells.

(c) Pyridoxine (B_6) is essential for haem and nucleoprotein production. Its deficiency in infants causes hypochromic microcytic anaemia which does not respond to iron therapy but only to pyridoxine.

(d) Riboflavin, nicotinic acid, pantothenic acid, biotin and thiamine are B vitamins that are involved in various intermediary metabolic processes. It is more common to find multiple deficiencies of these vitamins than deficiencies of one member. When deficiency occurs the result is almost always a normochromic normocytic anaemia which can be cured by administering the missing vitamin.

(e) Recently, vitamin E has been reported to be essential for the normal synthesis of erythrocyte membrane in infants and, when deficient, it may cause haemolytic anaemia due to excessive haemolysis, i.e. destruction of red cells.

4 *Trace elements.* Copper, cobalt, zinc, manganese, nickel and other minerals are present in the erythrocyte. There is no clear evidence in man to suggest that deficiency of any of these elements produces anaemia, but it is of interest to note that cobalt stimulates the formation of erythropoietin and, if given in quantities in excess of the normal dietary needs, results in the development of polycythaemia (increased number of circulating red cells).

Control of erythropoiesis

ERYTHROPOIETIN

Oxygen deprivation (hypoxia) or blood loss triggers the formation of a humoral factor, *erythropoietin*, which acts on the bone narrow to stimulate erythropoiesis.

Chemistry

Erythropoietin (or erythropoiesis-stimulating factor) is a glycoprotein (24% carbohydrate and 76% protein). Estimates of its molecular weight range from 71 000 to 100 000. Its protein is made of a single chain of about 340 amino acids.

Sources

In the past, many theories have been advanced claiming that the kidney, liver and other organs as the sources of erythropoietin. However, it has now been established that peritubular interstitial cells in the renal cortex and outer medulla are the site for the secretion of the hormone erythropoietin. The stimulus to erythropoietin production is hypoxia, caused either by anaemia or by other forms of impaired oxygen delivery to the kidney. The half-life of erythropoietin is 6.7 hours and the site of its degradation is not known.

Action

Erythropoietin stimulates the early stem cell (otherwise called the erythropoietin-sensitive stem cell or erythrocyte-committed stem cell) to differentiate into a proerythroblast. Erythropoietin seems to trigger the processes which eventually lead to haemoglobin synthesis. In addition, it also stimulates self-replication and proliferation of the erythropoietin-sensitive stem cells. Erythropoietin does not seem to influence the maturation stages in the erythrocyte production line.

Erythropoietin can be detected in the plasma and urine of normal persons and this is taken as evidence that the hormone is needed for normal erythropoiesis. But high levels occur in anaemia (except anaemia of renal failure) and in hypoxia due to living at high altitudes or as a result of heart or lung disease. In the latter situation, high levels of erythropoietin result in increased production of erythrocytes and consequently an increased number of circulating cells (erythrocytosis, polycythaemia) (Fig. 3.2).

In conclusion, erythropoiesis is determined by the rate of erythropoietin formation, which, in turn, is controlled by the relation of oxygen supply to oxygen demand. When oxygen supply does not meet demand, e.g. hypoxia, erythropoietin formation and thus erythropoiesis are increased, more red cells are produced and this corrects the decrease in oxygen supply to the tissues.

ANDROGENS, THYROID HORMONES, CORTISOL AND GROWTH HORMONE

Deficiency of any of these hormones usually results in anaemia, mostly normochromic or normocytic, which may be corrected by giving the missing hormone. This has been taken to support the importance of these hormones in normal erythropoiesis.

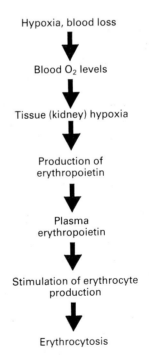

Fig. 3.2 The mechanism of production of erythropoietin.

Iron metabolism

DISTRIBUTION OF IRON IN THE BODY

The total amount of iron in the body of a healthy adult is $3-5$ g distributed as follows:

1 *Haemoglobin.* Most of the iron in the body ($65-75\%$, average: 3000 mg) is present in haemoglobin (in the red blood cells, 1 g haemoglobin contains 3.4 mg iron).

2 *Storage (available) iron.* Twenty per cent of the total body iron (average: 1 g) is present as the protein–iron compound ferritin in the liver, spleen and bone marrow.

3 *Cellular or tissue (non-available) iron.* This is the iron (average 150 mg) which is present in body cells as a major component of oxidative enzyme systems, e.g. cytochrome, and as part of myoglobin.

4 *Transport or plasma iron.* A very small amount of iron ($3-4$ mg) is found in the plasma in combination with a β-globulin, transferrin. This iron is being transferred from one point of use to another. A very minute amount of iron is present in blood in the form of ferritin. Measurement of serum ferritin is a reliable indicator of the level of iron stores in the body. Figure 3.3 shows a diagrammatic representation of iron metabolism.

IRON IN FOOD

The average diet provides $10-20$ mg iron/day.

Liver, being an important storage organ for iron, is the richest dietary source. Other good sources include: beef, mutton, fish, egg yolk, cereals, beans, lentils and green leafy vegetables. When food is cooked in iron utensils, some iron is added from the utensils.

Forms of iron in food

1 *Organic porphyrin* (haem) iron is the iron present in animal protein foods such as meats, fish and egg yolk.

2 *Inorganic (non-haem) iron* is the iron present in foods of vegetable origin.

Absorption

Most of the iron in food is in the oxidized ferric (Fe^{3+}) state; however, it is absorbed better when it is in the reduced ferrous (Fe^{2+}) state. This reduction is facilitated by gastric acidity and the presence of reducing agents in the diet, especially vitamin C and protein $-SH$ groups. The maximal absorption of iron occurs in the upper parts of the small intestine, especially the duodenum. Ferrous ions are carried into the mucosal cell by active transport. Once iron is inside the cell, it becomes attached to a non-ferritin protein carrier. Iron is then either transferred across the serosal border of the cells, where it is picked up by transferrin and carried to various sites in the body, or it remains within the cell, where it stimulates ferri-

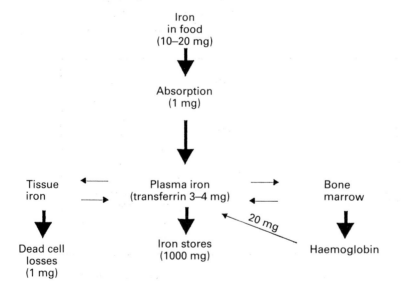

Fig. 3.3 Iron metabolism.

tin formation by combining with an intracellular protein apoferritin to form ferritin, which is deposited in the mucosal cell (Fig. 3.4).

Normally, 10–15% of the ingested iron is absorbed, i.e. 1–2 mg/day. Absorption depends on two important factors: the size of iron stores and the rate of erythropoiesis. If the former is small or the latter is high, then the rate of absorption will be increased. In addition, absorption of iron can be decreased by several factors: (i) the presence of phosphates, phytates or oxalates in the diet, as these anions bind and deposit iron in the gut and therefore iron will not be available in the free state for absorption; (ii) surgical removal of the stomach (gastrectomy) or achlorhydria (lack of hydrochloric acid), which decreases iron reduction; and (iii) malabsorption syndromes or any disease that causes steatorrhoea or chronic diarrhoea, which results in decreased absorption of iron.

TRANSPORT AND STORAGE

When the absorbed iron reaches the plasma it is converted back to the ferric (Fe^{3+}) state. It then combines with a specific iron-binding β-globulin called transferrin to form a ferric–protein complex which is the plasma transport form of iron. Transferrin conveys iron to the various body cells for storage or utilization.

Transferrin is a globular protein (molecular weight 86 000) and is normally 30–40% saturated with iron; the remaining 60–70% forms the unsaturated latent iron-binding capacity. The plasma iron level is normally 100–130 μg/ 100 ml. However, when the transferrin is fully saturated with iron the plasma iron level will be about 300 μg/100 ml. This is called the total iron-binding capacity (TIBC) of plasma (Fig. 3.3).

Iron is stored in the reticuloendothelial cells in the main storage organs for iron, which are the liver, spleen and bone marrow. Iron is stored in two forms: ferritin and haemosiderin. Ferritin consists of a protein shell of 24 subunits (apoferritin) formed of identical polypeptide chains arranged to form a hollow shell around a central cavity which contains a variable amount of ferric iron (about 2500 atoms), principally as hydrous ferric oxide-phosphate. Haemosiderin is believed to be an aggregate of ferritin molecules. In iron storage organs there is always a predominance of ferritin (65%) over haemosiderin (35%) molecules.

EXCRETION AND DAILY REQUIREMENTS

The body conserves iron very well and, under normal circumstances, little iron is lost. Total daily losses in men are 0.5–1.0 mg and these occur through several routes:

1 *Faeces.* Other than unabsorbed ingested iron, faecal iron comes from desquamated intestinal epithelial cells in addition to small amounts found in bile and saliva.

2 *Skin.* Skin loss consists of desquamated cells, lost hairs and nails and traces of iron in sweat.

3 *Urine.* Urinary loss of iron is negligible.

In women, the monthly menstrual losses are

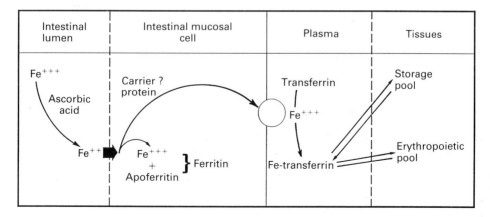

Fig. 3.4 A diagram illustrating the phases of iron absorption, its transport in plasma and its fate in the tissues.

about 15−28 mg iron, i.e. an additional 0.5−1 mg iron/day compared with men. About 400−900 mg of iron are lost by women during pregnancy and childbirth, i.e. approximately an extra 1−2 mg iron/day.

The daily requirements for iron are the amounts of iron needed to be absorbed to meet these losses, bearing in mind that women always have higher requirements than men due to menstruation, pregnancy and childbirth. Periods of growth during infancy, childhood and adolescence involve enlargement of the blood volume to maintain blood flow to the new tissues. During such periods the increased requirements for iron are met by increased absorption of iron from the intestine.

Haemoglobin

Oxygen and some carbon dioxide transport depend almost totally on the presence of the red respiratory pigment haemoglobin in the erythrocytes. Haemoglobin increases the ability of the blood to carry oxygen by 60-fold. Plasma carries 0.3 ml O_2/100 ml while whole blood carries 20 ml/ 100 ml. This vital function makes haemoglobin one of the most important constituents of higher organisms.

STRUCTURE

Haemoglobin (haem + globin) is a conjugated protein with a globular molecule (molecular weight 64 000) composed of four subunits (two α and two β chains); each unit contains a red iron−porphyrin or haem group bound to a polypeptide chain. The iron in the haem is in the divalent (ferrous, Fe^{2+}) state and can combine reversibly with oxygen to form oxyhaemoglobin without change in the valency of iron. The polypeptide chains are collectively called globin.

TYPES OF NORMAL HAEMOGLOBIN

Adult haemoglobin (HbA) All haemoglobins have one pair of amino acid chains in common, alpha (α) chains, each made up of 141 amino acids. In HbA, α chains are paired with beta (β) chains of 146 amino acids each. HbA constitutes 98% of the haemoglobin in healthy adults.

Haemoglobin A_2 (HbA$_2$) In HbA$_2$ the globin moiety is also composed of two α and two β chains. However, the β chains consist of 146 amino acids, 10 of which are different from those of the β chain in HbA (hence the name delta chain). HbA$_2$ is present in very small amounts at birth and reaches the adult level of about 2% during the first year of life.

Haemoglobin F (HbF) This is the major respiratory pigment during intrauterine life. It has two α chains and two gamma (γ) chains. Like the β chains, the γ chains are made up of 146 amino acids each. However, 37 amino acid residues are different from those of the β chain. At birth, 80−90% of the total haemoglobin is haemoglobin F. It then falls rapidly to about 5% at 6 months. A level of 1% may be detected at puberty and adulthood.

FUNCTIONS

1 *Carriage of oxygen* is the most important property of haemoglobin. When haemoglobin is exposed to oxygen two atoms of oxygen are taken up by each subunit molecule of haemoglobin to form oxyhaemoglobin.

2 *Carriage of carbon dioxide*. Reduced haemoglobin combines reversibly with carbon dioxide to form carbaminohaemoglobin:

$$Hb + CO_2 \rightleftarrows HbCO_2$$

3 *Buffer*. Haemoglobin, being a protein, acts as one of the buffers in blood.

Haemoglobin can also combine with gases other than oxygen and carbon dioxide. For example, Hb has greater (250 times) affinity to carbon monoxide than to oxygen. Treatment of carbon monoxide poisoning is to give pure O_2 at high pressure (hyperbaric oxygen).

Smokers of tobacco are reported to have about 5% of the haemoglobin or more combined with carbon monoxide.

BREAKDOWN

The life of the haemoglobin molecule ends with the end of the lifespan of erythrocytes. Haemoglobin is broken down into its components — globin and haem.

Fate of the haem group After the extrusion of iron the rest of the molecule is converted to biliverdin, which undergoes reduction to bilirubin. Both substances are named bile pigments, as they are excreted in bile. Bilirubin is poorly soluble in water. Once it is released from the reticulo-endothelial system into the plasma, it is bound to albumin, which greatly increases its solubility, and in the bound form bilirubin is carried to the liver. It is conjugated with glucuronic acid in the liver cells and this step renders bilirubin water-soluble. It is then excreted in bile into the small intestine, where it is converted by bacterial action to urobilinogen. Some of the urobilinogen remains in the small intestine and gives the stool its characteristic orange-brown colour. The rest is absorbed into the bloodstream and is eventually excreted by the kidneys in urine.

Jaundice Jaundice is a yellowish colour of the skin, sclera and other tissues due to the accumulation of excess bilirubin in the plasma and tissue fluids. The normal adult total plasma bilirubin range is 0.5−1.0 mg/100 ml. The yellow pigmentation is not obvious until the plasma concentration of bilirubin exceeds 2 mg/100 ml.

An excess of bilirubin in the plasma can result from three causes:

1 Excessive breakdown of erythrocytes — haemolytic jaundice (prehepatic).
2 Infective or toxic damage to the liver — hepatic jaundice.
3 Obstruction to the outflow of bile from the liver — obstructive jaundice (posthepatic).

ANAEMIA
Anaemia is defined as a decrease in the concentration of haemoglobin in the blood below the normal level for the same sex and age. Accordingly, anaemia may occur if the number of red cells (red cell count) falls below the normal level or if the erythrocyte count is normal but their load of haemoglobin (mean corpuscular haemoglobin) is less than normal. A patient suffering from anaemia usually complains of tiredness and easy fatigability on minor physical effort. These and other symptoms of anaemia are due to the decrease in the oxygen supply to the tissues.

Causes and classification
1 *Blood loss*, whether acute, as in haemorrhage following a car accident, or chronic, which occurs in chronic bleeding in the gastrointestinal tract, e.g. piles, peptic ulcer or the presence of worms, particularly hookworms, in the intestine.
2 *Diminished red cell production by the bone marrow.* This may occur in the following situations:

(a) Deficiency of a nutritional substance(s) required for erythropoiesis (nutritional anaemia). Iron deficiency causes the marrow to produce cells smaller in size than normal (microcytes) and with a small load of haemoglobin since iron is an integral component of the haemoglobin molecule. The cells therefore have a lighter red colour than normal (hypochromic). Iron-deficiency anaemia is described as microcytic hypochromic anaemia. It is the commonest type of anaemia in the developing countries and is mainly due to a deficiency of iron in the diet.

Deficiencies of vitamin B_{12} and folic acid cause megaloblastic anaemia. Folic acid deficiency in the diet is also common in tropical countries where the resulting anaemia is prevalent in malnourished children and in women during pregnancy, when the general body requirements for folic acid increase.
(b) Diminished red cell production may occur when the bone marrow space is occupied by secondary cancer cells or fibrous tissues, as in myelosclerosis, or as a result of destruction by radiation or drugs (aplastic anaemia).
3 *Excessive destruction of red cells (haemolytic anaemia).* The factors causing destruction may be within the red cells, e.g. presence of an abnormal haemoglobin such as HbS, due to which the cells take a sickle or crescent shape when Hb is in the reduced state. These sickle cells have a short lifespan as they are readily broken down and removed by the reticuloendothelial system. Sickle-cell anaemia is very common in the tropics and in people of negroid origin.

The cause of haemolysis may come from outside the red cells, e.g. blood group incompatibility, where an antibody may react with a blood group antigen present in the red cell membrane resulting in agglutination and haemolysis of red cells.

POLYCYTHAEMIA

Polycythaemia is an increase in the number of red blood cells per unit volume of blood.

Classification

1 *True or absolute polycythaemia*, characterized by an increase in the total number of circulating red cells (red cell mass). It is of two types:

(a) *Primary or polycythaemia rubra vera*, where bone marrow continues to produce red cells with no regard to the normal physiological negative feedback mechanism of the hormone erythropoietin.

(b) *Secondary polycythaemia*, due to hypoxia as in high altitudes and chronic respiratory or cardiac disease. This type of polycythaemia is characterized by high levels of erythropoietin released in response to hypoxia.

2 *Relative polycythaemia* occurs in cases of dehydration due to excessive fluid loss (vomiting or sweating) or decreased water intake. Plasma volume is decreased and consequently the red cells are concentrated (haemoconcentration). Thus, there is no increase in the red cell mass but the ratio of cells to plasma is high.

Leucocytes

CLASSIFICATION

The blood leucocytes (white blood cells) are a heterogenous population of nucleated cells lacking haemoglobin. There are five distinct morphological types classified into two groups on the basis of the presence or absence of granules in their cytoplasm:

1 Granulocytes (with cytoplasmic granules): these are the neutrophils, eosinophils and basophils.

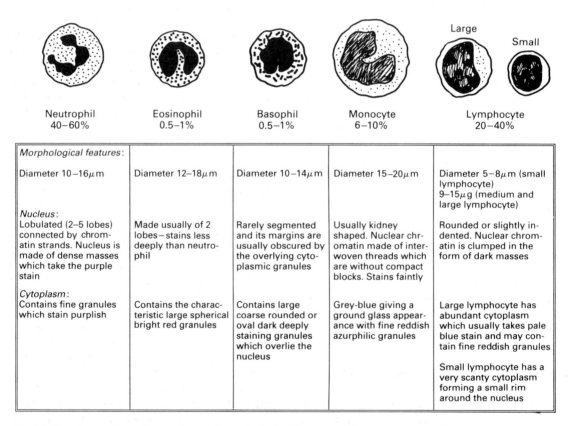

Neutrophil 40–60%	Eosinophil 0.5–1%	Basophil 0.5–1%	Monocyte 6–10%	Lymphocyte 20–40%
Morphological features:				
Diameter 10–16μm	Diameter 12–18μm	Diameter 10–14μm	Diameter 15–20μm	Diameter 5–8μm (small lymphocyte) 9–15μg (medium and large lymphocyte)
Nucleus: Lobulated (2–5 lobes) connected by chromatin strands. Nucleus is made of dense masses which take the purple stain	Made usually of 2 lobes–stains less deeply than neutrophil	Rarely segmented and its margins are usually obscured by the overlying cytoplasmic granules	Usually kidney shaped. Nuclear chromatin made of interwoven threads which are without compact blocks. Stains faintly	Rounded or slightly indented. Nuclear chromatin is clumped in the form of dark masses
Cytoplasm: Contains fine granules which stain purplish	Contains the characteristic large spherical bright red granules	Contains large coarse rounded or oval dark deeply staining granules which overlie the nucleus	Grey-blue giving a ground glass appearance with fine reddish azurphilic granules	Large lymphocyte has abundant cytoplasm which usually takes pale blue stain and may contain fine reddish granules
				Small lymphocyte has a very scanty cytoplasm forming a small rim around the nucleus

Fig. 3.5 The shapes, dimensions and special morphological features of the various types of leucocytes.

2 Agranulocytes (without cytoplasmic granules): these are the monocytes and lymphocytes.

Fig. 3.5 gives the dimensions and morphological characteristics of the leucocytes.

FORMATION OF LEUCOCYTES (LEUCOPOIESIS)

Sites of formation

1 Granulocytes: bone marrow.

2 Lymphocytes: bone marrow, thymus, lymph nodes and other collections of lymphoid tissues, e.g. wall of the intestine.

3 Monocytes: bone marrow.

FORMATION OF GRANULOCYTES (GRANULOPOIESIS)

The life-history of the granulocytes begins in the bone marrow, where there is progressive division and maturation from the earliest cell, the stem cell, successively through the cell types myeloblast, promyelocyte, myelocyte, metamyelocyte, band neutrophil and segmented neutrophil. The myeloblasts, promyelocytes and myelocytes are capable of mitotic division and cell replication; hence, these are collectively called the proliferating granulocyte pool. From the metamyelocyte stage onwards, no cell division occurs and therefore the metamyelocytes, band neutrophils and segmented neutrophils are together referred to as the maturation pool. Maturation takes the form of biochemical and morphological changes in both the nucleus and the cytoplasm. The nucleus becomes condensed and broken up into lobes. In addition, fine neutrophilic granules appear in the cytoplasm. The maturation pool is sometimes called the marrow granulocyte reserve, as it is believed to be the main source of extra neutrophils which enter the bloodstream in acute infections and other pathological states. The mature neutrophils, once released into the bloodstream, stay there for about 7–10 hours before they migrate to the tissues, where they function as mobile phagocytes.

NEUTROPHILS IN THE BLOODSTREAM

Mature neutrophils leave the bone marrow to enter the blood. Some of them join the blood circulation—the so-called circulating granulocyte pool. These are the cells available for blood sampling and counting. Others are deposited along the walls of the small vessels (marginal granulocyte pool), where they are in a state of rapid and continuous exchange with the circulating cells, and from this site they can be mobilized by exercise or by an adrenaline injection. The entry of these cells into the circulating pool accounts for the increased white cell count (leucocytosis) that accompanies exercise and other stressful situations.

FUNCTIONS OF LEUCOCYTES

The general function of leucocytes is defence against infection. However, the different types of leucocytes contribute to a different extent towards this general function.

FUNCTIONS OF NEUTROPHILS

The neutrophils are also called polymorphonuclear leucocytes because the nucleus is formed of 2–5 lobes. This cell is the most important cell in the cellular defences of the body against infection. To achieve this goal neutrophils execute several integrated functions: (i) the neutrophils must reach the site of infection (chemotaxis); (ii) they must ingest the foreign organism (phagocytosis); and (iii) they must kill or inhibit the multiplication of the micro-organism (microbial killing).

Chemotaxis

The neutrophils are actively motile cells; they can move more rapidly than any other cell in the body. Their movement is directed towards bacteria in a purposeful manner, being attracted to bacteria, site of infection or inflammation by a variety of chemotactic substances, e.g. products of certain bacteria, damaged leucocytes or other tissue components. The property of directed movement of the neutrophils is named chemotaxis. It accounts for the accumulation of neutrophils at sites where they are needed, e.g. infected wounds. Impaired chemotaxis can lead to increased susceptibility to infectious diseases, especially in children.

When neutrophils approach the infected site, they lie along the walls of the closest capillaries—a process called margination. Then individual neutrophils squeeze themselves between endothelial cells and gradually move out from the capillary—a process called diapedesis. Since neutrophils are motile cells, they move towards the bacteria.

Phagocytosis

Phagocytosis is the active process whereby a cell eats particulate matter (Fig. 3.6).

One of the remarkable features of neutrophils is their fine capacity to distinguish foreign cells like bacteria from homologous body cells and aged or damaged cells from fresh ones. This is due to the presence in plasma of certain substances (opsonins), such as γ-globulins (especially IgG) and complement C4, which coat bacteria and ageing cells, thereby making them 'palatable' to neutrophils. To opsonize means to prepare for eating. An opsonin is an agent in plasma which acts on foreign particles to increase their palatability to phagocytes.

Recognition is followed by close adhesion between the outer membrane of the neutrophil and the bacterium. This is followed by invasion of the neutrophil membrane and complete encirclement of the bacterium by pseudopodia. The pseudopodia fuse to enclose the bacterium in a phagocytic vacuole.

Microbial killing

Following the ingestion (phagocytosis) of the bacterium, the following sequence of events take place:

1 The fusion of the neutrophilic granules with the phagocytic vacuole.

2 Discharge of antimicrobial agents from the granules into the vacuole. These agents include lysozymes, myeloperoxidase and lactoferrin, which are capable of destroying a wide range of bacteria.

3 Killing and digestion of the ingested organism.

FUNCTIONS OF EOSINOPHILS (ACIDOPHILS)

The eosinophils are characterized by the presence of coarse, bright red granules in their cytoplasm. These granules contain an arginine-rich basic protein which attracts red acidic dyes like eosin. The eosinophil nucleus is often seen as two large lobes. Eosinophil functions are not very different from those of the neutrophil.

Chemotaxis

Unlike neutrophils, eosinophils are attracted more towards areas of chronic inflammation rather than acute inflammation. Chemotactic substances for eosinophils include histamine, antigen—antibody complexes, 5-hydroxytryptamine (5-HT), bradykinin and a specific 'eosinophil chemotactic factor.'

Eosinophils tend to accumulate at the sites of histamine release as is seen in allergic diseases of the skin or lungs.

Phagocytosis

Eosinophils are capable of ingesting a variety of particles ranging from bacteria and destroyed cells to antigen—antibody complexes. Phagocytosis involves the same sequence of events as already described for neutrophils. However, antimicrobial activity is considerably less than that of the neutrophil.

Eosinophils and inflammation

In inflamed tissues, eosinophils have been shown to be capable of antagonizing and inactivating

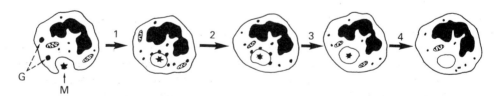

Fig. 3.6 Schematic diagram of the process of phagocytosis by a neutrophil. An opsonized microbe (M) after being recognized by the neutrophil is contained in an invagination of the neutrophil membrane. Thereafter, the particle is enclosed in a phagocytic vacuole. Some of the neutrophil granules (G) stick to the wall of the vacuole and then release their bactericidal substances, which induce killing and ultimate digestion of the microbe.

histamine and other chemical mediators of inflammation such as 5-HT and bradykinin. Through this function the eosinophil helps to limit and circumscribe the inflammatory process.

The eosinophilic response to an inflammatory stimulus is characterized by accumulation of eosinophils in the inflamed tissues with a simultaneous increase both in the production of eosinophils by the bone marrow and in the number of circulating eosinophils (eosinophilia), on their way from the bone marrow to the inflammatory sites. Chronic eosinophilia occurs in response to complex antigens, as in helminthic (worm) infestations (e.g. hookworm, ascaris and bilharzia) and in response to allografts (e.g. skin grafts).

The accumulation of eosinophils in inflammatory sites may be inhibited by high doses of corticosteroids, which also depress the chemotactic attraction of eosinophils.

Eosinophilia

From the foregoing account we may deduce that eosinophilia occurs in pathological states as a result of an antigen–antibody reaction.

Common causes of eosinophilia include:

1 Parasitic disease, e.g. worm infestations of the gut.
2 Allergic conditions:
  (a) Bronchial asthma.
  (b) Allergic rhinitis (hay fever).
  (c) Drug reactions, e.g. penicillin sensitivity.
3 Tropical eosinophilia, which represents a reaction to the filaria parasite.
4 Dermatological diseases.

FUNCTIONS OF BASOPHILS

The distinguishing morphological feature of the basophil is the large blue-black granules which appear to fill the cytoplasm, overlie the nucleus and tend to obscure nuclear configuration. Basophil granules contain abundant acid mucopolysaccharide, which accounts for their strong affinity for basic dyes such as methylene blue. Heparin is one of the important acid mucopolysaccharide constituents; other constituents include histamine, 5-HT and ribonucleic acid (RNA). The basophil is the carrier of histamine in the blood and, due to its being rich in both heparin and histamine, it bears a strong resemblance to tissue mast cells. The function of basophils is not known with certainty but they may have a role related to their content of the physiologically active substances such as heparin, histamine and 5-HT.

MONOCYTES (BLOOD MACROPHAGES)

The macrophage has its origin in the bone marrow monoblast and promonocyte. The mature monocyte reaches the bloodstream, where it stays for a variable period of time ranging from a few hours to 6 days. Then it leaves the circulation for the tissues, where it undergoes transformation to the larger and more effective phagocyte – tissue macrophage (histiocyte).

Functions

The macrophage contributes directly to the body defence systems by phagocytosis and killing of invading bacteria and, indirectly, by interacting and co-operating with lymphoid cells in both the afferent (or recognition of foreign material) and efferent (effector) limbs of the immune response. In the afferent limb, macrophages accumulate and retain antigens in immunological form.

Macrophages phagocytose damaged or altered host cells and microscopic debris, which justifies the descriptive name 'tissue scavengers'.

LYMPHOCYTES

Much of our knowledge about the cellular elements of the blood has been based on the concept that cells may be recognized and classified by morphological criteria. This concept, however, does not hold in relation to recognizing and classifying cells of the lymphoid series. The blood lymphocytes constitute a family of cells of different origins, migration patterns, sizes, staining characteristics, ultrastructure, lifespan and function.

Formation (lymphopoiesis)

Lymphocytes originate from the primitive unipotent stem cell (lymphoid-committed precursor) in the thymus, lymphoid tissues and bone marrow and then proceed along a known maturation line via the 'lymphocyte production pathway', which includes the following cellular stages:

1 *Lymphoblasts.* Normally these are only seen in lymphopoietic organs and almost never observed in peripheral blood.

2 *Intermediate and transitional forms* (large pyroninophilic blast cells).
3 *Small and medium—large lymphocytes* (blood lymphocytes).

These stages are not unidirectional. The process can, under certain circumstances, go in the reverse direction and small lymphocytes can grow into large lymphocytes and lymphoblasts. Such blastic transformation can be demonstrated *in vitro* by growing small lymphocytes in a suitable culture medium containing a non-specific mitogen, such as phytohaemagglutinin (PHA), or a specific antigen, e.g. tuberculin.

Lymphocytes in the bloodstream
Lymphocytes enter the peripheral blood either directly, by passing through the walls of blood-vessels in the various lymphopoietic organs, or indirectly, by entering the lymph stream and eventually reaching the bloodstream through the thoracic duct and other lymph ducts in the neck.

Classification
When seen under an ordinary light microscope, blood lymphocytes can be divided into small (5–8 μm diameter) and medium and large lympho-cytes (8–15 μm diameter). The majority of blood lymphocytes are of the small type.

Functions
Lymphocytes are the central cells in immunity. On the basis of this function, lymphocytes are divided into two types:
1 *Thymus-dependent lymphocytes (T-cells)* are so called because they originate in the thymus or bone marrow and migrate to the thymus. They have a lifespan of 100–300 days or even more (hence the name long-lived lymphocytes). This long lifespan is closely related to their property of constant movement from blood to tissues to lymph to blood again (recirculation of lymphocytes).

T-lymphocytes are the principal mediators of cellular immune responses such as rejection of tissue graft, e.g. kidney transplant, and delayed hypersensitivity reactions. They also play a minor role in the synthesis of immunoglobulins (antibodies).
2 *Thymus-independent lymphocytes (B-cells)*. In man, the B-cells develop in the bone marrow, the germinal centres of lymph nodes and the red pulp of the spleen. Their lifespan is 2–7 days (hence the name short-lived lymphocytes). They have been called B-cells because they are known as bursa cells. When the B-cells are properly stimu-lated by an antigen, they develop successively into large pyroninophilic lymphocytes and, lastly, plasma cells. The plasma cells are lymphoid cells which are capable of producing antibodies. Thus, the B-lymphocytes are the principal mediators of the humoral immune response.

Total leucocyte count
Although it is usually quoted in textbooks that the total leucocyte count is 4000 to 10 000 cells per cubic millimetre of blood, it should be em-phasized that this range applies more to Europeans than to residents of hot tropical countries. It is not uncommon to find a total leucocyte count among healthy students and blood donors in these geographical locations of between 2000 and 4000 cells/mm^3. Because there is a relatively low count of neutrophils, it is called neutropenia.

DIFFERENTIAL WHITE CELL COUNTS
The normal proportions of white blood cells are as follows:

Neutrophils	60–70%
Lymphocytes	20–30%
Monocytes	2–8%
Eosinophils	2–4%
Basophils	0–2%

LEUCOCYTOSIS AND LEUCOPENIA
An increase in the total leucocyte count above the normal is called leucocytosis. This may occur in health (physiological leucocytosis) or disease.

Physiological leucocytosis may occur under several conditions:
1 Diurnal variation: leucocyte counts are lowest in the morning and increase to a maximum in the afternoon.
2 After a protein meal.
3 Following physical exercise.
4 Stimulated by stress or an injection of adrenaline.

Disease states which commonly cause leuco-cytosis are bacterial infections (pyogenic infec-tions), e.g. tonsillitis, infected wounds or inflamed appendix. In these conditions, measurement of the total leucocyte count is essential for diagnosing the existence of the infection. The differential white cell count is also useful. In general, acute bacterial infections cause an increase in the neutrophils, while chronic and viral infections are associated with increased lymphocytes.

Leucopenia is a decrease in the total leucocyte count below the normal. It is often seen in conditions of malnutrition and is also an important feature of typhoid fever. Some drugs may depress the bone marrow and therefore result in leucopenia and, in particular, a decrease in the granulocyte count (agranulocytosis). Leucopenia can also be caused by vitamin B_{12} or folic acid deficiency.

LEUKAEMIA

Neoplastic growth of the stem cells in the bone marrow or lymphoid tissue leads to myeloid or lymphocytic leukaemia, respectively. Leukaemia can be acute or chronic depending on the time course of the disease. The total white cell count is usually very high (>50 000 mm^3). The leucocyte precursors in the bone marrow proliferate extensively and occupy more and more bone marrow space, resulting in a concomitant depression in the production of red cells, leading to anaemia, and of platelets, leading to thrombocytopenia and the risk of bleeding. The causes of the leukaemias are not fully understood, but among the known factors are chromosomal abnormalities, radiation and chemical agents, including drugs and some viruses.

The reticuloendothelial system (macrophage system)

The term macrophage was first used by Metchnikoff to refer to large mononuclear phagocytes to distinguish them from the smaller neutrophil polymorphs—microphages. He established the concept of phagocytosis and intracellular digestion as the essential primary mechanism of defence against bacterial agents. Afterwards, the term reticuloendothelial system was introduced by Aschoff and Landon in 1924 to refer to cells which take the vital stain trypan blue when injected into experimental animals. These cells included certain endothelial cells and reticular cells in addition to some macrophages. Thus, the concept was established that there is a system of cells within the body whose function is to collect, by a process of active phagocytosis, finely particulate material such as dyes or colloidal particles.

CELLULAR COMPONENT

1 Macrophages lining lymph sinuses in lymph nodes and blood sinuses in the liver (Kupffer cells), spleen, bone marrow, adrenal cortex and anterior lobe of the pituitary gland. It is important to realize that ordinary vascular endothelium (blood or lymph) is not phagocytic. This highly specialized function is reserved for the endothelium lining the sites mentioned above.
2 Reticular cells of lymph nodes, spleen and bone marrow.
3 Tissue macrophages (histiocytes): distributed widely throughout solid tissues.
4 Blood macrophages (monocytes).
Monocytes and histiocytes are sometimes referred to as mobile reticuloendothelial cells, i.e. the reticular and endothelial macrophages.

The reticuloendothelial system is, therefore, not a sharply defined system, either anatomically or physiologically, and its component cells have great powers of proliferation and regeneration. Furthermore, because of their wide distribution, reticuloendothelial cells have ready access to foreign particles that circulate in the bloodstream, as well as to materials which originate interstitially and find their way into lymphatics.

FUNCTIONS

1 Phagocytosis of bacteria and any foreign particles, the remains of dead tissue cells, erythrocytes, leucocytes, etc.
2 Breakdown of haemoglobin and the formation of bile pigments.
3 Assisting the immune system by the processing of antigens and the production of antibodies (see the immune response below).
4 Storage of iron.

The basis of immunity

The human body is in constant threat from foreign organisms that have a great potential to cause disease. To meet these threats there exists in the body a special defensive mechanism of immense potentiality which is mobilized when the body is invaded by foreign organisms and which is especially adapted to respond to the effects of the particular invader in question. This special mechanism is immunity. Its study (immunology) is of great importance in the understanding of disease processes and their prevention.

THE IMMUNE SYSTEM

The immune system consists of a series of specialized cells localized in tissues and organs throughout the body, e.g. the thymus, lymph nodes, spleen, bone marrow and the wall of the gastrointestinal tract. This system enables the human body to respond by means of cellular or humoral factors to the harmful effects of foreign macro-molecules, e.g. bacteria that get access to the internal environment and tend to alter its basic characteristics (the constancy of the internal environment), as is seen in disease conditions. The immune system is, therefore, one of the important homoeostatic mechanisms that maintain normal health and prevent disease.

In addition, the immunological defences operate against transplanted tissues and organs as well as malignant tumours or cancerous growth in the body. On all these fronts, the immune system specifically recognizes and selectively eliminates the foreign invader and neutralizes its harmful toxic products.

THE IMMUNE RESPONSE

Antigen is a substance capable of invoking a specific immunological response. It may be part of a virus, a bacterium or a transplanted or foreign tissue cell. Chemically, antigens may be proteins, polysaccharides, nucleic acids or a combination of these. It is now known that for any antigen there exists one or more genetically predetermined immunocompetent cells that can be stimulated to elicit an immune response. The antigen is recognized by the immunocompetent cells, either directly or after being processed by macrophages (Fig. 3.7).

Antibody (immunoglobulin) is a γ-globulin produced by special immunocompetent cells (plasma cells). Immunoglobulins have the property of interacting specifically with and binding on to the antigen in response to which they have been produced.

When an organism is first exposed to an antigen, a primary immune response is elicited, after a latent period due to division, proliferation and maturation of immunocompetent cells. It reaches a maximum in 1−2 weeks after exposure to the antigen. A good example is vaccinations, e.g. against smallpox or yellow fever. Usually the vaccination will be taken as effective only after the latent period has passed.

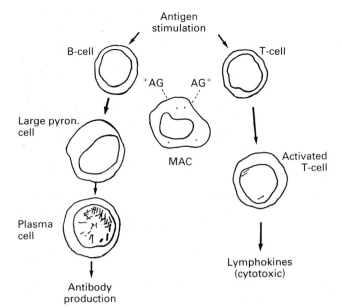

Fig. 3.7 Schematic representation of the interaction of an antigen (AG) with: (i) B-lymphocyte (B-cell), which is activated to proliferate to a large pyroninophilic (pyron.) cell which ultimately gives rise to an antibody-producing cell, i.e. humoral immune response; and (ii) T-lymphocyte (T-cell), which is stimulated to proliferate to become an activated T cell (effector cell or sensitized lymphocyte), the mediator of the cellular immune response. MAC = macrophage: helps in the processing of the antigen in the early stage of the immune response.

Upon re-exposure to the same antigen that provokes the primary response, the secondary immune response is obtained, which is more brisk. Its latent period is shorter and the antibody level rises faster and higher than in the primary response. This is because a large number of responsive cells become available to react with the antigen. The secondary immune response has given rise to the idea that the body has acquired some sort of immunological memory of the antigen due to the production of memory cells (Fig. 3.7). These memory cells have the capacity to produce both effector and memory cells upon subsequent stimulation by the original antigen.

Immunocompetent cells
Immunocompetent cells are cells capable of responding to contact with an antigen by manifesting or developing a specific cellular or humoral immunological capacity. These cells are equipped with cell surface receptors which are specific to given antigens. They also possess the mechanisms to respond to antigenic stimulation. Immunocompetent cells are lymphocytes, sometimes called virgin lymphocytes, of which there are two functionally distinct types: B-lymphocytes, responsible for humoral immune responses, and T-lymphocytes, which mediate the cellular immune responses.

Humoral immune response (Fig. 3.7)
B-lymphocytes have surface receptors for the interaction with the antigen. When the cell binds with the antigen, it is stimulated to proliferate and develop into large pyroninophilic cells, which may divide and ultimately give rise to the mature plasma cell. Plasma cells are individually capable of secreting one specific antibody or immunoglobulin. The antibody initiates the destruction of the antigen and promotes its removal by the phagocytic cells of the reticuloendothelial system. Antibody production is the manifestation of the humoral immune response.

ANTIGEN–ANTIBODY REACTION
The antigen–antibody reaction is designed to oppose the antigen as a threat to the human body. The reaction takes many forms, including neutralization, precipitation, agglutination and lysis.

Antibodies are responsible for this desirable reaction, which participates in the killing of the microbe and the neutralization of its harmful effects. However, they also have several forms of heightened reactivity to antigens collectively, named immediate-type hypersensitivity (or allergy). This is commonly seen in individuals allergic to dust, pollen of flowers or certain drugs such as penicillin.

While executing these functions, antibodies do not work on their own but get the help of the reticuloendothelial system and other phagocytic cells in the blood. The complement system is a complex group of enzymes in normal blood that work together with antibodies and play an important role as mediators of both immune and allergic reactions.

The cellular immune response
On contact with the antigen, T-lymphocytes proliferate and give rise to several types of effector cells with heightened reactivity towards the antigen. This reactivity is expressed in a variety of properties of these effector cells, among which are the following:

1 They are directly cytotoxic (hence the name killer cells) and are capable of the destruction of foreign cells, which are eventually phagocytosed by macrophages, monocytes and neutrophils.
2 They help B-lymphocytes to differentiate and proliferate — hence the name helper or T_H cells.
3 They secrete a variety of polypeptides called lymphokines which are capable of: (i) attracting macrophages and other phagocytic cells to the site of antigen–antibody reactions; (ii) stimulating further division and proliferation of lymphocytes, especially B-lymphocytes; (iii) inflicting damage on all cells except lymphocytes; and (iv) attacking a variety of viruses, thus preventing viral replication. The lymphokine with the latter ability, called interferon, has received much attention recently in view of its potential therapeutic use as an antiviral agent and also as an anticancer agent.

Cellular immunity is actively involved in the following:
1 Rejection of tissue or organ grafts, e.g. kidney transplant.

2 Delayed hypersensitivity reactions, e.g. tuberculin reaction.

3 Antitumour immunity, which involves the cellular immune mechanism in the defence against mutant or cancer cells.

4 Co-operation with B-cells in humoral immune responses.

THE ACQUIRED IMMUNE DEFICIENCY
SYNDROME (AIDS)

In a healthy person the cellular immune responses are performed with a predominance of circulating T-helper lymphocytes (T4 or CD4 cells) over T suppressor lymphocytes (T8 or CD8 cells); the normal ratio is $2:1$. The AIDS virus selectively destroys and therefore causes a decrease in the number of T-helper cells and a reversal in the T helper/T-suppressor cell ratio, with T-suppressor cells predominating. This results in generalized inhibition of the immune system and the affected person therefore loses resistance against infective agents, e.g. bacteria, viruses, etc.

Plasma proteins

Blood plasma contains 91% water and 9% solids, of which 7% are proteins and the rest are inorganic salts (mainly sodium and bicarbonate), in addition to substances being transported from one part of the body to another.

TYPES OF PLASMA PROTEINS

Plasma proteins are divided into three types with average concentrations (in g/100 ml) as follows:

Albumin	4.8
Globulins	2.3
Fibrinogen	0.3
Total	7.4

Albumin (molecular weight = 69 000)

Albumin is synthesized by the liver. Ten to 12 grams are produced each day and it has a half-life of 20 days. Because albumin is a relatively small molecule, plasma albumin accounts for 80–90% of the plasma colloid osmotic pressure (25 mmHg) although it makes up about 60% of the total plasma protein mass. Therefore, albumin is important in the formation and absorption of tissue fluids and in preserving a fluid balance

between blood and tissues. When the concentration of plasma albumin falls to low levels, less than 2 g/100 ml (hypoalbuminaemia), fluids leave the circulation readily and accumulate in excessive amounts in the interstitial compartment, leading to oedema. The commonest cause of hypoalbuminaemia in tropical countries is decreased intake of protein, as seen in children with malnutrition. Liver disease with gross damage to liver cells results in decreased synthesis of albumin. Kidney disease (nephrotic syndrome) results in excessive losses of albumin in the urine.

Albumin plays a role in the transportation of calcium, sulphonamides and bilirubin in the blood.

Globulins (molecular weight = 90 000–
1 300 000)

On the basis of electrophoretic mobility, globulins are subdivided into α_1-, α_2-, β- and γ-globulins. Alpha- and β-globulins are synthesized in the liver and γ-globulins in the reticuloendothelial system by plasma cells and lymphocytes.

Transport function of α- and β-globulins Alpha-globulins transport lipoproteins, lipids, hormones (cortisol-binding globulin, CBG; thyroid-binding globulin, TBG) and bilirubin in plasma.

Beta-globulins transport lipoproteins, lipids, cholesterol, iron (transferrin) and copper (ceruloplasmin).

The binding of vitamins A, D and K and some hormones by plasma proteins prevents them from being filtered in the kidney and lost in urine and also provides a stable blood reservoir of hormones on which tissues can draw.

Gamma-globulins (immunoglobulins) All the known antibodies are γ-globulins. The new-born child is protected by globulins which cross the placenta from the maternal to the foetal circulation. Local immunity is conferred to the gastrointestinal tract via the colostrum (first breast milk), which contains a special type of immunoglobulin (IgA). By 9 months the infant is able to produce his/her own immunoglobulins (IgG, IgM, IgD, IgE) which give active immunity against a variety of diseases. Solutions of γ-globulins are used therapeutically to give passive immunity.

A number of proteins with specific physiological functions have been isolated from the globulin fraction, e.g. prothrombin, transferrin, plasminogen, haptoglobins and many others. Haptoglobins are a group of globulins which have the distinctive property of binding free haemoglobin in the plasma. When there is excessive destruction of erythrocytes, free haemoglobin is picked up by haptoglobin, which can bind up to 150 mg Hb/100 ml blood and pass it to the reticuloendothelial system. When free haemoglobin levels exceed haptoglobin binding capacity, free haemoglobin appears in the urine (haemoglobinuria).

SUMMARY OF THE FUNCTIONS OF PLASMA PROTEINS

1 Transport functions (α- and β-globulins).
2 Defensive (immunoglobulins).
3 Reserve of body proteins.
4 Osmotic function (albumin) through control of the exchange of fluid between blood and tissues.
5 Viscosity of plasma is due mainly to fibrinogen and globulins.
6 Fibrinogen is important in blood clotting.

The blood groups

On the surface of human red blood cells are found a series of genetically determined glycoproteins and glycolipids that act as blood group antigens. They appear in early foetal life and remain unchanged throughout life. More than 100 blood antigens have been described, out of which at least 15 well-defined red blood cell group systems exist in most racial groups. Of these, only two are of major importance in clinical medicine — the ABO and rhesus (Rh) systems.

THE ABO SYSTEM

The ABO system depends on whether the *red cells* of an individual contain one, both or neither of the two blood group antigens A and B. Therefore, there are four main ABO groups: A, B, AB and O (Table 3.1).

These blood antigens are also called agglutinogens since, in the presence of the respective antibody (agglutinin), agglutination of red blood cells occurs.

The plasma may contain antibodies (agglutinins) against the A and B antigens: anti-A or alpha, anti-B or beta. These agglutinins are not present at birth but they appear between the 2nd and 8th month of life, most probably in response to A and/or B antigens taken in food of animal origin, especially meat, and in some bacteria. The anti-A and anti-B antibodies are described as naturally occurring antibodies. In the blood of any individual, the antibody present is the reciprocal of the antigen. Therefore, the ABO system can be defined with respect to both the characteristic antigen (agglutinogen) present on the red cells — and the characteristic antibody (agglutinin) present in the plasma or serum.

Inheritance of blood groups (Table 3.1)
The inheritance of the A and B antigens is dictated by the A and B genes. The O gene does not produce any demonstrable red cell antigen. This is the reason why group A genotype can be AA (homozygous) or AO (heterozygous). Similarly, for group B the possible genotype is BB or BO, while for blood group O the only possible genotype is OO. Group AB has both A and B genes and the only possible genotype is AB. Knowledge of these genotypes is useful in working out the probable

Table 3.1 Cell antigens, plasma agglutinins and genotypes of the ABO blood groups

Blood group	Antigen or agglutinogen in the red cells	Antibodies or agglutinins in plasma	Genotype
A	A	Anti-B (B)	AA, AO
B	B	Anti-A (A)	BB, BO
AB	A and B		AB
O	None	Anti-A and anti-B	OO

blood group of an offspring on the basis of the knowledge of the blood genotypes of the father and mother. It is also helpful in sorting out disputed parentage of the child.

Frequency distribution of blood groups
The frequency of the different blood groups in the ABO system varies widely in different ethnic groups, as is shown in Table 3.2.

THE RHESUS (Rh) BLOOD GROUP SYSTEM
The Rh system is described on the basis of the presence or absence of the rhesus antigen (D) on the surface of red blood cells. If present, the individual is said to be D-positive or Rh-positive; 85% of Europeans, 90–95% of Arabs and Africans and 98% of Asians are Rh- or D-positive. If absent, the individual is described as D- or Rh-negative.

Rhesus antigens
There are at least three sets of alternative antigens in the Rh system: D or d, C or c, E or e. However, D is a strong antigen and therefore clinically more important than the others. In blood banks, Rh grouping is performed with anti-D serum.

Rhesus antibodies
Rhesus antibodies are not naturally occurring antibodies but are described as immune antibodies, i.e. developed in response to a known antigenic stimulus. Anti-D antibody may be acquired as a result of immunization resulting from:
1 Transfusion of a Rh-negative individual with Rh-positive blood. This mishap usually occurs if the technician typing the blood in a blood bank makes a clerical error when determining the Rh blood group of either the donor or recipient or when issuing blood to a patient.
2 Presence of a Rh-positive foetus in a Rh-negative mother. Although foetal and maternal circulations are separate, entry of foetal Rh-positive cells into the maternal circulation may occur in the later weeks of pregnancy but mostly at the time of placental separation and delivery of the child. These Rh- or D-positive cells may provoke the production of anti-D.

Importance of blood groups in clinical medicine

BLOOD TRANSFUSION
The importance of blood groups becomes apparent when one considers the necessity of transfusing blood from one individual to another. For a safe blood transfusion, the following steps are followed:
1 The ABO and Rh blood grouping of both donor and recipient should be carried out.
2 When a suitable donor's blood for the recipient is found, a cross-matching test is done. The donor cells are mixed with the recipient's serum to find out whether the recipient's plasma contains an antibody which may react with the donor cells. For the same reason, another cross-match is carried out between the recipient's cells and the donor's plasma.

Table 3.3 summarizes the possible combinations that may arise in blood transfusion practice. It is clear that group AB recipients, because their plasma contains neither anti-A nor anti-B antibodies, can safely receive blood from donors of all ABO groups — thus the name universal recipient. Similarly, group O donors are referred to as universal donors, since their cells lack both antigens A and B and therefore their blood can be given safely to recipients of all ABO

Table 3.2 Distribution of the ABO blood groups in different ethnic populations (%)

	A	B	AB	O
Europeans	40	8	2	50
Arabs	27	20	4	49
Asians	21	28	7	44
West Africans	26	19	4	51

Table 3.3 Possible combinations that may occur in blood transfusion

	Donor red cell			
Recipient	AB	A	B	O
AB	−	−	−	−
A	+	−	+	−
B	+	+	−	−
O	+	+	+	−

+, reaction; −, no reaction.

groups. In the case of the universal recipients, the antibodies A and B in the plasma of group O, A or B donated blood are diluted in group AB recipients' plasma so that the antibodies are unlikely to react with the recipients' cells. For the universal group O donors, the anti-A and anti-B present in the donors' plasma are diluted in the recipients' plasma and therefore will not react with the recipients' cells. However, in recent years haemolytic reactions have arisen due to the use of universal donor blood containing anti-A or anti-B at a high titre. When used for transfusion of patients other than group O, these antibodies react with the recipient cells and cause a haemolytic reaction. The cardinal rule in blood banks nowadays is to transfuse patients with blood of the same ABO group; low-titre group O donors are used as universal donors only in emergency situations.

Similarly, when the universal recipients are transfused with group A, group B or group O blood with a high titre of anti-A or anti-B, a transfusion reaction may occur. Therefore, group AB recipients should always be transfused with group AB blood.

Complications of blood transfusion

Transfusing blood and blood products (fresh frozen plasma, platelet concentrate, etc.) into patients is common hospital practice in accidents and other emergency situations or during surgical operations. Strict precautions are followed before giving blood or any of its products to avoid certain reactions that may accompany the transfusion.

Most of the complications of blood transfusion result from antigen–antibody reactions between donor and recipient (incompatible blood transfusion). The outcome may be haemolytic, allergic or febrile reactions, which may occur at the time of giving blood or may be delayed for days or weeks. Other complications include transmission of disease from the donor to the recipient, e.g. malaria, syphilis, viral hepatitis and the AIDS virus. Nowadays in blood banks, donated blood is routinely screened for the presence of the infective agents that transmit syphilis, viral hepatitis and, more recently, AIDS.

In the case of a mismatched transfusion, agglutination of the donor cells and eventually their lysis occur. The donor cells are attacked by antibodies in the recipient's plasma. Donor plasma agglutinin rarely acts on recipient cells. This is because the donor's plasma is diluted by the recipient's plasma and therefore the concentration (titre) of the donor's agglutinin becomes too low to have any effect. Haemolysed red cells in the peripheral blood vessels are removed by phagocytic cells. The haemoglobin liberated can pass through the capillaries into the tissue fluid and can also filter through the glomeruli of the kidney and therefore appears in the urine (haemoglobinuria). Haemoglobin is phagocytosed and is converted to bilirubin. If the concentration of bilirubin in plasma is high enough, haemolytic jaundice results.

One of the most serious reactions to mismatched transfusions is acute kidney failure, which can be fatal. The kidney shuts down shortly following the transfusion. This results from circulatory shock and a marked drop in the renal blood flow and also from the precipitation of haemoglobin in the renal tubules, blocking a large number of tubules, which eventually leads to renal shut-down.

RHESUS INCOMPATIBILITY BETWEEN MOTHER AND FOETUS (HAEMOLYTIC DISEASE OF THE NEWBORN)

Rhesus incompatibility arises when a Rh-negative woman gets pregnant and the foetus is Rh-positive, having inherited the D gene from the father. As mentioned earlier, the mother may develop anti-D antibody as a result of such an incompatible pregnancy or if she has received incompatible (D+) blood before pregnancy. The first baby usually escapes the effects. Such immune antibodies, usually of the IgG type, may cross the placenta in future pregnancies. If the second foetus is Rh-positive, the immune anti-D may react with and destroy foetal red cells, resulting in the disease called haemolytic disease of the new-born or erythroblastosis foetalis. The foetus may be born with haemolytic anaemia, which, if severe, may necessitate exchange blood transfusion. Blood group O Rh-negative is infused in the new-born at one point while the haemolysed blood is gradually withdrawn from the infant at another point. Haemolytic disease of the new-born has an overall reported incidence of one in 200 of all pregnancies. This figure has decreased further with the use of prophylactic anti-D antibody, which is injected into Rh-negative mothers immediately after delivery of their first Rh-positive child. The injected antibody will react with and get rid of foetal Rh-positive cells before they immunize the mother.

Haemostasis

Haemostatic mechanisms prevent blood loss from intact vessels and stop excessive bleeding from severed vessels. Normal haemostasis is dependent upon a complex series of interactions between several components, which include: (i) local vascular factors; (ii) platelets; (iii) coagulation (fibrin formation); and (iv) fibrinolysis.

Bleeding seen in clinical practice often results from a defect in more than one of these components. Understanding of the integrated concept of the physiology of haemostasis is best achieved by considering the role of each of the above components.

VASCULAR FACTORS

Localized vasoconstriction is the immediate response of the vessel wall to injury. It is transient and lasts for less than 1 minute. The mechanism of vascular constriction is uncertain, but systemic or localized humoral and neural factors are believed to be important, particularly for the initial intense local constriction. The humoral factors involved include the localized release of the vasoconstrictor serotonin (5-hydroxytryptamine) from platelets and the systemic release of catecholamines from the adrenal medulla. However, the ability of the injured vessel to respond to these factors may be influenced by the general metabolic state of the patient. Acute hypoxia and metabolic acidosis diminish this important response.

PLATELETS

Platelets originate in the bone marrow, where the most recognizable primitive precursor, the megakaryoblast, matures through the stages of promegakaryocyte and megakaryocyte. This last cell has a diameter of 24–30 μm. Platelets of diameter 2–3 μm are formed by the breaking off of pieces of the cytoplasm of the megakaryocyte. Once released from the bone marrow, the platelets

pass through the spleen before reaching the bloodstream, where they survive for 3–5 days. The normal platelet count in peripheral blood is 100 000–400 000 platelets/mm^3 (average 250 000).

The functions of platelets can be summarized in the following processes:

1 The interaction of platelets with the vessel wall. Platelets in the circulation are in a quiescent state and are discoid in shape. A few seconds after injury to a blood-vessel, platelets rapidly adhere to the damaged area. Collagen exposed in the vessel wall attracts platelets to adhere to it and is also responsible for triggering the release of endogenous platelet adenosine diphosphate (ADP), which increases the stickiness of platelets.

2 The interaction of platelets with one another (platelet aggregation). In the presence of released ADP, platelets aggregate with one another to form a fragile and reversible primary haemostatic plug, which is not capable of satisfactory haemostasis.

3 Soon after aggregation, platelets start releasing the following substances:

(a) 5-HT, a potent vasoconstrictor.

(b) Platelet phospholipid or platelet factor 3 (PF3), which is essential in the process of blood coagulation.

(c) Proteins, e.g. platelet β-thromboglobulin (β-TG) and platelet factor 4 (PF4, heparin-neutralizing factor).

(d) Thromboxane A$_2$ (TXA$_2$), a prostaglandin formed from arachidonic acid. It is a potent vasoconstrictor and aggregating agent. Aspirin ingestion inhibits TXA$_2$ formation and prevents platelet aggregation. This is the basis of the use of aspirin in the prevention of thrombotic disorders, e.g. coronary heart disease.

The further stability of the platelet plug requires the availability of thrombin, which is generated as a result of the simultaneous activation of the coagulation system. Thrombin induces structural changes in platelets (so-called viscous metamorphosis). The platelets swell and interlock with each other. The simultaneous formation of fibrin, which infiltrates the platelet plug, results in a reinforced (concrete-like) structure. This final step is described as irreversible aggregation.

Platelets contribute in several ways to blood clotting. They contribute platelet phospholipids, without which coagulation fails. Platelet phospholipid (PF$_3$) is an integral part of the platelet membrane. Once the platelet is activated, the membrane phospholipid undergoes certain physical changes whereby it becomes exposed to the plasma. This phospholipid is essential for the process of blood coagulation. Clot retraction to about 10% of the original clot size is due to platelets. Platelet pseudopodia attached to fibrin contract, pulling the fibrin fibrils tightly together. This is due to the presence in platelets of large quantities of contractile proteins, the muscle proteins actin, myosin, tropomyosin and troposin. Moreover, there exists within the platelet a calcium ion-regulating mechanism resembling the sarcoplasmic reticulum of muscle cells. When internal contraction occurs in platelets, it facilitates the following:

1 Extrusion of the platelet release products.

2 Contraction of the aggregates.

3 Retraction of the blood clot.

4 Sealing of the porous haemostatic plug.

Blood coagulation

The coagulation mechanism is a complete set of biochemical reactions which terminate in the production of the enzyme thrombin, which, in turn, changes the soluble plasma protein fibrinogen into insoluble fibrin. A large number of chemical substances participate in this process; however, only 12 clotting factors have been identified and given numerical designations in the order of their discovery. A simplified schematic representation of the interactions of these clotting factors is shown in Fig. 3.8. This is based on the McFarlane cascade or waterfall hypothesis.

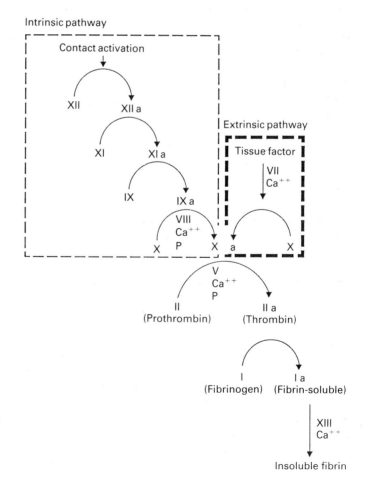

Fig. 3.8 A diagrammatic representation of the interaction of the various clotting factors in the intrinsic and extrinsic pathways of the activation of the coagulation mechanism.

ACTION OF THE COAGULATION SYSTEM

The clotting factors circulate in the bloodstream in the form of inactive precursors. Their activation may occur along two pathways.

Intrinsic pathway

The intrinsic pathway is dependent on certain clotting factors, all of which are present in the blood—hence the name intrinsic pathway. The trigger to this pathway is the activation of factor XII by contact with a foreign surface, i.e. any surface other than the smooth natural endothelial surface. Therefore, activation of factor XII may occur when blood comes in contact with exposed collagen when the vessel wall is injured. Activation also occurs when blood comes in contact with glass when coagulation is tested for in the laboratory. Once factor XII is converted into active factor XII (XIIa), it activates factor XI to form active factor XI (XIa), which, in turn, activates factor IX to form active factor IX (IXa). Active factor IX, in the presence of factor VIII (anti-haemophilic globulin), platelet phospholipid and calcium ions, activates factor X to activate factor X (Xa). Beyond this point the blood coagulation mechanism continues along a pathway common to both intrinsic and extrinsic mechanisms of the clotting system.

Extrinsic pathway

The extrinsic pathway is triggered by a material (tissue factor or tissue thromboplastin) from damaged tissues, which forms a complex with factor VII, which, in the presence of calcium ions, activates factor X to Xa. Thus, the extrinsic pathway takes a shorter time than the intrinsic pathway.

The common pathway for both the intrinsic and extrinsic mechanism of activation of the clotting system proceeds as follows (Fig. 3.8). Active factor X (Xa), in association with factor V, platelet phospholipid (platelet factor 3, PF3) and calcium ions, forms what is known as the thrombokinase complex, which converts prothrombin to thrombin. Thrombin is a proteolytic enzyme which splits two short peptide chains, fibrinopeptides A and B, from both ends of the fibrinogen molecule. The resulting residues, called fibrin monomers, join end to end to form long strands of fibrin

polymer, which aggregate to form a three-dimensional network of fibrin. As such, this aggregate is soft and pliable and is therefore not strong enough to give a satisfactory haemostatic plug. Final bonding and polymerization of these fibrin monomer strands to form a more stable fibrin gel requires the presence of factor XIII (fibrin-stabilizing factor) and calcium ions.

Summary

1 Both intrinsic and extrinsic pathways are necessary for normal haemostasis.
2 Both pathways are activated when blood leaves the blood-vessels for the tissues.
3 Thrombin is a key factor in both the intrinsic and extrinsic systems, in addition to its action on fibrinogen.
4 The activation of the clotting mechanism along the shorter extrinsic pathway results in the rapid formation of thrombin, which feeds back to activate the intrinsic pathway through factors VII and V. Factor VII can activate factor IX to active factor IX, and this forms an activation connection between both pathways.
5 Thrombin stimulates platelets to release ADP and TXA_2 and therefore enhances further aggregation of platelets.
6 Thrombin is essential for platelet morphological changes during haemostasis (viscous metamorphosis), which lead to the formation of the primary haemostatic plug.

Table 3.4 gives a summary of the roles of blood-vessels, platelets and blood coagulation in haemostasis.

Fibrinolysis

Fibrinolysis is the enzymatic breakdown of fibrin. Fibrinolysis is a normal function of the blood and it makes a major contribution to the maintenance of the patency of blood-vessels in health and disease. It can be considered as a counterbalance to the equally important clotting process, the former laying down fibrin and the latter removing it. If either side of the balance is tipped, the end result will be either a bleeding tendency due to excessive digestion of fibrin, or thrombosis and blocking of blood-vessels due to excessive clotting. Fibrinolysis, in the physiological sense, is produced by the enzyme plasmin, which is derived

Table 3.4 Summary of the mechanisms involved in haemostasis

Damage to blood vessels	Role of platelets	Blood clotting
Altered surface	Adhere to collagen	Activation of clotting factors
Exposed collagen	Release ADP, TXA_2	
Release of ADP	Platelet aggregation and seal small blood-vessels	Thrombin formation
Release of tissue factor	Release of 5-HT, vascular contraction	Fibrin formation
		Fibrin stabilization
Stimulation of vascular contraction	Release of platelet phospholipid (PF_3)	Sealing of large vessels

from a stable plasma protein precursor, plasminogen (a β-globulin) (Fig. 3.9). The conversion of plasminogen into plasmin requires activators which are present in the blood, tissues and urine. Plasmin digests unwanted intra- or extravascular deposits of fibrin, which is the desired physiological effect. However, it is a non-specific proteolytic enzyme, which can digest certain clotting factors, including fibrinogen, factor V and factor VIII. Under normal circumstances the fibrinolytic enzyme system is kept in check by inhibitors of plasminogen activators (antiactivators) and also by inhibitors of plasmin (antiplasmins) which are produced by the liver. The liver also inactivates circulating plasminogen activators. In recent years, tissue plasminogen activator (tPA) has been produced from cultured melanoma cells in large quantities and is currently used with some success in dissolving coronary clots (thrombi) in patients who suffer acute coronary thrombosis.

ANTICOAGULANTS

Anticoagulants are substances which inhibit the process of blood coagulation. The following anticoagulants are in common use in clinical practice:

1 *Substances which remove calcium ions from the blood:*

(a) Citrate, used in collection of blood in blood banks.

(b) Oxalate and ethylene diamine tetra-acetate (EDTA), used for blood collection in clinical laboratories.

2 *Heparin* is a mucopolysaccharide found in many tissues, with the highest concentration in the liver, lungs, granules of mast cells and basophil leucocytes. Heparin is a very potent anticoagulant. It interferes with the clotting process in two ways: first, it acts as a direct antithrombin and, second, it prevents the conversion of prothrombin to thrombin.

Heparin is commonly used for keeping blood in

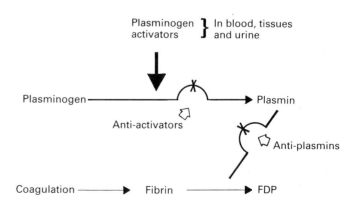

Fig. 3.9 The fibrinolytic enzyme system.

the fluid state during open-heart surgery and to prevent the occurrence of venous thrombosis in the postoperative period. It is also used for blood collection in laboratories.

3 *Warfarin and phenindione.* These substances suppress the synthesis of vitamin K-dependent factors (II, VII, IX and X) by the liver. Warfarin and phenandoin are widely used in clinical practice. Their advantage over heparin is that they are given by mouth while heparin can only be given by injection. Their effect continues for up to 48 hours while the action of heparin lasts for 6−8 hours.

Naturally occurring anticoagulants
Naturally occurring anticoagulants are present in small concentrations in normal plasma, where their main function is to keep the clotting mechanism in check. The best known of the naturally occurring anticoagulants are antithrombin III and protein C, which inactivate clotting factors VIII and V.

Haemorrhagic (or bleeding) disorders
Haemorrhagic disorders are disease conditions characterized by an excessive bleeding tendency due to a defect in the haemostatic system. Bleeding disorders may be caused by:
1 Vascular defects: either congenital or acquired.
2 Platelet defects: either a decrease in the number of circulating platelets (thrombocytopenia) or a defect in platelet function (thrombocytopathy).
3 Coagulation defects: either congenital or acquired.

CONGENITAL COAGULATION DEFECTS
The commonest congenital bleeding disorder is haemophilia A, due to factor VIII deficiency, and haemophilia B, or Christmas disease, due to deficiency of factor IX. Haemophilia is characterized by excessive bleeding, which may occur spontaneously in the skin, mucous membranes and joints, or after minor trauma or surgery, e.g. tooth extraction. Bleeding in these situations can only be stopped by replacing the missing clotting factor in the form of fresh blood transfusion, fresh frozen plasma or clotting factor concentrate.

ACQUIRED COAGULATION DEFECTS
Acquired defects may be due to diminished production of a clotting factor, as in liver disease and vitamin K deficiency, or to consumption of platelet and clotting factors, as in the serious condition disseminated intravascular coagulation (DIC), in which extensive activation of the clotting system occurs *in vivo*. This may end in fatal bleeding.

Further reading
1 Firkin, F., Chesterman, C., Pennington, D. & Rush, B. (eds) (1989) *de Gruchy's Clinical Haematology in Medicinal Practice*, 5th edn. Blackwell Scientific Publications, Oxford.

4: Excitable Tissues

Objectives

On completion of the study of this section, the student should be able to:

1 Understand the basic properties of cell membranes which underlie the process of excitation.

2 Comprehend the ionic basis of the resting and action potentials of the cell membrane, particularly that of nerve and muscle so as to make appropriate use of diagnostic techniques such as electrocardiography, electromyography and electroencephalography.

3 Explain the basis of the process of transmission at synaptic junctions.

4 Describe the molecular basis of contraction and relaxation of muscle fibres in order to explain the links between the electrical and mechanical events.

Introduction

Receptors in the skin and other sense organs act as transducers which change various stimuli such as heat, cold, light or sound into electrical signals. Nerve fibres are responsible for conduction of these signals, which are known as nerve impulses, to the central nervous system (CNS) and also from the various parts of the CNS to effector organs such as muscles and glands. Nerves and muscles are called excitable tissues because they respond to chemical, mechanical or electrical stimuli. Muscle cells demonstrate this excitation by producing mechanical contractions which take place in skeletal muscle, heart muscle and smooth muscle cells present in hollow viscera such as the stomach, intestine and bladder and in the blood-vessel walls. However, in a general sense all living cells are excitable because they respond to external stimuli and possess special properties which reside in the cell membrane. The most important of these properties of cell membranes necessary for excitation are:

1 They can maintain different concentrations of positively charged cations and negatively charged anions outside and inside their membranes (see Chapter 1).

2 Their permeability changes when they are appropriately stimulated.

3 When the permeability changes, only specific types of ions can pass in a certain direction.

A stimulus on the cell membrane first produces a change in permeability (or conductance). This is followed by movement of ions across the cell membrane. As the ions have different electrical charges, their movement from one side of the cell membrane to the other produces a difference in electrical charge between the outside and inside of the cell membrane. This difference in voltage is called the potential. The difference in voltage across the cell membrane at rest (i.e. not being stimulated) is called the resting membrane potential. When the cell membrane is stimulated the resting membrane potential is temporarily reversed and an action potential is produced. It is the action potential that is transmitted along nerves and causes muscles to contract and glands to release secretions.

In summary, a stimulus produces change in membrane permeability or conductance which leads to movement of ions across the cell membrane and thus results in a voltage difference or action potential. This chapter will explain the main mechanisms concerned with maintenance of the resting membrane potential and the production of the action potential. Furthermore, the mechanisms whereby the action potential causes muscle to respond will be described.

Ionic distribution across the cell membrane

The composition of the interstitial fluid (the internal environment) must be kept constant (as explained in Chapter 1). It is essential that the difference in ionic gradients between the inside of the cell and the outside is maintained for the resting and action potentials to develop. If no action potential can be produced by a maximal stimulus, then the cell is considered inexcitable or 'dead'.

Origins of the resting cell membrane potential

A number of forces act on cell membranes. These forces are responsible for: (i) the maintenance of the resting membrane potential; (ii) the development of the action potential; and (iii) bringing the cell back to its resting state after the action potential is over. These forces are diffusion, electrical gradients and active transport.

DIFFUSION

Diffusion is the movement of molecules from a place of higher concentration to a place of lower concentration. The rate of diffusion is proportional to the difference in concentration of the molecules in the two places. This is called the concentration gradient or chemical gradient. Ions are electrically charged (Na^+, K^+, Cl^-) and therefore positive ions move to the negative area and negative ions move to the positive area. This is called the electrical gradient.

ELECTRICAL GRADIENT

In Chapter 1 we have learned that there is a high concentration of non-diffusible protein anions in the intracellular fluid. The presence of such ions on one side of the membrane produces a form of equilibrium known as the Donnan effect. The negative charge of protein molecules hinders entry of diffusible anions and favours entry of cations (Fig. 4.1). The Donnan effect predicts that the distribution of diffusible ions in this case will be such that their concentration ratios inside (i) and outside (o) the membrane will be equal, as follows:

$$\frac{[K^+]_i}{[K^+]_o} = \frac{[Cl^-]_o}{[Cl^-]_i}$$

Also, the product of the concentrations of diffusible ions of the same valence on each side will be equal, i.e.:

$$[K^+]_i \times [Cl^-]_i = [K^+]_o \times [Cl^-]_o$$

Equilibrium potentials and the Nernst equation
If the cell membrane is permeable to only one ion, the potential produced by diffusion of that ion is called the equilibrium potential (E). If a cell is placed in a medium similar to the extracellular fluid (ECF) but containing K^+ only, K^+ will diffuse out of the cell (efflux). This occurs because K^+ concentration (chemical gradient) inside the cell is much higher (concentration gradient). As this outward movement continues, there is a gradual build-up of an opposing force acting in the opposite direction. This force, which tends to keep K^+ inside the cell, is the electrical gradient due to the attraction between the positively charged K^+ and the negatively charged intracellular protein and other intracellular anions. As more and more K^+ diffuses out, this electrical gradient will progressively increase until it ultimately reaches a value which balances the force of the chemical gradient and prevents further net K^+ efflux. At this point where the electrical gradient (potential) becomes equal to the chemical gradient, K^+ is said to be at electrochemical equilibrium and the potential is referred to as the K^+ equilibrium potential (E_{K^+}). The Nernst equation is used to calculate the equilibrium potential of individual ions.

In the case of positively charged ions such as K^+, the Nernst equation is:

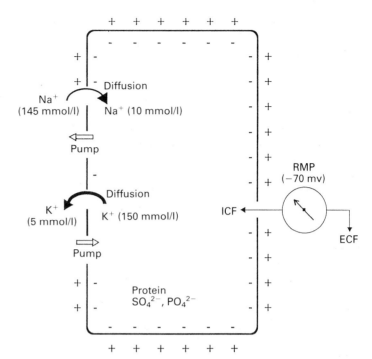

Fig. 4.1 Forces acting across the cell membrane and the resting membrane potential. ECF, extracellular fluid; ICF, intracellular fluid.

$$E_{K^+} = 61 \log \frac{[K^+]_o}{[K^+]_i} = 61 \log \frac{5 \text{ mM/l}}{150 \text{ mM/l}} = -90 \text{ mV}$$

and in the case of negatively charged ions such as Cl^-:

$$E_{X^-} = 61 \log \frac{[X^-]_i}{[X^-]_o} \text{ (in mV)}$$

where E_x is the equilibrium potential of the ion X, $[X_i]$ is the concentration of X inside the cell, and $[X_o]$ is the concentration of X in the extracellular medium. The Nernst equation depends on two assumptions:

1 The membrane is freely permeable to the ion being examined.
2 There is no other ion exerting an opposing influence in the medium.

The equilibrium potential for Na^+ or Cl^- can be calculated as follows:

$E_{Na^+} = 61 \log 150/15 = +61 \text{ mV}$
$E_{Cl^-} = 61 \log 125/9 = -70 \text{ mV}$

The sign of the potential refers to the polarity of the inside of the membrane.

Equilibrium potentials are never reached in normal cells because, when one ion is moving in one direction, other ions are moving in the opposite direction and this tends to prevent the development of the equilibrium potential. The equilibrium potential can be measured for a particular ion only under experimental conditions.

ACTIVE TRANSPORT
Active transport requires energy because it transports ions against their concentration gradients. The most important active transport system is the sodium–potassium pump, which is responsible for transport of Na^+ to the outside and K^+ to the inside against their concentration gradients. There is an enzyme called Na^+–K^+ adenosine triphosphatase (ATP-ase) present on the cell membrane. This enzyme is activated by Na^+ and K^+ to hydrolyse ATP and release the energy needed to pump Na^+ to the outside of the cell and K^+ to the interior. The Na^+–K^+ pump is responsible for keeping the different concentrations of Na^+ and K^+ in resting cells and also for bringing the cell back to the resting state after excitation.

The resting membrane potential

If a microelectrode is inserted into a cell and another electrode is placed into the interstitial fluid surrounding the cell and the two electrodes are connected to a voltmeter, the inside of the cell will show a negative deflection of about -70 mV (-70 to -90) with respect to the outside (Fig. 4.1). This potential is called the resting membrane potential. It is negative inside the cell with respect to the outside because of the following:

1 The resting cell membrane is $10-100$ times more permeable to K^+ than to Na^+. K^+ tends to leak out of the cell down its concentration gradient, carrying positive charge with it, and is unable to carry Cl^- with it because Cl^- has a higher concentration outside. This tends to make the cell interior more negative.

2 The non-diffusible anions (protein, sulphate and phosphate ions) cannot leave the cell. According to the Donnan effect, there is a slight excess of cations outside the cells and a slight excess of anions inside. This leaves the inside negative with respect to the outside.

3 A very small amount of sodium also diffuses into the cell down its concentration gradient because the cell membrane is only slightly permeable to Na^+. If K^+ continued to leak to the outside and Na^+ to the inside, a state of equilibrium would be reached and there would be no potential difference between the two sides of the membrane. This does not happen because of the Na^+-K^+ pump.

The main function of the Na^+-K^+ pump is to maintain the concentration gradients between the inside and outside of the membrane. Leaking K^+ and Na^+ ions are returned. The pump returns two K^+ ions to the inside of the cell while it removes three Na^+ ions from the inside. In this way the Na^+-K^+ pump maintains the electro-negativity of the cell membrane. Inhibition of the Na^+-K^+ pump, e.g. by metabolic inhibitors, leads to a decrease in the resting membrane potential (RMP).

The resting membrane remains polarized because the movements of ions at rest do not favour either an inward or an outward flux. A state of equilibrium is maintained.

The Nernst equation, though useful for the determination of the equilibrium potential of individual ions, is not accurate for the calculation of the RMP. For this purpose, the Goldman equation is more suited. It takes into account the equilibrium potential of the three important ions (K^+, Na^+ and Cl^-) which exist together in the body fluids. It also takes into consideration the differential membrane permeability to these ions:

$$\text{RMP (mV)} = 61 \log \frac{P_{K^+}\,[K^+]_o + P_{Na^+}\,[Na^+]_o + P_{Cl^-}\,[Cl^-]_i}{P_{K^+}\,[K^+]_i + P_{Na^+}\,[Na^+]_i + P_{Cl^-}\,[Cl^-]_o}$$

where P_{K^+}, P_{Na^+} and P_{Cl^-} are membrane permeabilities to K^+, Na^+ and Cl^- respectively; $[K^+]_o$, $[Na^+]_o$ and $[Cl^-]_o$ are extracellular concentrations of K^+, Na^+ and Cl^-; and $[K^+]_i$, $[Na^+]_i$ and $[Cl^-]_i$ are intracellular concentrations of K^+, Na^+ and Cl^-.

Certain facts need to be pointed out:

1 Since RMP $= E_{Cl}$ (-70 mV), it becomes evident that Cl^- distribution across the membrane depends exclusively on physiochemical forces and needs no active transport to maintain this distribution (unlike K^+ and Na^+, which need the Na^+-K^+ pump). There is evidence that Cl^- does not contribute to the value of the RMP and changing the extracellular Cl^- concentration does not affect the RMP. Hence, Cl^- can be omitted from the above equation, which becomes:

$$\text{RMP (mV)} = 61 \log \frac{P_{K^+}\,[K^+]_o + P_{Na^+}\,[Na^+]_o}{P_{K^+}\,[K^+]_i + P_{Na^+}\,[Na^+]_i}$$

2 Because P_{K^+} is almost one hundred times P_{Na^+}, the RMP depends primarily on K^+ with very little contribution from Na^+. Increasing the extracellular potassium concentration causes a marked reduction in the RMP, whereas changing the extracellular sodium concentration does not significantly affect the RMP.

The action potential (AP)

The action potential is a sudden reversal of membrane polarity produced by a stimulus. Action potentials occur in living organisms to produce physiological effects such as:

1 Transmission of impulses along nerve fibres.

2 Release of neurosecretions or chemical transmitters in synapses.

3 Contraction of muscle.

4 Activation or inhibition of glandular secretion.

Many cellular functions in animals and plants are induced by action potentials. During stimulation of muscle or nerve, an action potential can be recorded by inserting a microelectrode inside a cell with a reference electrode in the extracellular fluid (Fig. 4.2). By this method a monophasic potential is recorded. Action potentials recorded from the surface of nerves or muscles using two-surface electrodes are called biphasic because they have two waves (Fig. 4.3). The different action potentials can be recorded by a cathode ray oscilloscope which transforms changes in voltage into the graphic records shown in Figs 4.2 and 4.3.

DEVELOPMENT OF THE AP

When a cell membrane is stimulated by a physical or a chemical stimulus, the cell membrane permeability to Na^+ is dramatically increased. Sodium channels open and the sodium ions rush through the channels to the inside of the cell, causing the inside of the membrane to become positive with respect to the outside. This is called depolarization because it changes the membrane polarity from its polarized negative state. The membrane potential actually becomes reversed and reaches +35 mV.

At the end of depolarization, sodium permeability stops and potassium permeability increases abruptly and K^+ ions leave the cell down their concentration gradient, causing the inside of the membrane to return quickly to its original potential. This is called repolarization and the membrane potential is brought back to −70 mV. The duration of the action potential in skeletal muscle and nerves (depolarization and repolarization) is about 1−5 ms (1/1000 s).

The Na^+−K^+ pump now starts to move sodium out of the cell and potassium to the inside against their concentration gradients so that the membrane regains its resting membrane potential and is ready for another action potential. Figure 4.4 gives a graphic representation of the changes in

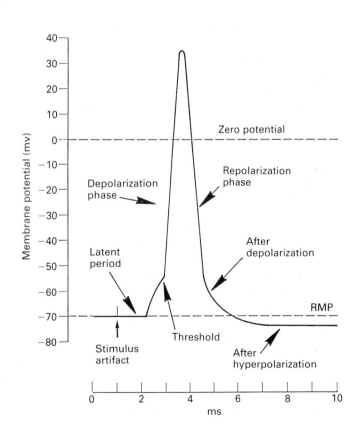

Fig. 4.2 Monophasic action potential recorded with one of the recording electrodes inside the cell.

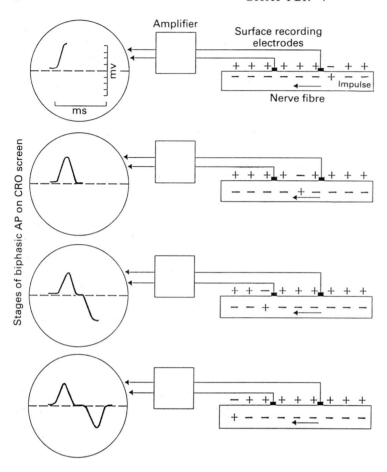

Fig. 4.3 Biphasic action potential recorded with both of the recording electrodes outside the cell. CRO, cathode ray oscilloscope.

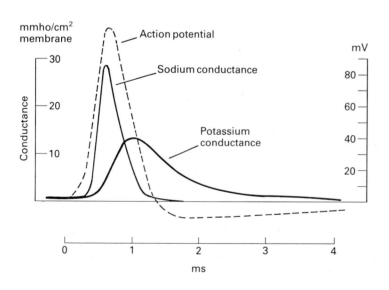

Fig. 4.4 Graphic representation of the changes in Na$^+$ and K$^+$ permeability (conductance) of the cell membrane during the action potential. (Reproduced with permission from Hodgkin, A.L. (1958) *Proc. R. Soc. Lond. (Biol).*, **148**, 1, based on Hodgkin, A.L. & Huxley, A.P. (1952) *J. Physiol.* **117**, 500–44.)

permeability of the cell membrane which take place during the action potential.

A stimulus which is just strong enough to move the RMP from its resting value (−70 mV) to the level (−55 mV) that leads to production of action potential is called a threshold stimulus (Fig. 4.5). Subthreshold stimuli result only in transient local responses which are proportional in amplitude (height) to the strengths of the corresponding subthreshold stimuli. In each of these instances the stimulus leads to the opening of some sodium channels, resulting in Na⁺ influx and membrane depolarization. This, however, is short-lived because it is promptly counteracted by local repolarizing processes, which overcome it and return the membrane potential (MP) to −70 mV. These local repolarizing processes, which tend to resist any change in the membrane potential, comprise potassium efflux and chloride entry into the cell.

The threshold stimulus, on the other hand, succeeds in taking the MP to the critical value of −55 mV at which the local repolarizing processes can no longer overcome and stop depolarization. From here onward, opening of Na⁺ channels leads to opening of more and more Na⁺ channels

and a chain reaction occurs. This progressive increase in Na⁺ conductance (permeability) is, of course, accompanied by Na⁺ influx, which causes depolarization and moves the MP toward E_{Na^+} (+60 mV). This value, however, is never reached because depolarization gets switched off at an MP value around +30 to +40 mV by two factors:

1 Inactivation and closure of Na⁺ channels, which occurs soon and automatically after their opening. Inactivated Na⁺ channels cannot be reactivated (opened) until the MP has returned to a value close to the RMP. This explains the occurrence of the absolute refractory period.

2 Delayed opening of K⁺ channels takes place and leads to K⁺ efflux, which is responsible for repolarization of the membrane. Sometimes a third component of the AP is observed: this is a hyperpolarization coming after repolarization. It is an inconsistent finding and is of variable duration. Finally, the Na⁺−K⁺ pump takes Na⁺ out of the cell and K⁺ into it and helps to restore the RMP, making the cell ready for another AP.

RECORDING OF APs
Action potentials are recorded by oscillographs and oscilloscopes. When the recording is obtained

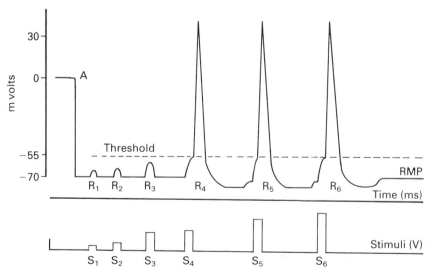

Fig. 4.5 Effect of strength of electrical stimulation. A, point of entry of microelectrode into the cell. Local graded responses (R₁, R₂ and R₃) are produced when subthreshold stimuli (S₁, S₂ and S₃) are applied. Response R₄ is due to a threshold stimulus. Responses R₅ and R₆ are due to supra threshold stimuli (S₅ and S₆). Note that the AP does not increase in amplitude with increased strength of stimulation.

between an intracellular active (or recording) electrode and an extracellular (or reference) electrode, the potential is called a monophasic potential, because one positive or one negative deflection is recorded. When both the recording and reference electrodes are placed on the surface of a muscle or nerve, two deflections or a biphasic potential will be produced by the stimulus. When the part of the nerve under the recording electrode is depolarized, its outside becomes negative and the recording electrode will record a negative deflection in relation to the reference electrode. When the depolarization wave reaches the reference electrode, the recording electrode will record a positive deflection in relation to the reference electrode (Fig. 4.3).

CHANGES IN CELL EXCITABILITY
DURING AP
During depolarization and for a short period following it, the nerve or muscle cannot be excited by even the strongest stimulus. This is called the absolute refractory period (ARP). A second stimulus given in the early part of ARP cannot produce a second AP over the first one because Na^+ channels are already maximally open and cannot be opened further. Similarly, if given in the latter part of ARP, a second stimulus cannot produce a second AP because Na^+ channels are now in the inactivation phase and cannot be opened. The ARP is followed by the relative refractory period (RRP), which covers repolarization and hyperpolarization. During the RRP, an action potential of a lower amplitude than normal can be elicited, but only by a stimulus stronger than the threshold stimulus (i.e. by a suprathreshold stimulus). The RRP may be as short as 0.4 ms in the big nerve fibres (A fibres) and as long as 4 ms in the smaller C fibres. Because bigger fibres have got shorter refractory periods, they can generate more action potentials per second than smaller nerve fibres (i.e. they are capable of higher firing frequency).

ALL-OR-NONE LAW
The action potential, unlike the local response (see below), is not graded but obeys the all-or-none law. This means that a given stimulus either elicits no AP or produces a full AP whose ampli-

tude (height) cannot be increased by further increase in the stimulus intensity. As shown in Fig. 4.5, subthreshold stimuli fail to elicit the AP and suprathreshold stimuli fail to make it bigger. Although normally the AP of a given cell has a fixed amplitude, it can become smaller if elicited in the relative refractory period of the cell or if the extracellular Na^+ concentration is low.

GRADED POTENTIALS
Graded potentials are local, non-propagated responses whose amplitude varies with the strength of the stimulus applied. A depolarizing local response leads to a reduction in the membrane potential, taking it closer to the threshold level at which the action potential develops and thereby making the cell more excitable. Excitability refers to the ability of the cell to produce action potentials. Such a local depolarizing potential can arise from movement of positive ions (e.g. Na^+) into the cell or exit of negative ions from the cell. It can also be produced by cathodal stimulation, which deposits negative charges on the outer side of the cell membrane, thereby reducing the membrane potential.

A hyperpolarizing local potential, on the other hand, makes the membrane more negative inside and thereby reduces excitability of the cell. It can result from entry of negative ions (e.g. Cl^-) into the cell, exit of positive ions (e.g. K^+) from the cell, or anodal stimulation. Graded potentials die out within 1–2 mm from their point of origin. They cannot transmit sensory or motor information for long distances. They are, however, very useful as receptor potentials, serving for the initial recognition of a stimulus at the sensory receptor. They are also of utmost importance for integration of information at synapses. Furthermore, they can summate (add up) to reach the threshold level necessary for generation of the action potential (Fig. 4.5). There are two ways for this summation to occur:
1 *Spatial summation* occurs when several subthreshold stimuli arrive simultaneously at the neurone, add up and become strong enough to depolarize the neurone to the threshold level.
2 *Temporal summation* occurs when several subthreshold stimuli are delivered rapidly and successively so that each is added on top of the

preceding one before that one dies out, until the threshold level is reached and an action potential is generated.

Examples of graded potentials that will be discussed later include excitatory postsynaptic potential (EPSP), inhibitory postsynaptic potential (IPSP) and muscle end-plate potential (EPP).

ACTION POTENTIAL OF CARDIAC MUSCLE

Depolarization of heart muscle is similar to that of skeletal muscle, i.e. it is due to increased membrane permeability to sodium ions. At the end of depolarization the membrane potential does not return to the resting level but stays positive (above zero) for about 200−250 ms (Fig. 4.6). This is called the plateau (flat top). The plateau phase is due to increased calcium permeability with consequent Ca^{2+} influx.

Towards the end of the plateau, potassium permeability starts and K^+ leaves the cells, producing repolarization. Due to the delay in repolarization, the absolute refractory period (ARP) of heart muscle lasts for nearly the entire duration of the contraction. The relative refractory period (RRP) is short and lasts from the end of the absolute refractory period until shortly after re-polarization. A second AP, triggered during the RRP, gives a weak contraction superimposed on the semirelaxed phase of the first contraction. It is weak because the shorter duration of the second AP results in less than normal Ca^{2+} entering the cell. Therefore, unlike skeletal muscle, ventricular muscle cannot develop summation and tetanus during high-frequency stimulation. The prolonged refractory period protects the ventricles from too rapid re-excitation and allows them to relax long enough to be filled with blood. It also allows AP conduction through the muscular network to be completed before a second excitation can occur. This prevents recycling of excitation.

ELECTRICAL PROPERTIES OF SMOOTH MUSCLE

There are three types of smooth muscle:

1 *Spontaneously active smooth muscle*. These muscles can contract spontaneously in the absence of external innervation. They generate electrical activity, which spreads by gap junctions and produces contractions. Two types of electrical activity can be seen:

(a) *Pacemaker potentials*: a focal pacemaker

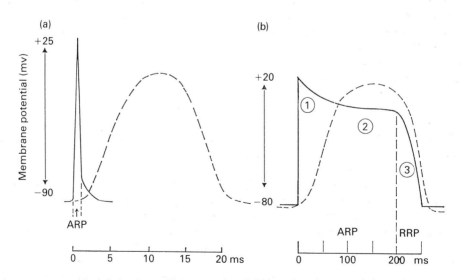

Fig. 4.6 (a) Action potential of skeletal muscle compared with (b) AP of cardiac muscle (note the time-scale). The broken line indicates the muscle contraction in each case. For cardiac muscle, (1) indicates depolarization due to Na^+ influx, (2) is the plateau phase due to Ca^{2+} influx, and (3) is repolarization due to delayed K^+ efflux. ARP, absolute refractory period; RRP, relative refractory period.

area depolarizes adjacent regions to produce action potentials. The pacemaker potential is due to slow influx of Ca^{2+}. The pacemaker region is not fixed and can shift from one location to another. This can be observed in the uterus and taeniae coli.

(b) *Slow wave depolarizations*: examples of these can be seen in the stomach and intestine. They are made up of slow, plateau-like waves, which last 2−8 s. Their frequency is determined by the cells whose rate is fastest. On reaching the threshold, they initiate action potentials. External innervation serves merely to regulate these spontaneous activities by stimulating them (parasympathetic, acetylcholine) or by inhibiting them (sympathetic, noradrenaline).

2 *Electrically inexcitable smooth muscle.* These do not generate electrical activity and depend upon external stimulation to contract. Examples are bronchial and tracheal smooth muscles.

3 *Intermediate smooth muscle.* This type shows spike-like electrical activity on stimulation. The force of contraction is proportional to the frequency of action potentials. Examples of these are found in the piloerector muscles, blood-vessels, vas deferens, seminal vesicles and iris.

PROPAGATION OF THE AP (IMPULSE TRANSMISSION)

Nerve fibres can be classified as myelinated and unmyelinated. An action potential developing at one point also stimulates adjacent parts of the membrane, causing new action potentials to spread in both directions. This process enables one stimulus to depolarize a whole cell and enables nerves to carry impulses from one part of the body to the other. The process of transmission is illustrated in Fig. 4.7.

Propagation of the action potential in unmyelinated (C) fibres occurs by continuous conduction. The depolarized area of the membrane depolarizes the area immediately next to it and itself becomes refractory for a brief period. This prevents propagation in the reverse direction and ensures unidirectional transmission (Fig. 4.7). In myelinated nerves, however, we get saltatory conduction whereby the impulse jumps from one node of Ranvier to the next. This is a much quicker way of impulse transmission.

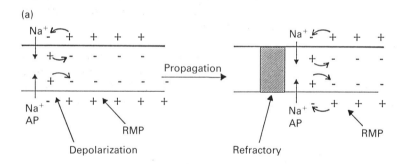

(a)

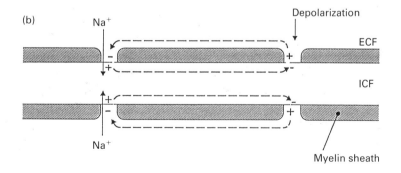

(b)

Fig. 4.7 (a) Continuous conduction in unmyelinated nerve. (b) Saltatory conduction in myelinated nerve. AP, action potential; RMP, resting membrane potential.

Synaptic transmission

A synapse is the junction between two neurones where the electrical activity of one neurone is transmitted to the other. Most synapses occur between the axon terminals of one neurone (presynaptic neurone) and the cell body or soma and dendrites of another, postsynaptic, neurone (Fig. 4.8). The presynaptic endings enlarge slightly to make the synaptic knob. The synaptic knob is separated from the postsynaptic membrane by the synaptic cleft. The synaptic knob contains vesicles which contain a transmitter substance. When the action potential arrives from the axon, it causes the calcium channels to open, i.e. increasing the membrane permeability to Ca^{2+}. Calcium attracts the vesicles to the membrane and once they are in contact they rupture (exocytosis). The released transmitter diffuses into the synaptic cleft and combines with specific receptors for that transmitter on the postsynaptic membrane. This changes the permeability of the postsynaptic membrane to specific ions and results in a postsynaptic potential. If this potential is depolarizing and is summated with others to reach the membrane threshold level, an AP results. The postsynaptic membrane usually contains no transmitter; this is why nerve conduction occurs only in one direction.

MECHANISM OF AP GENERATION IN A POSTSYNAPTIC CELL

Postsynaptic potentials (PSPs)
Postsynaptic potentials are produced in the post-synaptic membrane by the release of a transmitter. These potentials are integrated in the soma of the postsynaptic neurone. Depending on the type of synapse or transmitter, their potentials can be excitatory or inhibitory. Single PSPs do not usually cause any excitation or inhibition in the postsynaptic neurone because the change in membrane potentials does not reach the threshold to produce a response, but they may produce partial depolarization or hyperdepolarization. Threshold is reached by spatial or temporal summation and can be excitatory or inhibitory as explained below.

Spatial summation A number of synaptic knobs release their transmitter at the same time on a single postsynaptic membrane and each produces a partial depolarization of about a fraction of a millivolt. When small depolarization currents from transmitters released by the other knobs are added together, they depolarize the membrane by about 11 mV. This will cause an action potential to be generated in the postsynaptic cell, which is transmitted down the postsynaptic axon. If they do not reach threshold, these potentials decay and die away. This type of summation of EPSPs to reach threshold is called spatial summation, i.e. summation due to stimulation of a number of synaptic knobs.

Temporal summation If one synaptic knob is repetitively stimulated, its postsynaptic effects may add up and reach threshold to fire an action potential. This is called temporal summation.

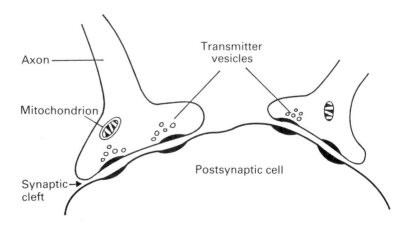

Fig. 4.8 A typical synapse.

Excitatory and inhibitory postsynaptic potential
If the transmitter released at the synapse opens Na⁺ channels and thereby depolarizes the small area immediately under the synaptic knob, the result will be an excitatory postsynaptic potential (EPSP).

If the transmitter opens Cl⁻ channels and thereby hyperpolarizes the postsynaptic membrane an inhibitory postsynaptic potential (IPSP) will result. EPSPs make the postsynaptic cell easier to excite. Conversely, IPSPs make the postsynaptic cell more difficult to excite.

Summation of several EPSPs causes the *generator zone* of the postsynaptic neurone to reach the threshold and an AP is elicited in the postsynaptic cell. The generator zone is situated at the origin of the axon, where it arises from the cell-body. It is the region where APs are generated. The generator zone is more excitable than the rest of the axon because it has a threshold potential of −59 mV (whereas the rest of the axon has a threshold potential of −55 mV). Therefore the membrane has to be depolarized by only 11 mV from its resting value of −70 mV in order to trigger an AP in the generator zone.

Synaptic delay
After maximal stimulation, the minimum time required for an impulse from a presynaptic knob to produce a response in the postsynaptic neurone is 0.5 ms. This is called the synaptic delay. This is the time taken for the transmitter to be released from vesicles and to produce an action potential in the postsynaptic membrane. A monosynaptic neurone pathway will have a delay of about 0.5 ms while a polysynaptic pathway will have a delay corresponding to the number of synapses.

Electrical transmission
Some synapses transmit signals electrically or without transmitters. In such synapses the synaptic cleft is replaced by a low-resistance bridge which allows the passage of the electrical impulse with ease. This type of transmission has a short synaptic delay and occurs only in the brain.

NEUROMUSCULAR TRANSMISSION
Transmission of impulses from nerves to muscles occurs at neuromuscular junctions, which are a special type of synapse. Each nerve fibre loses its myelin sheath, branches several times and supplies from three to several hundred muscle fibres. Each muscle fibre, however, receives innervation from only one neurone. The nerve endings enlarge slightly to make the synaptic knob, which fits into an invagination of the muscle cell membrane called the motor end-plate. At the motor end-plate the muscle postjunctional membrane makes numerous folds called the subneural or junctional folds. These, like villi, greatly increase the membrane surface area of the motor end-plate (Fig. 4.9).

Acetylcholine is the only transmitter at the synaptic knobs that is synthesized in the neural cytoplasm from choline and acetylcoenzyme A. The subneural clefts contain the enzyme acetylcholinesterase (AChE), which is capable of rapidly hydrolysing acetylcholine (ACh) to prevent its accumulation. Choline is released as a result of ACh breakdown and is taken up by the nerve and reutilized.

The events of neuromuscular transmission are similar to those occurring at synapses. The incoming action potentials increase the permeability of the nerve ending to calcium ions. Ca²⁺ enters the nerve endings, causing attraction and rupture of the acetylcholine vesicles. The acetylcholine diffuses in the subneural folds and combines with acetylcholine receptors on the motor end-plate. The permeability of the endplate to Na⁺ is increased, producing a depolarization called the end-plate potential. Therefore, neuromuscular junctions are excitatory only. The amount of acetylcholine released by an impulse is

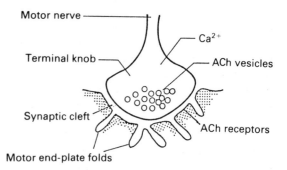

Fig. 4.9 Diagrammatic representation of the neuromuscular junction.

enough to activate 10 times the receptors needed to produce a full end-plate potential. This is not waste but a safety factor to ensure that the message gets across. The end-plate potential is a local, graded, non-propagated response which is present only in the end-plate region. It serves to depolarize the adjacent muscle membrane to its firing level. Action potentials are generated on both sides of the end-plate and are conducted by the local current mechanism to both ends of the muscle fibre. As we are going to see later in this chapter, the muscle action potentials will eventually produce muscular contraction.

Myasthenia gravis

Myasthenia gravis is a serious disease caused by a failure of transmission at the neuromuscular junction. It is characterized by tiredness and progressive paralysis of muscles; death commonly occurs due to failure of respiratory muscles. It is an autoimmune disease in which some of the acetylcholine receptors are destroyed by circulating antibodies. Acetylcholine released by the impulses cannot produce an immediate effect and is destroyed by AChE. Drugs, such as neostigmine, inhibit the action of AChE and allow accumulation of adequate amounts of acetylcholine to stimulate the remaining receptors. Such drugs are used for management of this disease.

Some drugs that modify neuromuscular transmission

1 *Muscle relaxants* are drugs widely used during general anaesthesia and in some other medical conditions to produce muscle relaxation. Curare is an example of non-depolarizing blockers, which occupy ACh receptors reversibly without producing depolarization. Succinylcholine, on the other hand, is an example of depolarizing blockers. It produces an initial depolarization of the sarcolemmal membrane, which prevents ACh from producing a response in the depolarized membrane. A few minutes later, as the membrane becomes repolarized, a second phase of decreased receptor sensitivity to ACh takes place.

2 *Alpha-bungarotoxin* is a chemical extracted from snake venom. It binds irreversibly to ACh receptors. The victim becomes totally paralysed.

When labelled with radioactive iodine, it is used in a histochemical technique to estimate the number of ACh receptors at the neuromuscular junction.

3 *Anticholinesterase drugs*, such as neostigmine, reversibly combine with AChE and thus allow ACh to accumulate in the cleft, giving it a more favourable opportunity to act on the receptors. They are used in the management of myasthenia gravis, as mentioned above.

4 *Organophosphate compounds* combine irreversibly with AChE, leading to long-term increases in neuromuscular transmission. They are used as insecticides and, regrettably, as a chemical weapon. They produce muscle fasciculations, abdominal cramps, respiratory distress, sweating and convulsions. They are a frequent cause of poisoning in agricultural areas when they are accidentally ingested.

NEUROTRANSMITTERS

Transmission of impulses in synapses and in neuromuscular junctions requires chemical transmitters. Transmitters are formed in the presynaptic knob and each transmitter has a specific receptor in the postsynaptic membrane. A transmitter can be excitatory or inhibitory depending on the ionic permeability that it will produce. Acetylcholine and catecholamines are usually excitatory transmitters because they increase sodium permeability, while gamma-aminobutyric acid (GABA) and glycine are inhibitory transmitters because they increase chloride permeability and therefore produce hyperpolarization. After producing their action, transmitters are either degraded by specific enzymes or taken up into the synaptic knob.

Contraction of muscle

Contraction is defined as the active process of generating a mechanical force in muscle. The force exerted by a contracting muscle on an object is known as muscle tension and the force exerted on a muscle by a weight is known as the load. To lift a load, muscle tension must be greater than the muscle load. To lift a load the muscle must shorten and this type of contraction is called an isotonic (constant tension) contraction because the load remains constant throughout the period of shortening. When muscle shortening is prevented by a load that is greater than the force of contraction, the development of tension occurs at constant muscle length and this is called isometric (constant length) contraction. Body movements involve mainly isotonic contractions, which involve movement. Supporting a weight in a fixed position or maintaining the erect posture (e.g. standing) involves isometric contractions. Both types of contractions are involved together in most muscular activities.

MECHANICS OF SKELETAL MUSCLE CONTRACTION

Simple muscle twitch
The mechanical response (contraction) to a single action potential is called a simple muscle twitch (Fig. 4.10a). Following the stimulus, there is an interval of a few milliseconds, known as the latent period, before contraction begins. This is the time taken for the action potential to develop. If the muscle is stimulated through its nerve, the latent period is longer because of the time taken in transmission at the neuromuscular junction. Figure 4.10b, c and d show the effect of two successive stimuli.

Summation of contraction
A single twitch is of no mechanical value because its duration is very short. To produce sustained and co-ordinated muscle movements, single twitches summate in two different ways:
1 *Spatial summation.* The muscle fibres, together with the motor neurone which innervates them, constitute a motor unit. In spatial summation, stimulation of numerous nerve fibres causes an increasing number of motor units to be excited. The responses of single motor units are therefore added together to produce a strong contraction by the muscle.
2 *Temporal summation.* Separate muscle contractions are caused by single action potentials when their frequency is low. When frequency of stimulation rises above 10 per second, the second stimulus causes a contraction to develop before the first one is over. As frequency increases, the degree of summation becomes greater, producing stronger contractions every time in a stepwise

fashion (Fig. 4.11). When muscle is stimulated at a progressively greater rate, a frequency is reached at which contractions fuse together and cannot be distinguished. This is called tetanization and the contraction is a tetanic contraction, which is smooth and maintained. Most contractions of muscles in everyday life are of this nature. They allow useful work to be done.

TYPES OF MUSCLE CONTRACTIONS

Isometric contractions
When the muscle is not allowed to shorten (isometric), the tension developed during contraction increases to reach a maximum (as shown in Fig. 4.12). The relationship shown in the diagram is known as a length–tension relationship. It is

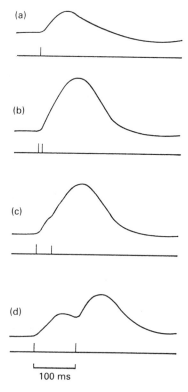

Fig. 4.10 (a) A single muscle twitch (contraction) due to a single short-lived stimulus; (b) complete summation of two contractions; (c) and (d) show incomplete summation as the interval between the two stimuli widens.

seen from Fig. 4.12 that there is an optimum length of the muscle at which it develops maximum tension when it is stimulated. This depends on the effect of stretch or relaxation of the muscle on the degree of overlap of the actin and myosin filaments in the sarcomere. Isometric contractions do not result in external work being performed. However, they are useful for maintenance of posture and supporting activities.

Isotonic contractions

When a muscle is allowed to shorten, it will exert mechanical work (e.g. lifting a load). In this case the tension developed is constant (isotonic) and is equal to the load. Figure 4.13 shows the force–velocity curve during isotonic contractions. It is seen from this relationship that the velocity de-

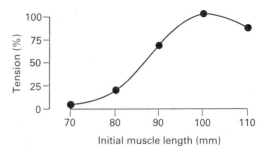

Fig. 4.12 Isometric contraction (length–tension curve). At point 100 mm, the length represents the resting length of the muscle, which gives the maximum strength of contraction.

creases as the load increases. The initial length of the muscle also affects the velocity of contraction. The power of contraction equals the force times velocity. Muscle power is usually maximum when only one-third of the maximum tension develops, i.e. isotonic contraction lifting one-third of the maximum load.

MOLECULAR BASIS OF MUSCLE CONTRACTION

Both skeletal and cardiac muscles show striations (alternating dark and light bands) under the light microscope and are therefore called striated muscles. The striated muscle fibre (cell) has a diameter ranging between 50 and 200 μm and can

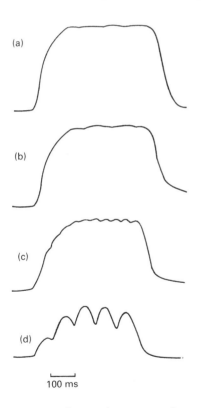

Fig. 4.11 Responses of a muscle to various frequencies of stimulation. Note that at (d) 20/s stimulation produces summation, at (c) 40/s incomplete fusion or clonus, and at (a & b) 80/s and 100/s respectively there is complete fusion or tetanus, which is a smooth maintained contraction.

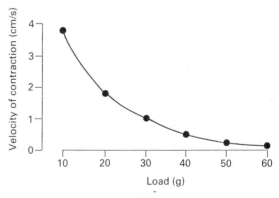

Fig. 4.13 Isotonic contraction (force–velocity curve). The velocity of contraction is inversely related to the load.

be as long as 15 cm. Each contains several hundred myofibrils in the cytoplasm. The myofibril is made up of contractile units called sarcomeres, whose length lies between 1.5 and 3 μm depending upon how much the muscle is stretched (Fig. 4.14).

The sarcomere consists of thick myosin filaments, thin actin filaments and two regulatory proteins called troponin and tropomyosin. It is bound on either side by a plate-like protein structure called Z line, to which actin filaments are attached. The dark A band lies in the middle of the sarcomere and is formed of overlapping myosin and actin (Fig. 4.14). The light I band is made of actin. The H zone is made of myosin only. During contraction actin and myosin slide upon each other, increasing the extent of their overlap and bringing the Z lines closer together. This leads to the narrowing and finally disappearance of the I band and H zones, but the A band remains unaffected.

The myosin molecule has a cross-bridge head and a rod-like tail (Fig. 4.15). The joint-like or neck region between the head and tail allows the head to swivel through an angle of 45−50° during contraction. The head has a site for binding ATP and it also contains an ATPase enzyme. This enzyme, however, gets activated only when the myosin head binds to actin. Each myosin filament (Fig. 4.16b) is made up of 150−360 myosin molecules arranged so that the heads project from the sides of the filament.

The actin molecule is a globular protein, 400 of which form a chain called F-actin. Two F-actin chains intertwine to form one actin filament (Fig. 4.16a). The actin molecules contain binding sites for the heads of myosin.

Tropomyosin is a relaxing or inhibitory protein which has a long thread-like molecule lying between the two F-actin chains, a bit to the side of the groove between them, and covering the binding sites on actin molecules. It thereby pre-

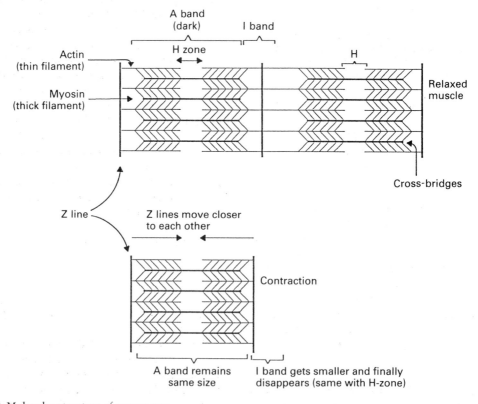

Fig. 4.14 Molecular structure of sarcomeres.

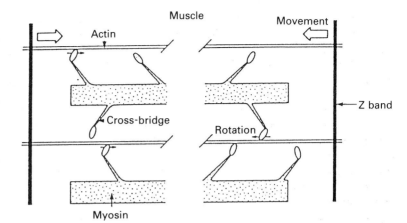

Fig. 4.15 Movement of actin and myosin filaments showing rotation of the cross-bridgehead of the myosin molecule. (Reproduced from Bray, J.J., Cragg, P.A., Macknight, A.D.C., Mills, R.G. and Taylor, D.W. (eds) (1989) *Lecture Notes in Physiology*, 2nd edn. Blackwell Scientific Publications, Oxford.)

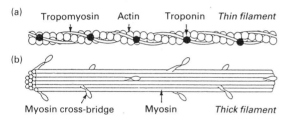

Fig. 4.16 Actin filament, formed of a spiral of two chains of actin molecules. The thread-like tropomyosin molecules lie between the two chains of actin molecules, slightly to one side of the groove between them, covering the binding sites for myosin. Troponin molecules are spaced at 40 nm intervals. (b) Myosin filament (made up of 150–360 myosin molecules). (Reproduced from Bray, J.J., Cragg, P.A., Macknight, A.D.C., Mills, R.G. and Taylor, D.W. (eds) (1989) *Lecture Notes in Physiology*, 2nd edn. Blackwell Scientific Publications, Oxford.)

vents the myosin head from combining with actin in the relaxed state of the muscle.

Troponin molecules have calcium-binding sites and are attached to the tropomyosin molecule at 40 nm intervals. When Ca^{2+} binds to troponin, this moves the tropomyosin threads sideways into the groove between the two F-actin chains, thereby uncovering the binding sites on actin. The myosin heads immediately get attached to these sites, forming cross-bridges.

THE EVENTS IN MUSCLE CONTRACTION
As mentioned earlier, acetylcholine released by the motor nerve at the neuromuscular junction

leads to formation of the end-plate potential (EPP). This, in turn, depolarizes the sarcolemma (muscle cell membrane), leading to formation of the muscle action potential. The muscle action potential spreads to the inside of the cell via a system of transverse tubules (T tubules), which are invaginations in the muscle cell membrane. Arrival of the AP at the sarcoplasmic reticulum, which serves as a Ca^{2+} store and contains a high concentration of Ca^{2+}, leads to opening of Ca^{2+} channels on its membrane. Consequently, Ca^{2+} diffuses out into the cell cytoplasm and combines with troponin. This makes troponin pull tropomyosin sideways into the groove between the two F-actin chains, thereby exposing the active sites on actin. Myosin heads, with ATP on them, get attached to actin, resulting in the formation of a high-energy actin–myosin complex. This binding of actin to myosin activates the ATPase enzyme present in the myosin head, leading to hydrolysis of ATP:

$$ATP \xrightarrow{\text{ATPase}} ADP \xrightarrow{\text{ATPase}} AMP$$
$$P_i + E \qquad P_i + E$$

where E represents energy and P_i represents inorganic phosphate.

This hydrolysis of ATP needs magnesium (3 μmol/litre). The energy thus liberated is used for movement of cross-bridges between actin and myosin so that these filaments slide upon each other, bringing the Z lines closer together and making the sarcomere shorter. Summing up of all increments of shortening occurring in all sarcomeres results in the overall shortening of the muscle. When ATP is hydrolysed and lost, what remains is called a rigor complex or a low-energy actin–myosin complex. When a new ATP molecule comes and occupies the vacant site on the myosin head, this triggers detachment of myosin from actin. The free myosin head now swings back to its original position and then gets attached to another actin molecule and the cycle gets repeated (i.e. hydrolysis of ATP, swivelling of cross-bridges over 45–50°, sliding of filaments, detachment of myosin from actin, and so on). The rigor complex is very stable and, unless an ATP molecule becomes available to occupy the vacant site on myosin and trigger the release of myosin from actin, the actin remains permanently attached to myosin. In this circumstance the muscle becomes extremely rigid, a situation referred to as rigor mortis. It occurs following death due to cessation of metabolism and lack of ATP.

Relaxation of muscle is an active process. ATP is used to pump Ca^{2+} back into the sarcoplasmic reticulum, calcium is detached from troponin, tropinin ceases pulling on tropomyosin, and this returns to its original position and covers the active sites on actin, preventing formation of cross-bridges, and the muscle relaxes. Therefore, ATP is needed for both contraction and relaxation:

1 ATP hydrolysis provides energy for movement of cross-bridges during contraction.

2 ATP is needed for breakdown of the rigor complex and disengagement between actin and myosin.

3 ATP is used for pumping Ca^{2+} back into its stores and thereby promoting muscle relaxation.

Further reading

1 Ganong, W.F. (1991) Excitable tissues: nerve. In *Review of Medical Physiology*, 15th edn, pp. 43–56. Appleton & Lange, San Mateo, California.
2 Ganong, W.F. (1991) Synaptic and junctional transmission. In *Review of Medical Physiology*, 15th edn, pp. 77–104. Appleton & Lange, San Mateo, California.
3 Keynes, R.D. & Aidley, D.J. (1981) *Nerve and Muscle*. Cambridge University Press, Cambridge.

5: The Autonomic Nervous System

Introduction

The whole nervous system can be anatomically divided into central and peripheral. The central is encased within the bones of the cranium and the vertebral column. Thus, the central nervous system (CNS) consists of the brain and the spinal cord. The peripheral nervous system consists of nerve connections between the CNS and the various organs and tissues. These connections include nerve trunks, plexuses and all the smaller nerve fibres which distribute themselves to the various muscles, skin and other organs.

On the other hand, there are physiological divisions of the nervous system. The somatic part of the nervous system mainly deals with the voluntary and conscious aspects of neurological control such as skeletal muscle movement and the various somatic sensory functions. The other division, which is the subject of this section, is the autonomic nervous system, which deals with involuntary control mechanisms such as regulation of the heart, the circulation and digestive functions, i.e. *visceral functions*. Each of the somatic and autonomic divisions has a central and peripheral component.

The control of visceral functions

The division of the nervous system into autonomic and somatic is not to say that they function

Objectives

On completion of the study of this section, the student should be able to:

1 Understand the functional relationships between the somatic and autonomic nervous systems in order to differentiate between somatic and autonomic disorders.

2 Understand the anatomical and functional aspects of the sympathetic and parasympathetic divisions of the autonomic nervous system so as to appreciate the integration of the control of visceral functions.

3 Understand the nature of chemical transmission and the types of autonomic receptors in order to explain the mechanism and site of action of drugs which modify the functions of organ systems.

4 Understand the roles of visceral sensation so as to explain visceral control mechanisms and the physiology of visceral pain.

5 Describe the higher control of the autonomic nervous system so as to appreciate the relationships between psychological factors and visceral disorders.

independently. There is a great deal of integration between the two. In fact, overlap occurs in the part dealing with emotions and the psyche. This is of particular importance in the manifestations of psychosomatic disorders, when emotional or psychological factors manifest themselves as visceral or somatic symptoms and signs.

Since the autonomic nervous system is directly concerned with regulating visceral or involuntary functions, it is responsible for many mechanisms which control the internal environment. Autonomic nerves also enable the body to react automatically to external environmental inputs such as heat or light. The main integrating station for autonomic functions is the part of the brain known as the hypothalamus. From the hypothalamus impulses pass down to the various autonomic nerves which supply the internal organs and tissues.

Divisions of the autonomic nervous system

The motor fibres (efferents) of the autonomic

nervous system can be divided, from a functional point of view, into two types: the sympathetic and the parasympathetic. Anatomically they also have different origins in the central nervous system. Sympathetic fibres leave the central nervous system in the thoracic and upper two or three lumbar spinal nerves; hence the name *thoracolumbar outflow*. These nerves distribute themselves widely to supply blood-vessels, the heart, lungs and abdominal viscera.

Parasympathetic fibres leave the central nervous system in some of the cranial nerves, namely the 3rd, 7th, 9th and 10th cranial nerves (oculomotor, facial, glossopharyngeal and vagus nerves, respectively). The first three of these nerves distribute themselves to structures in the head and neck while the vagus nerve distributes itself mainly to the thoracic and abdominal organs. Other parasympathetic fibres leave the central nervous system in the sacral spinal nerves; hence the other name of the parasympathetic division, the *craniosacral outflow.*

The sympathetic division enables the body to mobilize and spend energy and to prepare organ systems to face emergency situations. The parasympathetic division functions in the state of relaxation and therefore contributes to the regulation of visceral functions to enable the body to perform restorative and vegetative functions.

Organization of the autonomic nerves
In general, the autonomic nerves, whether sympathetic or parasympathetic, consist of two neurones. The first neurone has its cell body in the central nervous system (the brain nuclei or the grey matter of the spinal cord). The second neurone has its cell body in a ganglion which lies outside the CNS. Thus, the two neurones are called preganglionic and postganglionic respectively. In this respect the autonomic nerves are different from somatic nerves. Somatic nerve cell bodies are never found outside the CNS.

It is noteworthy that a single preganglionic neurone synapses with several postganglionic neurones, thus spreading the effect over a larger area. The sympathetic and parasympathetic parts have some differences and some similarities in their structure and function. Although they generally have opposite effects, there is often a degree of integration in their functions.

THE SYMPATHETIC DIVISION

Anatomical considerations
The preganglionic fibres of the sympathetic nerves originate in the lateral horn cells in the spinal grey matter. These lateral horns constitute a column of cells which characterizes the grey matter of all the thoracic and the upper two lumbar segments. The preganglionic fibres leave the spinal cord in the ventral roots. Outside the vertebral column these fibres leave the spinal nerves in the white rami communicantes (WRC) and synapse with the postganglionic neurones in the *paravertebral ganglion* (Fig. 5.1). These ganglion neurones have connections between them and therefore they form the sympathetic chain, which lies on both sides of the vertebral bodies on the posterior wall of the thorax. The sympathetic trunk formed by interganglionic connections allows synapses to be made several segments above or below the point of exit of the preganglionic fibre from the spinal cord. Some preganglionic fibres, however, do not synapse in the paravertebral ganglia but pass through them to synapse in *collateral ganglia*, e.g. the coeliac ganglion and the mesenteric ganglia.

The paravertebral ganglia give rise to postganglionic fibres in the grey rami communicantes (GRC) (see Fig. 5.1). The postganglionic fibres join somatic nerves and distribute themselves to the target organs. These sympathetic fibres supply the blood-vessels of the skin, sweat glands, hair follicles and the blood-vessels in the skeletal muscles.

Sympathetic fibres which do not synapse in the thoracic and upper lumbar paravertebral ganglia of the sympathetic chain leave the ganglia to synapse with postganglionic neurones in other ganglia (Fig. 5.2):
1 *Cervical ganglia.* These receive an upward extension of the sympathetic trunk from T1 to T4. Their postganglionic fibres distribute themselves to structures in the head, neck and upper limbs.
2 *Collateral ganglia.* These are found in the abdomen and their postganglionic fibres supply the abdominal viscera down to the proximal part of the colon.
3 *Sacral ganglia.* These receive a caudal extension of the sympathetic trunk from the lower thoracic and upper lumbar segments. They dis-

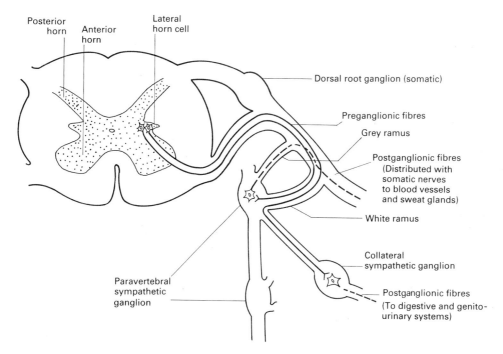

Fig. 5.1 The organization of pre- and postganglionic fibres of the sympathetic nervous system. Preganglionic fibres are shown as solid lines and postganglionic fibres as broken lines.

tribute their postganglionic fibres to the distal colon, pelvic organs and lower limbs.

The preganglionic sympathetic fibres are usually much shorter than the postganglionic ones.

Effects of sympathetic stimulation
The sympathetic part of the autonomic nervous system is activated during conditions of stress and emergencies (e.g. fear, anxiety, hypoglycaemia, severe pain, etc.). Under these conditions the body needs to be prepared for 'fight or flight'. Collectively, therefore, the effects of sympathetic stimulation will promote mechanisms which increase energy metabolism and the efficiency of the supply systems (circulatory and respiratory systems). Table 5.1 gives the various effects brought about by sympathetic and parasympathetic stimulation.

THE PARASYMPATHETIC DIVISION

Anatomical considerations
The parasympathetic fibres originate in the nuclei of the cranial nerves mentioned above and in the sacral segments of the spinal cord. The ganglia of parasympathetic fibres are usually situated near or within the tissues or organs supplied. Thus, the postganglionic fibres are very short compared with the preganglionic fibres. The distribution of parasympathetic fibres is given in Fig. 5.3.

Effects of parasympathetic stimulation
The physiological effects of parasympathetic stimulation are related to those activities associated with relaxed states connected with vegetative functions such as feeding, resting and sexual stimulation. The responses of the digestive system to food intake are due to parasympathetic stimulation, e.g. salivary secretion, secretions of the stomach, pancreas and intestine. Parasympathetic stimulation also increases the mobility of the gastrointestinal system and the urinary system (see Table 5.1).

Chemical transmitters
The effects of sympathetic and parasympathetic stimulation are produced by the release of chemi-

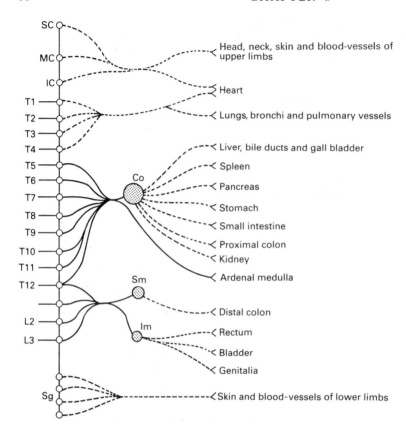

Fig. 5.2 Diagrammatic representation showing the origins and distribution of sympathetic nerves. The solid lines are preganglionic fibres and the broken lines are postganglionic fibres. SC, MC and IC are the superior, middle and inferior cervical ganglia, respectively. Co, Sm, Im and Sg are the coeliac, superior mesenteric, inferior mesenteric and sacral ganglia, respectively.

cal substances or transmitters which serve two functions: (i) to transmit the nerve impulse from the preganglionic fibre to the postganglionic neurone in the autonomic ganglia, and (ii) to combine with receptors in the target cells so as to produce the desired physiological effect.

In all the ganglia, whether sympathetic or parasympathetic, the transmitter substance is *acetylcholine*. Therefore, all the preganglionic fibres release acetylcholine at the synapse between the pre- and postganglionic neurones. The adrenal medulla is homologous with a sympathetic ganglion. The fibres reaching it come from preganglionic neurones and therefore release acetylcholine (see Fig. 5.3).

Acetylcholine and noradrenaline are the two main transmitters released by the postganglionic fibres. In general, all postganglionic fibres of the parasympathetic release *acetylcholine* as the main transmitter (with a few exceptions which will not be mentioned at this stage). Some post-

ganglionic cholinergic neurones also release vasoactive intestinal peptide (VIP). On the other hand, most of the sympathetic postganglionic fibres release *noradrenaline*. The postganglionic sympathetic fibres which supply the sweat glands and the blood-vessels in skeletal muscles are an exception; they release acetylcholine. Figure 5.4 gives a summary of autonomic transmitter substances and their sites of release.

The chemical transmitters act locally either at the synapse or at the target cell receptors. Their effect lasts only for a few seconds because they are either destroyed or removed from the site of their release. Acetylcholine is broken down by the enzyme cholinesterase, which is found in the terminal endings of all cholinergic nerves. Noradrenaline, on the other hand, is rapidly taken up by the nerve endings by active transport. This process accounts for the removal of most of the noradrenaline released. The remainder diffuses out with the surrounding extracellular fluid and

Table 5.1 Effects of sympathetic and parasympathetic stimulation

Organ	Sympathetic effects	Parasympathetic effects
The eye		
Iris	Dilatation of pupil	Constriction of pupil
Ciliary muscle	Relaxation (far vision)	Contraction (near vision)
Lacrimal glands	None	Lacrimation
The heart		
SA node	Increased heart rate	Decreased heart rate
Myocardium	Increased strength of contraction	No effect?
Coronary blood-vessels	Dilatation	Dilatation
Blood-vessels to skeletal muscles	Dilatation (cholinergic)	None
Skin	Vasoconstriction	None
	Sweating	
Bronchial muscle	Relaxation	Constriction
Gastrointestinal tract		
Salivary glands	Decreased secretion	Increased secretion
Pancreas	Decreased secretion	Increased secretion
Stomach	Decreased secretions and motility	Increased HCl secretions and motility
Intestinal secretions and motility	Decreased	Increased
Sphincters	Constriction	Relaxation
Ureters and urinary bladder		
Detrusor muscle	Relaxation	Contraction
Internal sphincter	Contraction	Relaxation
Male sex organs	Ejaculation	Erection
Adrenal medulla	Secretion of adrenaline and noradrenaline	None

is removed by the circulation, where it is quickly destroyed by enzymes.

Adrenergic receptors

Receptors are the site on the cell membrane with which the chemical transmitter combines to produce its effect on the cell. The effect may be on the cell membrane itself, e.g. change in permeability leading to electrical changes, or it may be on the cell cytoplasm, where biochemical reactions are stimulated. A good example of the latter effect is activation of adenylcyclase, which leads to cyclic adenosine monophosphate (AMP) formation. Cyclic AMP is important for the initiation of many intracellular reactions.

A quick review of the effects of sympathetic stimulation reveals that in some organs contraction of smooth muscle is produced, e.g. constriction of blood-vessels. In other organs, sympathetic stimulation produces relaxation of smooth muscle, e.g. coronary vessels, intestine and urinary bladder. The explanation of this is the presence of different receptors on the target cells. Two main types of adrenergic receptors, known as alpha and beta receptors, have been described. In general, alpha receptors are stimulatory and beta receptors are inhibitory, although this is not always the case (Table 5.2).

Other evidence for the presence of receptors comes from the effects of drugs on sympathetic activity. Certain drugs have been found to block or inhibit some of the effects of alpha or beta receptor stimulation. On the basis of this observation such drugs have been classified as α-blockers and β-blockers.

The distribution of adrenergic receptors in the various tissues can be deduced from the effects of drugs. Table 5.2 gives the main examples of alpha

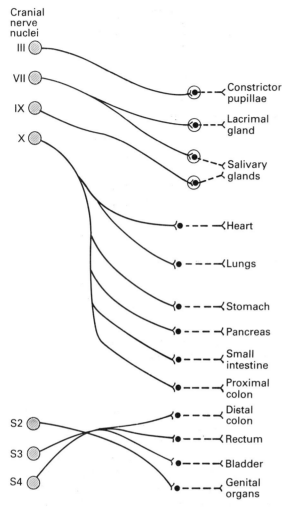

Cranial
nerve
nuclei

III

VII

IX

X

Constrictor
pupillae

Lacrimal
gland

Salivary
glands

Heart

Lungs

Stomach

Pancreas

Small
intestine

Proximal
colon

Distal
colon

Rectum

S2

S3

S4

Bladder

Genital
organs

Fig. 5.3 The origins and distribution of parasympathetic nerves. III, VII, IX and X are the cranial nerves which contain parasympathetic fibres. S2, 3 and 4 indicate the 2nd, 3rd and 4th sacral segments. Note the long pre-ganglionic fibres; postganglionic fibres are usually within the target organ. However, some of the cranial nerves have separate ganglia, denoted by the open circles.

Table 5.2 The effects of α- and β- adrenergic receptors

Sympathetic effects due to α-receptors	Sympathetic effects due to β-receptors
Dilatation of iris (mydriasis)	β_1-receptors
Constriction of blood-vessels in gut, skin, salivary glands, lungs and cerebral circulation	Cardio-acceleration
	Increased myocardial contractility
	Lipolysis
Constriction of sphincters in GI tract	Renin secretion
Increased motility of ureters	β_2-receptors
Contraction of trigone and internal sphincter of urinary bladder	Coronary vasodilatation
	Skeletal muscle vasodilatation
Ejaculation	Skeletal muscle tremor
Sweating, slight and localized	Bronchodilatation
Pilomotor muscle contraction	Relaxation of uterus
Decreased secretion of salivary glands	Intestinal relaxation
	Bladder relaxation
Decreased secretion of pancreatic juice	Calorigenic effect
	Glycogenolysis
Decreased insulin and glucagon secretion	Amylase secretion

Table 5.3 Some examples of adrenergic receptor blockers

Alpha receptor blockers	Beta receptor blockers
Phenoxybenzamine (Dibenyline)	*Non-selective*
	Propranolol (Inderal) (β_1 and β_2)
Phentolamine (Rogitine)	Oxprenolol (β_1 and β_2)
Prazosin (α_1)	
Yohimbine (α_2)	*Selective*
	Atenolol (Tenormin) (β_1)
	Metoprolol (β_1)
	Practolol (β_1)
	Butoxamine (β_2)

and beta receptor effects of sympathetic stimulation. It is evident that within one organ more than one type of receptor may be found. Recently, the alpha and beta receptors have been divided into further subdivisions, e.g. α_1-, α_2-, β_1- and β_2-receptors. New drugs are now available with more selective action on specific receptors. Table 5.3 gives examples of adrenergic receptor blockers. These drugs are becoming more and more important in therapeutics, especially in the treatment of hypertension and coronary heart disease.

Cholinergic receptors

Acetylcholine is released in all preganglionic nerve endings, i.e. in the sympathetic and parasympathetic ganglia. On the basis of the effects of

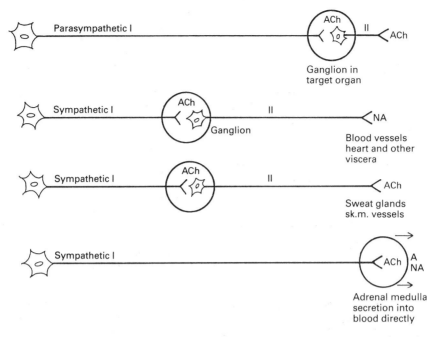

Fig. 5.4 Schematic representation of the various sites of release of autonomic transmitters. I and II refer to preganglionic and postganglionic fibres, respectively. A, adrenaline; ACh, acetylcholine; NA, noradrenaline; sk.m., skeletal muscle.

drugs which have similar actions to acetylcholine, it has been found that the receptors in the ganglia are different from those in peripheral tissues supplied by postganglionic fibres. Thus, nicotine produces effects similar to those of preganglionic stimulation (nicotinic actions of acetylcholine). On the other hand, the drug muscarine produces effects similar to postganglionic parasympathetic fibres (muscarinic effects of acetylcholine). Cholinergic receptors in the autonomic ganglia can be blocked by drugs such as *hexamethonium*. On the other hand, the muscarinic actions of acetylcholine at the postganglionic endings are blocked by *atropine*. In fact, atropine blocks all the autonomic cholinergic effects except those at the autonomic ganglia.

Apart from the autonomic nerves, acetylcholine is also the transmitter at the neuromuscular junction of skeletal muscle. The cholinergic receptors in skeletal muscle can also be stimulated by nicotine but they are not identical with those found in autonomic ganglia. At the neuromuscular junction, the action of acetylcholine is blocked by the drugs *curare* and *tubocurarine*. Furthermore, acetylcholine is an important neurotransmitter in the central nervous system.

Myasthenia gravis is a disease caused by autoimmune destruction of acetylcholine receptors at the neuromuscular junction of skeletal muscle (see Chapter 4).

Visceral afferent fibres

Visceral functions are continuously monitored by various sensory receptors situated in the organs or in the central nervous system. Examples of peripheral receptors are the chemoreceptors and baroreceptors which are concerned with the regulation of cardiovascular and respiratory functions. Examples of central receptors are the thermoreceptors and osmoreceptors in the hypothalamus which are concerned with the regulation of body temperature and water balance, respectively. There are also receptors for pain and stretch in the viscera.

Afferent fibres from the viscera travel in the fibres of the autonomic nervous system (see

Fig. 5.1). These afferent fibres are essential for the regulation of the various organ systems. The visceral afferent fibres have their cell bodies in the cranial nerve ganglia and in the dorsal root ganglia of the spinal cord. Thus, we find visceral nerves in the facial nerve, glossopharyngeal nerve and vagus. Visceral afferent fibres are also found in the thoracic and upper two or three lumbar nerves and the sacral nerves, i.e. all the spinal nerves which have sympathetic or parasympathetic components. Therefore, visceral pain goes to these segments of the spinal cord and is sometimes felt on the part of the body or skin supplied by somatic nerves of the same segment. This is called referred pain. A good example of this is shoulder pain due to diaphragmatic irritation or gall-bladder disease.

In the spinal cord, visceral afferent fibres travel in the spinothalamic tracts. In the brain they are found in the fibres of the thalamic radiation. Cortical representation is in the postcentral gyrus and is found to be intermixed with somatic representation.

Special features of autonomic functions
The following special features characterize the mechanisms responsible for autonomic functions. They are highlighted here for the sake of contrast with somatic functions.

1 The viscera have their own intrinsic functions and therefore extrinsic autonomic denervation does not result in complete loss of function or paralysis. Autonomic nerves seem to control the level of activity and its modulation according to requirements.

2 Autonomic nerves have a basal level of discharge or tone which can be increased or decreased

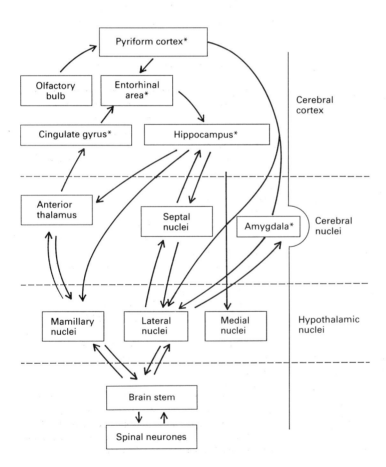

Fig. 5.5 Diagrammatic representation of the major connections of the hypothalamus. Parts of the cerebral cortex and cerebral nuclei are marked (*) to indicate that they are part of the limbic system.

as necessary. Sympathetic tone is the term given to the continuous basal discharge to blood-vessels which keeps them in a state of partial or tonic contraction at rest. Vasoconstriction is produced by increasing the sympathetic tone and vasodilatation is produced by decreasing it.

The parasympathetic or vagal tone to the heart at rest keeps the heart rate at the right level for maintenance of the resting cardiac output. An increase in the heart rate can be produced by reducing the vagal tone. Increasing the tone results in reduction of heart rate.

Similarly, abolition of vagal tone to the gastrointestinal tract results in temporarily decreased motility and slow passage of food, recovery occurring within 24 hours. However, gastric acid secretion is permanently decreased, which allows a peptic ulcer to heal, thereby providing a rationale for surgical section of the vagus nerve in the surgical management of peptic ulcers.

3 One of the characteristics of sympathetic functions is the mass discharge. This means that the sympathetic discharge occurs simultaneously to many parts of the body. Thus, in the 'fight or flight' response there are numerous effects occurring at the same time, e.g. increased cardiac output, increased blood flow in the skeletal muscle, decreased blood flow in the skin and splanchnic area, increased rate of metabolism, increased blood glucose, mental alertness, etc.

This is not the case with parasympathetic reactions, which tend to be localized specific reflexes, e.g. the heart rate decreases in response to a rise in blood-pressure by increasing vagal tone; salivary secretion in response to the smell and taste of food; secretion of gastric juice; emptying of the rectum and urinary bladder, etc.

Higher control of autonomic functions

There are several areas of the brain which are concerned with autonomic regulation. Neurones in the posterior part of the hypothalamus regulate sympathetic functions, such as blood-pressure and heart rate, while those in the anterior part regulate some parasympathetic effects, such as contraction of the urinary bladder. There is, however, considerable overlap in this distribution of hypothalamic nuclei. Other areas, such as the reticular formation of the medulla and pons,

exert influences on autonomic functions such as heart rate, blood-pressure and gastrointestinal secretions and motility.

Autonomic functions are initiated by reflexes which have their centres in the spinal cord or the cranial nerve nuclei in the brain stem. These reflexes are influenced by the higher centres mentioned above. Some of the reflexes, like the voiding of urine and defecation, are also controlled by higher voluntary centres.

The hypothalamus is situated at the base of the brain. It is closely related to the limbic system, of which it is considered by some as an integral part. There are numerous nuclei and areas in the hypothalamus. These have been given various names, the most well-known of which are the supraoptic and paraventricular nuclei (see Chapter 17). These nuclei and areas are not anatomically as well-defined as they might appear in diagrams. The hypothalamus receives information from various parts of the CNS and gives outputs both to

Table 5.4 Functions of various nuclei of the hypothalamus

Anterior hypothalamus[a]	Posterior hypothalamus[a]
Preoptic nucleus	*Posterior hypothalamic nucleus*
Bladder contraction	Dilatation of pupil
Decreased heart rate	Increased blood-pressure
Decreased blood-pressure	Shivering
Supraoptic nucleus	*Perifornical nucleus*
Antidiuretic hormone release	Hunger
	Increased blood-pressure
	Rage
Paraventricular nucleus	
Oxytocin release	*Ventromedial nucleus*
	Satiety
Anterior hypothalamic area	
Temperature regulation, sweating	*Dorsomedial nucleus*
	Stimulation of GI tract
Lateral hypothalamic nuclei[a]	*Mammillary bodies*
Thirst centre	Feeding reflexes
Feeding centre	

a The lateral hypothalamic nuclei overlap both the anterior and posterior nuclei on both sides. One of the main functions of the lateral nuclei is to provide connections with the cerebral nuclei, e.g. the thalamus and the limbic nuclei.

the cerebrum above and to the brain stem below (Fig. 5.5). It sends outputs to the anterior thalamus and the limbic cortex in the cerebrum. It also sends outputs to the reticular formation of the brain stem and the vital centres of the medulla.

The hypothalamus is largely concerned with vegetative or visceral functions (see Table 5.4). It mediates some of the effects of the autonomic nervous system. For this reason it is sometimes called the head ganglion of the autonomic nervous system. Section of the brain stem above the medulla, however, does not affect most basal functions of the autonomic nervous system. This emphasizes the importance of spinal and medullary centres in autonomic responses.

The limbic system consists of those parts of the brain which are responsible for instinctive behaviour and emotions. In lower animals this part of the brain represents the largest portion of the cerebral hemispheres. In man it is restricted mainly to parts around the hilum of the cerebral hemisphere (see Chapter 17). The close connections of the hypothalamus with the limbic system help to explain the causation of abnormal autonomic reactions. Such conditions have been collectively known as psychosomatic disorders, characterized by hyper- or hypoactivity of autonomic functions. Peptic ulcer, hypertension, ulcerative colitis and migraine are frequently encountered examples of these conditions. Almost all diseases have a psychological element in them. The patient's reaction to his complaints are influenced by his psyche and by social factors in his environment. The doctor is well advised to take account of these factors in communication with patients as well as in diagnosis and management.

Further reading

1 Ganong, W.F. (1991) The autonomatic nervous system. In *Review of Medical Physiology*, 15th edn, pp. 207–213. Appleton & Lange, San Mateo, California.
2 Guyton, A.C. (1991) The Limbic system. In *Textbook of Medical Physiology*, 8th edn, pp. 651–654. Saunders, Philadelphia.

6: The Cardiovascular System

Objectives

On completion of the study of this section, the student should be able to:

1 Acquire basic information on the properties of cardiac muscle and the electrical events taking place in the heart, so as to use the electrocardiogram (ECG) in order to identify conduction defects, arrhythmias, ischaemic heart disease, etc.

2 Describe the events of the cardiac cycle and correlate the electrical events with the contraction and relaxation phases and the heart sounds so as to explain the pathophysiology of valvular and congenital heart disease.

3 Comprehend the factors which influence and regulate the cardiac output so as to use this information in the diagnosis and management of heart failure.

4 Understand the relationships between pressure, flow and resistance in the circulatory system so as to explain the pathophysiology of blood-pressure abnormalities, heart failure and oedema.

5 Understand the circulation in various special regions (pulmonary, coronary, cerebral) so as to explain abnormalities of blood flow such as myocardial and cerebral ischaemia.

Introduction

The heart provides two pumps in series which drive the circulation of blood through the systemic and pulmonary vessels. The circulation of blood serves many functions, of which the most well-known is the provision of oxygen and nutrients to the tissues as well as removal of waste products. However, one major role of the cardiovascular system is homoeostatic regulation. The heart and blood-vessels respond to physical and chemical changes in the internal and external environments. Adjustments in the pumping action of the heart and the distribution of blood flow help to bring the altered parameters back to normal. For example, changes in blood O_2, CO_2 or pH give rise to important cardiovascular responses which correct these changes. A rise or fall of the skin or deep body temperatures also produce cardiovascular responses which help in the regulation of body temperature. For these reasons, the cardiovascular system can be regarded as a homoeostatic system.

ANATOMICAL CONSIDERATIONS OF THE HEART

Atria

The atria are thin-walled chambers which receive blood either from the systemic circulation or from the pulmonary circulation. The atria open into the ventricles via the atrioventricular valves (Fig. 6.1).

The ventricles

The ventricles are thick-walled chambers which pump blood into the pulmonary trunk and the aorta. The wall of the left ventricle is three to four times thicker than the right ventricle, because of the greater work it has to do against aortic impedance due to the arterial blood-pressure. The right ventricle works against much less pressure in the pulmonary arteries. However, each ventricle has the same capacity and pumps the same volume of blood in a given period of time. Even minor differences over long periods of time result in heart failure. This leads to congestion of the systemic veins in the case of right

ventricular failure, and congestion of the pulmonary circulation in the case of left ventricular failure.

There is a fibrous tissue ring separating the atria from the ventricles. It separates the myocardium of the atria from that of the ventricles. Therefore, electrical activity in the atria cannot go directly to the ventricles. The fibrous tissue also gives attachment to the muscle cells of the heart and the atrioventricular valves.

The valves of the heart

There are no valves where the great veins enter the atria. However, the heart itself has four sets of valves which ensure that the blood flows in one direction and that no regurgitation occurs between the chambers of the heart. The right atrioventricular opening is guarded by the tricuspid valve (three cusps). At the left atrioventricular opening there is the mitral valve or bicuspid valve (two cusps). The openings of the aorta and the pulmonary trunk are guarded by the semilunar valves.

The atrioventricular cusps are held by chordae tendineae to papillary muscles. These muscles contract as part of the ventricular myocardium. Therefore, during ventricular contraction the cusps are pulled down so that they do not flap into the atria causing back flow of blood.

The conducting tissues

The conducting tissues of the heart consist of specialized myocardial cells. The pacemaker cells

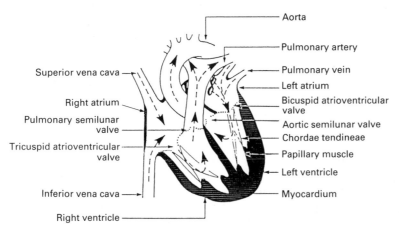

Fig. 6.1 The chambers and valves of the heart. The arrows indicate the direction of flow. (Reproduced from Bray, J.J., Cragg, P.A., Macknight, A.D.C., Mills, R.G. & Taylor, D.W. (eds), (1989) *Lecture Notes in Physiology*, 2nd edn. Blackwell Scientific Publications, Oxford.)

are found in the sinoatrial (SA) node in the right atrium near the junction with the superior vena cava. The atrioventricular (AV) node lies above the fibrous ring and represents the only myocardial connection between the atria and ventricles. The bundle of His starts in the AV node and gives rise to two branches (right and left), which descend into the interventricular septum. The terminal branches of the conducting tissue end in the subendocardial muscle of the ventricles.

Cardiac innervation

The heart receives extrinsic nerve supply from both sympathetic and parasympathetic divisions of the autonomic nervous system. The preganglionic neurones of the cardiac sympathetic nerves originate from the lateral horn cells of the upper thoracic segments. They relay in the sympathetic chain. The postganglionic nerves run in the cardiac sympathetic nerve and innervate the SA node, the atrial myocardium, the AV node and the ventricular myocardium. The parasympathetic supply to the heart originates in the neurones of the nucleus ambiguus of the medulla oblongata, also known as the cardioinhibitory area. The nerve fibres reach the heart in the vagus nerves. Preganglionic fibres synapse in parasympathetic ganglia near the heart. The postganglionic fibres are short and supply the SA node in the atrial muscle and the AV node.

Properties of cardiac muscle

To understand the properties of cardiac muscle, we must first take into account some basic anatomical considerations.

When seen under the light microscope, cardiac muscle appears striated but it differs in several ways from striated skeletal muscle. It consists of branching and interdigitating fibres which are closely attached to one another by structures known as intercalated discs (Fig. 6.2). These discs are specialized junctions. Physiologically they are of great significance because they offer little resistance to the passage of an action potential. Therefore, excitation will spread very easily from one fibre to another. The intercalated discs permit cardiac muscle to function as a syncytium which obeys the 'all or none law'. Stimulation of a single

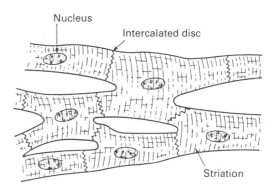

Fig. 6.2 Microstructure of cardiac muscle showing intercalated discs between muscle cells. (Reproduced from Bray, J.J., Cragg, P.A., Macknight, A.D.C., Mills, R.G. & Taylor, D.W. (eds) (1989) *Lecture Notes in Physiology*, 2nd edn. Blackwell Scientific Publications, Oxford.)

cardiac muscle fibre results in contraction of all the muscle fibres. Furthermore, cardiac muscle is capable of contracting without being triggered by a nerve impulse.

Cardiac muscle has four characteristic properties which will be considered below: contractility, rhythmicity, excitability and conductivity.

CONTRACTILITY

Contractility is the ability of the cardiac muscle to convert chemical energy into mechanical work. The heart acts as a muscular pump and the strength of its contractility determines its pumping power.

The mechanism of contraction depends on the contractile units, the myofibrils, which contain the protein molecules actin and fibrin. The shortening of the myofibrils (see Chapter 4) takes place with energy derived from adenosine triphosphate (ATP). Almost all the energy needed for contraction is derived from aerobic metabolism, so an adequate oxygen supply is essential for cardiac contractility.

Factors affecting myocardial contractility

Factors which affect myocardial contractility are said to have an inotropic effect. Those which increase contractility are called positive inotropic factors and those which decrease contractility are called negative inotropic factors. The factors which affect myocardial contractility are as follows:

1 *Starling's law of the heart.* The contractility of cardiac muscle depends on the initial length of its fibres. If the venous return is increased, this will stretch the cardiac muscle so that the initial length of the muscle fibres will increase. As a result of this increase in diastolic volume, the strength of contraction will increase provided that the fibres are not stretched above their elastic limits. Starling's law explains how cardiac muscle accommodates itself to the changes in venous return.

2 *Cardiac innervation.* Cardiac muscle is supplied by both divisions of the autonomic nervous system. Sympathetic stimulation produces an increase in contractility and vagal stimulation produces a decrease in contractility.

3 *Hormonal and chemical factors or drugs.* Adrenaline, noradrenaline, alkalis, digitalis and calcium ions are all positively inotropic. Acetylcholine, acids, ether, chloroform, some bacterial toxins (e.g. diphtheria toxin) and K^+ ions are all negatively inotropic.

4 *Physical factors.* Warming increases contractility while cooling decreases it.

5 *Oxygen supply.* This depends on the condition of the coronary vessels. Hypoxia (low oxygen content in the blood) depresses contractility.

6 *Mechanical factors:*

(a) Because cardiac muscle behaves as a syncytium, it obeys the 'all or none law', i.e. minimal or threshold stimuli lead to maximal cardiac contraction.

(b) Cardiac muscle shows the treppe or staircase phenomenon, i.e. if an isolated heart is stimulated by successive equal and effective stimuli, the first few contractions show a gradual increase in the magnitude of contraction.

(c) Cardiac muscle cannot be excited while it is contracting. This is due to the long absolute refractory period of cardiac muscle (see Chapter 4). Therefore, the muscle must relax before it can be restimulated. The length of the refractory period is vital to the functions of the heart as a pump. It prevents cardiac muscle from being tetanized like skeletal muscle. If it were possible to tetanize cardiac muscle,

the heart would be thrown into a state of continuous contraction, i.e. it would stop and the subject would die of circulatory failure.

(d) Cardiac muscle can perform both isometric and isotonic types of contractions.

RHYTHMICITY (AUTOMATICITY)

Rhythmicity is the ability of cardiac muscle to contract in a regularly constant manner. The rhythmicity is myogenic in origin, i.e. its origination has nothing to do with the nerves of the heart. It is not neurogenic. This is proved by: (i) the finding that, if the apex of the frog heart is cut off and placed in Ringer's solution, it will continue to contract in a rhythmic manner although it is devoid of any nervous tissue; (ii) the fact that the heart of the human or the chick embryo shows rhythmic contraction before the development of any nervous connections to it.

The pacemaker of the heart

The frog's heart is different from the human heart in that it has five chambers instead of four. The extra chamber is the sinus venosus (SV), which is similar to the sinoatrial (SA) node in the human heart. Experiments with frogs' hearts have proved that rhythmicity is not the same in the different chambers of the heart. The SV of the frog's heart or the SA node in man has the fastest rhythm of all parts of the heart. This is due to the fact that SA nodal fibres have an unstable resting membrane potential (Fig. 6.3). The slow depolarization (prepotential) seen between action potentials is due to a decrease in K^+ permeability to the outside, which causes the resting membrane potential (RMP) to decrease. These prepotentials are called pacemaker potentials and the SA node is therefore called the pacemaker of the heart. This is proved by the following:

1 If heat or cold is applied to the different chambers of the frog's heart, the rhythm is found to change only when the heat or cold is applied to the SV. No change is noted when the temperature stimulus is applied to the atria or ventricles.

2 If a string (ligature) is tied between the SV and the atria of the frog's heart (first Stannius ligature), the atria and ventricles will stop immediately, while the SV will continue to beat in its normal rhythm. After some time the ventricles

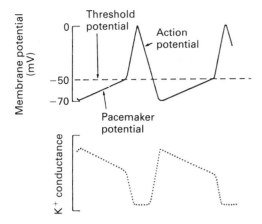

Fig. 6.3 Pacemaker potential and K$^+$ conductance in sinoatrial node. Note the prepotential indicating instability of the RMP. (Reproduced from Bray, J.J., Cragg, P.A., Macknight, A.D.C., Mills, R.G. & Taylor, D.W. (eds) (1989) *Lecture Notes in Physiology*, 2nd edn. Blackwell Scientific Publications, Oxford.)

will start contracting again but at a slower rhythm. This is called the idioventricular rhythm. This is due to the presence of an ectopic pacemaker, which may be in the AV node or any cardiac muscle. In humans ectopic pacemakers are normally dormant but begin to show activity if the SA node is inhibited or separated.

3 Measurement of the electrical potential changes over the different chambers of the heart has shown that the SA node is the first part of the heart to show electrical negativity. This means that the SA node is the first part to show activity.

Factors affecting rhythmicity
Nervous, physical and chemical factors affect rhythmicity as follows:

1 *Neural factors.* Vagal stimulation (or application of acetylcholine) increases the permeability of SA nodal tissue to K$^+$ ions and therefore increases the membrane potential. This decreases the slope of the prepotential and the rate of firing of the SA node and results in slowing of the heartbeat (bradycardia). Strong stimulation of the vagus nerve may lead to complete stoppage of SA nodal activity. Sympathetic stimulation (or application of noradrenaline) decreases the permeability of the nodal tissue to K$^+$ ions and so

decreases the resting membrane potential. This increases the slope of the prepotential and the rate of firing, resulting in acceleration of the heart (tachycardia).

2 *Physical factors.* Moderate warming of the SA node increases rhythmicity, while cooling decreases it.

3 *Effects of ions.* Application of Na$^+$ ions can initiate rhythmicity but cannot maintain it. A decrease in extracellular K$^+$ ions increases the slope of diastolic depolarization and therefore increases the rhythmicity. An increase in extracellular K$^+$ ions decreases the rhythmicity and may stop the heart in diastole (relaxed phase) by decreasing the slope of diastolic depolarization.

4 *Chemical factors.* Thyroid hormones and catecholamines increase the rhythmicity of the heart while acetylcholine, digitalis, some metabolites and oxygen lack decrease the heart rate.

EXCITABILITY (IRRITABILITY)
Excitability is the ability of the cardiac muscle to respond to adequate stimuli by generating an action potential followed by a mechanical contraction (Fig. 6.4).

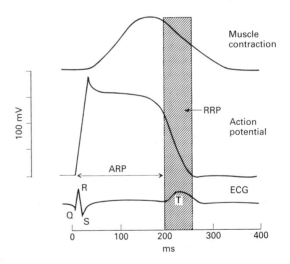

Fig. 6.4 Time relationship between the action potential and the contraction of ventricular muscle. ARP, absolute refractory period; RRP, relative refractory period. Note that summation of contraction is not possible.

Excitability passes through two phases:

1 *Absolute refractory period (ARP).* In cardiac muscle the ARP is very long compared with that in nerve or skeletal muscle cells. The ARP occupies the whole period of systole (contraction phase of cardiac muscle). The ARP corresponds to the period of depolarization, during which the excitability of the heart muscle becomes zero, i.e. no stimulus, however strong, can excite the muscle or produce any response. The difference between cardiac muscle and skeletal muscle in this phase is very important from a functional point of view. Skeletal muscle can be tetanized while cardiac muscle cannot, because the ARP occupies the whole contraction phase (systole).

2 *Relative refractory period.* This occupies the time of diastole of cardiac muscle and corresponds to repolarization. A strong enough stimulus can give rise to contraction during the relative refractory period. The refractory period can be affected by the heart rate, temperature, bacterial toxins, vagal stimulation, sympathetic stimulation and drugs.

CONDUCTIVITY

Conductivity is the ability of cardiac muscle fibres to conduct the cardiac impulses that are initiated in the SA node, which is the pacemaker of the heart. From the SA node the impulses are conducted to the AV node through atrial musculature and internodal pathways, which are three in number. They connect the SA node to the AV node. From the AV node the impulses pass to the AV bundle (bundle of His), which divides into two branches, right and left, which will transmit the impulses to the ventricles. From the apex of the heart the impulses are conducted to the base of the heart through the Purkinje fibres. Conduction velocity in the bundle of His is 1 m/sec and in the Purkinje fibres 4 m/s (fastest); it is about 0.4 m/s in ventricular muscle and 0.05 m/s in the SA node. It is slowest (0.01 m/s) in the AV node. The AV nodal delay gives time for the ventricles to be filled with blood before they contract (ventricular systole).

The electrocardiogram

The electrocardiogram (ECG or EKG) is a clinical test used as an indicator of cardiac function be-

cause all parameters of heart function tend to be represented in it.

WHAT IS AN ECG?

Associated with the passage of the cardiac impulse from the SA node to the AV node and down the AV bundles to the ventricles, there are changes of potential between the different points of the surface of the heart. The electric current generated by the ionic changes at the surface of the heart muscle spread into the tissue fluids, which form a continuous conductor between the heart and the surface of the body, to reach the skin surface. From there, this minute current can be picked up electronically, amplified and then allowed to drive an appropriate recorder to give an ECG record. The current is picked up by electrodes, which are metal plates applied to the wrists and ankle. Before placing the metal plates, electrode jelly is applied to the skin to ensure good conduction. The ECG can be defined as a record of the fluctuations in action potentials generated during one heartbeat.

LEADS

A pair of electrodes applied to appropriate parts of the body is called a lead. Different points of application, i.e. different leads, give information about the strength and direction, i.e. vector, of the electrical forces generated by the heart.

There are three different types of leads: bipolar leads, unipolar augmented leads and unipolar precordial or chest leads.

Bipolar leads

Bipolar leads are also called standard leads or conventional leads. They were the first leads used in 1903 by Einthoven, the father of modern electrocardiography.

Bipolar recording means recording the ECG using two active or exploring electrodes and recording the potential difference between them. The two electrodes are placed in two different positions on the body. These leads are three in number: lead I, lead II and lead III (Fig. 6.5).

Lead I This is the connection between the right arm (RA), designated as negative (−ve), and the left arm (LA), which is considered positive (+ve).

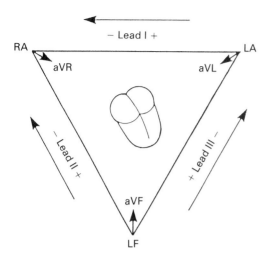

Fig. 6.5 Einthoven's triangle and the limb leads. RA, LA and LF are right arm, left arm and left leg, respectively; aVR, aVL and aVF are augmented limb leads. The arrows indicate the direction in which each lead detects the electrical activity of the heart. (Reproduced from Bray, J.J., Cragg, P.A., Macknight, A.D.C., Mills, R.G. & Taylor, D.W. (eds) (1989) *Lecture Notes in Physiology*, 2nd edn. Blackwell Scientific Publications, Oxford.)

Lead II This is the connection between the right arm (RA), which is made negative (−ve), and the left foot or leg (LF), which is considered positive (+ve).

Lead III This is the connection between the left arm (LA) which is designated as negative (−ve) and the left foot or leg (LF) which is positive (+ve).

Einthoven's triangle This is an equilateral triangle drawn arbitrarily around the area of the heart (Fig. 6.5). Einthoven's law states that, if an ECG is recorded through standard limb leads, then the total voltage of the QRS (ventricular) complex in leads I and III is equivalent to lead II. If the voltage of the QRS complex in lead III is +2 and in lead I is +3, then the voltage in lead II will be +5. In other words, if the voltage of QRS in any two leads is known, then the other one can be deduced from those two.

In order to record a normal ECG using bipolar recording, we use a machine called the electrocardiograph. We connect the three leads and a further electrode is applied to the right foot or leg (RL). This electrode is not used for recording. It is used to earth the subject and thus minimize any interference from the mains and electrical apparatus present in the room. The modern electrocardiograph, which was developed in 1930, is an amplifier-driven oscillograph provided with direct input recording. It replaced the original string galvanometer developed in 1903 by Einthoven. In order to make a recording using the modern electrocardiograph, two input connections must be made to the amplifier, and there is usually a switch on the apparatus which makes the appropriate connections internally when a particular lead is chosen. The detailed waveform produced by the ECG depends on the site of the recording electrodes.

Unipolar augmented leads
Unipolar augmented leads are the second type of leads used to record the ECG. Unipolar recording means recording an ECG using an active or exploring electrode connected to one limb (RA, LA or LF) and an indifferent electrode that joins any two of the other limbs at zero potential (i.e. two of the limbs are connected through electrical resistances to the negative (−ve) terminal of the electrocardiograph, while the third limb is connected to the positive (+ve) terminal). This type of recording between one limb and the other two limbs has the advantage of increasing the size of the potentials by 50% without any change in configuration from the non-augmented leads.

There are three types of augmented leads:
1 aVR: In this lead the right arm (RA) is +ve in reference to the left arm (LA) and left foot or leg (LF).
2 aVL: In this lead the left arm (LA) is +ve in reference to the right arm (RA) and left leg (LF).
3 aVF: In this lead the left leg (LF) is +ve in reference to the RA and LA (see Fig. 6.5).

Each augmented unipolar limb lead records the potential of the heart on the side nearest to the respective limb, i.e. lead aVR can be regarded as looking at the right side of the heart. Thus, when the record in the aVR lead is negative, this means that the side of the heart nearest to the right arm is negative in relation to the remainder of the heart. Lead aVL can be regarded as looking at the left side of the heart. Thus, when the record in aVL is negative, this means that the side of the heart nearest to the left arm is negative. Lead aVF can be regarded as looking at the inferior side of the heart. Thus, when the record in the aVF lead is positive, this means that the apex of the heart, which is the part of the heart nearest to the foot, is positive with respect to the remainder of the heart.

Unipolar precordial (or chest) leads
Unipolar precordial leads are the third or V type of lead used to record the ECG. In unipolar chest leads the ECG is recorded with one exploring electrode placed on the anterior surface of the chest over the heart. This electrode is connected to the +ve terminal of the electrocardiograph, and the −ve electrode, called the indifferent electrode, is normally connected simultaneously through electrical resistances to the right arm, left arm and left leg (Fig. 6.6). Usually six different

standard chest leads are recorded from the anterior chest wall, the chest electrodes being placed respectively at the following six points:

V_1 4th right interspace at sternal border
V_2 4th left interspace at sternal border
V_3 midway between V_2 and V_4
V_4 5th left interspace at the midclavicular line or at the apex of the heart
V_5 5th left interspace at the anterior axillary line
V_6 5th left interspace at the midaxillary line

The exploring electrode in the chest leads is in the form of a suction cup, applied to the chest by evacuating the air from the cup. The skin of the chest must be cleaned and rubbed with electrode jelly before applying the cup.

THE NORMAL ECG
The normal waveforms observed in an ECG record are labelled starting with the letter 'P' then 'QRS' then 'T'. The general appearance of an ECG record is shown in Fig. 6.7. This appearance varies from one lead to another, as will be seen later. Causes of the different waveforms of the normal ECG include:

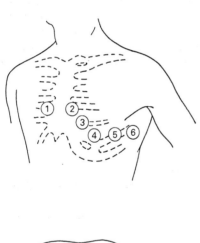

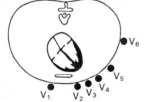

Fig. 6.6 Diagram showing the position of the unipolar chest leads. (Reproduced from Bray, J.J., Cragg, P.A., Macknight, A.D.C., Mills, R.G. & Taylor, D.W. (eds) (1989) *Lecture Notes in Physiology*, 2nd edn. Blackwell Scientific Publications, Oxford.)

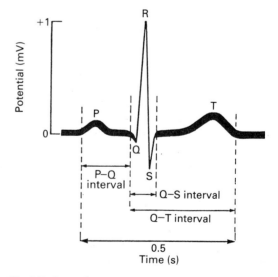

Fig. 6.7 General appearance of ECG record at rest. (Reproduced from Bray, J.J., Cragg, P.A., Macknight, A.D.C., Mills, R.G. & Taylor, D.W. (eds) (1989) *Lecture Notes in Physiology*, 2nd edn. Blackwell Scientific Publications, Oxford.)

1 *The P wave.* This corresponds to the spread of excitation from the SA node over the atrial muscle and therefore represents atrial systole, i.e. it is due to atrial depolarization.

The propagation of the excitation wave down the bundle of His does not produce any detectable external electrical change. It is demonstrated on the ECG by the isoelectric P–Q interval, where there is no deflection.

2 *The QRS complex.* This is due to spread of the excitation wave to the two ventricles, i.e. it is due to ventricular depolarization. The invasion of the two ventricles causes electrical changes which to a large extent cancel each other and the record returns to the base-line (S–T segment). The QRS complex at the beginning of the ventricular excitation and the T wave at the end are all that remains in the algebraic summation of electrical activity in the ventricles.

3 *The T wave.* This is due to ventricular repolarization. Should changes occur in the ventricular muscle, such as necrosis following coronary thrombosis, then the cancellation will not be so effective and the S–T segment will no longer be isoelectric but elevated or depressed.

The relaxation phase T–P is isoelectric.

TIME AND SPEED OF THE ECG
It is a fundamental principle of the ECG machines that they all run at a standard speed and they use paper with standard squares. Each large square (5 mm) is equivalent to 0.2 s, so there are five large squares (25 mm) per second and 300 large squares per minute. So an ECG event such as the QRS complex occurring once per large square is occurring at a rate of 300/min. The heart rate can be calculated by remembering this sequence. If the R–R interval is:

One large square, the heart rate is 300/min. This is calculated by dividing by the number of large squares, i.e. 300 over 1; 300/1 = 300.
Two large squares, the rate is 300/2 = 150.
Three large squares, the rate is 300/3 = 100.
Four large squares, the rate is 300/4 = 75.
Five large squares, the rate is 300/5 = 60.
Six large squares, the rate is 300/6 = 50.

Just as the length of the paper between two R waves gives the heart rate, so the distance between the different parts of the PQRST complex shows the time taken for conduction to occur through different parts of the heart. The P–R interval is of great clinical importance. The P–R interval is the time taken for the cardiac excitation (impulse) to spread from the SA node through the atrial muscle and the AV node, down the bundle of His and into the ventricular muscle. Most of the time is taken up by the delay in the AV node. The normal P–R interval is usually measured from the beginning of the P wave to the beginning of the QRS complex. In fact, it should have been called the P–Q interval (see Fig. 6.7) rather than the P–R interval. It normally lasts 0.12–0.20 s (three to five small squares). The prolongation of the P–R interval is an indication of a conduction defect in the AV system. If the P–R interval is very short, either the atria have been depolarized from close to the AV node or there is an abnormality of conduction from the atria to the ventricles.

The duration of the QRS complex shows how long excitation takes to spread through the ventricles. The QRS duration is normally 0.12 s (three small squares) or less; but any abnormality of conduction takes longer and causes widened QRS complexes.

SUMMARY OF THE ECG INTERVALS
P–R = 0.12–0.2 s
QRS = 0.12 s
Q–T = 0.4 s
S–T = 0.3 s
Standard rate (i.e. speed) of the ECG paper = 25 mm/s
Standard timing of the ECG paper = 0.2 s for each big square
One big square in the ECG paper = five small squares
1 mV = 1 cm of vertical deflection.

NORMAL ECG VARIATIONS
As we mentioned before when we discussed the general appearance of the normal ECG record, the appearance varies from one lead to another. We will now discuss these variations.

1 *The appearance of the ECG recorded using the three standard bipolar leads, i.e. leads I, II and III.* For any wave in the cycle, the largest deflection will be obtained in the lead which is parallel to the direction of the impulse (or *vector*) at the time. In most normal individuals the anatomical and electrical axes of the heart are directed downwards and to the left and hence all the waves should be tallest in lead II. Morever, due to the direction of the vector, the P and QRS principal deflections will be upwards in all the standard leads, as shown in Fig. 6.8.

2 *ECG recorded from the three augmented unipolar limb leads.* The augmented unipolar leads record activity from the part of the heart lying opposite the electrode. Note in Fig. 6.8 that the cardiac axis lies parallel to lead II and the highest deflection will be recorded in lead II and lead aVF. Therefore, lead aVR mainly records activity within the ventricular cavities and will show a predominantly negative RS deflection since the activation spreads from endocardial to epicardial regions, i.e. travels away from the electrode.

aVL records mainly from the upper left side of the heart and thus has a biphasic type of wave.

aVF is related to the lower surfaces of the ventricles and will record a steep R wave.

3 *ECG recorded from the six chest or precordial unipolar leads.* These leads look at the heart in a horizontal plane. Leads V_1 and V_2 are nearest to the right ventricle, while V_5 and V_6 are predominantly related to the left ventricle. V_3 and V_4 are intermediate in position. Because of its

thick wall, the left ventricle tends to dominate the electrical activity even in right ventricular leads; thus, V_1 and V_2 have substantial downward deflections in addition to the upward wave associated with proximal activation. They may therefore show biphasic RS complexes but with a predominant downward component. On the other hand, V_5 and V_6 are influenced by left ventricular activity alone and show steep R waves (see Fig. 6.9).

Two important points are worth mentioning here:

1 A general principle is that, if a wave of depolarization is approaching an exploring electrode, a positive voltage will be picked up by that electrode relative to an earthed reference electrode. If the wave of depolarization is receding from it, a negative voltage will be recorded. The best example is the record obtained from V_1 to V_6. In V_1 the R wave is smallest with a deep, negative S wave but, as we go from V_1 to V_6, then R wave height increases and the depth of the S wave disappears (this is known as R-wave progression).

2 The ECG record can reveal much information on cardiac abnormalities such as: tachycardia and bradycardia, abnormalities of sinoatrial node rhythmicity, beats originating in abnormal pacemakers, faults in transmission of the im-

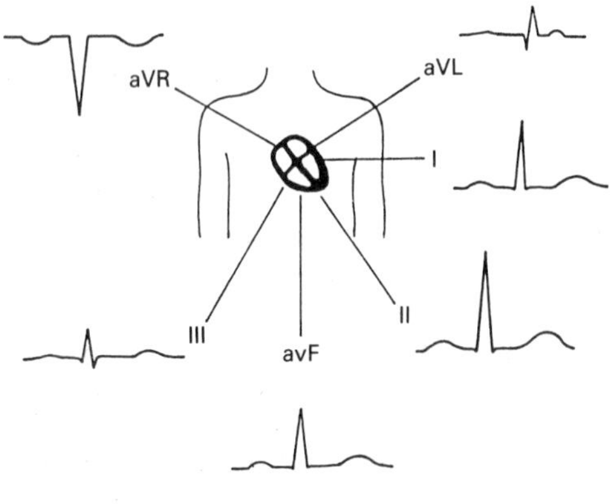

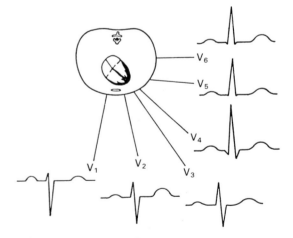

Fig. 6.8 ECG records from bipolar limb leads and unipolar augmented limb leads. (Reproduced from Bray, J.J., Cragg, P.A., Macknight, A.D.C., Mills, R.G. & Taylor, D.W. (eds) (1989) *Lecture Notes in Physiology*, 2nd edn. Blackwell Scientific Publications, Oxford.)

Fig. 6.9 ECG records from chest leads. (Reproduced from Bray, J.J., Cragg, P.A., Macknight, A.D.C., Mills, R.G. & Taylor, D.W. (eds) (1989) *Lecture Notes in Physiology*, 2nd edn. Blackwell Scientific Publications, Oxford.)

pulse in the conducting system, myocardial ischaemia, myocardial hypertrophy and electrolyte disturbances.

EXAMPLES OF ECG ABNORMALITIES

1 *Sinus tachycardia* is usually due to excessive stimulation of the SA node by means of the sympathetic nerves. It can also occur in hyperthyroidism. The ECG record shows regular sinus rhythm with a high frequency. Bradycardia can likewise occur due to vagal stimulation and hypothyroidism.

2 *Sinus arrhythmia*. This is a variation in the heart rate during respiration. It is more marked during deep inspiration due to impulses radiating from the respiratory centre inhibiting vagal tone and increasing the heart rate. It is physiological and does not signify heart disease. It is usually observed during childhood and early adolescence.

3 *Faults in the conduction system*. The most common site of block of conduction is the AV node. This commonly occurs as a result of ischaemic heart disease leading to ischaemia of the conducting tissues. In the early stages the impulses can pass but the rate of conduction between atria and ventricles (P−R interval) is prolonged. This is called *first degree heart block*. The P−R interval in such cases is greater than 0.2 s. When the P−R interval becomes more prolonged (0.35−0.45 s) then all the impulses cannot pass especially when the heart rate is fast. In this case one impulse may be conducted out of every 2 or 3 atrial beats (2 : 1 or 3 : 1 block). This is referred to as *second degree heart block*. Both first and second degree blocks are examples of incomplete heart block (Fig. 6.10a, b).

Complete heart block develops in severe AV ischaemia when none of the atrial impulses are conducted to the ventricles. In the ECG the P waves have no relation to the QRS complexes i.e. the atria and ventricles beat regularly but independently (Fig. 6.10c). The atria beat much faster than the ventricles which beat at their own rhythm (idioventricular rhythm). This is known as ventricular escape.

4 *Extrasystoles* (also known as premature or ectopic beats) occur when a QRS complex arises from an ectopic focus in the atria, AV node or

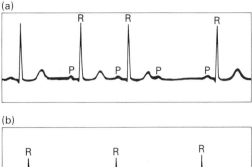

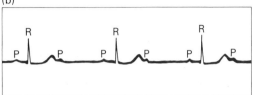

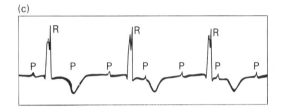

Fig. 6.10 (a) and (b) Incomplete heart block. (c) Complete heart block. See explanation in the text. (Reproduced from Emslie-Smith, D., Paterson, C.R., Scratcherol, T. & Read, N.W. (eds) (1988) *BDS Textbook of Physiology*, 11th edn. With permission from Churchill Livingstone, Edinburgh.)

in the ventricles, the extra systole is not preceded by a P wave in the usual manner and is followed by a compensatory pause (missed beat). The compensatory pause is due to the absence of the next normal beat falling in the refractory period of the extrasystole. Extrasystoles therefore lie between two normal QRS complexes. They may occur in a healthy heart or indicate early heart disease.

5 *Atrial fibrillation*. This is a form of arrhythmia characterized by uncoordinated contraction and relaxation occurring all over the atria. Atrial dilatation and slow conduction precipitate this condition (as in mitral stenosis). in the ECG, the P waves are absent because there is no coordinated depolarization of the atria. They are replaced by low voltage irregularities of the isoelectric line. The QRS complexes are completely irregular because impulses from the atria cross the AV node at random. The heart rate is rapid; usually more than 120/min. This condition does not cause a major reduction of the cardiac output. The reason is that co-ordinated atrial contractions are not essential for ventricular filling which occurs up to 70% passively. Although the ventricles beat irregularly, the cardiac output is maintained but at a decreased level.

6 *Atrial flutter.* When an ectopic focus fires repetitively at a higher rate than the SA node the result is atrial flutter. This is due to abnormal waves of excitation which cause semi-coordinated atrial contractions at a much higher rate (200–350). The ECG shows prominent P waves but only some of them are followed by a QRS complex. As the AV nodal tissue has a long refractory period, the P:QRS ratio is usually 2:1 or 3:1.

7 *Ventricular fibrillation.* This condition can be fatal within a very short time. As there is no coordinated ventricular beat the cardiac output drops to zero. The ECG shows a completely irregular incidence of waves with no rhythm. Eventually a low voltage irregular tracing results. Ventricular fibrillation is always associated with serious heart disease.

8 *Myocardial ischaemia* is one of the main causes of cardiac arrhythmias. Ischaemia usually shows as S–T segment depression and T wave invertion. The main ECG change after an acute myocardial infarction is elevation of the S–T segment of the recording over the infarcted area. This is due to leakage of K^+ from damaged muscle cells causing an increase in their resting membrane potential and rapid repolarization after each beat. Both these effects cause elevation of the S–T segment. Other changes in the different leads which help to locate the position of the infarct are outside the scope of this text.

9 *Effects of ionic changes* on the ECG. Abnormal Na^+ and K^+ concentrations affect the ECG record. In general, Na^+ deficiency in the extracellular fluid causes a low voltage ECG. However, the effect of K^+ is more serious. Hypokalaemia results in slowing of conduction. The P–R interval is prolonged and the S–T segment is depressed and the U wave appears prominent. Hyperkalaemia is far more serious and can cause death. Slight increases in K^+ cause the T waves to become tall and prominent due to change in repolarization.

The P–R interval and the QRS are normal. However, further increases in K^+ result in prolongation of the QRS complex with tall T waves. Further increases result in a decrease of the resting membrane potential and the myocardium does not respond. The atria become paralysed and eventually ventricular tachycardia and fibrillation occur.

Hypercalcaemia in clinical practice rarely reaches levels which show effects on the ECG.

Hypocalcaemia causes tetany. Effects on the ECG appear as prolongation of the Q–T interval and the S–T segment.

The cardiac cycle

The heart, as a pump, goes through phases of contraction and relaxation (systole and diastole). During these phases there are associated changes in pressure and volume which take place in the different chambers of the heart and in the great vessels. There are also vibrations, mainly generated by the closure of the valves. These vibrations result in the heart sounds. The correlation of these events with each other and their time sequence help us to understand dynamic cardiovascular functions.

Of special interest is the correlation of the mechanical changes with the electrical events (the ECG). The ECG demonstrates that the cardiac muscle of the atria acts as one unit and that of the ventricles as another. The timing of the electrical waves is seen to precede the mechanical contraction and relaxation phases.

The wave of contraction which occurs first in the atria and then in the ventricles is triggered off by depolarization of the pacemaker (SA node). The electrical signal results in depolarization followed by contraction of the atria. The signal is transmitted to the ventricles through the AV node. It cannot be transmitted through other sites because of the presence of fibrous tissue separating the cardiac muscles of the atria from that of the ventricles. There is a delay in conduction at the AV node and this allows time for the ventricles to fill with blood before their own contraction begins. The terms diastole and systole, if not qualified, usually refer to the phases of relaxation and contraction of the ventricles. But we can also speak of atrial diastole and atrial systole.

The cardiac cycle can be divided into three main parts for the sake of conveniently describing the various events:

1 Atrial systole.
2 Ventricular systole.
3 Diastole of both atria and ventricles.

Throughout the following description of the cardiac cycle, refer to Fig. 6.11.

ATRIAL SYSTOLE

The contraction of the atria starts towards the end of ventricular diastole and helps to complete the filling of the ventricles. About 70% of ventricular filling takes place passively through the AV valves due to the pressure gradient brought about by relaxation of the ventricles. Thus, the atrial systole only adds 30% of ventricular filling.

During atrial systole, there is slight regurgitation of blood into the venae cavae, which results in the generation of pressure waves in the veins near the heart (venous pulse).

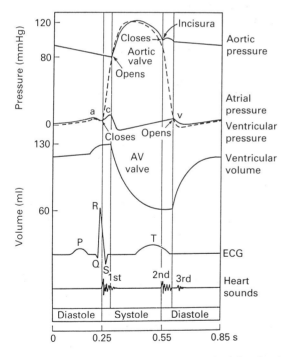

Fig. 6.11 Events of one cardiac cycle in the left side of the heart at rest. (Reproduced from Bray, J.J., Cragg, P.A., Macknight, A.D.C., Mills, R.G. & Taylor, D.W. (eds) (1989) *Lecture Notes in Physiology*, 2nd edn. Blackwell Scientific Publications, Oxford.)

VENTRICULAR SYSTOLE

The contraction of the ventricles follows immediately upon completion of atrial systole. The rise in ventricular pressure results in immediate closure of the AV valves. This causes the first heart sound, which marks the beginning of systole. The semilunar valves, however, do not open until ventricular contraction has raised the pressure in the ventricles enough to oppose the aortic and pulmonary artery pressure levels. This takes about 0.5 s. During this time the ventricles are closed chambers, because both the AV and the semilunar valves are closed, and therefore there is no change in the volume of the ventricles. This period is called the *isovolumetric contraction phase*. When the pressure reaches about 80 mmHg in the left ventricle, the aortic valve opens; and, when it reaches 10 mmHg in the right ventricle, the pulmonary valve opens. When both of these valves open, there follows the rapid ejection phase of systole. This soon slows down and causes the systolic pressure peak of about 120 mmHg in the aorta and 25 mmHg in the pulmonary artery.

At rest, during ventricular systole, each ventricle ejects about 70–90 ml of blood, which is known as the *stroke volume*. This is about 65% of the ventricular volume at the end of diastole. A residual reserve volume of about 50 ml is left in each ventricle at the end of systole (end-systolic volume). The term protodiastole refers to the last 0.04 s of systole. During protodiastole, the pressure in the ventricles falls, due to the flow of blood from the aorta into the systemic circulation.

DIASTOLE

Atrial relaxation precedes ventricular relaxation and continues through the greater part of ventricular diastole. At the beginning of ventricular diastole, the aortic and pulmonary valves close. This causes the second heart sound, which marks the beginning of diastole. The drop in pressure in the ventricles relative to that in the aorta and the pulmonary artery causes these valves to close immediately. The aortic valve closes when the pressure in the left ventricle falls to about 80 mmHg and the pulmonary valve closes when the pressure in the right ventricle falls to about 12 mmHg. Normally, the two valves close synchronously (at the same time).

In early diastole and after the closure of the semilunar valves, the AV valves do not open immediately. This is because the pressure in the ventricles is still higher than that in the atria. It takes about 0.04 s for ventricular pressure to fall to a level below that in the atria and only then do the AV valves open. This period between the closure of the semilunar valves and the opening of the AV valves is called the *isovolumetric relaxation phase*. The ventricle during this period is, a closed chamber and there is no change in volume; there is only change in pressure.

When the AV valves open, blood rapidly flows into the relaxed ventricles. This is the rapid filling phase of diastole. Towards the end of diastole, the atria contract and complete the filling of the ventricles, as previously described. Thus, one cardiac cycle is completed.

VARIATION IN THE DURATION OF SYSTOLE AND DIASTOLE

It is found that cardiac muscle can depolarize faster when the heart rate is rapid. The duration of the action potential can be decreased from 0.25 s to 0.15 s. At rest, systole lasts for 0.3 s but it can be shortened to 0.16 s when the heart rate is 190–200/min. Similarly, diastole is decreased from 0.53 to 0.14 s.

The refractory period is also affected by an increased heart rate. The absolute refractory period is shortened from 0.2 to 0.13 s and the relative refractory period from 0.05 to 0.02 s.

PRESSURE CHANGES DURING THE CARDIAC CYCLE

Aortic pressure

At the beginning of ventricular systole, there is no change in aortic pressure because the aortic (Ao) valve is still closed (isovolumetric phase of systole). When the Ao valve opens, there is a quick rise of pressure during the rapid ejection phase, which soon flattens at a peak of about 120 mmHg at rest (systolic pressure). The aortic pressure rises to a value slightly above that in the ventricles, due to the momentum of blood flow, i.e. kinetic energy becomes converted into pressure.

The last quarter of systole (protodiastole) is characterized by a fall in aortic pressure. This is due to slow ejection and the flow of blood from the aorta into the systemic circulation. At the beginning of diastole the pressure is about 80 mmHg (diastolic pressure). During diastole there is a slow and slight decrease in aortic pressure due to the continued flow of blood from the aorta into the systemic circulation.

Arterial pressure waves

The arterial pressure wave reflects the systolic peak and diastolic trough. However, these waves are sharper than those seen in the aorta (Fig. 6.12). There is an irregularity in the descending limb called the dicrotic notch. This has been attributed to a rebound of the aortic valves upon closure and

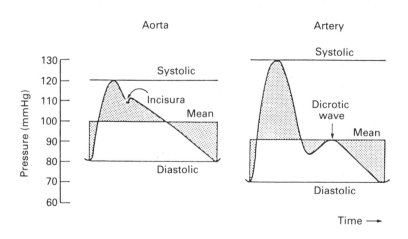

Fig. 6.12 Pressure waves in the aorta and a large artery. The shaded areas above and below the mean pressure are equal. (Reproduced from Bray, J.J., Cragg, P.A., Macknight, A.D.C., Mills, R.G. & Taylor, D.W. (eds) (1989) *Lecture Notes in Physiology*, 2nd edn. Blackwell Scientific Publications, Oxford.)

to oscillations in the aorta. Figure 6.12 also shows the calculation of the mean pressure from the arterial pressure wave.

Pressure changes in the pulmonary artery
Pressure waves in the pulmonary artery are similar to those in the aorta. The difference is only in magnitude. The systolic pressure is about 21–25 mmHg and the diastolic 5–10 mmHg.

Atrial pressure
The atrial pressure changes result in three waves (Fig. 6.13). The 'a' wave is due to atrial systole and is 4–6 mmHg in the right atrium and 7–8 mmHg in the left atrium. The 'c' wave is due to the rise in ventricular pressure causing the AV valves to bulge into the atria during the isometric contraction phase. It is also partly due to pulling of the atrial muscle during ventricular contraction. The 'v' wave is due to a build-up of pressure in the atria while the AV valves are closed. This wave falls once the AV valves open.

Jugular venous pulse waves
These waves are also named 'a', 'c' and 'v'. They are transmitted from the atria as there are no valves between them and the veins in the neck. Jugular venous pressure also rises during expiration and falls during inspiration due to changes in intrathoracic pressure (Fig. 6.13).

ELECTRICAL EVENTS
The waves of the electrocardiogram are changes in voltage detected on the surface of the body. They are conducted potential differences from cardiac muscle. Each wave is the sum of millions of action potentials occurring in individual cardiac muscle fibres. The characteristic pattern of the ECG shows waves named P, QRS complex and T. The details of the electrocardiogram and its clinical uses have been discussed earlier in this chapter.

The P wave is caused by spread of the depolarization wave in the atria. This electrical event is followed by the mechanical contraction of the atria about 0.02 s later.

The QRS complex occurs about 0.12–0.2 s after the P wave. This complex of waves is due to the spreading of depolarization in the interventricular septum and the ventricular muscle. It precedes the contraction of the ventricles by about 0.02 s.

The T wave represents the repolarization of the ventricles and occurs during the latter part of systole and before the onset of diastole.

THE HEART SOUNDS
The heart sounds can be detected over the anterior chest wall by means of a stethoscope (auscultation) or by the use of a sound recording device (phonocardiography). There are certain areas where the heart sounds can be heard best (Fig. 6.14). Usually, two sounds can be heard with ease and are known as the first and second heart sounds. Sometimes a third and fourth sound can be detected.

The first heart sound is due to the vibrations produced by closure of the mitral and tricuspid valves. It occurs at the beginning of systole. In fact, it is a good clinical indicator of the onset of

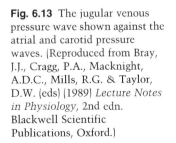
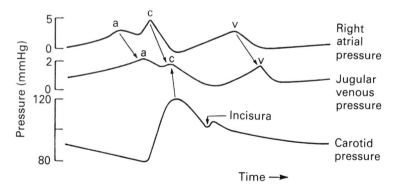

Fig. 6.13 The jugular venous pressure wave shown against the atrial and carotid pressure waves. (Reproduced from Bray, J.J., Cragg, P.A., Macknight, A.D.C., Mills, R.G. & Taylor, D.W. (eds) (1989) *Lecture Notes in Physiology*, 2nd edn. Blackwell Scientific Publications, Oxford.)

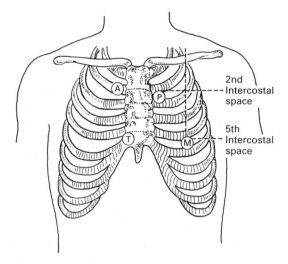

Fig. 6.14 Precordial auscultation areas for heart sounds. A, aortic area; M, mitral area; P, pulmonary area; T, tricuspid area.

ventricular systole. It is of a low pitch and is slightly longer in duration (0.15 s). The second heart sound is shorter (0.11 s) and higher in pitch. The two sounds can be simulated by the words 'lub dup'. The second heart sound is caused by the closure of the aortic and semilunar valves and it marks the beginning of ventricular diastole. The second heart sound may become split into two sounds during inspiration. This is called physiological splitting, brought about by the delay in closure of the pulmonary valve. The third heart sound is probably caused by the rush of blood during rapid ventricular filling in diastole. The fourth heart sound, which is rarely heard, occurs just before the onset of systole, i.e. just before the first heart sound, and has also been attributed to ventricular filling.

Auscultation of the heart is useful in clinical diagnosis of valve disease. Students of medicine are advised to familiarize themselves with the normal pattern and characteristics of the normal heart sounds so that they may be able to detect abnormal heart sounds (murmurs) or additional vibrations, which are called bruits and thrills. Murmurs are heard when there is abnormal flow, causing turbulence to occur. Abnormal vibrations can be caused by valve disease, e.g. narrowing

(stenosis) or incompetence (regurgitation). Bruits can be heard over a highly vascular organ, e.g. a highly vascular, enlarged and active thyroid gland, or over an arteriovenous fistula or shunt. Thrills are vibrations which can be felt by the hand when placed over the chest wall and they may accompany a murmur caused by valve disease.

Murmurs are described as systolic or diastolic according to their timing in the cardiac cycle. Thus, a murmur heard after the first heart sound and before the second is a systolic murmur, and one which comes after the second and before the first is a diastolic murmur. The valve source of the murmur can be discovered according to the location of the murmur in the areas of auscultation, as shown in Fig. 6.14. Thus, a murmur due to mitral stenosis occurs during diastole (because blood flows through this valve in diastole) and is best heard at the apex of the heart. Similarly, aortic stenosis will cause a systolic murmur, heard best at the base of the heart. A systolic murmur at the mitral area indicates abnormal back flow through the mitral valve (mitral incompetence), which should be closed during systole. A murmur due to aortic regurgitation is heard in diastole (because blood should not be flowing through this valve in diastole), but this time it is heard best at the base of the heart over the aortic area.

Cardiac murmurs are also produced by congenital malformations of the heart, such as interventricular septal defect, pulmonary atresia and patent ductus arteriosus.

CONCLUSION
Understanding the cardiac cycle helps to time the relationship between the heart sounds and the mechanical and electrical events. This is essential for timing abnormal changes in pressure, abnormal sounds and abnormalities of the ECG. It is also essential for the understanding of the haemodynamics of the pulmonary and systemic circulations.

The cardiac output
The cardiac output is defined as the volume of blood pumped by each ventricle per unit time. In healthy adults it is about 5 litres/min. The cardiac output (COP) depends on the heart rate and on

the stroke volume. In fact, it is equal to the product of these two parameters.

COP = Stroke volume × heart rate

Example
COP = 70 ml × 72/min
    = 5040 ml/min (approximately 5 litres)

This volume will vary with the size of the individual. Women have smaller cardiac outputs than men, and children have smaller outputs than adolescents or adults. However, it has been found that the cardiac output is normally about 3.2 litres/m^2 of body surface min. This is called the cardiac index. The surface area of the body is a function of both height and weight. During moderate muscular exercise the cardiac output may increase by two- or threefold. In athletes the cardiac output may reach 30 litres/min or more.

MEASUREMENT OF CARDIAC OUTPUT

From the definition stated above, it is clear that the COP is equal to the volume of blood that flows in the pulmonary artery or the aorta per min. *Fick's principle* states that the amount of a substance taken up by an organ in a given time equals the difference in concentration of that substance between arterial and venous blood, multiplied by the volume of blood flowing through the organ during the same period of time, i.e.:

$$\frac{\text{Amount of substance}}{\text{taken up by organ}} = \frac{\text{Arteriovenous}}{\text{difference}} \times \frac{\text{Blood}}{\text{flow}}$$

If we take O_2 uptake in the lungs as an example, we can use Fick's principle to calculate the cardiac output. Each minute the blood flow in the lungs is equal to the output of the right ventricle. We therefore need to measure the O_2 uptake per minute and the arteriovenous (AV) O_2 difference.

Example

O_2 uptake in the lungs	= 250 ml/min
AV oxygen difference	= 20/100
	− 15/100 ml
	= 5 ml/100 ml
COP (pulmonary blood flow)	= O_2 uptake
	÷ AV O_2 difference

By substitution in the above formula:

$$COP = \frac{250}{5} \times 100 = 5000 \text{ ml/min}$$

Another method used for measurement of cardiac output is known as the indicator or *dye dilution method*. This method depends on the observation that a dose of isotope or a dye injected into the right atrium will flow as a bolus in the pulmonary circulation and through the left side of the heart and then into the arterial system before it recirculates and completely mixes with the blood. This makes it possible to calculate the blood flow per unit time. Figure 6.15 shows the curve obtained by plotting the dye concentration in arterial blood against time during the first circulation of the dye bolus. This curve is used to obtain the time taken by the dye to circulate once through the heart.

The dye concentration is plotted on a logarithmic scale and time on a linear scale. The log−linear plot makes the descending limb of the curve a straight line, which is extrapolated to zero concentration. The point of intersection gives the theoretical time taken by the dye bolus to circulate through the lungs and to leave the heart.

The average concentration of the dye is calculated by integration of the area under the curve. Dividing the amount of dye (Q) by the average arterial concentration (C) gives the volume of blood (V) required to carry the dye, i.e.:

$$Q/C = V$$

This volume is the cardiac output during the time given by the curve for one circulation of the dye.

Example
5 mg of dye was injected and the average arterial concentration was found to be 1.5 mg/litre. A single circulation time, as shown in Fig. 6.6, is 40 s.

$$\text{Cardiac output in 40 s} = \frac{5 \text{ mg}}{1.5 \text{ mg/litre}}$$
$$= 3.33 \text{ litres}$$

$$\text{Cardiac output/min} = 3.33 \times \frac{60}{40} = 4.995 \text{ litres}$$

This is an example of the dye dilution method.

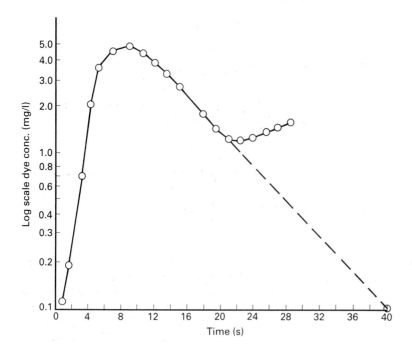

Fig. 6.15 The indicator or dye dilution method for determination of cardiac output. The broken line is the extrapolation drawn to find the circulation time.

There are other techniques for measuring cardiac output which are outside the scope of this text.

CONTROL OF CARDIAC OUTPUT

The control of cardiac output is essential for adjustments of the blood-pressure and blood flow requirements under different physiological and pathological states. For example, muscular exercise increases the demand for O_2 carriage and this, in turn, requires more blood flow to carry O_2 from the lungs and to deliver it at a higher rate to working muscles. In cases of fever or exposure to heat, there is dilatation of the vascular bed with consequent lowering of arterial blood-pressure. This requires a higher cardiac output so as to maintain the blood-pressure. Cardiac output is simply controlled by means of adjustments to the heart rate and the stroke volume. The regulation of these two important parameters will now be discussed in detail.

Regulation of the heart rate

The mechanisms which influence the heart rate are either neural or humoral in nature. The neural mechanisms depend on cardiac innervation. The sympathetic and the parasympathetic (vagus) nerves exert their influence on the SA node and therefore modify the heart rate by their effect on the pacemaker. The inherent rhythm of the SA node in the denervated heart is higher than the heart rate at rest (about 70/min). This is because the vagus exerts a continuous flow of impulses known as vagal tone, which tends to slow the heart rate. Inhibition of vagal tone causes the heart to beat faster. On the other hand, inhibition of the sympathetic nerves will cause a slight decrease in the heart rate, but their stimulation results in a great increase in sinus rhythm.

The neural regulation of the heart rate by means of autonomic nerves is achieved by means of cardiac centres, mainly in the medulla oblongata. Figure 6.16 shows a diagrammatic representation of the cardiac centres. The cardioinhibitory centre (depressor area) and the vasomotor centre (pressor area) receive inputs from various sensory receptors in the body. The most important of these are the baroreceptors and the chemoreceptors.

The baroreceptors are stretch receptors located in the arterial system, the most well-known of which are in the carotid sinus at the bifurcation of the common carotid artery and in the arch of the aorta. The chemoreceptors are special cells

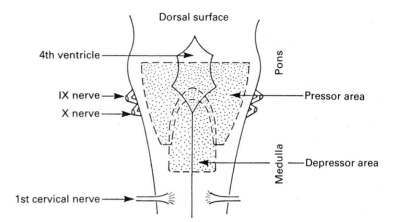

Dorsal surface

4th ventricle

IX nerve

X nerve

Pons

Pressor area

Medulla

Depressor area

1st cervical nerve

Fig. 6.16 Diagrammatic representation of the medulla and pons showing the location of cardiovascular centres. (Reproduced from Bray, J.J., Cragg, P.A., Macknight, A.D.C., Mills, R.G. & Taylor, D.W. (eds) (1989) *Lecture Notes in Physiology*, 2nd edn. Blackwell Scientific Publications, Oxford.)

(known as glomus bodies) situated in the carotid and aortic bodies in close proximity to the location of the baroreceptors and they send their afferent impulses through the buffer nerves (Fig. 6.17). The receptor cells are sensitive to changes in PO_2, H^+ concentration and PCO_2. The following is a description of some of the more important reflexes which regulate the heart rate.

The baroreceptor reflex Stimulation of the baroreceptors takes place when the blood-pressure rises. In fact, there is a direct relationship between the rate of firing of the baroreceptors and the blood-pressure, i.e. the higher the blood-pressure the higher the frequency of impulses generated in the baroreceptors (Fig. 6.18). There are very few impulses in the nerves coming from the baroreceptors when the pressure is low but this increases when the mean pressure rises. The baroreceptor reflex is found to be more sensitive to changes in pressure within the physiological range and less so when the pressure is too low or too high.

Stimulation of the baroreceptors by a rise in blood-pressure produces reflex slowing of the heart by excitation of the cardioinhibitory centre. This works through the vagus, which causes

Fig. 6.17 Afferent nerves of the baroreceptors and chemo-receptors and the cranial nerves which connect them to the cardiovascular centres in the medulla. For simplicity, the aortic bodies have been omitted. (Reproduced from Bray, J.J., Cragg, P.A., Macknight, A.D.C., Mills, R.G. & Taylor, D.W. (eds) (1989) *Lecture Notes in Physiology*, 2nd edn. Blackwell Scientific Publications, Oxford.)

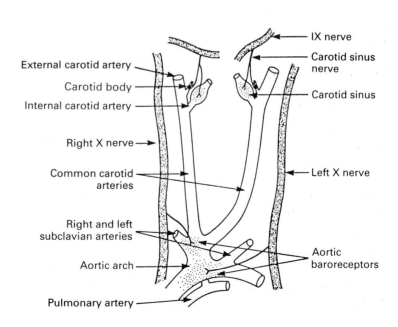

External carotid artery

Carotid body

Internal carotid artery

Right X nerve

Common carotid arteries

Right and left subclavian arteries

Aortic arch

Pulmonary artery

IX nerve

Carotid sinus nerve

Carotid sinus

Left X nerve

Aortic baroreceptors

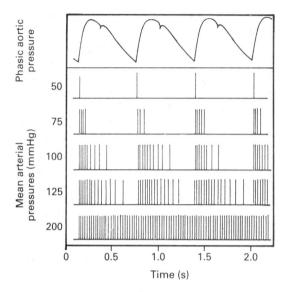

Fig. 6.18 Frequency of impulse traffic in a single fibre of the carotid sinus nerve in relation to mean arterial blood-pressure. (Reproduced from Berne, R.M. & Levy, M.N. (1986) *Cardiovascular Physiology*, 5th edn, by permission of C.V. Mosby Co., St Louis.)

slowing of the SA node, and results in decreasing the cardiac output. On the other hand, there is also reflex inhibition of the vasomotor centre. This causes a decrease in the sympathetic vaso-constrictor tone to blood-vessels, leading to vaso-dilatation in the peripheral circulation. Two effects result: a decrease in the peripheral resist-ance and an increase in the capacity of the per-ipheral circulation. In this way, the baroreceptor reflex causes not only a decreased cardiac output but also a decreased peripheral resistance, both of which contribute to decreasing the blood-pressure and bringing it back to normal.

Conversely, when the blood pressure falls, the baroreceptor impulses rapidly decrease in fre-quency. This initiates an immediate sympathetic response through the vasomotor centre. There is an increase in the heart rate, an increase in the strength of myocardial contraction and peripheral vasoconstriction. All of these result in rapidly elevating the blood-pressure to normal. The reflex constitutes an important life-saving mechanism in response to falling blood-pressure, which is

valuable for short-term regulation of blood-pressure. The responses of the baroreceptors will be discussed further when the regulation of ar-terial blood-pressure is considered.

Chemoreceptor reflexes The PO_2, H^+ and CO_2 concentrations in the blood are monitored by the peripheral chemoreceptor cells in the carotid and the aortic bodies. It has been found that these bodies have a very high rate of blood flow in relation to their size. This characteristic makes the glomus cells most suitable for monitoring the chemical composition of the blood. Oxygen, being one of the vital components, stimulates the per-ipheral chemoreceptors if its partial pressure falls (hypoxia). Also, if the pH falls, as in acidosis, stimulation of the chemoreceptors takes place. High PCO_2 can affect the peripheral chemore-ceptors, but only to a minor extent. As will be discussed in the chemical control of respiration (Chapter 7), PCO_2 is more effective on central chemoreceptors. The chemoreceptors are also stimulated when the blood-pressure falls. Their rich blood supply also makes them sensitive to a decrease in blood-pressure because it leads to a diminished O_2 supply (hypoxia).

Hypoxia also leads to anaerobic glycolysis, with excessive production of lactic acid. Lactic acidosis causes a drop in the blood pH, which stimulates the chemoreceptors. When the carotid and aortic chemoreceptors are stimulated by the above changes, they send impulses to the cardiac centres in the medulla. The result is a reflex increase in the heart rate due to stimulation of cardiac accel-eratory neurones, which work through sympath-etic stimulation. These responses increase the blood-pressure and improve the O_2 delivery and CO_2 uptake by the circulation.

The Bainbridge reflex It has been found that stretch of the right atrium by an increased venous return results in a reflex increase in the heart rate (tachycardia). However, when the heart rate is already high the reflex produces a decrease in the heart rate or may be absent. The sensory nerves of this reflex are in vagal afferent fibres from the right atrium. The reflex is abolished when the vagus is cut. The Bainbridge reflex acts as a safeguard against accumulation of blood in the right side of the heart and the venous system. The heart responds to an increase in venous return by increasing the cardiac output.

Receptors in joints and tendons Movement of skeletal muscles increases the heart rate. There are receptors in joints and tendons which detect the skeletal muscle movement. These are called proprioceptors. Impulses originating in these receptors will travel to the central nervous system, where they are relayed to hypothalamic and cardioacceleratory neurones. This reflex is partly responsible for the increased heart rate during muscular exercise.

Higher centres It is well known that emotional factors, working through the senses of hearing or vision or even initiated by thought or dreams, can affect the heart rate. These effects seem to work through the limbic system and the hypothalamus. There are neurones in the anterior part of the hypothalamus which cause the sympathetic nerves to accelerate the heart.

Regulation of the stroke volume

The stroke volume mainly depends on the amount of blood coming back to the heart (venous return), but it also depends on the condition of the myocardium and several factors which affect myocardial contractility.

The effect of Na^+, K^+ and Ca^{2+} ions on myocardial contractility Excess K^+ ions (hyperkalaemia) causes the myocardium to become hyperpolarized. The cardiac muscle fibres become relaxed and extremely weak. This is due to a decrease in the resting membrane potential and the action potential. As the action potential becomes smaller, the myocardial contraction becomes weaker. Excess K^+ can block conductions at the AV node. The heart comes to a standstill with the myocardium relaxed.

The effect of excess Ca^{2+} is the opposite of that of excess K^+. Excess Ca^{2+} causes the myocardium to contract more strongly and the heart may go into spasm (Ca^{2+} rigor). This is probably due to a direct effect of Ca^{2+} on the contractile process. On the other hand, low Ca^{2+} causes relaxation of cardiac muscle. Under physiological conditions,

however, serum Ca^{2+} is not allowed to vary widely. A decrease in serum Ca^{2+} will cause death due to skeletal muscle and nerve excitation resulting in tetany before the heart is affected. Similarly, high levels of Ca^{2+} usually result in deposition of Ca^{2+} in tissues rather than hypercalcaemia which affects the heart.

Slow calcium channel-blockers such as nifedipine and verapamil suppress Ca^{2+} entry into the cardiac muscle cell and the conducting tissues. These drugs can also inhibit the contraction of vascular smooth muscle. They therefore cause vasodilatation and decrease arterial blood-pressure and thus help the myocardium by decreasing the afterload on the left ventricle (see below).

Excess Na^+ ions depress myocardial contractility. This is because the Na^+ ion competes with Ca^{2+} receptors in the contractile process. Low Na^+, on the other hand, may result in cardiac fibrillation.

Mechanical factors These are related to the state of stretch of the cardiac muscle at the beginning of contraction and also to the resistance offered to the muscle during contraction. The stretch of the ventricular muscle is related to cardiac filling, which depends on the venous return. This is often referred to as *preloading* of the cardiac muscle. On the other hand, the resistance to myocardial contraction is proportional to the pressure in the aorta and pulmonary arteries. This is referred to as impedance to flow and also as afterloading of the cardiac muscle.

1 *The length–tension relationship (preloading).* The stretch of the myocardium by blood causes the muscle to contract more strongly. The famous physiologist, Starling, stated that the strength of myocardial contraction is proportional to the initial length of the muscle fibres. This statement became known as Starling's law of the heart. Thus, the myocardial strength increases as the filling of the heart increases. Therefore, it can be regarded as an intrinsic autoregulation of the myocardium, which enables the heart to cope with increased venous return. This autoregulatory mechanism also ensures that the output of the right and left ventricles are matched for each beat. It safeguards against blood accumulating

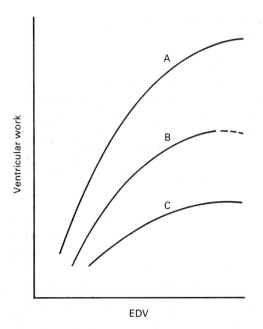

Fig. 6.19 Relationship between myocardial stretch (EDV) and tension (ventricular work). Curve A shows the effect of sympathetic stimulation, B shows the relationship at rest, and C the effect of myocardial depressants (e.g. parasympathetic stimulation, hypoxia, hypercapnia and acidosis).

in the venous system or in the pulmonary circulation.

The curves shown in Fig. 6.19 illustrate the relationships between the stretch of cardiac muscle and the strength of contraction (length–tension curve). The degree of stretch is given by the end-diastolic volume (EDV). The strength of contraction (tension) is given by myocardial work (myocardial performance). The curves have become known as the Frank–Starling curves. It should be noted, however, that the degree of stretch works positively up to a certain limit or optimum. If this limit is exceeded, then further stretching produces a negative effect on contractility.

2 *Aortic and pulmonary impedance (afterloading).* The pressure in these great arteries affects the force of contraction by offering resistance to blood flowing out of the ventricles. High pressure in the aorta or pulmonary artery causes

the contractile force to decrease because the ventricles have to contract against increased resistance and therefore the stroke volume is decreased.

3 *The heart rate* indirectly affects the force of contraction. As the heart rate increases, the duration of diastole becomes shorter and the EDV smaller. Thus, the stroke volume, according to Starling's law, should be decreased. Only when the heart rate exceeds 120/min does the diastolic filling become significantly affected. During muscular exercise, this is compensated for by an increased sympathetic stimulation, which increases the strength of contraction.

Effects of neural regulation on stroke volume
1 *Sympathetic stimulation.* The main regulatory mechanism of the stroke volume is the inotropic effect of catecholamines. During muscular exercise there is increased sympathetic discharge, which increases the force of ventricular contraction through β_1 adrenergic receptors. This increases the stroke volume by decreasing the residual or end-systolic volume of the ventricles. Similarly, during stimulation of the baroreceptor reflex by hypovolaemia, sympathetic discharge is responsible for increasing the stroke volume.

2 *Effects of parasympathetic stimulation.* There is little negative inotropic effect on the ventricle while the atrial contraction is decreased. The effects of sympathetic and parasympathetic stimulation can be seen in Fig. 6.19 superimposed on the length–tension relationship as shown by the Frank–Starling curves.

THE VENOUS RETURN
Several mechanisms help to return the blood through the venous system to the heart:
1 The most important of these is the venous pressure gradient, due to the pumping action of the heart, which is usually less than 10 mmHg. The pressure in the right atrium is zero or negative. This pressure gradient is enough to ensure a venous return equal to the cardiac output at rest. The blood volume and the state of constriction or dilatation of the systemic veins determine the pressure gradient and, in particular, the central venous pressure (i.e. pressure in the great veins and the right atrium). In the case of haemorrhage,

venoconstriction ensures adequate venous pressure.

2 During walking or other forms of muscular work, contracting muscles squeeze the capillaries and veins within them. The deep veins lying between muscles are also squeezed. Due to the presence of valves in the veins, the blood flows towards the heart. This mechanism is known as the *muscle pump*. Standing for long periods of time without movement can result in pooling of blood in the lower extremities, decreasing the venous return and the cardiac output. Fainting can occur if the blood flow to the brain is decreased.

3 During inspiration the intrathoracic pressure becomes negative relative to atmospheric pressure. The thin walls of the atria and great veins allow the central venous pressure to become negative as well. Therefore, a suction force, known as the *respiratory pump*, draws blood into the venae cavae and the right atrium. This is assisted by the descent of the diaphragm, which increases the pressure in the abdominal cavity. The opposite of the respiratory pump occurs during the Valsalva manoeuvre, i.e. forced expiratory effort against a closed glottis as during coughing or straining. The venous return decreases and so does the cardiac output and blood-pressure. As a result, an increase in the heart rate is brought about by the baroreceptor reflex.

4 Gravity also plays a role in the venous return from the head and neck. Conversely, if the body is erect, pooling may result in the lower limbs. On lying down, venous return is improved. This is why the head should be kept down and the legs raised as a first-aid procedure in cases of hypotension, e.g. after severe blood loss. In cases of congestive heart failure, lying flat on the back causes dyspnoea due to more venous return to the heart and therefore pooling of blood in the lungs causing breathlessness.

The effect of gravity may become exaggerated in patients receiving drugs which block the sympathetic ganglia. Postural hypotension may occur on standing up from the lying position due to vasodilatation and pooling of blood in the lower extremities, together with failure of the sympathetic response.

SUMMARY OF NERVOUS REGULATION OF THE HEART

Sympathetic stimulation

Sympathetic stimulation causes an increase in the heart rate through β_1 adrenergic receptors in the SA node. There is also a shortening of the conduction time in the AV node and a positive inotropic effect on the atrial and ventricular muscle. These effects cause an increase in the cardiac output and the arterial blood-pressure.

Parasympathetic stimulation

Parasympathetic stimulation releases acetylcholine and slows the heart rate by suppressing the pacemaker potential. Conversely, suppression of vagal tone or blocking acetylcholine by atropine causes an increase in the heart rate up to 150–180/min. Acetylcholine causes slowing or blocking of conduction in the junctional myocardium of the AV node. There is also a negative inotropic effect on the atria but little or no such action on the ventricles. This effect is brought about by hyperpolarization of the resting membrane potential and shortening the duration of the action potential.

Indirect effects on the ventricles include a decrease in the stroke volume due to poor atrial contraction, which leads to a decrease in ventricular filling and therefore a negative effect on Starling's law. Together with a decreased heart rate, this effect leads to a decrease in cardiac output and arterial blood-pressure.

Cardiac centres

The cardioinhibitory area consists mainly of the nucleus ambiguus of the vagus nerve in the medulla oblongata. Cardioinhibitory neurones are also found in the nucleus of the tractus solitarius (NTS). A rise in blood-pressure stimulates arterial baroreceptors and produces bradycardia through excitation of the cardioinhibitory area. Conversely, a fall in blood-pressure will produce tachycardia by inhibiting vagal tone and increasing sympathetic discharge. Emergency situations produce tachycardia by sympathetic stimulation, but there is no 'cardioacceleratory centre' as such.

The vasomotor centre (VMC) or pressor area is situated in the ventrolateral part of the medulla oblongata, including the cardioinhibitory area. It sends fibres to the preganglionic sympathetic nerves of the lateral horn cells of the spinal cord. Both excitatory and inhibitory neurones from the medulla converge on the sympathetic preganglionic neurones. Excitation will bring about an increase in the heart rate, vasoconstriction and a rise in blood-pressure. Inhibition of sympathetic vasomotor tone causes vasodilatation and a decrease in arterial blood-pressure.

Reflexogenic areas

Numerous receptors and reflexes have been described in the cardiovascular system. Only well-studied receptors and reflexes will be described here. Arterial blood pressure is monitored by the systemic *baroreceptors*. Those in the carotid sinus send impulses via the sinus nerve, which is a branch of the glossopharyngeal nerve, and those in the aortic arch send impulses in the afferent fibres of the vagus nerve. Fibres from the baroreceptors synapse in the vicinity of the nucleus of the tractus solitarius (NTS), from which interneurones make connections with the nucleus ambiguus and the vasomotor centre. The latter controls vasomotor tone via the lateral horn cells of the spinal cord. An increase in blood-pressure sends excitatory impulses to the cardioinhibitory centre and inhibitory impulses to the vasomotor centre. The total response will therefore be slowing of the heart rate and vasodilatation (i.e. a fall in the peripheral resistance). Both effects will decrease the blood-pressure.

Systemic arterial *chemoreceptors* are situated in the carotid bodies near the bifurcation of the common carotid artery and in aortic bodies near the aortic arch. The cells of these bodies receive a rich blood supply. They are stimulated by a decrease in P_{O_2} and H^+ and an increase in P_{CO_2}. A decrease in the perfusion of the chemoreceptor bodies (e.g. after a fall in arterial blood-pressure) will also result in local hypoxia and stimulation of the chemoreceptors. Afferent nerve fibres from the chemoreceptors reach the NTS in the same nerves as those from the arterial baroreceptors. In the medulla these afferent fibres converge on the VMC.

Stimulation of the chemoreceptors by one or more of the above stimuli causes bradycardia, a negative inotropic effect on the ventricular myocardium and vasoconstriction. However, separate stimulation of the aortic chemoreceptors has been found to cause tachycardia, a positive inotropic effect and an increase in peripheral resistance. Chemoreceptor stimulation by hypoxia also results in secretion of adrenaline by the adrenal medulla, which causes tachycardia and an increase in cardiac output.

Low-pressure stretch receptors are found in the right atrium. Stimulation of the atrial receptors type B, which respond to an increase in venous return, results in tachycardia mediated by sympathetic nerves. These receptors may be responsible for the Bainbridge reflex. Other atrial receptors, when stimulated, result in vasodilatation. Both the above responses can be considered as safety mechanisms against excessive increases in venous return or central venous pressure.

Pulmonary receptors In the pulmonary circulation (with uncertain location) there are receptors stimulated by vascular distension which mediate reflex bradycardia and a fall in systemic arterial blood-pressure. Their physiological significance and that of possible pulmonary chemoreceptors are still uncertain.

METABOLISM OF CARDIAC MUSCLE

At rest more than 60% of the energy supply of the myocardium comes from free fatty acids. The remainder is supplied mainly by carbohydrate and small amounts of lactate, pyruvate and ketone bodies. As cardiac muscle has a high protein turnover, amino acid uptake is mainly used for protein synthesis and maintenance of cardiac muscle.

The immediate source of energy for myocardial contraction is high energy phosphate from ATP. Stores of ATP are resynthesized from creatine phosphate. This indicates that an adequate supply of O_2 is essential for myocardial contractions. In fact myocardial oxygen consumption (MVO_2) is closely related to the work of the heart. In hypoxia (as in severe muscular exercise) the heart reverts to anaerobic glycolysis supplying limited energy and resulting in formation of lactate, which cannot be utilized as a main source of energy. In severe myocardial ischaemia ATP is soon depleted resulting in the formation of AMP and adenosine. The latter leaks out of the cell. This curtails resynthesis of ATP. If the ischaemia persists myocardial cellular death begins in about 30 min. Medications used to dissolve coronary clots are best used as early as possible, preferably within the first half-hour of the attack.

THE WORK OF THE HEART

The work of the heart as a pump is converted into (a) pressure (pressure work) and (b) flow (kinetic work). By far the greater part goes to the building up of blood-pressure in the arterial system. The energy used for this pressure work by the left ventricle is about seven times that of the right ventricle. This is due to the higher impedance in the aorta and the thickness of the left ventricle wall. As a result the oxygen consumption of the left ventricle is far greater than that of the right. The kinetic work of each ventricle is only about 1–2% of the total work done at rest, but increases significantly during exercise due to the increase in the velocity of flow or cardiac output.

CARDIAC RESERVE

In healthy persons at rest the heart uses only a fraction of its maximum output. This margin of difference is called the cardiac reserve. In a sedentary adult the cardiac reserve is five times the resting cardiac output while in athletes it can be sevenfold. The cardiac reserve is called upon during stress such as muscular exercise or disease conditions which require cardiovascular adjustments such as heart failure, valvular disease and blood loss. Many disease conditions result in decreased cardiac reserve. In such cases the patient fatigues quickly e.g. feels breathless, the heart rate becomes too fast or the muscles become painful due to inadequate oxygen supply.

The physiological mechanisms activated to achieve the cardiac reserve are *cardiac acceleration* and the increase in *stroke volume*. These are discussed in detail in the appropriate parts of this chapter. In chronic disease such as hypertension and aortic stenosis myocardial hypertrophy contributes to the cardiac reserve.

In clinical practice measurement of the cardiac reserve is a useful assessment of the condition of the heart. However, the cardiac output may be difficult to measure in most hospitals. Indirect methods such as exercise tolerance tests (the so-called stress test) are carried out instead (see Chapter 14).

Applied physiology

HEART FAILURE

The condition of heart failure is not a disease entity. It can be defined as failure of the heart to pump adequate amounts of blood to maintain the various functions of the body. It can be due to a variety of disease processes which can be grouped into three categories (a) those due to myocardial failure, such as myocardial infarction, chronic overload, or toxins and poisons, (b) mechanical abnormalities such as valvular disease, congenital cardiac anomalies and increased impedance to forward flow (e.g. hypertension and aortic stenosis), and (c) cardiac arrhythmias such as fibrillation and flutter.

The pathophysiology of heart failure varies with the underlying pathology. In all cases there is a decreased cardiac output for a given end-diastolic volume (EDV), i.e. depression of the Frank–Starling curve.

Congestive heart failure (CHF)

As a result of decreased cardiac output left atrial pressure increases leading to back pressure and pulmonary congestion. This is the cause of shortness of breath or *dyspnoea*, which increases when the patient lies down (orthopnoea).

After myocardial infarction the immediate adjustments that take place in heart failure are due to decreased cardiac output and systemic venous congestion. Activation of cardiovascular reflexes takes place mainly due to the baroreceptors, chemoreceptors and cerebral ischaemia. There is an increase in the heart rate and strength of myocardial contraction due to sympathetic activation. This results in partial compensation for the damaged myocardium and an almost normal cardiac output is maintained.

In chronic states in which both ventricles are involved, there is activation of the renal mechanisms that result in fluid retention. This is due to (a) decreased glomerular filtration brought about by decreased renal blood flow, (b) secretion of renin and activation of the renin–angiotensin system, and (c) secretion of aldosterone. These effects cause Na$^+$ and water retention. The increased blood volume increases the venous return and right atrial pressure leading to an improvement of the cardiac output. At rest the patient may have no symptoms but becomes breathless on effort because of the decreased cardiac reserve.

When the cardiac output cannot be maintained for adequate renal function, more fluid is retained. This leads to a progressive rise in the *right atrial* pressure, venous congestion and increased capillary pressure leading to *oedema* of the lower limbs or lower back in recumbent patients. Venous congestion also leads to *liver enlargement* and congestion of the abdominal viscera. These are the classical manifestations of CHF.

Further retention of fluid leads to stretch of the myocardium and oedema of the cardiac muscles and eventually leads to death of decompensated heart failure. In the terminal stages oedema of the lung takes place with extreme shortness of breath. This state calls for immediate therapy, otherwise the patient soon dies.

Left ventricular failure
Can occur in acute myocardial infarction, mitral or aortic valve disease or as a result of severe hypertension. In such cases the left ventricle fails without involvement of the right ventricle. The result is severe pulmonary congestion brought about by the normal right ventricular output and severe rise of pressure in the left atrium. Pulmonary oedema occurs and the patient coughs up frothy sputum, often blood-stained. Left ventricular failure can be fatal in less than half an hour unless appropriate measures are taken.

Right ventricular failure
Rarely occurs due to pulmonary hypertension or mitral stenosis without involvement of the left ventricle. In such cases systemic venous congestion is a prominent feature. The liver becomes enlarged and systemic oedema and ascites occur (i.e. CHF). In this case dyspnoea is due to low pulmonary perfusion leading to hypoxia and respiratory stimulation by the chemoreceptors.

High cardiac output failure
This is a term given to describe conditions where the cardiac output is higher than normal but still inadequate to meet demands of oxygen consumption or haemodynamic adjustments, e.g. thyrotoxicosis, vitamin B$_1$ deficiency (beriberi) and cases of large arteriovenous shunts.

The systemic circulation
Oxygenated blood is pumped by the left ventricle into the aorta, from which systemic arteries supply the various organs and limbs. Figure 6.20 shows that each main artery supplies a defined vascular bed, which is drained by a main vein. The veins join the venae cavae. Thus organ or regional blood supply is arranged in parallel. This makes it possible to regulate parts of the systemic circulation independently to meet local needs, without affecting the general circulation.

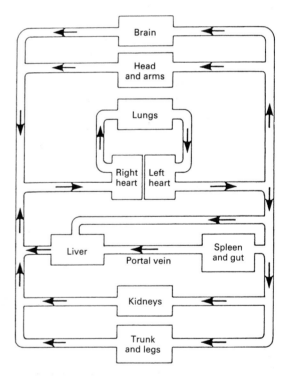

Fig. 6.20 Schematic diagram of the circulatory system showing the main vascular beds. (Reproduced from Bray, J.J., Cragg, P.A., Macknight, A.D.C., Mills, R.G. & Taylor, D.W. (eds) (1989) *Lecture Notes in Physiology*, 2nd edn. Blackwell Scientific Publications, Oxford.)

The arteries are large-diameter vessels. The thick walls are composed mainly of circular smooth muscle with a good proportion of elastic tissue. This permits stretching during systole and recoil during diastole and gives the damping effect which converts pulsatile blood flow into a more steady stream. From small arteries the blood flows into the arterioles, whose walls contain a thick layer of smooth muscle and much less elastic tissue than arteries. This permits control of the diameter of the arterioles and therefore the peripheral resistance to blood flow. For this reason the arterioles have been called the *resistance vessels*. An arteriole has a diameter of about 20 μm. The total cross-sectional area of the arterioles in the whole body is about 400 cm^2.

Blood flows from the arterioles into the capillaries, which have thin walls made of one layer of flat endothelial cells. There are precapillary smooth muscle sphincters, which are not innervated but which respond to chemical substances in the blood or interstitial fluid. The final part of an arteriole (metarteriole) appears to be connected to a venule by a main capillary channel (preferential channel). True capillaries form a network around this main vessel (Fig. 6.21). The diameter of a true capillary is about 5 μm at the arteriolar end and about 9 μm at the venular end. Therefore, the red cells have to squeeze through the capillaries and become convex discs with their convexity towards the direction of flow and their edges closely applied to the capillary wall. This favours exchange of O_2 taking place between the interstitial fluid and the haemoglobin in the red blood cells. The capillaries are called *exchange vessels*.

At any one time, only 5% of the blood volume is in the systemic capillaries, but this proportion is the most vital part of circulating blood because it is across the capillary membrane that gas and nutrient exchange takes place. The transit time from the arteriolar end to the venular end of the capillary is 1–2 s. In an adult the total cross-sectional area of the capillaries is about 4500 cm^2 and their surface area, which is available for exchange, is about 6300 m^2. The pressure in the capillaries varies considerably. Typical values in human nail-bed capillaries are 32 and 15 mmHg at the arteriolar and venular ends, respectively. There is a small pulse pressure of about 5 mmHg in capillaries.

Capillary pores at the junctions between endothelial cells allow molecules of up to 10 nm in diameter to pass through the capillary wall. These gaps are much smaller in brain capillaries and thus contribute to the blood–brain barrier. Larger gaps (20–100 nm) or fenestrae are found in the intestinal villi and some of the kidney capillaries.

The veins and venules are thin-walled vessels with much less muscle in them than in the arteries. However, they are still capable of a certain degree of contraction. The total cross-sectional area of the venules is 4000 cm^2 while that of the

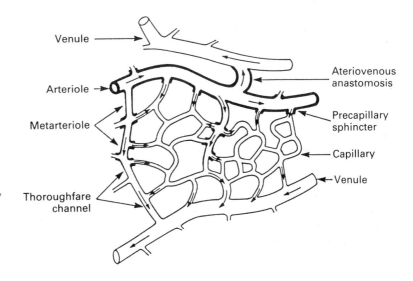

Fig. 6.21 Diagrammatic representation of microcirculation. (Reproduced from Bray, J.J., Cragg, P.A., Macknight, A.D.C., Mills, R.G. & Taylor, D.W. (eds) (1989) *Lecture Notes in Physiology*, 2nd edn. Blackwell Scientific Publications, Oxford.)

Venule

Arteriole →

Metarteriole

Thoroughfare channel

Ateriovenous anastomosis

Precapillary sphincter

Capillary

Venule

veins is about 40 cm². However, the venules and veins together hold more than 60% of the blood volume. This is why they are called the *capacity vessels.* Changes in blood volume can be compensated for by venoconstriction or dilatation and provide an important vascular response to circulating noradrenaline or sympathetic stimulation. The veins have valves, which are made of crescent-shaped folds of intima. These valves play an important role in venous return, as already described.

Normally, the fluid leaving the capillaries is more than that returning to them. The extra fluid does not accumulate because it is drained by the lymphatics. Lymph flows back through the thoracic duct into the great veins in the thorax. The lymph is propelled by rhythmic contractions in the walls of the right lymphatic duct and the thoracic duct. The flow of lymph is helped by the respiratory pump and by the muscle pump in the same way as they help the venous return. The lymph vessels and the large ducts also have valves which prevent back flow.

The arterial blood-pressure

To ensure a steady blood flow or perfusion to the tissues, the mean arterial blood-pressure is maintained at about 100 mmHg. This is achieved by control of the cardiac output and by variations in the diameter of the arterioles, which are responsible for regulation of the peripheral resistance. Local and central mechanisms help to regulate the blood flow through specific parts of the circulation. This section deals with the relationships between pressure, flow and resistance (i.e. haemodynamics) and the regulation of arterial blood-pressure under physiological conditions. The above terms will be explained and defined. The applied physiology of arterial blood-pressure will also be considered.

BASIC CONCEPTS OF HAEMODYNAMICS
The perfusion of tissues depends on the pressure generated by the pumping action of the heart. The difference in mean blood-pressure between the aorta and the right atrium (i.e. effective perfusion pressure) determines the blood flow in the systemic circulation. Flow is defined as the volume of blood which passes a given point in the

circulation per unit time in ml/sec. Flow should not be confused with velocity, which is defined as displacement of blood from one point in the circulation to another in cm/sec.

Resistance (R)
Resistance to flow is generally determined by the diameter (radius, r) of the tube, the viscosity (η) of the fluid and the length (L) of the tube, as given by the following formula (Poiseuille–Hagen formula):

$$R = \frac{8\eta L}{\pi r^4}$$

This relationship is accurate for rigid tubes (e.g. glass or steel). In blood-vessels (which are not rigid) this mathematical relationship is applicable within limits. However, it is important to note that the main influence on variation of resistance is the diameter of the blood-vessels (expressed as r^4). Adjustment of the diameter of the arterioles is the primary mechanism for the regulation of the peripheral resistance.

In summary, the relationship between pressure, flow and resistance is similar to that in Ohm's law, which applies to electricity:

$$\text{Ohm's law: } I \text{ (current)} = \frac{V \text{ (voltage)}}{R \text{ (resistance)}}$$

$$\text{Similarly: } F \text{ (flow)} = \frac{P \text{ (pressure difference)}}{R \text{ (resistance)}}$$

and $P = F \times R$

Blood flow
Blood flow can be laminar or turbulent (Fig. 6.22). Turbulence is associated with the generation of sounds in the cardiovascular system (e.g. heart sounds and murmurs).

The degree of turbulence depends on the density of the fluid, the diameter of the vessel, the velocity of flow and the viscosity of the fluid. Turbulence is given by Reynold's number (Re):

$$Re = \frac{\text{Density} \times \text{diameter} \times \text{velocity}}{\text{viscosity}}$$

The greater the Reynolds number, the greater the turbulence. Murmurs are heard over the heart if there is narrowing (stenosis) or insufficiency of

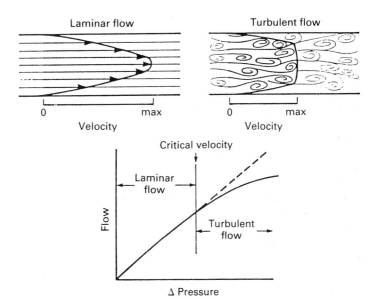

Fig. 6.22 The upper drawing shows laminar and turbulent flow. The lower (graph) gives the relationship between pressure and velocity of flow. Laminar flow becomes turbulent when a certain critical velocity is reached. (Reproduced from Bray, J.J., Cragg, P.A., Macknight, A.D.C., Mills, R.G. & Taylor, D.W. (eds) (1989) *Lecture Notes in Physiology*, 2nd edn. Blackwell Scientific Publications, Oxford.)

the valves, which affects the diameter of cardiac orifices between the heart chambers. Viscosity may be greatly decreased in anaemia, which is also associated with tachycardia and therefore greater velocity. This may explain the systolic murmurs commonly heard in anaemic patients.

MEASUREMENT OF ARTERIAL BLOOD-PRESSURE

If a glass tube is connected with an artery, blood will rise into the tube to a height of about 2.5 m. The top of the blood column will be seen to move up and down with each contraction and relaxation of the left ventricle. These were the basic observations made in AD 1732 by one Stephen Hales in an experiment on a horse. We can now measure the blood-pressure indirectly but conveniently by using a sphygmomanometer (Fig. 6.23). A rubber bag is wrapped around the upper arm and inflated by a hand-pump. The cuff is connected to the mercury manometer. When the cuff is inflated to a pressure above that in the brachial artery, blood flow to the forearm stops and the radial pulse cannot be felt by palpation of the radial artery at the wrist. The cuff is now deflated slowly until the pulse is felt again. The pressure reading at this point is the systolic pressure. This is called the palpation method. When a stethoscope is placed on to the antecubital fossa and the cuff

inflated to a pressure higher than the systolic pressure, there is no flow and no sounds can be heard (Fig. 6.24). On slow deflation, sounds are heard when the pressure falls to the systolic pressure. These are known as Korotkov sounds. They are due to turbulent blood flow because the artery opens for a limited period of time only when the blood-pressure exceeds that inside the cuff. On further deflation the sounds become louder and sharper. Suddenly they fade or stop. In adults at rest the point at which the sounds disappear is taken as the diastolic pressure. In children and in adults after exercise, the point at which the sounds fade is more accurate. Thus, by using the auscultation method, we can obtain readings of both the systolic and the diastolic arterial blood-pressure.

Normal blood-pressure is often quoted as 120 mmHg systolic and 80 mmHg diastolic. However, it is difficult to define normal values because of the errors of sphygmomanometric measurement, age variation and the state of relaxation (basal state) or otherwise of the individual.

REGULATION OF ARTERIAL BLOOD-PRESSURE

The neural regulation of arterial blood-pressure is closely linked with that of the heart. Normal arterial pressure depends on the cardiac output,

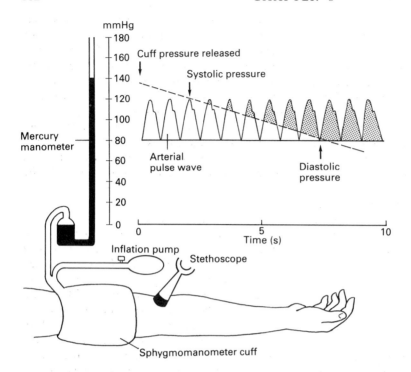

Fig. 6.23 Measurement of systolic and diastolic blood-pressure using the sphygmomanometer.

the peripheral resistance and the blood volume. It is the regulation of these three variables which determines the arterial blood-pressure. Variation in cardiac output mainly determines the systolic blood-pressure. The regulation of cardiac output, namely the heart rate and stroke volume, has

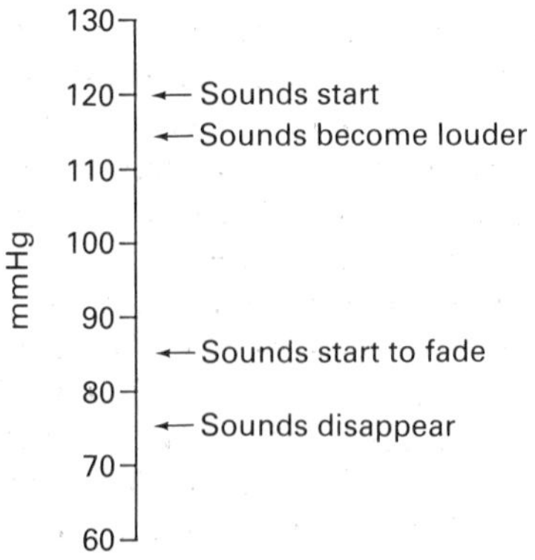

Fig. 6.24 Using the auscultatory method, systolic arterial blood-pressure is taken at the point when the sounds start and the diastolic is read when the sounds either start to fade or disappear completely.

been described. The following will deal with the other two parameters.

Regulation of peripheral resistance
Variations in the peripheral resistance determine the diastolic blood-pressure. The diameter of the arterioles is regulated by varying the sympathetic vasomotor tone. There are several reflex mechanisms which regulate the vasomotor tone, the most important of which are the baroreceptor and chemoreceptor reflexes:

1 The *baroreceptors* send a continuous flow of signals to the vasomotor centre and the vagal centre. The frequency of these signals is directly proportional to the arterial blood-pressure. When blood-pressure falls, there is a decrease in the signals reaching the vasomotor centre, resulting in a reflex increase in sympathetic vasomotor tone (vasoconstriction) and consequently an increase in the peripheral resistance. As mentioned before, there is also an increase in the heart rate. Conversely, an increase in blood-pressure leads to inhibition of vasomotor tone and reflex vagal slowing of the heart.

2 The *chemoreceptors* are stimulated by low P_{O_2},

high P_{CO_2} and low plasma pH. Although the main function of the chemoreceptors is in respiratory control, they also exert important effects on the vasomotor centre. They can be stimulated during hypotension (e.g. after haemorrhage), which causes hypoxia due to decreased blood flow to the chemoreceptor cells. The result is reflex vasoconstriction, with an increase in the blood-pressure. There is also an increased output of catecholamines from the adrenal medulla, leading to tachycardia. Increased P_{CO_2} has a direct effect on the vasomotor centre, producing reflex vasoconstriction. In the tissues, however, hypoxia and hypercapnia produce local vasodilatation. Therefore, the central and local effects oppose each other and there is no general effect on blood-pressure.

Cerebral ischaemia or local hypercapnia, produced by increased intracranial pressure or diminished cerebral blood flow, has a direct effect on the vasomotor centre, which tends to increase the blood-pressure. A reflex bradycardia usually results, due to baroreceptor stimulation.

The reflex vasomotor effects mentioned above are most active in the skin, splanchnic area and muscle. These tissues can tolerate decreased O_2 supply without damage to the cells. The brain and heart tissue cannot function without a constant blood supply and therefore their blood-vessels are only partly under neural control.

3 Other vasomotor reflexes can be initiated by receptors in various organs. *Atrial stretch receptor* stimulation (e.g. by increased venous return) produces reflex vasodilatation and a fall in blood-pressure. *Thermoreceptors* in the skin or hypothalamus produce skin vasodilatation on exposure to heat and vasoconstriction on cooling. These responses help to maintain body temperature. Pulmonary receptors have been described which produce vasoconstriction as a result of lung inflation.

4 *Hormonal agents* also play an important role in the regulation of peripheral resistance. However, both resistance and capacity vessels can be involved.

Adrenaline and noradrenaline are secreted from the adrenal medulla on sympathetic stimulation (e.g. in emergency or stressful situations, asphyxia and hypoglycaemia). Noradrenaline is vasoconstrictor, while adrenaline is vasoconstrictor in most parts of the circulation but vasodilator in skeletal muscle. Therefore, noradrenaline increases the peripheral resistance, while adrenaline has no marked effect on total peripheral resistance, which may even decrease if the skeletal muscle vasodilatation is great.

Other hormonal agents include angiotensin, kinins, prostaglandins and histamine. Angiotensin II has a powerful vasoconstrictor effect but the others have, so far as is known, no role in the normal regulation of blood-pressure.

Regulation of blood volume
The mechanisms which regulate blood volume are mainly renal. There are low-pressure *volume receptors* in the walls of the right atrium, which send impulses to the brain via the vagus. Some of these receptors may also be responsible for secretion of the hormone known as atrial natriuretic peptide (ANP). The volume receptors and ANP have an important role in regulation of aldosterone and sodium excretion. Retention of sodium is one of the main mechanisms for maintaining blood volume. The macula densa cells of the juxtaglomerular apparatus in the kidney may also be considered as volume receptors, because they are stimulated by decreased blood flow in the kidney. They secrete renin in response to decreased renal blood flow and therefore trigger off the renin–angiotensin system, the most important mechanism for aldosterone regulation (see Chapter 10). Angiotensin II has a powerful vasoconstrictor effect, which contributes to the hormonal control of peripheral resistance.

Antidiuretic hormone (ADH) or vasopressin, of the posterior pituitary, is released when volume receptors are stimulated, e.g. by hypovolaemia. This hormone is also released in response to osmoreceptor stimulation and therefore helps in the regulation of body water and the thirst mechanism (see Regulation of ADH, Chapter 10).

Large blood volume variations may affect the cardiac output and therefore the arterial baroreceptors. This leads to vasomotor stimulation, which results in an increase of cardiac output and peripheral resistance as well as venoconstriction, which compensates for the volume loss by decreasing the capacity of the venous system.

Summary

The mean arterial pressure is determined by the cardiac output and the peripheral vascular resistance. The systolic pressure is a function of the stroke volume and therefore is determined mainly by the cardiac output. The diastolic pressure depends more on the peripheral resistance. Regulatory mechanisms depend on fast acting reflexes which control the heart rate, stroke volume and peripheral resistance. Long-term mechanisms are mainly concerned with regulation of the body fluids and blood volume.

Fast acting reflex mechanisms
1 The baroreceptor reflex regulates the heart rate and peripheral resistance through vasomotor tone.
2 Chemoreceptor reflex regulates the heart rate and vasomotor tone.
3 Central nervous system ischaemic reflex.
4 Adrenal medullary hormones.
5 Vasopressin vasoconstrictor effect.

Long-term regulatory mechanisms Renal water and electrolyte regulation by:
1 Renin−angiotensin system.
2 Aldosterone secretion.
3 Vasopressin antidiuretic effect.
4 ANP secretion.

Local regulation of blood flow
As the main goal of the circulation is to maintain adequate perfusion of the tissues, the blood flow (volume per unit tissue mass per unit time) determines the supply of O_2 and nutrients. It has been explained that regulation of the cardiac output and the peripheral resistance maintains a pressure gradient for adequate blood flow in the various organs. However, the activities and demands of different organs may vary from time to time. For this reason, local control of the blood flow in a particular organ or region of the body can take place without affecting the entire circulation. The following mechanisms are responsible for local circulatory adjustments of blood flow in organs or tissues:
1 There are intrinsic mechanisms which enable blood-vessels to react to local conditions. An increase in perfusion pressure stretches the arteriolar wall, which reacts by contraction of its smooth muscle and therefore decreases the diameter. Thus, an increase in perfusion pressure results in a myogenic response, which tends to keep the flow constant. This mechanism is particularly well-developed in the kidney and the brain and is referred to as *autoregulation*. It enables these organs to keep their blood flow constant, despite wide variations of the perfusion pressure. However, this is not true of very low or very high perfusion pressures. At a certain low pressure, called the critical closing pressure, the arterioles collapse and blood flow stops (Fig. 6.25).

In addition to the myogenic response described above, another theory of autoregulation holds that an increase in perfusion pressure causes an increase in tissue fluid formation and therefore a rise of extramural pressure, causing the vessels to be compressed. This may partly explain autoregulation in an encapsulated organ like the kidney.
2 A second intrinsic mechanism depends on tissue metabolites. Accumulation of products of metabolism in active tissues causes vasodilatation. This explains the increased blood flow, or reactive hyperaemia, in a limb after brief ischaemia. The factors known to produce vasodilatation are CO_2, hydrogen ions, potassium excess and oxygen lack. This is called metabolic autoregulation; it enables tissues to regulate their blood flow according to their activity and metabolic rate. It is particularly effective in the regulation of blood flow in cardiac and skeletal muscle.

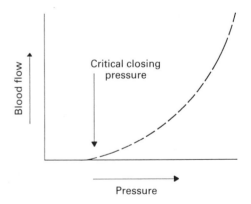

Fig. 6.25 The graph shows the relationship between blood flow and the pressure in blood-vessels. Note the critical closing pressure.

An increase in other metabolites, such as ATP, ADP, adenosine, pyruvate and lactate, has also been found to cause vasodilatation.

3 A third local regulatory mechanism depends on humoral agents released in the tissues. The release of the enzyme kallikrein by active salivary and sweat glands and by other exocrine glands converts plasma kininogens into active vasodilator substances or kinins such as bradykinin and kallidin. Kinins are also formed in inflammation and in response to allergic reactions, and they contribute to the vasodilatation seen in these conditions. Other humoral vasodilators include histamine, which is produced by mast cells and granulocytes in allergy and inflammation. Some prostaglandins are vasodilators while others are vasoconstrictors. These are produced in inflamed tissues and their synthesis can be suppressed by anti-inflammatory agents such as aspirin.

4 A fourth local mechanism is the effect of temperature. Warming of an organ produces local vasodilatation, while cooling results in vasoconstriction. This is an essentially local effect, which can occur in denervated limbs. The effect may be explained partly by a direct effect on vascular smooth muscle and partly by local changes in tissue metabolite production.

Applied physiology

HYPERTENSION AND HYPOTENSION

Hypertension

Elevated arterial blood-pressure (hypertension) is diagnosed when the diastolic pressure is higher than 90 mmHg. Usually, both systolic and diastolic pressures are raised. It is a disease with no known cause in about 90% of cases. These are diagnosed as having essential or *benign hypertension*, which means we do not know the primary cause. The remaining cases can be secondary to renal disease, hypersecretion of aldosterone (Conn's syndrome), excess glucocorticoids (Cushing's syndrome) or a catecholamine-secreting tumour of the adrenal medulla (phaeochromocytoma). Hypertension is also a symptom of toxaemia of pregnancy, which may be due to a pressor peptide released by the placenta.

Patients with essential hypertension initially have periodic elevations of blood-pressure which are exaggerated physiological responses to physiological stimuli such as exercise, cold or excitement. These responses can be managed by sympatholytic drugs. Eventually the blood-pressure elevation becomes continuous, which suggests failure of the baroreceptors to respond or that their normal range of response becomes readjusted (or set) at a higher level. Increased arterial blood-pressure is due to an increase in either cardiac output or peripheral resistance. Since the cardiac output is usually normal in chronic established benign hypertension, the main cause of the sustained increase in blood-pressure is the peripheral resistance. The underlying mechanism is either excessive arteriolar constriction and/or structural abnormalities in the resistance vessels.

Chronic hypertension may continue for many years as benign hypertension with few complications, but it may rapidly develop into *malignant hypertension*, due to unknown triggering factors, more in men than in women. This causes swelling of the optic disc (papilloedema), renal failure and cerebral symptoms.

Hypotension

Hypotension may occur in normal subjects as a result of the effect of gravity on venous return. On prolonged standing, pooling of up to 500 ml of blood occurs in the lower limbs. Interstitial fluid also accumulates in the lower extremities, due to increased hydrostatic pressure. These effects may cause a decrease in venous return and cardiac output. If the blood flow to the brain decreases to a critically low level, fainting occurs. However, compensatory mechanisms involving the baroreceptors and volume receptors lead to an increase in the heart rate and activation of the renin–angiotensin system. These help to maintain normal cardiac output and the state of constriction of blood-vessels. Local autoregulation in the cerebral circulation also helps to maintain normal cerebral blood flow.

Attacks of fainting or syncope may be precipitated by hypotension due to a reflex vasomotor inhibition, usually when the patient is standing. Consciousness is recovered quickly on lying down. Severe emotions can cause syncope due to fall of blood-pressure mediated by the limbic system and the hypothalamus. A possible mechanism is skeletal muscle vasodilatation, brought about by release of adrenaline and activation of the sympathetic cholinergic vasodilator system.

Orthostatic or postural hypotension may occur in some subjects on standing up from a lying down position. It causes transient dimness of vision and dizziness or a fainting attack. The mechanism seems to involve failure of the sympathetic compensatory mechanisms which readjust vascular tone on standing. There is also failure of adrenal medullary secretion of noradrenaline in response to baroreceptor stimulation. Patients with hypovolaemia due to a variety of causes are more likely to develop postural hypotension.

Hypotension may be due to heart failure, when the heart is unable to maintain a systolic pressure of 100 mmHg. This occurs in coronary artery disease due to myocardial damage and in mitral stenosis and aortic stenosis. Attacks of fainting are common in such patients. Peripheral vasodilatation, as caused by exposure to heat or vasodilator drugs, may precipitate such attacks. If a fainting attack is prolonged (coma), cerebral ischaemia and brain damage may occur.

HAEMORRHAGE

Acute blood loss is serious because it leads to a sudden drop in venous return, a decrease in end-diastolic volume and therefore a reduction in cardiac output and arterial blood-pressure. If the blood loss is less than 20% of the blood volume, rapid compensatory mechanisms are usually adequate, and spontaneous readjustment of the blood-pressure takes place. However, if the blood loss is more than 20%, the compensatory mechanisms are not enough and blood transfusion must be given as early as possible.

The rapid compensatory mechanisms increase the peripheral resistance and the cardiac output to restore the arterial blood-pressure as follows:
1 Responses initiated by the arterial baroreceptors. Their rate of discharge falls and this releases the pressor area of the vasomotor centre to increase its discharge. The result is generalized vasoconstriction due to increased sympathetic vasomotor tone. This is accompanied by increased venoconstriction, especially in the splanchnic area. Therefore, both the peripheral resistance and the central venous pressure are increased. The baroreceptor reflex also causes tachycardia by means of increased sympathetic discharge as well as inhibition of vagal centres (cardioinhibitory neurones).
2 The above mechanisms are augmented by stimulation of the peripheral chemoreceptors. These are stimulated by lack of O_2 due to diminished blood flow (stagnant hypoxia) and decreased O_2 carrying capacity of the blood. H^+ concentration also rises due to increased anaerobic glycolysis and lactic acid formation.

Cerebral blood-vessels are not constricted because they have little vasomotor control. The coronary vessels are dilated due to increased myocardial metabolism. However, renal blood flow may be compromised due to constriction of afferent and efferent arterioles. This causes decreased urine formation (oliguria), sodium retention and increased blood urea.
3 Responses due to the adrenal medulla. Sympathetic stimulation due to the baro- and chemoreceptor reflexes stimulates the adrenal medulla to release catecholamines into the bloodstream. Noradrenaline also reaches the circulation from sympathetic postganglionic fibres. The catecholamines augment the vasomotor tone and also increase the stroke volume and dilate the coronary vessels. Other effects of catecholamines after haemorrhage increase secretion of adrenocorticotrophic hormone (ACTH) and cortisol and enhance blood coagulation by releasing fibrinogen and prothrombin from the liver and by activation of factor V.

The rapid compensatory mechanisms described above explain the signs of hypovolaemia: pale, cool skin and tachycardia with low-volume (thready) pulse. There is also an increase in the respiratory rate, mediated by O_2 lack and acidosis. Tachypnoea helps increase the venous return and the efficiency of oxygenation of blood. Restlessness (increased skeletal muscle movement), in some cases, is probably due to stimulation of the reticular formation. Muscle movements increase the venous return through the 'muscle pump'.

Thirst is also a prominent symptom of hypovolaemia. It is due to cellular dehydration. This is explained as follows. When the arterioles constrict, the capillary pressure falls. This leads to net movement of interstitial fluid into the bloodstream. The decreased interstitial fluid volume causes net movement of fluid out of the intracellular compartment, leading to thirst.

4 Angiotensin II is a powerful vasoconstrictor. Its formation in response to hypovolaemia is due to activation of the renin–angiotensin system. Angiotensin II exerts a generalized vasoconstrictor effect, which contributes to the increase in peripheral resistance. It also contributes to the thirst mechanism by acting centrally. Angiotensin II is an important stimulus for aldosterone secretion and therefore it helps to retain Na^+ and increase the extracellular fluid (ECF).

The long-term or delayed mechanisms are mainly concerned with the restoration of blood volume and red blood cell (RBC) formation, as follows:

1 Restoration of plasma volume. The movement of ECF into the circulation leads to haemodilution and a fall of plasma protein concentration. Preformed albumin is released from the liver and is eventually followed by increased plasma protein synthesis. The plasma proteins are restored in 3–4 days.

As long as hypovolaemia exists, stimulation of aldosterone and ADH (vasopressin) secretion continues to play an important role in the restoration of ECF volume (see Chapter 10).

2 Red blood cell formation is stimulated by the renal hormone erythropoietin, which is released in response to hypoxia (see Chapter 3). Anaemia following haemorrhage is associated with an increase in the concentration of 2,3-diphosphoglycerate (2,3-DPG). This is also stimulated by hypoxia and favours the delivery of O_2 from haemoglobin (Hb) to the tissues.

SHOCK

Shock can be defined as a state of the circulation in which decreased blood flow disturbs the functions of organ systems. It is usually the result of heart failure or a severe decrease in peripheral resistance, both of which produce a sharp fall in blood-pressure. The patient presents with a pale, cold and sweaty skin due to sympathetic stimulation. There is also tachycardia, restlessness, thirst, decreased formation of urine and metabolic acidosis. If not treated adequately it may become irreversible, with brain damage, respiratory failure and death.

When the main problem lies with the cardiac output, as in coronary artery occlusion, this is called cardiogenic shock or central circulatory failure. When the fault is in the blood volume or peripheral resistance, it is known as peripheral circulatory failure. The systolic blood-pressure is usually less than 80 mmHg. The main disturbance is either blood loss or dehydration (hypovolaemic shock), together with a large vascular bed. Other causes of peripheral circulatory failure include severe bacteraemia, severe allergy (anaphylaxis), extensive burns (leading to massive plasma loss from the skin) and adrenal cortical failure.

Circulation in special regions

This section will deal with the adjustments of the blood flow in special vital regions of the body. Aspects of blood flow in other organs will be found in the various chapters dealing with organ systems.

PULMONARY CIRCULATION

The pulmonary arterial system has low resistance (about one-tenth that in systemic circulation). The pulmonary vessels are thinner-walled, shorter and more compliant. The blood-pressure in the pulmonary arteries is about 25 mmHg systolic and 10 mmHg diastolic (mean 15 mmHg). In the pulmonary capillaries it is 10 mmHg and in the left atrium it is about 8 mmHg during diastole. However, the total blood flow, the cross-sectional area and the velocity of flow in the pulmonary circulation are similar to those in the systemic circulation. At rest, a red blood cell takes about 0.75 s to pass through a pulmonary capillary and much less than that during exercise. The amount of blood in the lungs during standing is about 450 ml; on lying down it may increase to nearly double this amount. This causes a decrease in the vital capacity and explains dyspnoea (breathlessness) on lying flat (orthopnoea) in patients with cardiac failure or pulmonary disease.

The regulation of pulmonary blood flow is mainly passive, due to the high distensibility of the pulmonary vessels. When the cardiac output increases, as in muscular exercise, the blood flow in the lungs increases, pulmonary arterial pressure increases and parts of the lungs normally with little perfusion receive increased blood flow. There are, however, important vascular responses to alveolar O_2. Hypoxia causes vasoconstriction. Note that this is the opposite effect to that in the systemic circulation. Therefore, hypoxic alveoli tend to have less blood flow. This helps to divert blood flow to well-ventilated alveoli and thus gaseous exchange is enhanced. However, during exposure to hypoxia of high altitude or in lung disorders causing decreased ventilation, pulmonary vasoconstriction has obvious disadvantages, such as increased vascular resistance. This causes an increase in pulmonary blood-pressure and increases ventricular workload. Accumulation of CO_2 in underventilated areas leads to a rise in hydrogen ion concentration, which also causes vasoconstriction.

The pulmonary blood-vessels are supplied by sympathetic nerves, which produce vasoconstriction, and parasympathetic nerves, which are vasodilators. Adrenaline administration or sympathetic stimulation can decrease the pulmonary blood flow by about one-third.

CORONARY CIRCULATION

The coronary arteries are the first branches arising from the aorta, just above the aortic valves. The left coronary artery supplies most of the left ventricle and part of the anterior surface of the right ventricle as well as the interventricular septum. The right coronary artery supplies most of the right ventricle and the posterior wall of the left ventricle. The sinoatrial node is supplied by both arteries. The atrioventricular node and bundle are supplied by the right coronary artery. The right bundle branch is supplied by the right coronary artery while the left terminal bundle branch is supplied by both arteries.

The coronary arteries supply the myocardium with blood mainly during diastole (Fig. 6.26). About 80% of coronary blood flow takes place during diastole. During systole the aortic cusps are kept away from the coronary orifices by eddy currents and therefore the flow is maintained. When the myocardium is in a state of contraction,

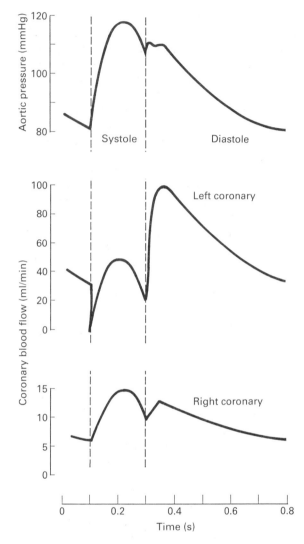

Fig. 6.26 A diagram showing phasic blood flow in the coronary arteries. (Reproduced from Berne, R.M. & Levy, M.N. (1988) *Physiology*, 2nd edn, by permission of C.V. Mosby Co., St Louis.)

the blood-vessels are compressed and flow is decreased. Subendocardial myocardial tissue in the left ventricle is further compressed by the high pressure inside the ventricle. This part of the heart muscle is most affected by ischaemic damage (myocardial infarction). In the right ventricle and both atria, coronary flow is better due to the greater pressure gradient between these chambers and the aorta.

The venous drainage of the left ventricle is mainly through venules which run with the small arteries and converge to form the coronary sinus and the anterior cardiac veins. This is known as the superficial drainage system. There is a deep system which drains the remaining parts of the heart. It is made up of arteriosinusoidal vessels which open directly into the heart chambers.

Coronary blood flow is mainly regulated by *metabolic autoregulation*. The degree of coronary dilatation is proportional to myocardial O_2 consumption, i.e. hypoxia is a potent vasodilator. Cardiac muscle has a high oxygen consumption relative to its mass. At rest it receives 250 ml of blood/min and extracts about 30 ml of O_2/min (i.e. 60%). Other tissues extract about 25% of arterial O_2 supply. Therefore, during exercise there is little further O_2 extraction and the increased O_2 uptake has to come from an increased coronary blood flow. Hypoxia is thought to release a vasodilator substance, probably adenosine. Hypoxia can increase the coronary blood flow fivefold. Reactive hyperaemia occurs when blood flow is temporarily obstructed. This increases blood flow in capillaries rather than in arteriovenous shunts. Other chemical factors involved in autoregulation include P_{CO_2}, H^+, K^+ and prostaglandins.

Adrenaline causes vasodilatation in the coronary arteries by acting on β_2-receptors. Therefore, in conditions where adrenaline is released from the adrenal medulla by sympathetic stimulation, the coronary blood flow is increased.

Neural regulation depends on sympathetic and parasympathetic nerves, which play a smaller role in regulation of coronary blood flow than autoregulation. Parasympathetic cholinergic nerves produce vasodilatation. Noradrenaline produces mild vasoconstriction by acting on α-adrenergic receptors. However, after sympathetic stimulation of the cardiac nerves, the increase in the heart rate and myocardial contractility increases the O_2 uptake and the net result is vasodilatation (by metabolic autoregulation) rather than vasoconstriction. Similarly, during exercise the coronary blood flow is maintained by autoregulation despite the increase in the heart rate and the shorter duration of diastole. When there is narrowing of the coronary arteries (athero-

sclerosis), the myocardial blood flow on exertion may not be sufficient, and myocardial ischaemia gives rise to chest pain (angina pectoris).

The pressure in the aorta, compared with that in the coronary arteries, significantly affects the coronary blood flow, especially during systole, when both pressures fall simultaneously. Also, the mean and systolic aortic pressure affects the coronary blood flow indirectly by changing aortic impedance, which, in turn, affects the work of the heart and its O_2 consumption.

CEREBRAL CIRCULATION

The brain receives its blood supply from the two internal carotid arteries and the vertebral arteries. These two sources communicate through the circle of Willis (Fig. 6.27), ensuring adequate circulation even when a major artery is occluded. The cerebral arteries have less-developed muscle layers. The veins drain into the venous sinuses, which are kept patent by the attachments of the dura mater. This ensures venous drainage even when there is an increase in intracranial pressure.

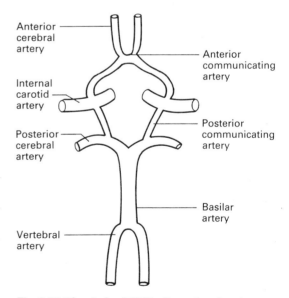

Fig. 6.27 The circle of Willis. (Reproduced with permission from Emslie-Smith, D., Paterson, C.R., Scratcherol, T. & Read, N.W. (eds) (1988) *BDS Textbook of Physiology*, 11th edn. Churchill Livingstone, Edinburgh.)

The blood flow to the brain is about 50 ml/100 g, i.e. if the brain is about 1400 g, then it receives 700 ml, which amounts to 15% of the cardiac output.

The capillaries in the brain have special endothelial tight junctions and lack the fenestrae commonly found in other capillaries. This is believed to be the structural basis of the *blood–brain barrier*, one of the protective mechanisms of the brain and the spinal cord.

The cerebral blood flow is kept constant mainly by *myogenic autoregulation*. Thus, the arteries dilate when the blood-pressure decreases and constrict when the blood-pressure is high. This is effective over a wide range of mean arterial blood-pressure (50–150 mmHg). Significant hydrostatic pressure changes occur on changing the position of the head in relation to the heart, e.g. performing a handstand. The camel and the giraffe have to lower the head by about 2–4 m when drinking water. Autoregulatory mechanisms maintain a constant blood flow despite the effect of gravity. Venous congestion and oedema are prevented by ECF passing to the cerebrospinal fluid, which is reabsorbed in the cerebral ventricles.

Metabolic autoregulation also plays an important role. The cerebral arterioles are particularly sensitive to P_{CO_2} changes and a low blood pH. Marked vasodilatation can cause up to a twofold increase in cerebral blood flow when the P_{CO_2} rises from 40 to 100 mmHg. Conversely, a marked vasoconstriction occurs when the P_{CO_2} falls below 40 mmHg. There is a decrease in cerebral blood flow, which may cause dizziness, when the P_{CO_2} decreases, e.g. due to hyperventilation (as in high altitudes) or due to excessive laughter or crying (as in hysterical attacks). The P_{O_2} plays a smaller role than P_{CO_2}. Hypoxia produces little vasodilatation when P_{O_2} decreases from 100 to 50 mmHg.

The cerebral blood-vessels have both sympathetic and parasympathetic supply but they seem to play a minor role in blood flow regulation.

Cerebral blood flow can be affected by change of the intracranial pressure. An increase in the intracranial pressure occurs in cases of hydrocephalus, brain tumours or meningitis. The cerebral blood flow is decreased due to compression of the intracranial blood-vessels. This causes hypoxia and hypercapnia in the cerebral circulation and directly stimulates the vasomotor centres, leading to an increase in the blood-pressure. There is also local metabolic vasodilatation. The increase in perfusion pressure and the vasodilatation helps to keep the blood flow constant. The increase in arterial blood-pressure is responsible for reflex bradycardia, usually observed in cases of increased intracranial pressure. It is mediated by the baroreceptors and is known as Cushing's reflex.

Further reading

1 Guyton, A.C. (1991) Cardiac output, venous return, and their regulation. In *Textbook of Medical Physiology*, 8th edn, pp. 221–233. Saunders, Philadelphia.

2 Guyton, A.C. (1991) Dominant role of the kidneys in long-term regulation of arterial pressure and in hypertension; the integrated system for pressure control. In *Textbook of Medical Physiology*, 8th edn, pp. 205–220. Saunders, Philadelphia.

3 Guyton, A.C. (1991) Muscle blood flow and cardiac output during exercise; the coronary circulation: and ischemic heart disease. In *Textbook of Medical Physiology*, 8th edn, pp. 234–244. Saunders, Philadelphia.

4 Guyton A.C. (1991) Muscular distensibility and functions of the arterial and venous systems. In *Textbook of Medical Physiology*, 8th edn, pp. 159–169. Saunders, Philadelphia.

5 Guyton, A.C. (1991) Nervous regulation of the circulation, and rapid control of arterial pressure. In *Textbook of Medical Physiology*, 8th edn, pp. 194–204. Saunders, Philadelphia.

6 Guyton, A.C. (1991) Overview of the circulation and medical physics of pressure, flow, and resistance. In *Textbook of Medical Physiology*, 8th edn, pp. 150–158. Saunders, Philadelphia.

7 Little, R.C. (1981) *Physiology of the Heart and Circulation*, 2nd edn. Year Book Medical Publishers, Chicago.

8 Smith, J.J. & Kampire, J.P. (1984) *Circulatory Physiology—the Essentials*, 2nd edn. Williams & Wilkins, Baltimore.

7: The Respiratory System

Introduction

The respiratory system serves two major functions in the human body: (i) it provides mechanisms for exchange of oxygen and carbon dioxide between the environment and tissues, and (ii) it plays a main role in the regulation of pH of the extracellular fluid. To achieve these two functions, the respiratory system provides the

---

> **Objectives**
>
> On completion of the study of this chapter, the student should be able to:
>
> **1** Understand the mechanics of breathing leading to normal ventilation.
>
> **2** Understand the relationship between volume and pressure in the lungs so as to explain abnormalities of lung compliance.
>
> **3** Describe the processes of gas exchange in the pulmonary alveoli and tissues so as to explain defects of these mechanisms leading to hypoxia or hypercapnia.
>
> **4** Understand the neural and chemical control of *ventilation* so as to explain the variations that occur in physiological and pathological states.
>
> **5** Describe the role of the respiratory system in the regulation of blood pH so as to explain the mechanisms of respiratory acidosis and respiratory alkalosis.
>
> **6** Become aware of the pulmonary function tests used in the diagnosis and management of respiratory disorders.

mechanical movements which result in *ventilation*, i.e. movement of the air in and out of the lungs during breathing. The lung alveoli provide mechanisms for the exchange of gases between the blood and the alveolar air. The regulation of breathing is closely linked with this exchange of gases and the breathing is stimulated or depressed according to the need to get more oxygen and to remove excess carbon dioxide. The removal of CO_2 provides a mechanism for getting rid of carbonic acid, which is a major factor in the regulation of blood pH.

The respiratory system has other subsidiary functions, which will be mentioned in the discussion of other body functions.

Ventilation

Ventilation is the process dealing with the air movement between the lung and atmospheric air.

LUNG VOLUMES AND CAPACITIES

For convenience, definitions of certain volumetric parameters commonly used in respiration are presented before the description of the mechanics of breathing. It should be explained at this stage that, when a volumetric parameter comprises more than one volume, it is referred to as a capacity.

1 *Tidal volume* is the volume of air inspired or expired at each breath. At rest (quiet breathing) the tidal volume is about 500 ml in a normal adult male. Its volume varies with body size, sex, age and activity. It is measured by a spirometer or gas meter.

2 *Respiratory minute volume or pulmonary ventilation* is the volume of air breathed in or out of the lungs each minute. In other words, it represents the volume of air which ventilates the lung every minute. If the resting tidal volume is 500 ml and if the resting respiratory rate is 12/min, then the pulmonary ventilation rate or lung minute volume is $500 \times 12 = 6000$ ml/min.

3 *Inspiratory reserve volume* is the extra volume of air which can be inspired by a maximal inspiratory effort over and above the tidal inspiration. It is defined as the volume of air inspired by a maximal inspiratory effort after normal inspiration. It is also measured by a spirometer or gas meter.

4 *Inspiratory capacity*. Since inspiration starts after the end of expiration, then the amount of air that can be inspired is equal to the tidal volume plus the inspiratory reserve volume. This is referred to as the inspiratory capacity (it is called capacity because it is formed of more than one volume, i.e. tidal volume plus inspiratory reserve volume). It is defined as the volume of air inspired by a maximal inspiratory effort after normal expiration.

5 *Expiratory reserve volume* is the extra volume of air expired by a maximal expiratory effort over and above the tidal expiratory volume. It is defined as the volume of air expired by a maximal expiratory effort after normal expiration. It is measured by a spirometer or gas meter.

6 *Vital capacity* is the total volume of air that can be exchanged between the lungs and the environment in one respiratory cycle. It is defined as the volume expired by a maximal expiratory effort after maximal inspiration. It comprises tidal volume plus inspiratory reserve volume plus expiratory reserve volume. The vital capacity also depends on the size, age and sex of the individual. It is a good test for assessment of lung function and can be measured by a spirometer or gas meter.

7 *Residual volume* is the volume of air that remains in the lung after maximal expiration. Since it is the volume of air that cannot be expired, it is not available for measurement by a spirometer or gas meter; other methods have to be used (see Lung function tests). Assessment of the residual volume is a good index of the state of the lung. It is usually increased by age and lung disease when the lung elasticity or bronchial diameter is affected.

8 *Functional residual capacity (FRC)* is the volume of air remaining in the lung after normal expiration. It includes the expiratory reserve volume and the residual volume. Like the residual volume, it cannot be measured by a spirometer or by a gas meter. As the FRC is large (2.2 litres) and the part of inspired tidal air which reaches the alveoli is small (0.35 litre), the composition of alveolar air remains relatively constant during the inspiration–expiration cycle.

9 *Total lung capacity* is the maximal volume of air that can be accommodated in the lungs. It comprises tidal volume, inspiratory reserve volume, expiratory reserve volume and residual volume.

The various volumes and capacities are shown graphically in Fig. 7.1 and their normal values are shown in Table 7.1.

Age, sex and body size have a significant effect on some of the lung volumes and capacities. For example, both residual volume and functional residual capacity increase with age while vital capacity decreases with age. Similar changes are usually observed in lung diseases affecting elasticity and expansibility (compliance) of the lungs.

MECHANICS OF BREATHING

Anatomical considerations
The lungs are enclosed in an airtight compartment formed of the thoracic cage and the diaphragm. Its only connection with the atmosphere is through the mouth, nose and respiratory conducting system of tubes — airways. The lungs are surrounded by a minute space filled with a film of

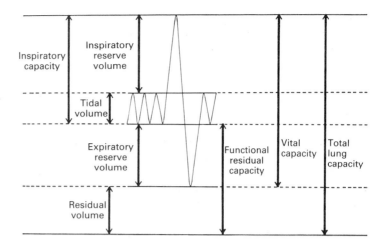

Fig. 7.1 Graphic representation of lung volumes and capacities during normal breathing, maximum inspiration and maximum expiration.

fluid, the pleural space, lying between the visceral pleura attached to the lung surface and the parietal pleura which lines the inner surface of the thoracic cage and the thoracic surface of the diaphragm. The thin film of fluid in the pleural space lubricates the movements of the two surfaces during breathing. The chest wall is formed of muscles, ribs and vertebrae. On the outside it is covered by skin and subcutaneous tissue. The diaphragm is a muscle attached to the lower portion of the thoracic cage. It separates the thorax from the abdominal cavity. The oesophagus passes through the diaphragm to the stomach. The diaphragm, when relaxed, assumes a dome shape with its convexity towards the thoracic cavity. When it contracts during inspiration, it descends and becomes less convex.

The ribs are connected by two layers of muscles: external and internal groups of intercostal muscles. The two groups of muscles differ in the direction of their muscle fibres from their origin to their insertion. The external intercostal muscles pass downwards and forwards from their origin to their insertion while the internal intercostal muscles pass downwards and backwards from their origin to their insertion (Fig. 7.2). The ribs are articulated posteriorly to the fixed vertebral column by socket joints and anteriorly to the mobile manubrium and sternum by cartilage joints. From their attachment posteriorly to the vertebral column the ribs curve downwards, laterally, forwards and then medially.

Table 7.1 Normal lung volumes and capacities in litres for adult males and females

	Volume (litres)	
	Male	Female
Tidal volume (TV)	0.5	0.5
Inspiratory reserve volume (IRV)	3.3	1.9
Expiratory reserve volume (ERV)	1.0	0.7
Residual volume (RV)	1.2	1.1
Inspiratory capacity (IC)	3.8	2.4
Functional residual capacity (FRC)	2.2	1.8
Vital capacity (VC)	4.8	3.1
Total lung capacity (TLC)	6.0	4.2

All pulmonary volumes and capacities are 20–25% less in females than in males.

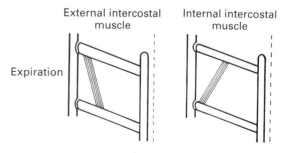

Fig. 7.2 Diagrammatic representation showing the arrangement of fibres of external and internal intercostal muscles and position of ribs during expiration.

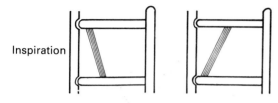

Inspiration

Fig. 7.3 Movements of the ribs and sternum during inspiration to increase the anteroposterior diameter of the thoracic cavity. Compare with their position in expiration.

Chest movements

The arrangement of the ribs and intercostal muscles is such that movements of the ribs cause alterations in the dimensions of the chest. When the external intercostal muscles contract they tend to raise the upper ribs and sternum. This movement will displace the sternum and ribs forwards, increasing the anteroposterior diameter of the thoracic cavity. Moreover, as the arched ribs are raised, they rotate outward in a 'bucket handle' movement, increasing the transverse diameter of the chest slightly. Contraction of the internal intercostal muscles tends to lower the ribs and sternum, reducing the anteroposterior diameter of the thoracic cavity (Fig. 7.3).

When the diaphragm contracts, it becomes less convex and pushes the abdominal viscera down-wards. This increases the vertical diameter of the thoracic cavity (Fig. 7.4).

During normal resting inspiration, the increase in the vertical diameter of the chest, brought about by the descent of the diaphragm, contributes about 70% of the increase in the volume of the thoracic cavity. The increase in the anteroposterior diameter, brought about by upward movement of the sternum and ribs, contributes about 25–30% of the volume change while the increase in the transverse diameter, brought about by the rotation of the ribs, contributes about 1–2%.

Intrapleural pressure

A manometer connected to a needle inserted into the pleural space will record subatmospheric pressure (negative pressure) throughout the respiratory cycle during normal quiet breathing. The pressure will vary between −2 mmHg at the end of normal expiration to about −6 mmHg during inspiration (Fig. 7.5). During forcible inspiration, it may fall to −70 mmHg. During forcible expiration against a closed glottis (Valsalva manoeuvre), the intrapleural pressure becomes positive, rising to 50 mmHg or more above atmospheric pressure.

The changes in oesophageal pressure detected by a balloon inserted into the oesophagus reflect

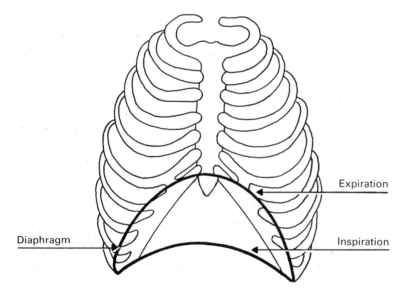

Diaphragm

Expiration

Inspiration

Fig. 7.4 Movement of the diaphragm during inspiration to increase the vertical diameter of the thoracic cavity.

Intrapleural pressure mmHg

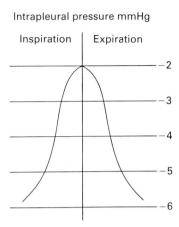

Fig. 7.5 Changes of intrapleural pressure during inspiration and expiration in normal resting breathing.

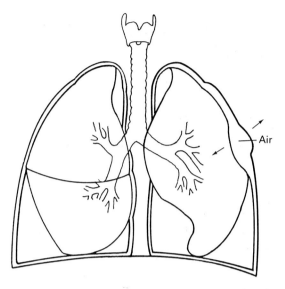

Fig. 7.6 The effect of introducing air into the pleural cavity through the chest wall. The lung will recoil inward (collapse) and the chest wall will recoil outward.

the changes in intrapleural pressure fairly accurately.

Causes of subatmospheric (negative) intrapleural pressure Several factors contribute to the creation of a negative intrapleural pressure:

1 *The lungs.* The walls of the lungs are rich in elastic tissue, which tends to collapse the lungs because the elastic recoil of the lung acts in an inward direction. Thus, with advancing age due to loss of elastic tissue, intrapleural pressure rises towards the atmospheric pressure. The lung recoil is also potentiated by the surface tension of the fluid lining the alveoli and alveolar sacs. However, the surface tension forces are normally reduced by surfactant, a lipoprotein secreted by type II cells of the alveolar epithelium. Congenital deficiency of surfactant (hyaline membrane disease of the new-born) may be one of the causes of the failure of the lungs to expand at birth.

2 *The chest wall.* Formed of ribs, cartilage and muscle, the chest wall has a tendency to recoil outward, i.e. in an opposite direction to the elastic recoil of the lungs. At equilibrium, these two opposing forces lead to a negative pressure in the pleural cavity. If air is introduced into the pleural cavity by opening the chest wall, the lungs will collapse (recoil inward) while the chest wall will recoil outward and the intrapleural pressure becomes atmospheric (Fig. 7.6).

3 *Pleural capillaries and lymphatics.* The intrapleural space is rich in blood capillaries and lymphatics, which tend to absorb fluid from the pleural cavity. This adds to the negativity of the intrapleural pressure.

Intra-alveolar (intrapulmonary) pressure
Pressure within the alveoli of the lungs varies with the different stages of the respiratory cycle. When recorded at the end of expiration with the glottis open, it is atmospheric. During inspiration, because the lungs are expanding with the increasing size of the thoracic cage, the pressure tends to fall below atmospheric. If the glottis is open, air will be drawn into the lungs. During quiet inspiration, the pressure falls to about −1 mmHg but with forcible inspiration it may fall much lower, to about −70 mmHg. During expiration, as the lungs recoil, the intra-alveolar pressure rises above atmospheric and forces the air out of the lungs if the airway is open. With normal quiet expiration, intra-alveolar pressure usually rises to 1 mmHg, but during forceful expiration with the glottis closed (Valsalva manoeuvre) it may rise up to 100 mmHg.

Mechanism of air flow between the lung and the atmosphere

The external intercostal muscles and the diaphragm are innervated by somatic nerves, the intercostal and the phrenic nerves, respectively. Stimulation of the nerves leads to contraction of the external intercostal muscles, increasing the anteroposterior diameter and, to a lesser extent, the transverse diameter of the chest, while contraction of the diaphragm increases the vertical diameter. The resulting increase in the volume of the chest tends to lower the intrapleural pressure. Consequently, because of their expansile properties, the lungs expand. The expansion of the lung lowers the intra-alveolar pressure below atmospheric so that, if the glottis is open, air will be drawn into the lungs.

When inspiration comes to an end by cessation of impulses to the inspiratory muscles, the muscles relax. The diaphragm ascends to assume a more convex shape, reducing the vertical diameter of the chest. On relaxation of the external intercostal muscles, the sternum and ribs descend and the thoracic wall recoils. Consequently, the intrapleural pressure rises (becomes less negative). The elastic lung then recoils and compresses the air in the alveoli, leading to the rise of the intra-alveolar pressure above atmospheric. If the airways are open, air is forced out of the lungs.

It is evident from the above description of the mechanisms of inspiration and expiration that during quiet breathing inspiration is an active process, involving muscular contraction, while expiration is a passive process, brought about by the relaxation of muscles of inspiration, recoil of the chest wall and elastic recoil of the lungs.

Accessory muscles of respiration

Distinction should be made between an increase in inspiration and an increase in pulmonary ventilation (minute volume).

During quiet breathing only about one-tenth of the muscle fibres of the external intercostal muscles and the diaphragm are active at any one time. With more powerful inspiration, more muscle fibres are recruited into activity until all fibres become active during maximal inspiration. This alone can increase pulmonary ventilation 10-fold, up to about 60 litres/min. However, there are other muscles involved in increasing pulmonary ventilation. These groups of muscles are designated accessory muscles of respiration; some are inspiratory and others are expiratory. They act by increasing the depth of inspiration even further, and by decreasing the airway resistance to air flow to allow quick and easy flow of air between the lungs and the atmosphere.

A group of muscles attached to the thoracic cage can be brought into action during maximal ventilation or whenever there is difficulty in breathing from any cause. When these muscles contract, they lift the entire thoracic cage upwards. This chest movement tends to facilitate inspiration by a further increase in the anteroposterior diameter of the chest, but a more important effect is that it straightens the airways, decreasing their resistance to air flow. The most important accessory muscles of inspiration are the sternocleidomastoids, the anterior serrati and the scaleni.

Normally, expiration is a passive process that does not involve active muscular contraction. However, during maximal expiratory efforts, as in severe exercise or when there is difficulty in breathing, e.g. bronchial obstruction in bronchial asthma, expiration becomes an active process with active muscular contraction. A number of muscles can be designated as accessory muscles of expiration. The internal intercostal muscles, by depressing the sternum and ribs, decrease the anteroposterior diameter of the chest. When the abdominal recti contract, they depress the thoracic cage, while also reducing the anteroposterior diameter of the chest. At the same time, they compress the abdominal viscera, pushing them upwards against the diaphragm, which will rise further and reduce the vertical diameter of the chest. The internal intercostal muscles and the abdominal recti are the main accessory muscles of expiration.

Fatigue of the respiratory muscles

Breathing continues throughout life. It involves voluntary muscular contraction. One of the properties of voluntary muscles is fatigability. The question may be asked whether the diaphragm and external intercostal muscles get fatigued since they are voluntary muscles. In fact they do, but, because during quiet breathing only one-tenth of the fibres are actively contracting, their activity is taken up by other fibres when they get fatigued. This will ensure that quiet breathing will continue indefinitely as only 10% of the fibres are contracting at any one time, to be replaced by others when they get fatigued.

Work of breathing

During quiet breathing, inspiration is an active process requiring energy. This amounts to about 2–3% of the total energy expenditure of the body under resting conditions. During severe exercise both the total energy expenditure of the body and that required for the work of breathing increase. Although the absolute total energy expenditure of the work of breathing increases significantly during activity, in the region of 15–20 fold, its percentage of the total body energy expenditure may not change significantly.

There are several components of breathing that require energy:

1 Energy is required by the contracting muscles of breathing, the diaphragm and external intercostal muscles during quiet breathing, and by accessory muscles of respiration during maximal or laboured breathing.

2 Energy is required to overcome the viscosity of the expanding lung tissue and of the tissues of the thoracic cage. This is designated as the non-elastic tissue resistance. When the lungs become diseased, e.g. consolidated or fibrotic, the non-elastic resistance increases.

3 Energy is required to stretch the thoracic and lung elastic tissues (elastic tissue resistance) and to overcome the surface tension in the alveoli and alveolar sacs during inspiration. This energy requirement will increase when the surfactant is deficient.

4 Energy is required to overcome airway resistance to air flow. The narrower the airways, the greater is the resistance to air flow and the more will be the energy requirement, as in the case of bronchial asthma or obstructive emphysema.

Compliance

Compliance means the ability to expand or stretch. It is the reciprocal of elasticity. The compliance of the lung is a useful measurement for the diagnosis of respiratory disease.

DEFINITIONS

Elastance

Elastance can be defined as the stretching force which must be applied to a certain body to produce a unit change in length.

Compliance

Compliance is the change in length or volume per unit change in the stretching force.

It has already been stated that, when the intra-alveolar pressure falls below atmospheric, air is drawn into the lungs. Conversely, when the intra-alveolar pressure rises above atmospheric, air is forced out of the lungs. The changes in intra-alveolar pressure are determined by changes in intrapleural pressure, which closely correspond to changes in oesophageal pressure.

Normally, when the lungs are *in situ* within the chest, the combined compliance of the lungs and thorax is measured; it is about 0.11 litre/cm H_2O pressure (i.e. when the intra-alveolar pressure increases by 1 cm H_2O, the thorax and lung expand by a volume of 0.1 litre). If the measurement is done on the lungs alone after being removed from the chest, the compliance of the lungs alone is recorded, which is greater than the combined compliance of the lungs and thorax. The compliance of the lungs alone is around 0.2 litre/cm H_2O pressure. Because the lungs and chest wall act in series, their combined compliance add as reciprocals, i.e.:

$$\frac{1}{\text{Combined compliance}} = \frac{1}{\text{Lung compliance}} + \frac{1}{\text{Chest wall compliance}}$$

MEASUREMENT OF COMPLIANCE OF THE LUNGS AND THORAX COMBINED

Combined compliance can be measured using a mouthpiece fitted with a flow metre and a valve. A balloon is introduced into the oesophagus to record the oesophageal pressure (intrapleural pressure). The nose is closed by a nose-clip. The subject inspires in increments of approximately

100 ml at a time. At the end of each increment of inspiration the valve is shut and the subject relaxes his inspiratory muscles while the oesophageal pressure is recorded. The process is repeated until the inspired increments of air equal the normal tidal volume. The same processes are repeated again during a tidal expiration. Now the subject expires in increments of about 100 ml of air at a time till the end of tidal expiration. The oesophageal pressure is noted each time. The volume change is then plotted against the oesophageal pressure change for inspiration and expiration. Figure 7.7 shows a normal compliance curve. From this curve the work of breathing can be calculated. The total work of breathing is indicated by the area OAECFO. Of this, area OADCFO represents work done to overcome elastic resistance (compliance work). The shaded area AECDA represents work to overcome airway and tissue resistance (viscous resistance). Area ADCBA represents the work done against airway and tissue resistance during expiration. This is normally part of the work done to overcome elastic resistance during inspiration. On expiration, the work is passively done by the elastic recoil of the tissues expanded during inspiration.

Usually a high compliance indicates that a given change in pressure moves a large volume of air into the lungs. Lung compliance is decreased when there is fibrosis, congestion, oedema or bronchial obstruction or when the surface tension of fluid lining the alveoli is increased. The compliance is usually small in the newborn but it increases gradually with age. In adults it decreases with old age. Therefore, the main factors which affect compliance are lung congestion, lung size and surface tension, which is mainly related to surfactant.

Lung congestion leads to decreased compliance and is well documented clinically. Lung size is directly proportional to compliance. For this reason, the term *specific compliance* was introduced. It is the compliance per unit volume of lung.

Surfactant is a substance which lowers the surface tension of the alveoli. Surfactant contains phospholipids (dipalmitoyl-phosphatidyl-choline) neutral lipids, protein and carbohydrates. It is produced by type II alveolar epithelial cells. The importance of surfactant lies in the fact that it prevents the collapse of alveoli during expiration when their volume is small (law of Laplace). Another important function is that surfactant increases the compliance of the lungs and therefore decreases the work of breathing. By decreasing the surface tension around lung capillaries, surfactant also decreases the formation of tissue fluids in the lungs. In the absence of surfactant, the lungs become rigid, oedema is formed and the alveoli collapse. These features are characteristically seen in *respiratory distress syndrome* (RDS) of the new-born.

Distribution of pulmonary ventilation

Pulmonary ventilation is the volume of air breathed in or out per minute. Inspired air is distributed to distinct spaces in the lungs:

1 Part of it occupies the air-conducting system of tubes which extends from the mouth and nose down to the terminal bronchioles. In this specific anatomical space there is no gas exchange with blood. It is therefore referred to as the *anatomical dead space*.

2 The rest of the inspired air occupies the space distal to the terminal bronchioles, namely the respiratory bronchioles, the alveolar ducts, alveoli and alveolar sacs, where gas exchange takes place (Fig. 7.8).

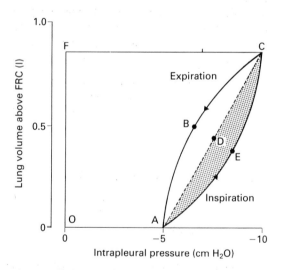

Fig. 7.7 The lung compliance curve. FRC, functional residual capacity. See text for explanation.

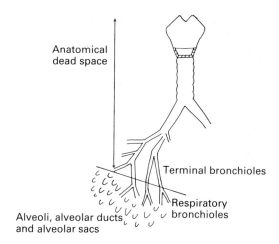

Anatomical
dead space

Terminal bronchioles

Respiratory
bronchioles

Alveoli, alveolar ducts
and alveolar sacs

Fig. 7.8 Diagram to show the anatomical dead space, where no gas exchange takes place, and the parts of the lungs where gas exchange does take place.

Each minute a certain volume of air equivalent to:

(Tidal volume − anatomical dead space)
× respiratory rate

undergoes exchange with the blood. This volume is referred to as the alveolar ventilation rate, as distinct from the total respiratory minute volume or pulmonary ventilation rate. In a normal subject the anatomical dead space is approximately one-third of the normal resting tidal volume.

If, in a normal subject, the resting volume is 500 ml and the respiration rate is 12/min, then:

1 Minute volume (pulmonary ventilation rate) is equal to:

$$500 \times 12 = 6000 \text{ ml/min}$$

2 Alveolar ventilation rate is equal to:

$$\tfrac{2}{3} \times 500 \times 12 = 4000 \text{ ml/min}$$

3 Dead space ventilation is equal to:

$$\tfrac{1}{3} \times 500 \times 12 = 2000 \text{ ml/min}$$

It should be pointed out that, if tidal volume is significantly reduced (although the pulmonary ventilation rate can be maintained by increasing the respiration rate by shallow rapid breathing), the alveolar ventilation rate will be significantly diminished. This is detrimental to gas exchange

in the lungs. The following example will illustrate this fact.

In an adult male the following results were obtained:

Tidal volume $= 300$ ml
Respiratory rate $= 20/\text{min}$
Pulmonary ventilation $= 300 \times 20$
$= 6000$ ml/min
Alveolar ventilation $= [300 - (\tfrac{1}{3} \times 500)] \times 20$
$= (300 - 167) \times 20$
$= 133 \times 20$
$= 2660$ ml/min
Dead space ventilation $= \tfrac{1}{3} \times 500 \times 20$
$= 3340$ ml/min

It is evident that, although the minute volume is normal, alveolar ventilation is significantly decreased. This is because the dead space is a constant volume and its share of pulmonary ventilation has increased (from 2000 ml/min to 3340 ml/min) as the respiratory rate increased. Therefore, a small reduction in tidal volume may therefore lead to a considerable decrease of gas exchange in the lung.

MEASUREMENT OF ANATOMICAL
DEAD SPACE

Expired air is a mixture of air coming from the alveoli (alveolar air) and air in the anatomical dead space, which is atmospheric. Alveolar air normally contains about 6% CO_2, while expired air contains around 4% CO_2. If, for practical purposes, the atmospheric CO_2 (which is 0.04%) is taken as 0%, then:

Tidal volume × 4% = Alveolar air × 6%
= (Tidal volume
− anatomical dead space)
× 6%

Therefore,

Anatomical dead space = Tidal volume
$$\times \frac{6 - 4}{6}$$

Or, in other words,

Anatomical dead space =
$$\frac{\text{Tidal volume} \times (\text{alveolar } CO_2 - \text{expired air } CO_2)}{\text{Alveolar } CO_2}$$

This is Bohr's equation. This method of measuring the anatomical dead space assumes normal perfusion of the lung. In certain lung diseases this assumption is not valid. When some alveoli do not receive adequate perfusion with blood, they become non-functional. The volume or anatomical dead space plus that of unperfused alveoli is known as the *physiological dead space*. In normal persons the anatomical and physiological dead spaces are nearly equal. In certain diseases of the lungs the physiological dead space may be 10 times the anatomical dead space or more, i.e. 1000–2000 ml.

UNEVEN VENTILATION

In the upright subject the bases of the lungs are found to be better ventilated than the apices. This can be demonstrated by breathing radioactive xenon. The uneven ventilation is due to the effect of gravity. Similarly, a subject in the supine position will have better ventilation of the posterior parts of the lungs than the anterior parts. Uneven ventilation can significantly affect gas exchange in the lungs.

Gas exchange in the lungs

Oxygen is continuously taken up from the lungs by the blood, and carbon dioxide is likewise continuously removed from the blood by the lungs. Air reaching the alveoli during inspiration will mix with air already in the alveoli. Consequently, alveolar air contains less oxygen and more carbon dioxide than inspired air. Air expired will constitute a mixture of alveolar air and the dead space air, which is atmospheric. Table 7.2 gives

approximate normal values for atmospheric air, expired air and alveolar air.

The exchange of oxygen and carbon dioxide between the alveoli and blood is a purely passive process of gas diffusion. However, the rate of diffusion is determined by several factors.

ALVEOLAR–CAPILLARY MEMBRANE

The alveolar–capillary membrane is a semipermeable membrane which separates alveolar air from the pulmonary capillary blood. The electron micrograph in Fig. 7.9 shows alveoli, capillaries and RBCs in a section of rat lung. Several layers separate the red cell membrane from the alveolar air. These are the fluid film lining the alveoli, the alveolar membrane, interstitial fluid and the capillary wall (Fig. 7.10).

Naturally, change in thickness of the alveolar–capillary membrane will affect the rate of gas diffusion. The thickness of the membrane cannot be diminished significantly, but with deep breathing, as in severe exercise, it may be decreased. In disease it can become much thicker, e.g. by fibrosis or accumulation of fluid in the alveoli, as in pulmonary oedema. This may lead to alveolar–capillary block.

PARTIAL PRESSURE GRADIENT OF GASES ACROSS THE ALVEOLAR–CAPILLARY MEMBRANE

In a gas mixture like air, the partial pressure of a gas is the pressure exerted by that gas in relation to the total pressure of the gas mixture. Therefore, the *partial pressure* is the product of the total pressure of the mixture and the fractional con-

Table 7.2 Composition of respiratory gases

	Atmospheric air		Alveolar air		Expired air	
	%	PP (mmHg)	%	PP (mmHg)	%	PP (mmHg)
O_2	20.84	159	13.6	104	15.7	120
CO_2	0.04	0.3	5.3	40	3.6	27
N_2	78.62	597	74.9	569	74.5	566
H_2O	0.50	3.7	6.2	47	6.2	47
	100	760	100	760	100	760

PP, partial pressure.

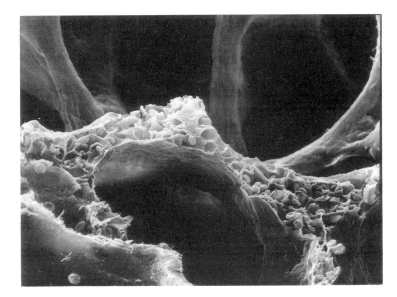

Fig. 7.9 Electron micrograph showing the alveoli, capillaries and red blood cells in a section of rat lung. (Courtesy of Dr Michael Scott, King Fahad Medical Research Centre, King Abdulaziz University, Jeddah, Saudi Arabia.)

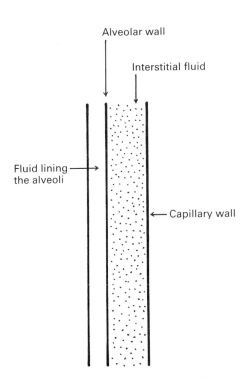

Fig. 7.10 Components of the alveolar–capillary membrane.

centration of the gas under consideration. Table 7.2 gives the composition and partial pressures of inspired air (atmospheric air), alveolar air and expired air. The partial pressure of oxygen in the mixed venous blood reaching the alveoli is about 40 mmHg, while the partial pressure of oxygen in the alveolar air is around 100 mmHg. Oxygen therefore diffuses from the alveolar space to the capillary blood along a partial pressure gradient of about 60 mmHg.

The partial pressure of carbon dioxide in the mixed venous blood reaching the alveoli is about 46 mmHg, while in the alveolar air it is about 40 mmHg. Thus carbon dioxide diffuses from the capillary blood to the alveolar space along a partial pressure gradient of about 6 mmHg.

THE PHYSICAL PROPERTIES OF GASES

The solubility and molecular weight of gases are a factor in the rate of diffusion. The more soluble the gas, the faster it will diffuse through the alveolar–capillary membrane, which is a fluid membrane. The higher the molecular weight of the gas, the slower it will diffuse through. Carbon dioxide is 23-fold more soluble than oxygen.

The solubility of a gas and its molecular weight determine its diffusion coefficient. This is defined as the rate of diffusion through a unit area of a given membrane per unit pressure difference:

$$\text{Diffusion coefficient} = \frac{\text{Solubility}}{\sqrt{\text{Molecular weight}}}$$

If we take the diffusion coefficient of oxygen to be 1.0, that of carbon dioxide is approximately 20.0. Carbon dioxide can therefore diffuse through the alveolar capillary membrane 20 times faster than oxygen. Consequently, diffusion failure will affect oxygen long before carbon dioxide diffusion is impaired.

SURFACE AREA

As the surface area of the alveolar–capillary membrane increases, the total volume of a gas exchanged will be proportionately increased. The alveolar–capillary membrane surface, in adult males, is approximately 70 m^2. During severe exercise, the depth of breathing is significantly increased, which increases the alveolar ventilation. The cardiac output is likewise increased several-fold. This normally leads to the opening up of small pulmonary capillaries, which results in an increase of the capillary surface area. What is important for diffusion purposes is not an increase of the alveolar surface alone or of the capillary surface alone; what is important is the effective surface area, i.e. the functioning alveoli in contact with functioning capillaries. It is reasonable, therefore, to expect an increase in the effective surface area during severe exercise. In diseases which reduce alveolar ventilation, e.g. emphysema and obstructive lung diseases, and in diseases which interfere with normal alveolar perfusion, e.g. in pulmonary embolism and restrictive lung diseases, the effective surface area is decreased.

VENTILATION/BLOOD FLOW RATIO

As mentioned above, the rate of diffusion of gases is determined by the effective surface area, i.e. the functional alveoli in contact with functioning capillaries, where the alveolar air comes in contact with capillary blood.

$$\frac{\text{Ventilation/}}{\text{blood flow ratio}} = \frac{\text{Alveolar ventilation}}{\text{Blood flow}}$$

In a normal adult male at rest:

Alveolar ventilation = 4 litres/min

$$\frac{\text{Pulmonary blood flow}}{\text{(cardiac output)}} = 5 \text{ litres/min}$$

Ventilation/blood flow ratio = $\frac{4}{5}$ = 0.8

As an extreme example of disturbed ventilation/blood flow ratio, no exchange of gases is expected to take place in functioning alveoli not perfused with capillary blood (wasted ventilation); nor would any gas exchange take place if blood is perfusing non-functioning alveoli (wasted perfusion).

Perfusion of the lungs is also uneven. The bases of the lungs receive more blood than the apices. The ventilation/perfusion ratios are 3.3 at the apices and 0.63 at the bases. This shows that perfusion is more uneven than ventilation. During exercise, however, lung perfusion becomes more uniform.

In normal individuals, an average ventilation/perfusion ratio of 0.8 is ideal for effective exchange of gases because each lung will have the same ratio as the overall ratio for both lungs together.

Example 1 Even ventilation with even blood flow:

	Alveolar ventilation (l/min)	Blood flow (l/min)	Ventilation/ perfusion ratio
Right lung	2	2.5	2/2.5 = 0.8
Left lung	2	2.5	2/2.5 = 0.8
Both lungs	4	5.0	4/5.0 = 0.8

However, if there is uneven ventilation or perfusion, the overall ventilation/perfusion ratio (of both lungs combined) may be normal but the ratio is abnormal for individual lungs, leading to wasted ventilation or wasted perfusion.

Example 2 Uneven ventilation with even blood flow:

	Alveolar ventilation (l/min)	Blood flow (l/min)	Ventilation/ perfusion ratio
Right lung	1	2.5	1/2.5 = 0.4
Left lung	3	2.5	3/2.5 = 1.2
Both lungs	4	5.0	4/5.0 = 0.8

In the right lung there is wasted blood flow, leading to a low ventilation/perfusion ratio in that lung, while in the left lung there is wasted ventilation, resulting in a high ventilation/perfusion ratio. It is clear that, although the overall ventilation/perfusion ratio is normal, the ratio is significantly disturbed in the individual lungs, leading to ineffective exchange of gases.

Example 3 Even ventilation with uneven blood flow:

	Alveolar ventilation (l/min)	Blood flow (l/min)	Ventilation/ perfusion ratio
Right lung	2	1	2/1 = 2.0
Left lung	2	4	2/4 = 0.5
Both lungs	4	5	4/5 = 0.8

In the right lung there is wasted ventilation whereas in the left lung there is wasted perfusion. Although the overall ventilation/perfusion ratio is normal, it is significantly disturbed in the two lungs separately, leading inevitably to reduced gas exchange between the blood and alveolar air.

CHEMICAL REACTIONS

Haemoglobin readily combines reversibly with oxygen and carbon dioxide. If blood contains the normal value of 150 g of haemoglobin per litre of blood, and if haemoglobin is fully saturated with oxygen, i.e. at partial pressure of O_2 of 100 mmHg, blood will carry $150 \times 1.34 = 200$ ml O_2/litre of blood (since 1 g of haemoglobin, when fully saturated, combines with 1.34 ml O_2 at standard temperature and pressure, dry (STPD)). The O_2 dissolved in 100 ml of blood at an oxygen partial pressure of 100 mmHg is only $0.003 \times 100 = 0.3$ ml, since 100 ml of blood carries only 0.003 ml O_2/1 mmHg partial pressure. About 98.5% of the total oxygen transported in arterial blood is combined with haemoglobin. If there is no haemoglobin, the blood volume needs to increase about 70 times to transport the same volume of oxygen. Conversely, a partial pressure of O_2 of about 7600 mmHg (7 atmospheres) will be needed for 100 ml plasma to hold 20 ml O_2 in the dissolved form.

The situation is not so critical with carbon dioxide, as it is about 20 times more soluble than oxygen. Haemoglobin binds about 20% of the total carbon dioxide transported in blood.

TEMPERATURE

The rate of diffusion of gases is normally dependent on temperature, but since body temperature is normally maintained within narrow limits, it will have no influence on the rate of exchange of gases across the alveolar–capillary membrane.

DIFFUSION CAPACITY

The *diffusion capacity* of a gas is defined as the total volume of the gas that diffuses across the alveolar–capillary membrane per unit time per unit partial pressure difference, usually measured in ml/min/mmHg. All the above factors affect the diffusion capacity of any particular gas. One would therefore expect that, all other factors being the same, carbon dioxide, because of its greater solubility, would have a higher diffusion capacity than oxygen – about 20 times more.

During exercise, both alveolar ventilation and blood flow increase. The alveolar membrane is stretched, leading to increased surface area and decreased thickness. Previously dormant pulmonary capillaries open up and existing capillaries dilate, leading to increased surface area of the capillary membrane. Consequently, the effective surface area for gas exchange will increase, leading to increased diffusion capacity for both oxygen and carbon dioxide (Table 7.3).

Diseases that affect any of the factors concerned with diffusion across the alveolar–capillary membrane will lower the diffusion capacity of oxygen more than that of carbon dioxide. Impaired diffusion of oxygen reaches fatal levels long before the diffusion capacity of carbon dioxide is significantly reduced.

Gas transport

TRANSPORT OF OXYGEN

Oxygen is transported in the blood in two forms:
1 The main mode of transport is in the form of oxyhaemoglobin. Over 98% of the total oxygen

Table 7.3 Diffusion capacity for O_2 and CO_2 at rest and during severe exercise

	Diffusion capacity	
	At rest	During severe exercise
O_2	20 ml/min/mmHg P_{O_2}	65 ml/min/mmHg P_{O_2}
CO_2	400 ml/min/mmHg P_{CO_2}	1300 ml/min/mmHg P_{CO_2}

carried in arterial blood is in this form. Under normal resting conditions the pulmonary capillary blood is fully equilibrated with alveolar air, i.e. the P_{O_2} in the capillary blood is nearly 100 mmHg. However, P_{O_2} in systemic arterial blood is usually below 100 mmHg (about 95 mmHg) even though it may be 100 mmHg in pulmonary capillary blood. The reason for this is that some venous blood mixes with arterial blood. There are two main sites for this physiological shunt:

(a) Some of the bronchial venous blood drains into the pulmonary veins. The deep bronchial veins eventually end in a main pulmonary vein or the left atrium. The superficial bronchial veins also communicate with the pulmonary veins.

(b) Some of the venous blood from the coronary circulation drains directly into the left side of the heart via venae cordis minimae, which drain into all four chambers of the heart.

2 Less than 2% of oxygen in arterial blood is found in dissolved form. At P_{O_2} of 100 mmHg, about 0.3 ml O_2 dissolves in 100 ml of blood. In venous blood (P_{O_2} of 40 mmHg) about 0.12 ml O_2/100 ml of blood is found in dissolved form.

Oxyhaemoglobin
Haemoglobin has great affinity for oxygen. It combines loosely and reversibly with oxygen by a process of oxygenation (not oxidation). The re-action is very fast, requiring less than 10 ms. The combination of oxygen with haemoglobin is a function of the partial pressure of oxygen, in a relationship outlined by the oxygen–

haemoglobin dissociation curve (Fig. 7.11). This relationship forms the basis of oxygen uptake in the lungs and its release to the tissues. At zero partial pressure of oxygen, the percentage satu-ration of haemoglobin with oxygen is zero, while at P_{O_2} of 100 mmHg the saturation is about 97.5%.

The oxygen dissociation curve (plotting per-centage saturation of haemoglobin with oxygen against partial pressure of oxygen) is character-istically sigmoid in shape (S-shaped).

Haemoglobin is formed of four haem units attached to polypeptide chains. Oxygenation of one haem unit leads to changes in the configur-ation of the haemoglobin molecule which increase the affinity of the second unit, and oxygenation of the second unit increases the affinity of the third and so on. This is the reason why the dissociation curve starts slowly but gains quickly, assuming the sigmoid shape.

The other characteristic of the curve is that there is a steep rise in the percentage saturation of haemoglobin between P_{O_2} of 0 mmHg and 75 mmHg. Beyond P_{O_2} of 75 mmHg there is a slow rise in the curve, becoming more or less flat at P_{O_2} of 80 mmHg or above.

The partial pressure of oxygen in arterial blood is about 95 mmHg, while in the tissues it drops to about 40 mmHg.

At P_{O_2} of 40 mmHg, haemoglobin is 75% saturated; at P_{O_2} of 95 mmHg, haemoglobin is 97% saturated. When 1 g of haemoglobin is fully saturated with oxygen, it binds up to 1.34 ml of oxygen at STP. Therefore, if haemoglobin con-centration is 150 g/litre of blood and arterial blood is 97% saturated, then the oxygen content of arterial blood will be about:

$$1.34 \times 150 \times \frac{97}{100} = 195 \text{ ml/litre of blood}$$

In venous blood, saturation of haemoglobin with oxygen is 75%. Therefore, the oxygen con-tent of venous blood will be approximately:

$$1.34 \times 150 \times \frac{75}{100} = 150 \text{ ml/litre of blood}$$

This means that the arteriovenous oxygen differ-ence is approximately 45 ml/litre of blood. Indeed, under resting conditions in a normal

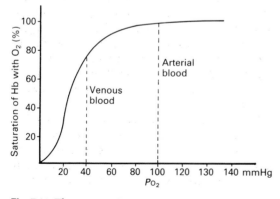

Fig. 7.11 The oxygen–haemoglobin dissociation curve.

person, oxygen uptake by the tissues is about 40–50 ml/litre of blood.

During exercise, however, oxygen expenditure by the active muscles is considerably increased. The P_{O_2} in the tissues may drop to 15 mmHg (or even lower), corresponding to a percentage saturation of haemoglobin of about 20%, or oxygen content of about:

$$1.34 \times 150 \times \frac{20}{100} = 40 \text{ ml/litre of blood}$$

The arteriovenous oxygen difference may then rise to about 150 ml/litre of blood during severe exercise.

It is evident that the quantity of oxygen carried in a volume of blood is dependent on the P_{O_2} as well as the haemoglobin concentration. The percentage saturation of haemoglobin with oxygen is dependent on P_{O_2} and totally independent of haemoglobin concentration. If oxygen content (instead of percentage saturation of haemoglobin with oxygen) is plotted against P_{O_2}, the level of the curve will be dependent on the haemoglobin concentration of the sample of blood (Fig. 7.12). But when plotting percentage saturation against P_{O_2}, as is usually done, the curve will always be the same, whatever the haemoglobin concentration is, if other factors remain the same.

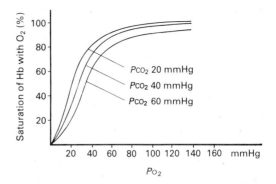

Fig. 7.13 The effect of P_{CO_2} on oxygen–haemoglobin dissociation curve (Bohr effect).

Factors which affect the oxygen–haemoglobin dissociation curve There are several factors that may affect the oxygen–haemoglobin dissociation curve. They may shift the curve to the right, indicating lowered affinity of haemoglobin for oxygen, or shift it to the left, indicating an increased affinity for oxygen.

1 *Partial pressure of carbon dioxide.* A rise in P_{CO_2} decreases the affinity of haemoglobin for oxygen and hence shifts the oxygen dissociation curve to the right (Fig. 7.13). This is called the *Bohr effect.* In the tissues, P_{CO_2} is high, and the more active the tissues are, the higher it will be. Therefore, more oxygen will be available for the tissues to meet this extra activity. In the lung, P_{CO_2} is lower than in the tissues—about 40 mmHg. The affinity of haemoglobin for oxygen in the lungs will be increased, leading to easier uptake of oxygen by haemoglobin in the lungs.

2 *pH.* Fall of pH (rise in hydrogen ion concentration) shifts the curve to the right, indicating a lowered affinity of haemoglobin for oxygen. During activity, pH of the tissues drops, due to the production of greater amounts of CO_2 and organic acids like lactic acid. This will facilitate the release of oxygen to the active tissues (Fig. 7.14).

3 *Temperature.* Rise of temperature also shifts the dissociation curve to the right. In active tissues, mainly due to oxidative processes, heat is generated and more oxygen is released from haemoglobin to provide the extra requirements of the tissues (Fig. 7.15).

4 *2,3-diphosphoglycerate (2,3-DPG).* 2,3-DPG is

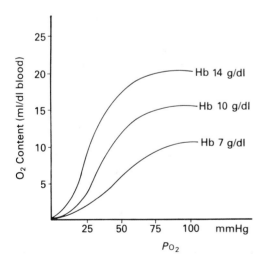

Fig. 7.12 Effect of anaemia on oxygen content of the blood at different P_{O_2} values.

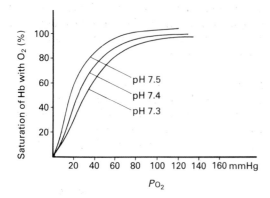

Fig. 7.14 The effect of pH on oxygen–haemoglobin dissociation curve.

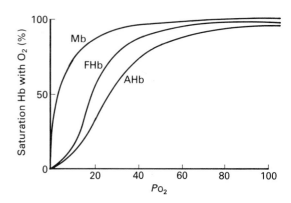

Fig. 7.16 The oxygen dissociation curves for: (a) adult haemoglobin (AHb); (b) foetal haemoglobin (FHb); and (c) myoglobin (Mb).

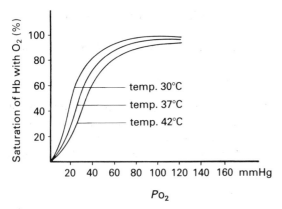

Fig. 7.15 The effect of temperature on oxygen–haemoglobin dissociation curve.

Myoglobin This is an iron pigment found in red muscle. Because it is formed of one haem unit, its oxygen dissociation curve does not exhibit a sigmoid shape; it is a rectangular hyperbola. Myoglobin has greater affinity for oxygen than adult or foetal haemoglobin. Its oxygen dissociation curve is further to the left than that of foetal haemoglobin (Fig. 7.16). Myoglobin releases its oxygen at a much lower P_{O_2} than haemoglobin. It is therefore very useful in muscles that are subjected to sustained contractions, during which blood supply is usually markedly decreased. Myoglobin will then provide the necessary oxygen supply for these muscles, where P_{O_2} may drop nearly to zero.

P_{O_2} in tissues during severe exercise drops considerably, to about 10 mmHg or less. This will result in greater release of oxygen from blood in tissues (Fig. 7.17).

Dissolved oxygen

There is a direct relationship between P_{O_2} and dissolved oxygen (Fig. 7.18). As oxygen is poorly soluble, within the normal values of P_{O_2} in blood, only a minute fraction of oxygen is transported in the dissolved form. In 100 ml of blood at body temperature, 0.003 ml of O_2 dissolves/1.0 mmHg P_{O_2}. Therefore, in arterial blood, dissolved oxygen does not normally exceed 0.3 ml/100 ml, and in venous blood it is about 0.12 ml/100 ml. The dissolved O_2 is in equilibrium with the O_2 combined with haemoglobin. It is

found in red blood cells bound to haemoglobin. It tends to decrease the affinity of haemoglobin for oxygen and therefore shifts the dissociation curve to the right. It helps the release of oxygen to the tissues. Its concentration in the red blood cells tends to increase in conditions of hypoxia at a high altitude.

Foetal haemoglobin has a greater affinity for oxygen than that of adult haemoglobin, probably because it binds less effectively with 2,3-DPG. Its oxygen dissociation curve is more to the left than that of adult haemoglobin, but, because it is formed of four haem units, it still posseses the sigmoid shape.

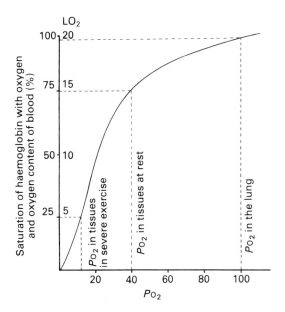

Fig. 7.17 Effect of severe exercise on P_{O_2} in tissues. Note the corresponding saturation of blood (25–75%), which indicates greatly enhanced oxygen release.

the dissolved O_2 which gets transferred to the tissues and becomes replaced from O_2 carried by haemoglobin.

Dissolved oxygen, although it constitutes less than 2% of the total oxygen transport in blood, is essential for tissues that do not have blood supply, like the cornea and cartilage, which depend on

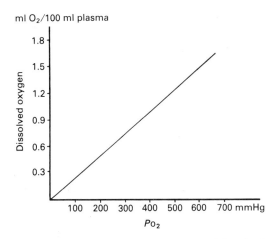

Fig. 7.18 The relationship between P_{O_2} and dissolved oxygen in plasma.

oxygen dissolved in the tissue fluids. Dissolved oxygen can be considerably increased by breathing pure or *hyperbaric oxygen*. This is part of the rationale for oxygen therapy.

CARBON DIOXIDE TRANSPORT

Carbon dioxide is transported from the tissues, where it is produced by metabolism, to the lungs, where it is unloaded to be removed to the atmosphere. It is transported by the plasma and by the red blood cells.

Transport of carbon dioxide by the plasma
Carbon dioxide produced by active cells diffuses by concentration gradient into the tissue fluid and the blood capillaries to reach plasma, where it is carried in several forms:
1 Some of it dissolves in the plasma and is transported in this form.
2 A small amount reacts with water to form carbonic acid:

$$CO_2 + H_2O \rightarrow H_2CO_3$$

The reaction proceeds very slowly because in the plasma there is no *carbonic anhydrase*, which catalyses the reversible reaction. The carbonic acid then dissociates into hydrogen ions and bicarbonate:

$$H_2CO_3 \rightleftharpoons H^+ + HCO_3^-$$

3 Carbon dioxide in the plasma reacts with the amino group of plasma proteins to form *carbamino proteins*.

Transport of carbon dioxide by the red blood cells
1 Most of the carbon dioxide that diffuses from the tissues into plasma enters the red blood cells and dissolves in the intracellular fluid.
2 Some of the dissolved CO_2 reacts with water to form carbonic acid. The reaction is a fast one due to the presence of *carbonic anhydrase* in red cells.

In the tissues, due to the high concentration of carbon dioxide, the reaction proceeds to the right, i.e. formation of carbonic acid. The acid formed dissociates into:

$$H_2CO_3 \rightleftharpoons H^+ + HCO_3^-$$

The bicarbonate ions accumulating in the red blood cell diffuse into the plasma by a concentration gradient. Since the main cation within the cell is potassium, which cannot freely diffuse out of the cell, electrical equilibrium is maintained by diffusion of chloride ions, the main extracellular anion, into the cell. This is called the *chloride shift*.

The hydrogen ions produced are buffered by haemoglobin which has unloaded its oxygen. The plasma bicarbonate is mainly the bicarbonate that diffuses out of the cell (Fig. 7.19).

3 A portion of the dissolved CO_2 reacts with the amino group of the reduced haemoglobin to form *carbamino haemoglobin.*

In summary, carbon dioxide is transported in the blood in three forms: dissolved form, as bicarbonate, and as carbamino compounds. Under resting conditions about 4 ml of CO_2 is taken up by 100 ml of blood draining the tissues and is consequently released in the lungs. Of this amount, approximately 2.8 ml (70%) is transported as bicarbonate, 0.8 ml (20%) as carbamino compounds and 0.4 ml (10%) in dissolved form.

Plasma transports more than 60% of the carbon dioxide which diffuses in the blood in the tissues, but the chemical reaction taking place in the red blood cells, due to the presence of the enzyme carbonic anhydrase, provides practically all of the bicarbonate ions transported in plasma.

Carbon dioxide dissociation curve
The volume of carbon dioxide carried in the blood is determined by the partial pressure of carbon dioxide (PCO_2). The dissolved form is directly proportionate to PCO_2 (0.06 ml dissolved in 100 ml of blood/1 mmHg PCO_2).

Figure 7.20 shows the relationship of the carbon dioxide content of blood to PCO_2 (carbon dioxide dissociation curve). The curve is affected by the saturation of haemoglobin with oxygen (Haldane effect). Oxyhaemoglobin shifts the curve to the right, i.e. in the lungs carbon dioxide is released from the blood. Reduced haemoglobin shifts the curve to the left, i.e. more carbon dioxide is taken up by the blood in the tissues. Thus, PCO_2 influences oxygen saturation of haemoglobin (Bohr effect) and PO_2 influences the CO_2 dissociation curve (Haldane effect). However, the Bohr effect is much more potent and more important than the Haldane effect.

Control of ventilation
Since the main functions of respiration are to supply oxygen to the tissues, to remove CO_2 from the tissues and to play a major role in the regulation of blood pH, it has to be capable of readjustment in states of hypoxia or hypercapnia or during metabolic disturbance of the acid–base balance. The readjustment is achieved by several mechanisms involved in the regulation of ventilation. These mechanisms can be grouped into two main categories, referred to as nervous and chemical control mechanisms, which are closely integrated.

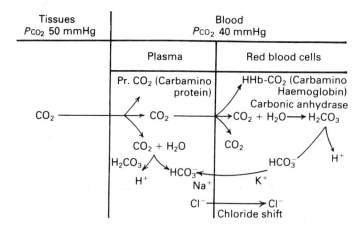

Fig. 7.19 The diagram illustrates the mechanism of carbon dioxide transport in blood, namely, as bicarbonate (70%), as carbamino compounds (20–25%) and as dissolved CO_2 (5–10%). It also shows the movement across the red cell membrane, which explains the chloride shift.

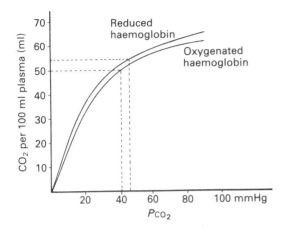

Fig. 7.20 The carbon dioxide dissociation curve with oxygenated and reduced haemoglobin (Haldane effect).

THE RESPIRATORY CENTRE

The respiratory centre is composed of several groups of neurones spread throughout the entire length of the medulla and pons. It can be divided into four major groups of neurones:

1 *The dorsal respiratory group* is located in the entire length of the dorsal aspect of the medulla. It comprises inspiratory neurones, which discharge rhythmically during resting and forced inspiration; hence the name *rhythmicity centre*. Thus, they are almost entirely responsible for inspiration.

2 *The ventral respiratory group* lies ventrolateral to the dorsal respiratory group along the entire length of the medulla. They are inactive during quiet breathing but become activated during increased pulmonary ventilation, as in exercise. They are mainly expiratory neurones with some inspiratory neurones, both of which are activated when expiration becomes an active process. At rest, when expiration is passive, they remain inactive.

3 *The apneustic centre* is situated in the lower pons. It sends excitatory impulses to the dorsal respiratory group of neurones and potentiates the inspiratory drive. It receives inhibiting impulses from the sensory vagal fibres of the Hering–Breuer inflation reflex and inhibitory fibres from the pneumotaxic centre in the upper pons. Severing of the vagal fibres and abolition of impulses from the pneumotaxic centre by mid-pontine section result in apneustic breathing (prolonged inspiration), while a section between the medulla and pons to remove the inspiratory drive of the apneustic centre results in gasping breathing, i.e. shallow inspiration and a prolonged period of expiration.

4 *The pneumotaxic centre* is located in the upper pons. It transmits inhibitory impulses to the apneustic centre and to the inspiratory area to switch off inspiration.

Both inspiratory and expiratory areas are influenced by impulses from the pneumotaxic and apneustic centres in the pons and indirectly by higher centres (Fig. 7.21). However, the dorsal respiratory group is most probably the integrating site of the various inputs (from chemoreceptors, baroreceptors, proprioceptors and lung stretch receptors) that reflexly affect breathing.

NERVOUS CONTROL OF VENTILATION

The rhythmicity centre sends excitatory impulses via the intercostal and phrenic nerves to the external intercostal muscles and the diaphragm, respectively. It receives impulses directly from higher brain centres, from other centres in the brain stem and from special receptors which give rise to respiratory reflexes.

Higher brain centres

1 Impulses from the cerebral cortex bring about voluntary modification of breathing. This is indicated by the ability to hyperventilate voluntarily and by temporary cessation of breathing (voluntary apnoea). One can only modify breathing voluntarily but cannot completely control it.

2 Impulses from the cerebellum co-ordinate breathing with other activities such as swallowing, talking and coughing.

3 Impulses from the hypothalamus and the centres for emotion and temperature regulation modify breathing. Breathing is modified during emotional stress, such as fear, excitement or anger, and by change in body temperature. Panting in dogs (shallow hyperventilation) is a mechanism for dissipation of heat from the body, controlled by hypothalamic centres.

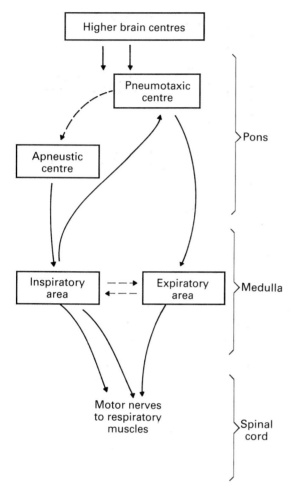

Fig. 7.21 Diagrammatic representation of the localization of respiratory neurones in the medulla and pons. The solid arrows indicate stimulation and the broken arrows indicate inhibition.

Centres in the medulla and pons
1 The rhythmicity centre is interconnected with the cardiac and vasomotor centres located in the medulla oblongata. Impulses radiate between the three centres.
2 The rhythmicity centre receives excitatory impulses from the apneustic centre located in the lower part of the pons. This area tends to produce deep inspiration and to stop breathing in inspiration when stimulated. Arrest of breathing in inspiration is known as apneusis.
3 The pneumotaxic centre is located in the upper

part of the pons. The main role of the pneumotaxic centre is to inhibit inspiration. When it is active, inspiration is shallow, and when it is less active the depth of breathing is greater. It acts directly on the inspiratory centres to suppress their discharge. On the other hand, the pneumotaxic centre has a suppressing effect on the apneustic centre. Figure 7.22 is a schematic diagram illustrating the arrangement of the various inputs controlling respiration.

Special receptors
The rhythmicity centre also receives nervous impulses from peripheral parts of the body which control ventilation reflexes. The following are the main examples of receptors associated with modification of breathing:
1 Sensory vagal fibres arising from stretch receptors in the smooth muscles of bronchi and bronchioles send inhibitory impulses to the apneustic centre, which accordingly modulates its discharge to the rhythmicity centre. When the lungs are inflated during inspiration the stretch receptors are stimulated, and an increasing volley of impulses is sent up to the apneustic centre to inhibit it, and inspiration ultimately comes to an end. This reflex mechanism (Hering–Breuer inflation reflex) protects the lungs from over-inflation. Cutting vagal fibres arising from the lungs prolongs inspiration. The breathing becomes deep and slow. It has also been demonstrated that there are vagal fibres arising from the lungs which are stimulated by deflation. They tend to bring expiration to an end, and allow inspiration to start. This reflex is therefore referred to as the Hering–Breuer deflation reflex. It is a much weaker reflex than the inflation reflex.
2 It has also been found that active or passive movement of muscles and joints will stimulate breathing, even when their blood supply is cut off, while their nerve supply is left intact. These are proprioceptive impulses arising from muscle and joint receptors all over the body, including the diaphragm, which is rich in muscle spindles. Proprioceptive stimulation of breathing may be the main stimulus for ventilation at the onset of muscular exercise.
3 Stimulation of the skin receptors by noxious

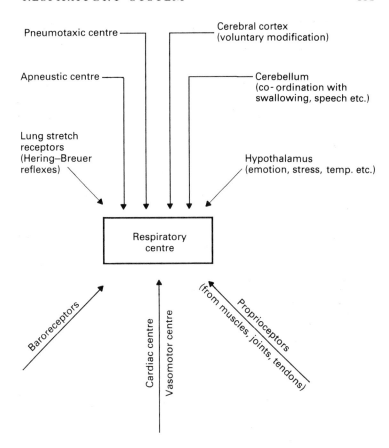

Fig. 7.22 Diagrammatic representation of the various central and peripheral inputs to the medullary respiratory centre.

stimuli may also stimulate ventilation. Stimulation of baroreceptors in the aortic arch and carotic sinuses may give rise to reflex modification of breathing.

CHEMICAL CONTROL OF VENTILATION

The rhythmicity centre is affected by chemical changes in the blood via two types of chemoreceptors: peripheral and central (Fig. 7.23).

Peripheral chemoreceptors

Peripheral chemoreceptors are located mainly in the carotid and aortic bodies, but may be found anywhere in the circulatory system. When stimulated, the carotid bodies and aortic bodies send excitatory impulses to the rhythmicity centre, via the glossopharyngeal and vagus nerves, respectively.

Peripheral chemoreceptors are highly sensitive to changes in arterial P_{O_2} and, to a lesser extent,

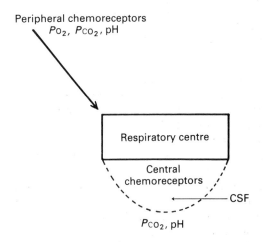

Fig. 7.23 Chemical regulation of ventilation.

to P_{CO_2} and pH. A fall in P_{O_2}, together with a rise of P_{CO_2} and a fall in pH, stimulates the chemoreceptors, bringing about reflex increase of ventilation. When arterial levels of P_{O_2}, P_{CO_2} and pH are normal, there is a low grade of tonic activity in the nerves arising from the peripheral chemoreceptors. With low P_{CO_2} and high pH, the tonic activity will be decreased and ventilation is depressed. When pH falls due to causes other than respiratory, e.g. metabolic acidosis of diabetes mellitus, there will be reflex hyperventilation to wash out CO_2 and bring the pH up to the normal range. Metabolic alkalosis (alkalosis not due to a respiratory cause, e.g. due to excessive vomiting or ingestion of large quantities of bicarbonate) depresses ventilation. Carbon dioxide will be retained in the blood, and eventually will compensate for the alkalosis. Voluntary hyperventilation will wash out CO_2, lead to respiratory alkalosis and increase the period of voluntary apnoea. Breathing high concentrations of CO_2 results in high arterial P_{CO_2} and low pH and therefore stimulates ventilation.

Central chemoreceptors
Experimental findings confirm the presence of chemoreceptors other than the carotid and aortic bodies. They also indicate the nature of the chemical changes in the blood that stimulate these receptors and bring about ventilatory responses. Denervation of the carotid and aortic bodies will abolish ventilatory responses to hypoxia (oxygen lack), while responses to high P_{CO_2} and low pH persist. This confirms the presence of other chemoreceptors. Lack of oxygen with denervation of peripheral chemoreceptors depresses ventilation by its direct effect on the respiratory centre. This suggests that hypercapnia (high P_{CO_2}) stimulates ventilation by its direct effect on the respiratory centre. This also indicates that the only mechanism by which hypoxia stimulates ventilation is by its effect on the peripheral chemoreceptors. There is greater ventilatory response to hypercapnia than to metabolic acidosis for equivalent changes in arterial pH, indicating that the effect of CO_2 is separate from the influence of hydrogen ion concentration. Assuming that the hydrogen ion is the unique stimulus for these central chemical

receptors, these observations also suggest the presence of a diffusion barrier between the blood and the central receptors (blood–brain barrier), a barrier that is freely permeable to CO_2 but poorly permeable to hydrogen and bicarbonate ions.

These central chemoreceptors are most probably located on the ventrolateral surfaces of the medulla oblongata (maybe in other sites as well), which are bathed with cerebrospinal fluid. They are highly sensitive to the hydrogen ion concentration of the cerebrospinal fluid evoked by arterial P_{CO_2}, since CO_2 can freely cross the blood–brain barrier into the cerebrospinal fluid while the barrier is relatively impermeable to hydrogen and bicarbonate ions.

Hypoxia
Hypoxia refers to conditions in which there is deficient oxygen supply to the tissues. Disturbances which interfere with oxygen uptake in the lungs, its transport in blood and its delivery to the tissues may lead to hypoxia. Accordingly, hypoxia can be divided into four main categories:
1 *Hypoxic hypoxia* results from any disturbance which interferes with normal oxygenation of the arterial blood, leading to low arterial P_{O_2}. It can therefore be due to: (a) factors in the atmosphere, such as high altitude; (b) factors interfering with normal O_2 diffusion in the lung; or (c) mixing of arterial blood with venous blood as in venoarterial shunts as in congenital heart disease. In hypoxic hypoxia, arterial P_{O_2} will be low and the percentage saturation of haemoglobin with oxygen in arterial blood will be low. Consequently, the oxygen content per unit volume of blood will be low.

Clinical causes of hypoxic hypoxia also include ventilation defects, e.g. paralysis of respiratory muscles, airway obstruction and poisoning with drugs that inhibit the respiratory centre, e.g. morphine and barbiturates. Hypoxia due to hypoventilation is always associated with high arterial P_{CO_2}.

Shunting of venous blood can also occur when there is wasted perfusion, i.e. when a larger number of alveoli do not get adequate ventilation. This leads to a large amount of venous blood passing through the lungs without being oxygenated.

Diffusion defects can occur in conditions which cause thickening of the pulmonary membrane, e.g. lung fibrosis, oedema and consolidation, as in pneumonia. In this case, there would be wasted ventilation.

Therefore, the concept of the *ventilation/perfusion ratio* ($\dot{V}/\dot{Q}$; where $\dot{V}$ = alveolar ventilation and $\dot{Q}$ = cardiac output, both in litres/min) is most important in the causation of hypoxic hypoxia due to pulmonary or cardiac disease. In a situation where there is wasted ventilation, $\dot{V}/\dot{Q}$ will be more than 1.0. In this case, the physiological dead space is increased because some ventilated alveoli have inadequate perfusion. In conditions where there is wasted perfusion, certain alveoli, although well perfused, do not get adequate ventilation. In this case oxygen therapy does not give the desired effect.

2 *Anaemic hypoxia.* This is due to lowering of the oxygen-carrying capacity of the blood. It may therefore be caused by anaemia, by abnormal haemoglobins or by rendering haemoglobin unavailable for oxygen uptake, as in carbon monoxide poisoning.

In this type of anaemia the arterial P_{O_2} is normal; the percentage saturation of haemoglobin with oxygen is normal, except in carbon monoxide poisoning. However, the oxygen content of arterial and venous blood per unit volume of blood is lowered, because there is insufficient haemoglobin to carry it.

3 *Stagnant hypoxia.* In this type of hypoxia there is slow or no circulation to the tissues. It may be a localized disturbance of circulation to a certain part of the body, e.g. the limbs, or it may be generalized or central, as in heart failure.

In this type of hypoxia, arterial P_{O_2} is normal, percentage saturation of haemoglobin in arterial blood is normal and O_2 content per unit volume of arterial blood is normal. The only disturbance can be seen in the venous blood, where P_{O_2}, percentage saturation of haemoglobin and oxygen content are all lowered significantly.

4 *Histotoxic hypoxia.* The disturbance is in the uptake of oxygen by the tissues, either due to poisoning of cellular enzymes or due to a long diffusion distance between the capillaries and cells, as may occur in tissue oedema. In this type of hypoxia, all parameters in the arterial blood are normal. As there is little or no utilization of O_2 by the tissues, the P_{O_2}, percentage saturation of haemoglobin and oxygen content of venous blood will all be higher than normal.

Table 7.4 shows blood gas findings in the various types of hypoxia.

Hypoxia is best treated with oxygen therapy and by correcting the underlying cause.

Cyanosis

Cyanosis is a blue coloration of the skin and mucous membranes. It starts to show when the level of reduced haemoglobin in capillary blood is more than 5 g/100 ml of blood. Reduced haemoglobin is dark in colour and, when its concentration exceeds 5 g/100 ml, its colour can overshadow the colour of oxygenated haemoglobin, no matter what the total haemoglobin concentration is.

Cyanosis can be localized (peripheral) or it can be generalized (central).

Table 7.4 Comparison of blood gas findings in the various types of hypoxia

Type of hypoxia	Arterial blood			Venous blood		
	P_{O_2}	% Sat	% O_2	P_{O_2}	% Sat	% O_2
Hypoxic	Low	Low	Low	Low	Low	Low
Anaemic	Normal	Normal (except CO poisoning)	Low	Low	Low	Low
Stagnant	Normal	Normal	Normal	Low	Low	Low
Histotoxic	Normal	Normal	Normal	High	High	High

% Sat, % saturation of haemoglobin; % O_2, O_2 content in ml/dl blood.

The peripheral type is most commonly seen at the tips of fingers when the person is exposed to severe cold. Cold leads to vasoconstriction, which will decrease the blood flow through peripheral parts. This allows more time for extraction of oxygen from the blood, which results in a higher level of reduced haemoglobin.

The generalized type of cyanosis is due to a central cause, such as venoarterial shunts, or severe hypoxia, as may occur in the new-born or in persons exposed to atmospheric air at very high altitudes.

Since cyanosis occurs when the capillary blood contains more than 5 g/100 ml of reduced haemoglobin, it is more commonly seen with polycythaemia and very rarely with anaemia, especially severe anaemia. It is therefore more evident in conditions associated with polycythaemia, e.g. polycythaemia vera, high altitudes or congenital heart diseases with right-to-left shunts, and in the new-born.

The role of the respiratory system in acid−base balance

DEFINITIONS AND NORMAL VALUES
The normal range of arterial pH is 7.35−7.45, with an average of 7.4. Acidosis is defined as any condition in which arterial pH is less than 7.35, while alkalosis is defined as any condition in which arterial pH is more than 7.45. Values of pH below 7.0 or above 7.8 are incompatible with life.

The pH of the blood is mainly determined by the ratio of concentration of bicarbonate to carbonic acid, i.e. the ratio of salt to acid as given by the Henderson−Hasselbalch equation:

$$pH = pK + \log \frac{salt}{acid}$$
pK for carbonic acid $= 6.1$

In normal persons, the concentration of bicarbonate is about 24 mmol/litre, while that of carbonic acid is approximately 1.2 mmol/litre, giving a ratio of 20 : 1:

$\log 20 = 1.3$
$pH = 6.1 + 1.3 = 7.4$

The concentration of carbon dioxide in the body fluids is determined by the rate of CO_2 production in the tissues and the rate of its removal by the lungs. There is a time-lag between its production and removal, leading to its accumulation in the body fluids to a concentration of about 1.2 mmol/litre under resting conditions. If the rate of production in the tissues, as in very severe exercise, exceeds the rate of removal by the lungs, its concentration in the body fluids will increase.

It is evident that, if the ratio of salt to acid (in this case bicarbonate to carbonic acid) increases, either by increase of the salt or decrease of the acid, pH will rise, and vice versa. For example:
1 In the case of hyperventilation, CO_2 is washed out, leading to lowered carbonic acid concentration in the blood. The ratio of bicarbonate to carbonic acid will therefore increase and pH will rise. This is referred to as respiratory alkalosis.
2 Conversely, if CO_2 is retained in the body, as in respiratory failure or impaired removal of CO_2 for any reason, the ratio of bicarbonate to carbonic acid will decrease and pH will consequently fall. This is referred to as respiratory acidosis.

The body, however, tends to resist extreme changes in pH. In the case of respiratory acidosis, the kidney will excrete more acid urine in the form of non-carbonic acids, e.g. acid phosphate, and retain more bicarbonate, thus raising the level of bicarbonate in the blood. This will approximate the bicarbonate/carbonic acid ratio towards the normal ratio; therefore, salt and acid are raised. This is a form of compensatory mechanism. The reverse happens in respiratory alkalosis, i.e. the kidney reabsorbs less bicarbonate and secretes fewer hydrogen ions.

When the acidosis is not due to carbonic acid, it is referred to as metabolic acidosis, as in diabetes mellitus, where acidosis is due to excess keto acids, or in cases of severe diarrhoea, where there is excessive loss of bicarbonate and potassium coming from pancreatic and intestinal secretions.

When there is excess bicarbonate in the blood as a result of excessive ingestion, or if there is excessive loss of hydrochloric acid in persistent vomiting, metabolic alkalosis will result.

In conclusion, the role of the lung in pH regulation depends on the sensitivity of the respiratory centre to CO_2 and H^+. Metabolic acidosis stimulates respiration and consequently more CO_2 is washed out of the blood and the pH rises. On the other hand, alkalosis of metabolic origin depresses respiration and therefore accumulation of CO_2 brings the pH back to normal. These responses represent respiratory compensatory mechanisms in the regulation of the acid–base balance.

Investigation of pulmonary disease

When investigating lung disease, a certain protocol is usually followed:

1 History taking and clinical examination.
2 Radiological examination.
3 Laboratory tests of lung function.
4 Other laboratory tests and procedures, e.g. bacteriological examination of blood, sputum or pleural fluid.

These tests help to locate and identify the pathology of the lesion. However, in this text only the tests for lung function will be considered. The question arises: 'Why do we perform lung function tests?'

1 They help in the diagnosis and management of patients with lung disease.
2 They are useful in preoperative assessment of fitness for general anaesthesia and surgery.
3 They assist in the follow-up of the progress of disease and therapy.
4 They are utilized in epidemiological surveys, e.g. industrial hazards and occupational lung diseases.
5 They are used for insurance purposes and when assessing compensation for disability.
6 They are useful in clinical research.

THE SCOPE OF LUNG FUNCTION TESTS

1 They supplement, but do not replace, other clinical diagnostic procedures.
2 They cannot tell where the lesion is and much less what the lesion is. They can make neither a clinical nor a pathoanatomical diagnosis. For example:

(a) They can reveal impaired diffusion across the alveolar–capillary membrane but cannot distinguish between interstitial oedema and intra-alveolar fluid accumulation. Nor can they distinguish between reduced alveolar surface area and thickened alveolar–capillary membrane.

(b) They can reveal the existence of right-to-

left shunt but cannot indicate whether the shunt is intracardiac or intrapulmonary.

LUNG FUNCTION TESTS

The lung function tests used for the diagnosis, management and prognosis of pulmonary diseases basically assess the following:

1 Ventilatory functions.
2 Gas diffusion (across the pulmonary membrane).
3 Elastic properties of the lungs.
4 Airways resistance.
5 Ventilation/perfusion relationship.

Ventilatory functions

Measurement of lung volumes and capacities (see Fig. 7.1) Using a bell-type spirometer, the subject breathes normally and records tidal volume. Following a normal inspiration, the subject is asked to take a deep inspiration. This records the inspiratory reserve volume (IRV). Following a normal expiration, the subject is again asked to make a deep expiration. This records the expiratory reserve volume (ERV). The lung volumes and capacities have been explained at the beginning of this chapter under the heading Ventilation.

Forced vital capacity The simplest and the easiest test of ventilatory functions is the *forced expiratory volume* (FEV). The subject is asked to breathe out forcibly after a maximal inspiration and the volume expired is measured. The volume expired in the first 1 sec of the expiration is designated FEV_1 and the total volume expired is the *forced vital capacity* (FVC). It is called forced vital capacity because of the forced expiration.

The volume exhaled can be measured by recording the volume expired against time, using a bell-type spirometer. The record obtained is called a spirogram (Fig. 7.24).

In a normal subject the FEV_1 is about 4.2 litres and FVC about 5.0 litres. This means that FEV_1 is 84% of the FVC (Fig. 7.24 – A). In general, this ratio is always more than 80%. An FEV_1 less than 80% of FVC is abnormal (Fig. 7.24 – B).

FEV_1, FVC and FEV_1% give very useful information for the assessment of obstructive and

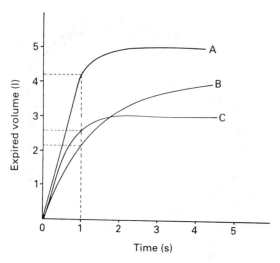

Fig. 7.24 Spirograms recorded by a 'vitalograph' for three patients. The maximum volume expired is the FVC for each patient. A: FVC = 5 l; FEV_1 = 4.2 l; FEV_1 = 84% – normal; B: FVC = 4 l; FEV_1 = 2.2 l; FEV_1 = 55% – obstructive lung disease; C: FVC = 3 l; FEV_1 = 2.7 l; FEV_1 = 90% – restrictive lung disease.

restrictive impairments, e.g. as are seen in patients suffering from bronchial asthma, chronic bronchitis and emphysema (obstructive lung diseases) and from fibrotic conditions of lungs, including pneumoconiosis (restrictive lung diseases).

Another important piece of information can be obtained from this spirogram. The time taken for the expiration of FVC is not 2 s (Fig. 7.24 – B) but 4 s. This is due to the increased resistance offered by the bronchi to the flow of air, because of their narrowing, which is the primary pathology in obstructive lung diseases. The measurement of time for a particle of gas to flow out of the lungs is called transit time. The mean transit time (MTT) in normal persons is always less than 3 s. Measurement of MTT is suggested by some to be a sensitive indicator of airways disease.

Figure 7.24 – C shows the spirogram of another patient. The FEV_1 in this patient is 2.7 litres, FVC is 3.0 litres and FEV_1% is 90%. This is typical of restrictive lung disease. In this patient the FVC is comparatively less. Although FEV_1 is less than the normal, FEV_1% of the FVC is normal.

FEV_1 and FVC measurements are repeated in patients suspected of obstructive lung disease

after giving the patient a bronchodilator (salbutamol 1% for 3 min through a nebulizer). Improvements in both FEV_1 and FVC indicate reversibility and confirm the occurrence of bronchospasm in these patients.

Peak expiratory flow rate (PEFR) Another important forced expiratory manoeuvre, which is rapid and can be performed at the bedside, is the measurement of PEFR. The subject, after making a full inspiration, exhales forcibly through a flowmeter (e.g. Wright flow meter, Fig. 7.25). In normal persons with a maximum effort, the PEFR goes up to 8–12 litres/s. In athletes and trained young men it can go as high as 20 litres/s. PEFR is decreased in patients suffering from bronchial obstruction. If airway obstruction is suspected, reversibility is tested by repeating the test after giving a bronchodilator. If reversibility is positive, airway spasm is confirmed. A PEFR record of the patient gives useful information on the progress or deterioration of the patient's condition.

Forced expiratory flow 25–75 (FEF25–75) This pulmonary function index is calculated from the spirogram obtained by using the bell-type spirometer. The patient, after full inspiration, forcibly exhales. The volume exhaled is recorded against time on a moving drum, in the same way as done to record the FEV. The middle half of the total spirogram (25–75%) is marked out and the time taken for this volume to be exhaled is measured (Fig. 7.26). The FEF25–75 is the volume divided by time in seconds. Normally FEF25–75 is 3–5 litres/s.

There is generally good correlation between FEV_1 and FEF25–75 in patients suffering from obstructive lung diseases. FEF25–75 has the advantage of avoiding measurements during the effort-dependent first quarter of FVC. Like other spirometric indices, FEF25–75 is modified by age, sex and body size.

Flow–volume loops Special electronic equipment is required to record the flow volume loops. The subject performs the same forced expiratory manoeuvre after full inspiration as for the FEV. Instead of using the bell-type spirometer, the flow and volume are simultaneously recorded on the

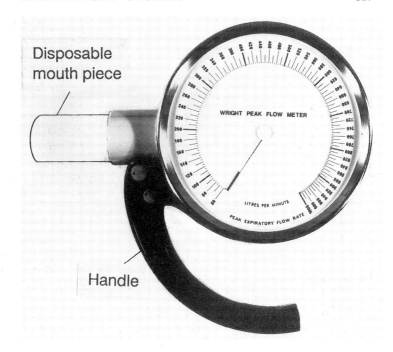

Fig. 7.25 Wright peak flow meter.

$x-y$ recorder. Usually flow is recorded along the y axis and volume along the x axis (Fig. 7.27). An important feature of the flow–volume loop is that it is impossible to get outside the loop no matter how much and how long the expiratory effort. It also shows the inspiratory loop, which is often not shown in spirograms.

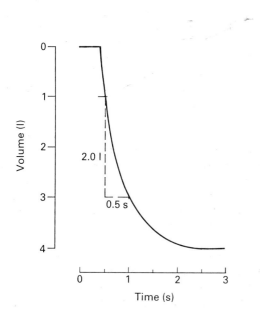

Fig. 7.26 Peak expiratory flow 25–75 recorded by a bell-type spirometer. FEF25–75 in this case = 4 l/s.

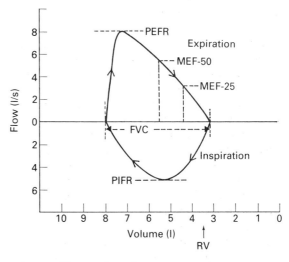

Fig. 7.27 Flow–volume loop in a normal subject. FVC = forced vital capacity; PIFR = peak inspiratory flow rate; PEFR = peak expiratory flow rate; MEF50 = maximum expiratory flow at 50% of FVC–normal = 4–6 l/s; MEF25 = maximum expiratory flow at 25% of FVC–normal = 2–4 l/s; RV = residual volume.

Recording of flow–volume loops is one of the more recent ventilatory function tests, which is routinely used in most pulmonary investigation hospital laboratories. It measures the following:

1 Forced vital capacity expiration (FVC).
2 Vital capacity inspiration (FCI).
3 Peak expiratory flow rate (PEFR).
4 Maximum expiratory flow at 50% of FVC (MEF50). Normal value: 4–6 litres/s.
5 Maximum expiratory flow at 25% of FVC (MEF25). Normal value: 2.4 litres/s.
6 Peak inspiratory flow rate (PIFR).
7 Maximum inspiratory flow rate at 50% of vital capacity.
8 Maximum inspiratory flow rate at 25% of vital capacity.

MEF50 and MEF25 are effort-independent flow rates. In obstructive airway diseases these parameters are significantly decreased (Fig. 7.28). In restrictive lung diseases the loop becomes elliptical. The FVC and FEV_1 are decreased. All the flow parameters are decreased but their ratio to the FVC does not change.

Inspiratory flow loops are useful in upper airways obstruction such as tracheal obstruction, laryngeal obstruction due to inflammation or neoplasm or obstruction of the larynx by inflammation of the epiglottis. In all these conditions the inspiratory loops are flattened.

In upper airways obstruction the expiratory loops are also flattened. This reduces the utility of the inspiratory loops in diagnosis as compared with the expiratory loops.

Patients who record obstructive types of flow–volume loops are given bronchodilators (salbutamol 1% for 3 min) and repeat the test to see reversibility.

MEF50 and MEF25 reduction in the presence of normal FVC and FEV_1 also indicates small or peripheral airways obstructive disease.

Test of uneven ventilation and closing volume
Uneven ventilation is best tested by the radioactive xenon technique. This test requires a gamma camera and radioactive xenon; it is used for research and a very few selected patients. The routine test is the single-breath nitrogen test. The advantage of this test is that, besides measuring the uneven ventilation, the closing volume can also be measured simultaneously. To perform this test, pure oxygen, a rapid nitrogen analyser and a recorder are used; the latter records nitrogen concentration against lung volume. The subject takes a full inspiration of pure oxygen, followed by slow expiration (the flow should not be more than 0.5 litre/s). The rapid nitrogen analyser analyses the concentration of nitrogen near the mouth. Four phases are recorded (Fig. 7.29). Phase 1 is expired from the anatomical dead space so it contains pure O_2. It is very small. The 2nd phase is also short. In this phase nitrogen concentration rises rapidly as the dead space and alveolar gases

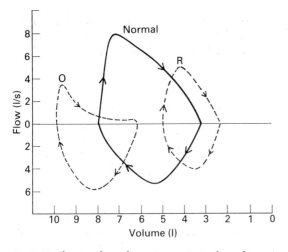

Fig. 7.28 Flow–volume loops in restrictive lung disease (R) and obstructive lung disease (O), compared with a normal subject.

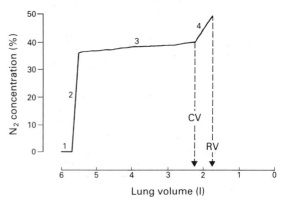

Fig. 7.29 Single-breath N_2 method. CV, closing volume; RV, residual volume.

are expired. Phase 3 is a plateau of N_2 concentration because it is the alveolar gas. Phase 4 shows a rapid rise in N_2 concentration. This signals the onset of closure of airways.

Phase 3 in normal subjects is almost flat. In patients with uneven ventilation, this phase rises steadily and the slope is the measure of inequality of ventilation. It is expressed as the percentage increase in nitrogen per litre of expired volume (0.70% increase per litre).

In patients with weak airways due to destruction of lung parenchyma, as in emphysema and obstructive airway disease, the closing of the airways takes place early and the closing volume, which is the volume of phase 4, is increased.

Helium dilution technique (closed circuit method) (Fig. 7.30) This method is used for measurement of total lung capacity (TLC) and residual volume. Helium is poorly diffusible through the alveolar capillary membrane. The subject rebreathes into a spirometer containing a known volume of a known concentration of helium in oxygen for a few minutes. The final concentration of helium in the spirometer is measured by a helium analyser.

Example
If the initial volume of helium–oxygen mixture = 2 litres,

Initial concentration of helium = 8%,
Final concentration of helium = 4%,
TLC to be measured = x litres

$$2 \times \frac{8}{100} = (x + 2) \times \frac{4}{100}$$
$$16 = 4x + 8$$
$$4x = 8$$
$$x = 2 \text{ litres}$$

The residual volume (RV) can be obtained by subtraction (TLC−FEV = RV).

In obstructive lung disease the TLC and RV are normal or higher than normal, but in restrictive disease TLC is decreased. In emphysema the RV is much higher than normal.

Body plethysmography (body box) (Fig. 7.31) The techniques for measurement of residual volume and functional residual capacity assume free communication of all parts of the lung with the atmospheric air for efficient and complete dilution and wash-out of either helium or nitrogen. If there is obstruction to any part of the lung, these techniques will give low values. To obtain true values of all the air in the lungs, whether or not they are in free communication with the atmosphere, the body plethysmograph is used. The principle used is based on Boyle's law for gases which relates gas pressure to gas volume:

$$P^1 V^1 = PV$$

The subject sits within the box and the door is tightly closed. The subject breathes air around him through a mouthpiece mounted with a

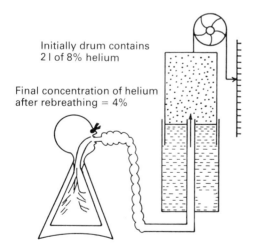

Initially drum contains 2 l of 8% helium

Final concentration of helium after rebreathing = 4%

Fig. 7.30 Helium dilution technique using the closed-circuit method.

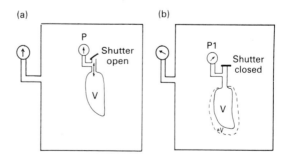

(a) (b)

P   P1

Shutter open   Shutter closed

V   V

Fig. 7.31 Sequence of events in body plethysmography (body box). (a) Before the mouthpiece shutter is closed. (b) Inspiration with the mouthpiece shutter closed.

shutter that is electrically controlled. At the end of normal expiration (when measuring the functional residual capacity), the shutter is closed and the subject inspires against this obstruction. Figure 7.31 demonstrates the events before and after closing the mouthpiece shutter; at the end of expiration the intra-alveolar pressure, P, is atmospheric as there is no gas flow. The thoracic gas volume (V) at the end of expiration, or functional residual capacity, is unknown. The mouthpiece shutter is then closed and the subject inspires against the obstruction. The new thoracic volume will then become $\Delta V + V = V^1$. The pressure of the new gas volume, V^1, recorded in the mouthpiece at the end of the inspiration, will drop to P^1.

The increase in thoracic volume ΔV is determined by the rise in the plethysmographic pressure, which is calibrated with volume change.

Since P, P^1 and ΔV are known, V (the functional residual capacity) can be worked out from Boyle's law:

$$PV = P^1 (\Delta V + V)$$

The total lung capacity is measured, using the helium dilution technique, and is usually measured simultaneously with the single-breath method used for calculation of the gas transfer factor ($TLCO$) (see below).

Gas diffusion

Diffusion capacity is defined as the total volume of a gas that diffuses through the lung alveolar–capillary membrane per unit time per unit partial pressure difference of the gas. It is recorded in ml/min/mmHg partial pressure (see Table 7.3).

The rate of transfer of gas across the pulmonary membrane depends upon the surface area of the pulmonary membrane, partial pressure difference across the pulmonary membrane, the pulmonary membrane thickness, the molecular weight and the solubility coefficient of the gas. All these factors operate while diffusion studies are conducted. Other factors which offer serious limitations to the study of O_2 diffusion across the pulmonary membrane are the pulmonary blood flow and the haemoglobin concentration in the blood. In order to minimize the effects of these latter factors, O_2 is not used for the study of

diffusion; instead, carbon monoxide (CO) is used, because CO transfer is only diffusion-dependent and is not affected by perfusion.

The diffusion capacity of lungs is measured by determining the CO transfer factor ($TLCO$):

$$TLCO = \frac{\text{Amount of CO breathed per minute}}{\text{Mean alveolar CO partial pressure}}$$

Two techniques are used for TLCO determination:

1 *Single-breath technique.* The subject takes a full inspiration of gas mixture (10% helium, 0.3% CO in air) and holds his breath for 10 s and then exhales. The first 1 litre or so is discarded (dead space volume) and, out of the remaining, 1 litre is collected (alveolar air) and analysed for CO and He concentrations. From the difference of helium concentration in the inspired and expired air and from the volume of air inspired, the total lung capacity (TLC) is calculated:

$$C_1 \times V_1 = C_2 \times (V_1 + V_2)$$

where C_1 and V_1 are concentration of He and volume of inspired air, C_2 is the concentration of He in expired air, and V_2 is the total lung capacity. From the TLC and concentration of CO in inspired and alveolar air, the TLCO is calculated.

Breath-holding may be difficult for some patients. In such cases, the steady-state technique is used.

2 *Steady-state technique.* The principle of the technique is the same as for the single breath. In this technique the subject breathes 0.25% CO and 6% He in air steadily until the rate of CO uptake becomes constant. When this steady state is reached, the alveolar air is collected and analysed and TLCO calculated as for the single-breath method. On average, the TLCO in young adults is 25 ml/min/mmHg. This can be changed into O_2 diffusing capacity by multiplying the TLCO by a factor of 1.23.

TLCO is the total amount of CO transferred per minute. As the surface area of the pulmonary membrane is different in different individuals and in some pulmonary diseases, the transfer coefficient (KCo) is calculated, which corrects the effects of the size of the lungs and gives the diffusing capacity per litre of lung volume:

$$KCo = \frac{TLCO}{\text{Alveolar volume}}$$

Arterial blood gases Full investigations of a pulmonary patient are never satisfactory without the determination of the arterial blood gases. Arterial blood is taken by arterial puncture, e.g. the radial, brachial or femoral artery, with a syringe in which the dead space is filled with heparin. The blood obtained is immediately analysed in an arterial blood gas analyser. If the blood cannot be analysed promptly after collection, it must be kept in ice-water at 4 °C for not more than half an hour. In no case should it be kept for more than an hour before analysis.

The most important outcome of pulmonary diseases is hypoxia. One of the important causes of hypoxia is hypoventilation, which is well illustrated by the CO_2 gas equation:

$$P_{CO_2} = \frac{\dot{V}CO_2}{VA} k$$

where $\dot{V}CO_2$ is CO_2 output, VA is alveolar ventilation and k is a constant. This relationship shows that, if alveolar ventilation is halved, P_{CO_2} is doubled. Similarly, the relationship of pH to P_{CO_2} can determine the degree of hypoventilation. A crude but reliable relationship has been observed. For every 10 mmHg rise in P_{CO_2}, the pH drops roughly by 0.1. Thus, from looking at pH only, one can guess the degree of hypoventilation if the drop in pH is entirely due to hypoventilation.

Determination of arterial blood gases before and after exercise or after treatment with 100% O_2 can give a fairly good idea about other causes of hypoxia. If, after moderate physical exercise, hypoxia is increased, it indicates either a diffusion defect or ventilation/perfusion imbalance or both. Hypoxia due to diffusion defects and hypoventilation are easily corrected by 100% O_2 therapy. If P_{O_2}, after therapy, fails to rise to normal levels, hypoxia is most probably due to a shunt.

Test for elastic properties of the lungs
Compliance is the change in volume of the lung per unit change in pressure. In order to measure compliance, one has to know the intrapleural pressure. Direct measurement of the intrapleural pressure is difficult. Instead, intraoesophageal pressure (which is not identical to the intrapleural pressure but reflects intrapleural pressure changes fairly well) is measured.

The subject swallows a small balloon at the end of a rubber tube which is placed half-way between the pharynx and the stomach. The balloon is filled with a few millilitres of air. The rubber tube is connected to a pressure-measuring device. The subject inspires air from a spirometer to total lung capacity and expires in steps of 500 ml through a valve connected with an automatic shutter. After each step the lungs are relaxed with the glottis open and the oesophageal pressure is measured. The pressure measured is plotted against the volume to get the pressure–volume curve. The slope of the curve represents the compliance (see Fig. 7.7).

One of the factors which affect the compliance of the lungs is the lung size. Compliance is directly related to the size of the lungs. For this reason the compliance per unit volume of lung, which is known as the *specific compliance*, is calculated. Large lungs show large compliance; if the elastic properties of the lungs are affected, the specific compliance of the same large lung is decreased.

Sometimes compliance is measured during inspiration and expiration by measuring the oesophageal pressure at the end of inspiration and expiration. This measurement, which is known as dynamic compliance, is sensitive and decreases significantly in patients with increased airway resistance.

Test of airway resistance
Airway resistance is measured by finding the pressure difference between the mouth and the alveoli and the rate of air flow, using the pressure, resistance and flow relationship expressed by Ohm's law:

$$\text{Airway resistance} = \frac{\text{Mouth pressure} - \text{alveolar pressure}}{\text{Flow of air}}$$

Flow of air and mouth pressure are easy to measure, but the measurement of alveolar pressure requires the use of the body plethysmograph. The subject sits in the body box connected to the pneumotachograph head, which measures the airflow. A side tube from the mouthpiece is connected to a transducer, which measures instantaneous mouth pressure. Before inspiration the pressure in the box is atmospheric. When the subject inspires, the alveolar pressure falls as the gas expands due to change in volume of the chest. This increases the pressure in the box, from which the change in the volume of the lungs can be determined. The change in the volume of the lungs due to inspiration can be converted to alveolar pressure change, using Boyle's law, if the initial lung volume is known.

Test of ventilation/perfusion relation
Unequal ventilation can be tested by the nitrogen wash-out test described earlier. The distribution of blood in the lungs is tested by injecting albumin labelled with radioactive iodine or technetium. The radioactivity of the lung is measured by a gamma camera. The photograph of radioactivity shows directly the distribution of blood in the lungs. Instead of radioactive albumin, radioactive xenon dissolved in saline can also be used. From the numerical data of radioactivity, the results can be expressed as blood flow per unit volume of the lungs.

The ventilation can be measured by using radioactive xenon gas. The subject takes a single breath or series of breaths of radioactive xenon, after which the radioactivity is measured by a gamma camera, and the results are expressed as ventilation per unit lung volume.

Further reading

1 Guyton, A.C. (1991) Respiration. In *Textbook of Medical Physiology*, 8th edn, pp. 402–462. Saunders, Philadelphia.
2 Levitzky, M.G. (1990) *Pulmonary Physiology*, McGraw-Hill, New York.
3 West, J.B. (1990) *Respiratory Physiology–The Essentials*, 4th edn, Williams & Wilkins, Baltimore.

8: Nutrition and Metabolism

Objectives

On completion of the study of this section, the student should be able to:

1 Understand the functions of essential nutrients required for maintenance of health and identify their sources.

2 Become familiar with major diseases caused by nutritional deficiencies so as to further his knowledge and skills in their diagnosis and management during his clinical studies.

3 Understand the concept of the metabolic rate so as to explain variations of energy expenditure and to describe the energy requirements of individual patients.

4 Comprehend the basic metabolic concepts which underlie different physiological states so as to predict nutritional needs.

5 Describe the effect of injury and infection on nutritional requirements, qualitatively and quantitatively, so as to be able to institute appropriate nutritional management.

6 Appreciate the importance of nutritional considerations in the management of disease during drug therapy and convalescence.

Introduction

Maintaining the human body in a constant state of health requires a continuous supply of nutrients. Some of these nutrients are consumable, i.e. they are used up. Examples of these are fats and carbohydrates, which are mainly catabolized to provide energy, normally in the form of adenosine triphosphate (ATP), for work as well as for vital life processes at the cellular level (e.g. ion transport, maintenance of nerve potentials and chemical synthesis). Fats (triacylglycerols) and carbohydrates (glycogens) are also forms of energy storage in the body, which are mobilized when required. Other nutrients, such as proteins, are essential components of the structure of living cells. Many important enzymes and hormones are made up of protein or polypeptide molecules.

To a certain extent, these sources of energy are interchangeable, but excess quantities of fat cannot be oxidized without the provision of minimal amounts of carbohydrate or protein.

In adults and in children, cells are continuously renewed and in this process some protein losses are incurred. On the other hand, growing children and pregnant mothers are forming new cells for new tissues and growth of organs. Following injury or disease, cells which have been damaged will have to be replaced. Proteins are made of amino acids; however, not all amino acids can be synthesized in the human body (i.e. essential amino acids). It is clear, therefore, that a continuous supply of dietary protein is important to supply the essential amino acids. It is also pertinent to note that some specialized carbohydrates and lipid molecules are components of cell membranes and some hormones and enzymes.

In addition to the above, the body requires a supply of minerals and trace elements, which take part in the structure of specialized molecules in the human body. We also need a continuous supply of a group of substances known as vitamins. These substances, more often than not, cannot by synthesized in the human body and must be present in the diet we consume. They play numerous biochemical roles and their deficiencies cause disease conditions (such as corneal opacity or xerophthalmia, rickets, beriberi and scurvy) characterized by tissue damage, failure of growth and impairment of specific metabolic processes.

A proportion, possibly large, of any meal can be described as inert, having no nutritive value. This is because it is not completely, or only very incompletely, absorbed. Many plant polysaccharides, such as cellulose and other woody material (lignin), fall into this category since they are not attacked by the host's digestive enzymes, even though some digestion by bacterial enzymes occurs. However, recent evidence suggests that these dietary fibres (e.g. cellulose) are of considerable value in the diet. How and why does fibre in the diet exert its beneficial effects? This is not entirely clear, but it is established that molecules of the fibre polysaccharide are strong adsorbing agents. They adsorb large quantities of water and also possibly toxic molecules which may be produced by the flora of the digestive tract. The adsorption of large amounts of water is also very important, and this leads to a much larger faecal mass and therefore easier transit through the intestines and easier expulsion.

Nearly all food contains small quantities of toxic substances. They can be added during processing, for example, to preserve the food or gain entry to the food incidentally, for example during crop spraying or treatment with fertilizers.

Water is not strictly a nutrient but it is required as a solvent for body fluid components. Water intake is vital for replacement of losses through the skin and urinary system. Water is also a by-product of metabolism. A balance between water intake and water loss is essential for maintenance of the constancy of the internal environment.

The delicate balance is of particular importance in hot environments, where evaporative water losses in sweating are an essential mechanism for heat dissipation.

This chapter deals with the nutrients and their sources, physiological roles and requirements. The main focus is to correlate the physiological state with variations in requirements and adjustments in underlying metabolism, so as to understand the changes which occur under physiological conditions, stress and relevant clinical situations.

Physiology of feeding

Food intake is regulated over long periods of time with such precision that a person can maintain his/her body-weight for 20 years or more within a few kilograms. The precision of this regulation becomes clear if it is realized that during the same period the individual would have consumed at least 4 tons of the essential nutrients, not considering the water intake. If his/her variation of body-weight was within 4 kg, then the precision of the control system is 10^{-3}. These calculations are based on a diet giving about 10 MJ/day (2400 kcal) and consisting of about 400 g carbohydrate, 60 g fat and 80 g protein.

APPETITE AND HUNGER

Appetite can be defined as the desire to eat. It may be influenced by the taste and smell of food and may be greater for some food items than others. There is reason to believe that appetite is increased for substances deficient in the body at any one time. An example of this is salt appetite under conditions of excessive sweat losses of NaCl. There is evidence in animals and in man that appetite for foods already eaten during a meal is less than that for new food items. Preference for new items is probably a biological mechanism to ensure that a variety of foods is consumed. In prosperous societies, however, it can lead to overeating due to availability of a large variety of food items.

The feeling of hunger is one of the important physiological stimuli for the appetite. The feeling of satiety is experienced after feeding. There are neurones in the hypothalamus which have been designated as the *feeding centre* in the lateral hypothalamus and as the *satiety centre* in the ventromedial nucleus. In hungry animals these neurones are stimulated by the sight and smell of food and they also display specific responses to different food items. Stimulation of the feeding centre neurones results in promotion of eating behaviour, while injury or damage results in severe loss of appetite (anorexia). On the other hand, stimulation of the satiety centre results in loss of appetite and its destruction causes overeating (hyperphagia). The latter is the underlying pathology in what is known as hypothalamic obesity. There is evidence that the hyperphagia of hypothalamic obesity is partly explained by hypersecretion of insulin, due to stimulation of the vagi during the cephalic phase of gastro-intestinal stimulation.

Hypoglycaemia is usually associated with hunger and hyperglycaemia with satiety. However, hunger can occur with hyperglycaemia, as in diabetes mellitus. A theory which has been advanced for the regulation of feeding is known as the glucostatic hypothesis. The cells in the satiety centre have a high rate of glucose utilization and, unlike other parts of the nervous system, their glucose uptake can be influenced by insulin. When their glucose utilization is high, satiety is experienced; but, when utilization falls, hunger is felt. Thus, a high arteriovenous concentration difference of glucose (high utilization) is associated with satiety, while a low concentration difference is associated with hunger. This hypothesis explains the presence of hunger with hyperglycaemia in diabetes. In this case, there is decreased utilization of glucose by the neurones in the satiety centre. However, the precise mechanism for the maintenance of body-weight is not fully understood. There seems to be a relationship between a set point for body-weight and feeding behaviour in the normal control mechanism.

The nutrients

SOURCES AND PHYSIOLOGICAL ROLES
The main nutrient sources of energy are carbohydrates and fats. Although proteins can also be used to supply energy, it is good dietary planning to avoid this. Proteins are more expensive than carbohydrates. It is important to ensure that chil-

dren receive adequate energy in order to use the dietary protein for growth. Excess energy intake is the main cause of overweight and all the attendant diseases associated with it, such as diabetes mellitus, hypertension and coronary heart disease.

CARBOHYDRATES
The body tissues require a constant daily supply of carbohydrates, in the form of glucose, in all metabolic reactions. Comparatively little is stored. Approximately 100 g of glycogen is stored in the liver and about 225 g in the muscles, and there is about 10 g of glucose in the blood. For most individuals, the supply available from storage depots would be insufficient for one day's need. The principal function of carbohydrate is to serve as a major source of energy for the body. When metabolized, carbohydrates yield 4 kcal or 17 kJ gram.

Carbohydrates are available in the form of starches and sugars. Starch is contained in cereals, roots and stems (Table 8.1). Pulses and seeds also have appreciable amounts of carbohydrate (see Table 8.2). Sugar is found in sugar-cane and sugar-beet. Sugar-cane contains 10% sugar (sucrose) while sugar-beet roots contain 6% sugar (sucrose and fructose). Refined sugar intake is blamed for the causation of overweight, dental caries, hyperlipidaemia and atherosclerosis. However, the evi-

Table 8.1 Carbohydrate contents of cereals and cereal foods

	Percentage by weight
Wheat	69
Sorghum (durra)	74
Rice	87
Millet (*Pennisetum* species)	75
Maize	64
Barley	69
Cassavas	35
Potatoes	18
Sweet potatoes	28
Yams	27
Bread	45–60
Porridge (sorghum species)	17
'Kisra' (sorghum species)	42

Table 8.2 Carbohydrate contents some pulses and seeds

	Percentage by weight
Beans	54
Lentils	57
Peas	50
Soya beans	24
Ground-nuts	19

dence for the latter is inconclusive. Fruits contain various amounts of sugar. Fructose and sucrose are contained in many fruits (see Table 8.3).

FATS

Butter, animal fat and vegetable oils are almost pure fat; they provide the highest amount of energy per gram (9 kcal = 38 kJ), which is more than twice the amount of energy supplied by each gram of carbohydrate. Because of the high energy content and low solubility of fats, they are used as a storage form of energy. Not only ingested fat but carbohydrate and amino acids not immediately used by the tissues are converted to fat and stored in the adipose tissue. Fat sources provide 5–20% of the energy content of diets in the poorer parts of Africa, Asia and Latin America. They may provide up to 80% of the energy intake in more affluent societies. This is simply explained by the fact that fats and oils are expensive and their intake tends to be influenced by the income in various communities. Excessive fat consumption, especially of the saturated types, results in hyperlipidaemia and the risks of atherosclerosis, coronary heart disease and hypertension. Some plant foods with a high fat content are shown in Table 8.4.

Energy can also be obtained from ingested alcohol (7 kcal = 30 kJ/g).

PROTEIN

The primary structure of proteins is the number, type and order of amino acids. Proteins vary in size from relatively small polypeptides to very complex molecules with several hundred thousand amino acid units. There are nine amino acids that are classified as essential, since they must be supplied in the food. Body synthesis of these amino acids is lacking or so limited as to be insufficient to meet metabolic needs (see Table 8.5). Histidine was first found to be required by infants; however, recent work in humans suggests that it may also be essential for adults. Without an adequate supply of the essential amino acids, protein cannot be synthesized or body tissue maintained.

The other amino acids, which can be syn-

Table 8.4 Examples of plant food with high fat content

	Percentage by weight
Ground-nuts	44
Sesame	51
Soya beans	24

Table 8.3 Fructose and sucrose contents various fruits

	Percentage by weight
Apples	15
Apricots	6
Bananas	16
Dates, fresh	38
Dates, dried	63
Grapes	15
Guavas	16
Mangoes	16
Oranges	12
Pears	27

Table 8.5 Essential and non-essential amino acids

Essential amino acids	Non-essential amino acids
Valine	Glycine
Lysine	Alanine
Threonine	Serine
Leucine	Cystine
Isoleucine	Tyrosine
Tryptophan	Proline
Phenylalanine	Hydroxyproline
Methionine	Citrulline
Histidine	Arginine
	Glutamine
	Aspartate
	Glutamate

thesized by the body in adequate amounts for normal function, are termed non-essential (see Table 8.5). This is not to suggest that these amino acids are not essential constituents of the proteins, but rather that it is not essential to include them in the diet since the tissues can synthesize their own supply from carbohydrate, fat and other amino acids.

The main sources of animal protein are meats (mainly muscle). The protein contents of some animal foods are shown in Table 8.6. All meats have large amounts of water, from 60 to 75% of their weight. Dried meat and dried fish have very little water and therefore their protein content is very high, up to 70%. The protein contents of some commonly used pulses and legumes are shown in Table 8.6.

Although the protein content of cereals is low compared with other seeds, it is important to note that large amounts of cereals are usually consumed for their carbohydrate content. Therefore, the amount of protein consumed in cereal foods constitutes a significant amount of the total daily protein intake. This is of particular importance in poorer parts of the world, where 80–90% of the energy comes from cereal foods. It

has been estimated that in some developing countries 90% of the protein intake comes from staple cereal foods.

Proteins of plant origin, however, are deficient in one or more of the essential amino acids. They are therefore of lower quality than animal proteins. Amino acid (or chemical) scores can be calculated for the most deficient (or most limiting) amino acid to describe the quality of any protein source. The amino acid score is calculated as follows:

Amino acid score =

$$\frac{\text{mg of amino acid in 1 g of test protein} \times 100}{\text{mg of amino acid in 1 g of reference protein}}$$

This score is a crude way to evaluate the percentage of adequacy of the protein because it does not take into account the digestibility of the protein, the availability of the amino acids, the utilization of those amino acids by the human body or the ability of that protein to support cellular synthesis. However, it can be used to calculate dietary protein requirements of a mixture of proteins compared with a reference protein; for example, if egg or milk protein requirement is 30 g/day and the amino acid score of the dietary proteins available is 60%, then the required amount of this mixture would be $[(30 \times 100)/60] = 50$ g/day. As will appear later in this chapter, other factors also influence the requirements of dietary protein.

VITAMINS

In addition to adequate energy and protein intake, there are certain substances which are required in minute amounts to achieve normal growth of the young and to maintain health in adults. These substances, which are known as vitamins, mainly act as coenzymes or a prosthetic group of enzymes responsible for promoting essential chemical reactions in tissue metabolism. They are often called accessory food factors in view of the fact that they do not supply calories or contribute appreciably to body mass. The exact mode of action of some vitamins is not yet clear but clinical conditions due to deficiency are well known. Vitamins regulate metabolism, help convert fat and carbohydrate into energy and assist in the formation and maintenance of bones and

Table 8.6 Protein contents of some animal and plant foods

	Percentage by weight
Animal foods	
Beef	18–32
Veal	21
Lamb	18–29
Pork	13–33
Poultry	17–30
Fish	17–19
Egg	13
Milk (cow)	3.5
Plant foods	
Beans (various types)	24
Peas	22
Soya beans	37
Chick-peas	19
Ground-nuts	26
Sesame	20
Cereals	7–16

tissues. As these substances cannot by synthesized in the body, a constant supply of dietary vitamins is essential. Table 8.7 gives a list of the main vitamins required in human diets and their sources.

Vitamins are complex chemical substances with known chemical structure and most of them have been synthesized. Specific biochemical roles have been described for most vitamins. For specific biochemical actions, a textbook of biochemistry should be consulted. The following is a brief description of their roles in metabolism and associated deficiencies.

Water-soluble vitamins

Vitamin B₁ The active form, thiamine pyrophosphate, is an important coenzyme in mammalian tissues and required in the oxidative decarboxylation of pyruvate, oxoglutarate and 2-oxo acids derived from methionine, threonine, leucine, isoleucine and valine. In addition, it acts as a coenzyme for the transketolase reaction in the pentose phosphate pathway. Thiamine is needed for the metabolism of carbohydrates, fats and proteins. However, all the evidence from the effects of thiamine deficiency link it with disturbance of carbohydrate metabolism, especially in the brain. Thiamine requirement is linked to carbohydrate intake. Its deficiency causes dry or wet beriberi and gastrointestinal disturbances.

Vitamin B₂ This vitamin maintains the health of mucous membranes. It is required for the synthesis of two coenzymes: flavin mononucleotide (FMN) and flavin adenine dinucleotide (FAD). These, in turn, form the prosthetic groups of several different enzyme systems (the so-called flavoproteins), which catalyse oxidation–reduction reactions in the cells and function as hy-

Table 8.7 Water-soluble and fat-soluble vitamins required in the human diet and their sources

	Sources
Water-soluble vitamins	
Vitamin B₁ (thiamine)	Thiamine is mainly found in the germ of cereals and seeds. Yeast, liver and meats are rich sources
Vitamin B₂ (riboflavine)	Milk, organ meats, eggs, cereals
Vitamin B₆ (pyridoxine)	Cereals, liver, yeast
Nicotinamide (or niacin)	Meats, liver, pulses
Pantothenic acid	Liver, kidney, egg yolk, milk
Biotin	Egg yolk, liver, kidney, meat
Vitamin B₁₂ (cobalamin)	Liver, meat
Folic acid (folacin)	Green leafy vegetables, meat, liver
Vitamin C (ascorbic acid)	Green peppers, chili (fresh), guavas, citrus fruits and parsley
Fat-soluble vitamins	
Vitamin A (retinol)	Milk and butter are good sources; also liver, kidney, egg yolk
Carotene (vitamin A precursor)	A yellow pigment found in vegetables and fruits. It can be converted in the body to retinol. One microgram of retinol equals 6 μg of carotene
Vitamin D (calciferol)	A rich source is fish-liver oil. Diet is not an important source of vitamin D. Most of it is synthesized in the skin from 7-dehydrocholesterol (provitamin D₃) by the effect of ultraviolet light. The liver hydroxylates the vitamin to 25-hydroxycholecalciferol and the kidney makes the active form 1,25-dihydroxycholecalciferol under the effect of parathyroid hormone. For this reason, vitamin D₃ is now viewed as a prohormone
Vitamin E (α-tocopherols)	Vegetable oils, meats, eggs, leafy vegetables
Vitamin K (several naphthoquinones)	Fresh leafy vegetables, liver, fish-liver oil

drogen carriers in the mitochondrial electron transport system. They are also coenzymes of several dehydrogenases and enzymes of oxidative deamination of amino acids. Vitamin B_2 deficiency leads to inflammation of the tongue (glossitis) and ulcerations at the angle of the mouth (angular stomatitis).

Vitamin B_6 This is rapidly converted in the body into the coenzymes pyridoxal-5'-phosphate and pyridoxamine phosphate. They form the prosthetic groups of enzyme−coenzyme systems in the pathways of energy production and amino acid and fat metabolism. In addition, vitamin B_6 catalyses the synthesis of tryptophan and its conversion into nicotinamide. It is also concerned with the synthesis of haemoglobin. It is therefore important for growth. Deficiency causes anaemia, irritability and convulsions. Deficiency in some patients can be associated with oral contraceptives and some drugs.

Nicotinamide This is necessary for formation of nicotinamide adenine dinucleotide (NAD) and nicotinamide adenine dinucleotide phosphate (NADP) (oxidative coenzymes). Deficiency causes pellagra, with skin, neurological and gastrointestinal manifestations. Its role in the prevention of pellagra is not fully understood.

Pantothenic acid This is incorporated in coenzyme A (CoA) and is thus essential in the intermediary metabolism of carbohydrate, fat and protein. No deficiency has been described in humans. It is widespread in foods and is synthesized by intestinal bacteria. There is no known deficiency disease.

Biotin This forms part of several carboxylase enzymes. Absorption is inhibited by raw egg white (avidin). It is synthesized by intestinal bacteria and daily requirements are minute. It is an essential coenzyme for fatty acid synthesis and oxidation. Its deficiency results in lethargy, loss of appetite, nausea, dermatitis and muscle pain.

Vitamin B_{12} This vitamin is essential for normal function in the metabolism of all cells, especially for those of the gastrointestinal tract, bone mar-

row and nervous tissue, and for growth. It is involved in protein, fat and carbohydrate metabolism. It is necessary for the maturation of red blood cells and maintains the myelination in the nervous system. It is the only vitamin not yet synthesized. Failure to absorb vitamin B_{12} because of the absence of the intrinsic factor in the gastric secretion (e.g. due to surgical resection of the intrinsic factor-secreting portions of the stomach or the absorbing surfaces of the ileum) results in a deficiency state. The latter produces a megaloblastic maturation of the red cells in the marrow, with a resulting macrocytic anaemia and a leucopenia with hypersegmented polymorphonuclear leucocytes.

Folic acid This is important in the synthesis of DNA, purines, haem and some amino acids; hence it has an important role in the growth and maturation of red and white blood cells. Deficiency causes poor growth, megaloblastic anaemia and sprue.

Vitamin C This vitamin functions as a cofactor or a coenzyme and is required for a number of hydroxylation and oxidative reactions. It is required for the production and maintenance of collagen, a protein substance found in all fibrous tissues (connective tissue, cartilage, bone matrix, tooth dentin, skin and tendon). The integrity of cellular structure depends on it. It maintains the health of capillaries via the synthesis of collagen; the latter promotes healing of wounds, fractures and bleeding gums and decreases liability to infection. Vitamin C also facilitates removal of iron from ferritin. Its deficiency causes scurvy.

Fat-soluble vitamins

Vitamin A Retinol is required for the formation of the pigment rhodopsin (visual purple), which is needed for vision in dim light. The vitamin is also required for normal growth and for maintenance of healthy epithelial tissue. In addition, vitamin A also appears to play a role in fertility and various metabolic pathways (e.g. cholesterol synthesis, mucopolysaccharide synthesis). Deficiency causes night-blindness and xerophthalmia, which may progress to corneal opacity and blindness.

Vitamin D The active form of vitamin D_3 (1,25-dihydroxycholecalciferol) is essential for normal growth and development and is important for the formation of normal bones and teeth. Along with parathyroid hormone and calcitonin, vitamin D has an important role in the maintenance of appropriate serum levels of calcium and phosphorus to support normal mineralization of bone. The active form of vitamin D_3 acts on the gut to promote absorption of calcium and phosphate and on bone to increase calcium and phosphate mobilization. Its deficiency causes rickets and osteomalacia.

Vitamin E This vitamin acts on the germinal epithelium, and deficiency causes infertility, red cell fragility and muscular dystrophy in animals. It protects cell membrane lipids from oxidation, thus maintaining low cellular peroxide levels. In man, deficiency can develop in association with defective lipid absorption or transport. Selenium also acts as an antioxidant and therefore decreases vitamin E requirements. Deficiency takes months to develop because vitamin E stores are high.

Vitamin K Normal fat absorption is necessary for vitamin K absorption. Adequate intake is required for synthesis of prothrombin and other coagulation factors (VII, IX and X) in the liver. Deficiency is characterized by tendency to bleed. Patients with biliary obstruction, which leads to failure of fat absorption, may develop vitamin K deficiency and therefore tend to bleed. Such patients require vitamin K injections before surgery. A water-soluble analogue (K_3) is available for clinical use. Vitamin K also plays a significant role in calcification.

MINERALS

Calcium and iron have well-known roles in human and animal nutrition. Other minerals are needed in varying amounts. Those required in minute quantities (less than 100 mg/day) are referred to as trace elements. Some trace elements have recently attracted attention and acquired significance in human nutrition. The following account describes the main sources and physiological roles of minerals and trace elements.

Calcium

Calcium is the most abundant mineral in the human body and is required throughout life, but especially during periods of growth, pregnancy and lactation. The main source of calcium in the human diet is milk (160 mg/100 g) and cheese (400–1200 mg/100 g). Smaller amounts of calcium are also found in cereals and seeds. Sesame is a rich source, containing about 1200 mg/100 g. High phytic acid, oxalic acid and fibre content in the diet interferes with absorption. Calcium is absorbed mainly in the duodenum and jejunum, because of the lower pH of intestinal contents. Vitamin D is essential for its absorption, which is also enhanced by dietary protein and the presence of lactose.

Calcium is required for mineralization of bone. It is also an important constituent of body fluids, where it mediates several physiological functions in the extracellular fluid as well as intracellularly. These functions are discussed in relation to the physiology of the different organs. The calcium turnover is shown in Fig. 8.1. The regulation of the plasma calcium level will be discussed with parathyroid function (Chapter 10).

Phosphorus

Milk and its products are the main source of phosphorus in the human diet. Cow's milk has a higher phosphorus concentration than human milk. When diluted, cow's milk has a lower calcium/phosphorus ratio than human milk. This can lead to lower calcium absorption in infants fed on diluted cow's milk. Its absorption is in the jejunum and is affected by the same factors as calcium.

Phosphorus is needed in the formation of bone. Its reabsorption in the kidney is reciprocally related to that of calcium. A constant Ca : P ratio in tissue fluids is maintained by parathyroid hormone acting on bone and kidney. Hyperphosphataemia lowers the Ca : P ratio and can lead to tetany. Failure to reabsorb phosphorus in the renal tubules leads to hypophosphataemia, which can cause osteomalacia and rickets. Phosphorus plays

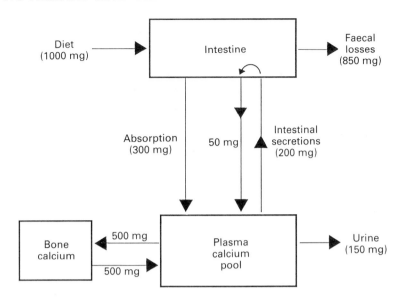

Fig. 8.1 An overview of calcium turnover and exchange in the body.

important roles in energy metabolism (ATP), in phosphate buffers and in phospholipids.

Iron

The adult male and non-menstruating female loses approximately 1.0 mg/day of iron. Of this amount, 50–60% is lost from the gastrointestinal tract and excreted in the bile or lost by shedding of mucosal cells. The remaining 40–50% is lost from the skin and the hair and in the urine. Iron is obtained from meat, cereals, pulses and vegetables. Red meats contain 2–3 mg/100 g in the form of haem. White meats contain 0.5–1 mg/100 g and liver has 7–10 mg/100 g. 'Hilba' (fenugreek seed) has a very high content – 22 mg/100 g of dry seeds – and pulses have 2–9 mg/100 g. Iron is absorbed mainly in the duodenum and upper jejunum in the ferric state. The oxidation of ferrous to ferric iron is accomplished by means of vitamin C, HCl and an NADH-dependent flavoprotein. Absorption depends on a protein receptor, apoferritin, found in the intestinal mucosa. The rate of absorption seems to be controlled by the amount of iron accepted by the intestinal mucosa in response to the body's requirement. The latter is reflected by the amount of available iron in the blood. There is no specific mechanism for adjusting the rate of excretion of iron; therefore, the iron content of the body is regulated by the rate of absorp-

tion. Normally only about 10% of dietary iron is absorbed. The absorption of iron may be interfered with in the presence of phytic acid and oxalate, which are usually found in whole cereal flour and other plant foods. That is why iron from plant sources (non-haem iron) is poorly absorbed.

Iron is required for synthesis of haemoglobin, myoglobin and the cytochrome oxidase enzymes. Haemoglobin contains about 67% of body iron, and about 30% is found in the storage forms of ferritin and haemosiderin. Figure 8.2 gives an overview of iron turnover.

Zinc

Zinc is available in meats, cereals, shellfish, kidney and liver, which are rich sources. The rate of absorption by the small intestine is somewhat related to zinc status, being greater than normal in zinc deficiency. The availability of dietary zinc is also a factor in determining absorption.

Zinc is a constituent of many metalloenzymes, such as carbonic anhydrase, superoxide dismutase, carboxypeptidases of the pancreatic juice, alkaline phosphatase and nerve growth factor (NGF). It is important for the synthesis of nucleic acids and protein. It is also required in the production of insulin and maintenance of normal immunocompetence.

Only about 20% of dietary zinc is absorbed in the intestine. Absorption from plant foods is poor,

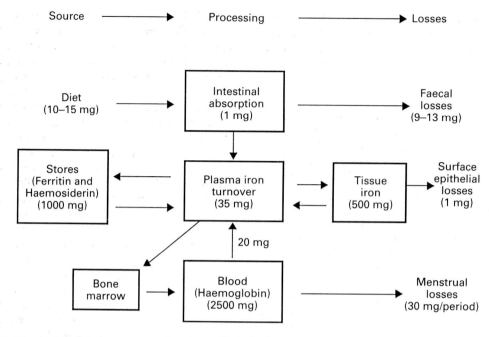

Fig. 8.2 An overview of iron turnover and exchange in the body.

especially if the diet has a high phytic acid content. Zinc is carried in the plasma bound to an α-2-macroglobulin, which binds about two-thirds, while albumin binds almost one-third. Only 2–5% zinc is free. Meats, grains and legumes are good dietary sources. Apart from pancreatic and biliary excretion, zinc is lost in sweat, hair and skin, as well as urine. Lactation imposes additional losses as does the transfer of zinc to the foetus during gestation. Figure 8.3 gives an overview of zinc turnover.

Serum zinc is decreased in hypoalbuminaemia and in patients with liver disease or nephrotic syndrome and patients receiving total parenteral nutrition (TPN). Patients susceptible to the development of zinc deficiency are those who are in a rapid growth period: infants, adolescents and pregnant mothers. Acute zinc deficiency results

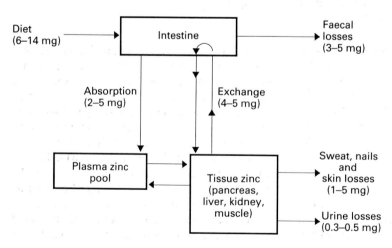

Fig. 8.3 An overview of zinc turnover.

in loss of appetite and dysfunction of taste and smell. Chronic zinc deficiency is characterized by retarded growth, anaemia, impaired wound-healing, hepatosplenomegaly and hypogonadism.

Chromium

Trivalent chromium is biologically active and can be obtained from all plant and animal foods. Vegetable oils, margarine, meat and milk products are good sources. The body contains about $100-200$ μmol of chromium. Following absorption, chromium is probably transported on transferrin. This trace element seems to be of importance in nutrition in relation to maintenance of glucose tolerance. It is a constituent of the glucose tolerance factor (GTF), the exact nature of which is not yet certain. This factor, together with insulin, is responsible for the entry of glucose into cells. If administered, it reverses glucose intolerance in animals fed on chromium-deficient diets. There is also evidence that some children with protein energy malnutrition (PEM), who develop glucose intolerance, can be treated by chromium chloride therapy. In animals, chromium deficiency results in diabetes mellitus, retarded growth and atherosclerosis. Patients on TPN can develop glucose intolerance, which responds to chromium therapy.

Cobalt

Meat, seafood, cereals and grains are good dietary sources of cobalt. It is the trace metal found in vitamin B_{12} (cobalamin). In the plasma, inorganic cobalt is distributed attached to albumin; 14% is deposited in the bone, 43% in muscle and smaller portions in other tissues, especially the kidney. Excretion of inorganic cobalt is mainly in the urine, the rate of which is thought to be important in the maintenance of cobalt homoeostasis. Cobalt is required in ruminants for bacterial synthesis of vitamin B_{12}. Man and non-ruminants need ready-made B_{12}. There is no known dietary deficiency of this trace element *per se*.

Copper

Copper is widely distributed in plant and animal foods. Vegetables, fish and liver are particularly good sources. Nutritional copper deficiency is rare. Milk is a poor source and therefore children may develop anaemia and neutropenia if copper deficiency occurs. This is particularly likely to happen in cases of chronic diarrhoea or malabsorption syndrome. Intake is about $2-4$ mg/day. The total amount of copper in the body is about 2 mmol. The absorption of copper and particularly its transfer from the intestinal mucosa into the blood is a regulated process influenced by changes in the physiological state (e.g. oestrogen levels, presence of cancer). Evidence from studies on rats suggests that the concentration of metallothionein in cells of the intestinal epithelium determines how much of the entering dietary copper is free to proceed into the blood or must stay behind attached to this high-cysteine protein. In the blood, copper is initially bound to albumin and transcuprein (a transport protein) and carried to the liver, where it is either incorporated into ceruloplasmin and other specific hepatic proteins or lost via the bile. Ceruloplasmin is secreted into the plasma and, apart from its possible enzymatic functions, transports the copper to cells throughout the body with the help of albumin and transcuprein. Figure 8.4 gives an overview of copper turnover.

Copper is a constituent of many enzymes (e.g. cytochrome oxidase). Its deficiency leads to inability to utilize iron stored as ferritin in the liver. Serum copper is normally increased during pregnancy and in women on oral contraceptives. This is due to an increase in the plasma α-2-globulin (ceruloplasmin), which binds 90% of plasma copper.

Selenium

Selenium is a trace element found in cereals and meat. Fish is particularly rich in selenium. Daily intake is $60-150$ μg/day. It is essential in human and animal nutrition because it is a component of the enzyme glutathione peroxidase and possibly some other enzymes. It is also a component of teeth. Selenium is thought to be absorbed fairly readily, although bioavailability differs with the source. It is transported from the gut mainly on very-low-density lipoproteins (VLDL) and low-density lipoproteins (LDL) and achieves its highest tissue concentrations in red cells, liver, spleen, heart and nails. Selenium is

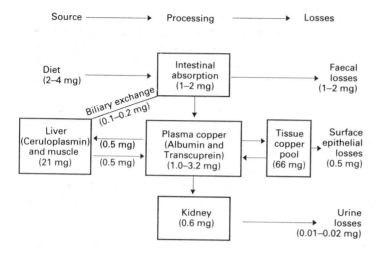

Fig. 8.4 An overview of copper turnover and exchange in the body.

lost from the body mainly in the urine, following its metabolism.

A cardiomyopathy has been associated with selenium deficiency in Chinese children. Deficiency and toxicity have been reported in cattle and sheep grazing lands with low or high selenium levels. This trace element acts as a synergistic antioxidant with vitamin E. Selenium deficiency has been described in patients receiving total parenteral nutrition (TPN). They developed weakness, pain and tenderness in skeletal muscles. Some patients also developed cardiomyopathy. These patients responded to selenium replacement.

Iodine

Iodine is available in cereals and other plant foods when the water and soil have adequate quantities. Seafood is particularly rich in iodine. The body contains about 10–20 mg of iodine, up to three-quarters of which is found in the thyroid gland. Iodine is efficiently absorbed in the inorganic form. Daily requirements are 100–200 µg. Certain substances can interfere with its absorption (goitrogens), e.g. found in cabbage and other vegetables. Iodides are readily absorbed from the intestinal tract and rapidly transported in the bloodstream to the thyroid gland, where they are oxidized to iodine and utilized in the production of the hormones. The iodine is linked with tyrosine molecules to form the thyroid hormones T_3 and T_4 (tri-iodothyronine and thyroxine). About one-third of the absorbed iodine is utilized by the thyroid gland, while two-thirds are excreted in the urine. Iodine in the faeces comes mainly from the bile.

Dietary deficiency causes enlargement of the thyroid to compensate (endemic goitre). People living in inland and mountainous regions where the soil and water are poor in iodine can be provided with iodized salt or iodized oil to prevent endemic goitre. Endemic goitre can also be found in places were goitrogenic substances are found in the staple diet.

Fluorine

Fluorine is mainly obtained from drinking-water. It is also available in tea and sea-fish, both of which have high concentrations. The recommended daily allowance is 1.5–4 mg/day. The only known function of fluorine is in bone and tooth enamel formation. Its deficiency is associated with dental caries. Fluoridation of the water-supply has been used to control dental caries where fluoride deficiency was established. Its use has been described in the management of osteoporosis, to decrease pain and increase bone density. Excess fluoride produces brown mottling of the teeth, which may later lead to pitting of the enamel.

Classification of foods

Foods can be classified in categories, each of which has its own characteristics, e.g. type of nutrients which it mainly supplies, water content and contribution to taste and variety. Knowledge of locally available foods will enable the physician to give appropriate nutritional advice to his patients.

The following is a widely accepted classification:

1 Cereals and starchy roots.
2 Pulses, legumes, seeds and nuts.
3 Meats.
4 Milk and eggs.
5 Fats and oils.
6 Vegetables and fruits.
7 Sugar and syrups.
8 Beverages.

The following is a résumé of the important nutritional contribution of each class.

Cereals and starchy roots
The main cereals consumed by man are wheat, rice, maize and barley. In some parts of Africa and Asia, varieties of sorghum (durra) and millet (bulrush, 'dukhunn', *Pennisetum typhoidium*) are used extensively. Cereal seeds contain starch as 70–80% of their weight. Therefore, cereals are the most important source of carbohydrate in human nutrition.

The grain contains protein as 7–12% of its dry weight. In poor economies, a person consumes about 500 g of cereals/day. He will also have consumed about 30–60 g of protein. Most cereal proteins, however, are deficient in one or more amino acids, e.g. wheat and sorghum protein are deficient in lysine and maize protein is deficient in tryptophan. If cereal foods are eaten with other seeds (e.g. ground-nuts, lentils, chick-peas, peas and beans), milk or meat, this deficiency will be supplemented. Cereals are also a good source of minerals and the B group of vitamins.

Starchy roots, such as cassavas and yams, and stems, such as potatoes, are mainly composed of starch. In many parts of Africa and Asia, cassavas, yams and other starchy roots are eaten for their high energy content, mainly in the form of starch. They have very little protein. Dependence on these food items after weaning can lead to protein deficiency (kwashiorkor).

Edible seeds (pulses, legumes, nuts and seeds)
There are many varieties of beans, peas, ground-nuts and other edible seeds which are popular in different cultures. The main nutrient ingredients of these are starch (about 60% of dry weight) and a high content of protein (about 20% of dry weight). If eaten with a cereal food, they will supply the deficient amino acids of the cereal protein. Legumes are generally deficient in the sulphur-containing amino acids but they have a high content of lysine. Most legumes have deficient or marginal contents of phenylalanine. Ground-nuts (peanuts) contain a higher fat content and they are therefore high in energy. Table 8.8 gives the composition of some major edible seeds.

Table 8.8 The composition of common edible seeds (per 100 g)

Edible seeds	Energy			Protein (g)	Fat (g)	CHO (g)	Ca (mg)	P (mg)	Fe (mg)
	H$_2$O	kcal	kJ						
Bean, broad ('fool')	10.6	354	1451	25.0	1.8	53.7	77	374	6.0
Bean, white ('fasulia')	12.0	349	1431	22.6	1.6	55.9	86	247	7.6
Chick-pea (hummus)	11.5	376	1542	19.2	6.2	56.7	376	324	7.3
Cow-pea	10.6	353	1447	23.2	1.2	57.2	77	420	7.0
Fenugreek seed ('hilba')	8.6	365	1497	29.0	5.2	50.0	180	186	22.0
Peanut (ground-nut)	6.0	589	2415	25.5	44.0	18.8	66	393	3.0
Lentil (peeled)	9.9	344	1410	25.8	1.8	58.8	24	271	10.6
Lupin ('turmos')	9.0	420	1722	40.0	13.0	26.0	90	545	6.3
Pea ('basilia')	13.0	386	1583	22.0	1.3	50.0	–	300	4.7
Pumpkin seed ('lub quraa')	4.5	602	2468	30.3	47.0	11.4	40	1064	9.2
Water-melon seed (roasted)	3.6	581	2382	27.1	50.3	16.3	44	696	13.0
Sesame seed ('simsim')	5.6	622	2550	20.0	51.4	13.9	1200	620	10.4
Sesame seed (ground) (tahina)	2.5	692	2837	21.5	62.0	10.2	100	840	9.0
Soya bean	7.0	447	1833	37.0	24.0	24.0	210	600	6.9

Meats

Beef, lamb, pork, fish and poultry are the main meats consumed by the various people of the world, to varying extents depending on availability, local culture and religious prohibitions. The nutritional importance of meat lies, first, in the high protein content and, second, the high quality of this protein. Meats contain all essential amino acids, i.e. they have high biological value.

Meats are muscle, which contains 18–24% protein. Red meats are also a good source of iron on account of their myoglobin content. The iron is present as haem, which makes it easier to absorb. Internal organs (offal), e.g. liver, spleen and gut, are also rich in protein and iron. Red meats and offal also have a high cholesterol content.

Vegetarians have to consume good mixtures of vegetable protein. The iron supply, especially for adolescents and women during pregnancy and lactation, is rather difficult to satisfy on a vegetable diet alone. Furthermore, vegetable diets tend to contain appreciable amounts of phytic acid, which interferes with the absorption of iron, zinc and other trace elements.

Milk and eggs

Human milk is the first food a baby gets. It provides all the essential ingredients of a balanced diet except for iron and vitamin C. Up to the 6th month, babies may depend entirely on milk without risk of nutritional deficits. Henceforth, supplements have to be given to satisfy mineral and vitamin requirements.

Table 8.9 gives a comparison of human milk with milk from other sources.

The energy content of milk varies mainly with its fat content. An approximate value is 270 kJ (65 kcal/100 ml).

It is important to note that milk is one of the richest human foods in calcium. Cheese is about 10 times the concentration of milk and provides a good storage form of the protein, fat and calcium content of milk. It enables surplus milk to be preserved and transported long distances. Various types of sour milk (e.g. yoghurt or 'laban zabadi' and 'robe') are eaten in various places. They, too, provide a way of keeping milk for moderate durations of time.

Eggs are a useful source of protein, vitamins and minerals. A medium-size egg (60 g) will provide 6 g of protein, 6 g of fat, 1 g of carbohydrate, 330 kJ (80 kcal) of energy, 1.5 mg of iron and 30 mg of calcium.

The egg yolk, in addition to vitamins and minerals, has a high cholesterol content (300 mg), second only to brain tissue.

Fats and oils

Some fats can occur in solid form when kept at low temperatures, and these are generally saturated lipids. Oils, however, do not solidify at moderately cool temperatures and are usually composed of unsaturated lipids. Such oils can be made saturated by hydrogenation to give margarine.

It can be observed from Table 8.10 that most edible fats are almost pure fat. The value of fats lies in two functions: (i) they are a good source of energy; and (ii) fats are a vehicle for absorption of the fat-soluble vitamins. Fats also make food more tasty and palatable. However, animal fat and the fat in dairy products have a high cholesterol content.

Table 8.9 The composition of milk (all values are per 100 g of milk)

	Human	Cow	Goat	Ewe	Camel
Protein (casein) (g)	1.5	3.5	3.7	6.5	3.7
Fat (g)	4.5	3.5	4.8	6.9	4.2
CHO (lactose) (g)	6.8	5.0	4.5	4.9	4.1
Iron (mg)	0.7	0.1	0.2	—	—
Phosphorus (μg)	14	91	129	—	—
Calcium (μg)	34	120	150	—	—

Table 8.10 Percentage composition of edible fats and oils. (Vitamin A in µg of retinol equivalents)

	Butter	Margarine	Ghee ('samin')	Lard/ animal fat	Seed oils	Red palm-oil
Fat	83	81	99	99	99.9	98.8
Water	16	15	1	1	0	0.7
Protein	1	0.6	0	0	0	0
Minerals	0	0	0	0	0	0.1
Vitamin A	840	900	800	0	0	4000

Fried foods absorb fat and this markedly increases their energy content. Red palm-oil, widely used in West Africa, is a good source of carotene and can therefore supply an adequate amount of vitamin A where milk is scarce. Its contents can be as high as 100 000 µg of retinol equivalents. Palm-oil and coconut-oil also have appreciable amounts of cholesterol. In some parts of the world where milk is scarce, vitamin A deficiency is common, particularly in children between the ages of 2 and 5 years. Serious effects on the eye result in permanent blindness.

Vegetables and fruits
Fresh vegetables have a high water content and are eaten mainly for taste and flavour. They contain, for practical purposes, no energy or protein. Vegetables, however, are an important source of minerals and water-soluble vitamins. Furthermore, vegetable matter provides fibre, which assists movement of food in the gut and prevents constipation and other diseases of the gastrointestinal tract.

Leafy vegetables are generally rich in minerals, particularly calcium and iron, and vitamin A and C. They also provide appreciable amounts of folic acid. Green peppers have a particularly high vitamin C content. On the other hand, vegetables are a poor source of energy due to their high water content. An exception to this is the green pulses and seeds which may be used as a vegetable. When used in reducing diets, vegetables are valuable for their filling effect, providing bulk and supplying vitamins and minerals.

Fruits have a high water content and rather low energy. Some fruits contain good supplies of calcium and phosphorus, but they are generally a poor source of iron. Some fruits contain significant amounts of carotene. All fruits contain vitamin C but only a few have a high content, such as guavas, strawberries, mangoes and the citrus fruits.

Sugar and syrups
Cane-sugar is a relatively recent addition to the human diet. Before the development of this industry, one had to chew 1–2 kg of cane to obtain three teaspoonfuls of sugar. Sugar contains sucrose, which gives energy (about 4 kcal/g) and nothing else. In contrast, cereal foods provide an important source of energy together with some protein, vitamins, minerals and fibre. Feeding excessive amounts of sugar to children, together with reduction in consumption of cereal foods and milk, can lead to serious malnutrition. Sugar is used in the preservation of canned fruits, jams and syrups. Because of their sugar content, sweets and cakes have a high energy content.

Beverages
Tea and coffee are widely consumed beverages. The active ingredient of these drinks is caffeine. The usual dose in a cup of tea is about 60 mg and in a cup of coffee 90 mg. Cola drinks may contain 50–200 mg of caffeine per litre. The energy in these drinks comes from added sugar. Alcohol-containing drinks provide 30 kJ (7 kcal)/g of alcohol. Beverages contain no other nutritional ingredients to any significant level. Bottled soft drinks contain either sugar or artificial sweeteners or both. The energy content, on the average, is about 17 kJ (150 kcal)/litre. Regular consumption of beverages can add significant energy intakes to a normal diet and may lead to overweight.

Nutritional requirements

In the previous pages, we described the nutrients, their roles and their sources. The requirements of the human body for each of the nutrients vary with many factors. Body size is the main determinant of energy and protein requirements. However, other factors, such as age, sex and physiological state (such as periods of rapid growth, pregnancy, lactation and old age), all have an influence on nutritional requirements.

The term minimum nutritional requirement refers to the physiological amount of a nutrient used daily by the body to maintain health, growth and a general state of well-being compatible with work and leisure. Over short periods of time this can come from body stores and should therefore be replaced from the diet. If the diet does not supply these minimum requirements, the stores become depleted and nutritional deficiency diseases appear. However, recommended nutritional allowances are estimates based on food intake surveys and experimental studies. There are wide variations between individuals and populations. Therefore, an average estimate is given for a large group (e.g. children in the age-group 1–3 years, male adolescents 13–15 years, etc.). These allowances are recommended by governments and international organizations and are usually 20% above the average for any group, allowing for a safety margin in nutritional advice and planning.

Recommendations in the United States and Europe greatly exceed surveyed nutrient intakes in developing countries. This brings up the concept of adaptation to the levels of nutrient intake available in any given community. In some societies children and adults do not attain their maximum growth potential (height and weight). They therefore adapt to the lower intake by decreasing the attained weight and height but manage to maintain health compatible with work and reproduction. Therefore, physiological nutritional requirements are not always possible to attain in developing countries. Increasing the energy intake in such communities has been found to increase productivity. Whether the recommended allowances are true requirements is difficult to say. These estimates can be considered as guidelines for nutritional planning and decision-making by the physician. They give a good margin of safety. The needs of an individual may depart signifi-

cantly from the recommended average for his weight, age and sex. Table 8.11 gives the recommended allowances as given by FAO and WHO (United Nations Food and Agriculture Organization and World Health Organization).

Energy requirements

The requirements of an individual depend on five variables: body-weight, physical activity, body composition, age and climate.

1 *Body-weight.* A large person needs more energy to move about, to climb stairs, etc. Energy expenditure experiments show that the energy required for a given task varies directly with body-weight. Furthermore, the energy needed for maintenance of body functions at resting or basal conditions varies directly with body-weight, which is representative of active cell mass.

2 *Physical activity.* Occupational and recreational work are more variable than other daily activities. A person working as a manual labourer needs more energy than an office clerk. A person who swims or plays football in his leisure time spends more energy than one whose hobby is reading or music. Table 8.12 gives the energy cost of various occupational and leisure activities.

3 *Body composition.* Lean body tissues have a higher metabolic rate than fat. Therefore, the higher the fat content, the lower the energy requirement. This factor is responsible for the sex difference in basal metabolic rate (BMR) often described in textbooks.

4 *Age.* Two factors are responsible for the effect of age: physical activity and the basal metabolic rate, both of which decrease with age. Children are very active, while elderly people tend to decrease their physical activity and eventually retire. The basal metabolic rate is high in children and decreases with age. It depends on many factors, such as body composition and active cell mass.

5 *Climate.* Extremes of climatic conditions tend to restrict physical activity (behavioural adaptation). The BMR is also found to vary with climate. It is lower in people living in hot environments. This means that at rest they produce less heat. We therefore consider this as a physiological adaptation.

Table 8.11 Recommended daily allowances (RDA)

	Body-weight (kg)	Energy (kcal)	Energy (MJ)	Protein (g)	Vit. A (µg)	Vit. D (µg)	Thiamine (mg)	Riboflavine (mg)	Niacin (mg)	Folic acid (µg)	Vit. B_{12} (µg)	Ascorbic acid (mg)	Calcium (g)	Iron (mg)
Children														
1	7.3	820	3.4	14	300	10.0	0.3	0.5	5.4	60	0.3	20	0.5–0.6	5–10
1–3	13.4	1360	5.7	16	250	10.0	0.5	0.8	9.0	100	0.9	20	0.4–0.5	5–10
4–6	20.2	1830	7.6	20	300	10.0	0.7	1.1	12.1	100	1.5	20	0.4–0.5	5–10
7–9	28.1	2190	9.2	25	400	2.5	0.9	1.3	14.5	100	1.5	20	0.4–0.5	5–10
Male adolescents														
10–12	36.9	2600	10.9	30	575	2.5	1.0	1.6	17.2	100	2.0	20	0.6–0.7	5–10
13–15	51.3	2900	12.1	37	725	2.5	1.2	1.7	19.1	200	2.0	30	0.6–0.7	9–18
16–19	62.9	3070	12.8	38	750	2.5	1.2	1.8	20.3	200	2.0	30	0.5–0.6	5–9
Female adolescents														
10–12	38.0	2350	9.8	29	575	2.5	0.9	1.4	15.5	100	2.0	20	0.6–0.7	5–10
13–15	49.9	2490	10.4	31	725	2.5	0.9	1.4	16.4	200	2.0	30	0.6–0.7	12–24
16–19	54.4	2310	9.7	30	750	2.5	0.9	1.4	15.2	200	2.0	30	0.5–0.6	14–28
Adult man (moderately active)	65.0	3000	12.6	37	750	2.5	1.2	1.8	19.8	200	2.0	30	0.4–0.5	5–9
Adult woman (moderately active)	55.0	2200	9.2	29	750	2.5	0.9	1.3	14.5	200	2.0	30	0.4–0.5	14–28
Pregnancy (latter half)		+350	+1.5	38	750	10.0	+0.1	+0.2	+2.3	400	3.0	50	1.0–1.2	
Lactation (first 6 months)		+550	+2.3	46	1200	10.0	+0.2	+0.4	+3.7	300	2.5	50	1.0–1.2	

Based on *Human Nutritional Requirements* (1974) WHO Monograph Series No. 61.

Table 8.12 Energy cost of physical activity

Classification of workloads	kJ	kcal	W
Very light	<10	<2.5	<170
Light	10−20	2.5−4.9	170−350
Moderate	21−30	5.0−7.4	350−500
Heavy	31−40	7.5−9.9	500−650
Very heavy	41−50	10−12.5	650−800
Exceedingly heavy	>50	>12.5	>800

Very light work	*Light work*	*Moderate work*
Most office jobs	Driving	Digging
Sitting, writing or reading	Electrical work	Shovelling
Needlework	House cleaning	Mechanical repairs
Cooking	Operating machines	Walking
Playing a musical instrument	Carpentry	Cycling up to 16 kph
		Tennis

Heavy work	*Very heavy work*	*Exceedingly heavy work*
Carrying loads	Fast swimming	Felling trees
Swimming	Sprinting	High performers in competitive athletics, e.g. long-distance running
Cycling at >16 kph	Cross-country running	
Walking uphill	Hill climbing	
Jogging		
Football		

Definitions

Kilocalorie (kcal) = amount of heat energy required to raise the temperature of 1 kg water by 1 °C

1 kcal = 4.184 kJ

Kilojoule (kJ) = amount of energy required to move 1 kg mass 1 m by a force of 1 newton (N)

Watt (W) = the power resulting from the exertion of 1 kJ energy for 1 second

1 W = 0.06 kJ/min
= 0.0144 kcal/min

Protein requirements

The economy of protein metabolism is such that daily protein turnover is about seven times the amount of amino acids required per day. This emphasizes that proteins are constantly broken down and the amino acids are reutilized. The reutilization is incomplete and this is why dietary amino acids are required. There is an obligatory loss of amino acid nitrogen, which has to be replaced to maintain nitrogen balance. The obligatory (or endogenous) nitrogen loss in a 65 kg man is 3.5 g/day, which is equivalent to 22 g of protein. This can be taken as the minimum protein requirement. Estimates of protein requirements are largely based on nitrogen balance studies. Additional amounts of protein are required for normal growth in infants, children and adolescents and during pregnancy and lactation. During the first 3 months of life, an infant needs 2.4 g/kg body-weight. Between the ages of 3 and 12 months, it falls to between 1.85 and 1.44 g/kg. Adequate energy is required for protein utilization in growth. In adolescents, increasing the energy intake was found to decrease oxidation of amino acids and enhance protein synthesis.

INFANTS AND CHILDREN

Infancy is defined as the 1st year of life. During this period, the infant doubles its weight at best and at 1 year his/her weight may be three times the birth-weight. This rapid growth rate requires relatively high nutritional requirements. Energy needs are determined by his or her basal metabolism, rate of growth, body size, age and activity. Enough calories must be provided to ensure growth and to spare protein from being used as energy. The best way to determine the adequacy of infants' energy intakes is to carefully monitor their gain in height and weight. Energy needs are supplied by carbohydrate and fat in milk during the first 6 months. Thereafter, energy also comes from supplementary feeding in the form of cereals and other strained and cooked foods, including meat, vegetables and fruits, which also supply additional protein, minerals and vitamins.

There is a relative decrease in the energy requirements with age. Obviously, the quantities recommended increase with age because of the increase in body-weight. Protein requirements during infancy, when there is rapid growth, are higher on a per kilogram basis than those of the adult or older child. Nitrogen from protein must be provided for the formation of new tissues, the maturation of tissue and the maintenance of tissues. Recommended protein requirements show a similar pattern to energy (see Table 8.13).

These protein requirements are based on high-quality protein, e.g. egg white. If poorly digestible or lower-quality protein is used, then corrections will have to be made. It is essential to supply adequate energy intake for protein utilization.

Minerals are essential for new tissues, in particular calcium and iron. Milk provides the

Table 8.13 The relationship between infant's age, energy requirements and protein requirements

Age (months)	Energy requirements		Protein requirements	
	(kJ/kg)	(kcal/kg)	(g/kg)	(g/day)
0–3	485	116	2.20	11
3–6	415	99	1.85	13
6–9	400	95	1.65	14
9–12	420	101	1.50	14

required calcium. After the 6th month, supplementary iron is necessary. It should be remembered that cow's or goat's milk has a high phosphorus content compared with human milk; this may lead to hypocalcaemia. To treat this, the Ca : P ratio in the feeds should be raised to 4 : 1. Hypocalcaemia is also associated with Mg deficiency and may give rise to tetany. Low serum Mg is also encountered in protein–energy malnutrition (PEM). Copper deficiency may be due to prolonged parenteral treatment, chronic diarrhoea and other diseases. It causes anaemia, loss of appetite and retardation of growth. Zinc deficiency has been reported in association with protein depletion. The daily requirement for infants is 3–5 mg.

Sodium and potassium are found in milk and other infant foods. These can easily be lost in diarrhoea and vomiting. Gastroenteritis (inflammation of the gastrointestinal tract) is a common condition in out-patients and paediatric wards. Therefore, dehydration is a common occurrence. In addition to the fluid loss, there is also loss of electrolytes and shifts of fluid between the intracellular and extracellular compartments. To prevent the occurrence of hypertonic dehydration or hypokalaemia, attention should be given to replacement fluid therapy, either oral or intravenous. Giving normal infant foods may be dangerous, as this can increase the tonicity of extracellular fluid, which may lead to convulsions and loss of consciousness.

Vitamin requirements during infancy are met by milk and supplementary feeding; therefore, vitamin deficiencies are uncommon. However, some infants who are breast-fed may develop vitamin K deficiency in the first few days after birth, giving rise to haemorrhagic disease of the new-born. Vitamin D is needed for calcium metabolism and skeletal growth in children. Since this nutrient is also synthesized in the skin, the amount required from dietary sources is dependent on factors such as geographical location and time spent out of doors. Children living in tropical areas may need no dietary vitamin D. In temperate zones, however, some dietary source is needed. Other fat-soluble vitamin deficiencies are seen with children who have malabsorption of fat, e.g. steatorrhoea. Recommended daily intakes are given in Table 8.11. Of particular importance is vitamin C. Milk is a poor source of this vitamin. Therefore, a mixed diet should aim at providing it for infants and children.

Protein−energy malnutrition (PEM) (Fig. 8.5) occurs in developing countries due to food shortages, poor nutritional education, recurrent infection or a combination of these factors. Typically it affects children under 5 years old.

Marasmus is the name given to protein and energy deficiency. This is seen in infants and children when there is a general shortage of food. It can be seen in adults also in famine situations. Poverty and poor housing conditions, as in slums and unauthorized settlements around big cities, predispose to PEM. Repeated infections, es-

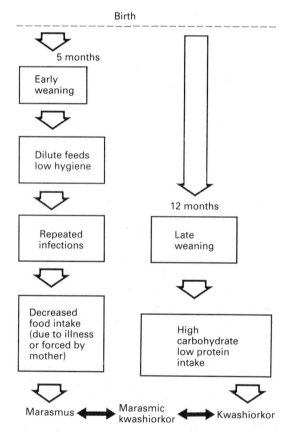

Fig. 8.5 Diagrammatic scheme of the causation of protein−calorie malnutrition. (Based on Maclaren, D.A. (1966) A fresh look at protein−calorie malnutrition. *Lancet* **ii**, 485−8.)

pecially of the respiratory system and the gastrointestinal tract, early weaning and enforced dietary restrictions during such illnesses are among the common causes of this disorder. Misinformed mothers may stop breast-feeding too early if a second pregnancy occurs.

The victims of this disease are emaciated, i.e. they lose subcutaneous fat and suffer severe muscle-wasting. They have a typical skin-and-bone appearance. Infants exhibit an anxious facial expression. These children get diarrhoea and gastroenteritis, leading to severe dehydration. The abdomen is often distended with gas. The child is usually severely underweight. Height is less affected (see Fig. 8.6a).

Kwashiorkor is the name given to PEM when the disease is due to protein deficiency associated with adequate energy intake. It affects children of 2−4 years old when the food supplements contain energy-rich but low protein contents. The typical example of kwashiorkor is the sugar baby of Jamaica. An abundance of sugar combined with poor protein foods leads to the development of a typical kwashiorkor picture (see Fig. 8.6b).

The child usually has oedema, enlargement of the liver and muscle-wasting but abundant subcutaneous fat. Although growth is retarded, weight is less severely affected because of the oedema. The child is apathetic and does not cry often, unlike the anxious marasmic child. In kwashiorkor there are skin lesions and the hair is thin, depigmented and easy to pluck. There is anaemia. Impaired intestinal absorption may lead to vitamin and mineral deficiencies.

The term marasmic kwashiorkor is given to a mixture of the above two extremes of PEM when various degrees of protein and energy deficiencies occur. A PEM marasmic child fed on sweetened drinks develops kwashiorkor signs, while a protein-deficient child who is too weak or apathetic to take adequate amounts of food develops marasmic signs. In clinical practice, then, these children are simply diagnosed as PEM.

NUTRITION IN PHYSIOLOGICAL STRESS
There are times when the physiological nutritional requirements are increased; therefore, dietary deficiencies are more likely to occur. Three such states will be presented: adolescence,

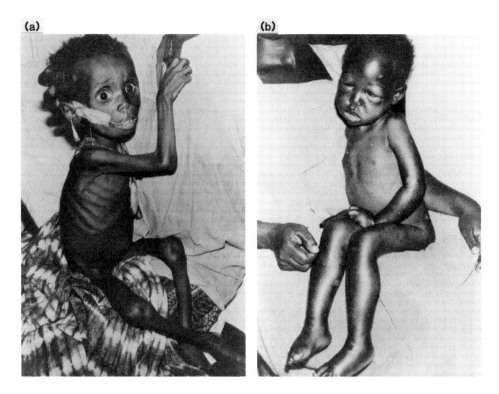

Fig. 8.6 (a) A child with marasmus. (b) A child with kwashiorkor. (Reproduced with permission from R.G. Whitehead, Director, Dunn Nutrition Centre, Cambridge, UK.)

pregnancy and lactation. Average recommended allowances for these states will be found in Table 8.11. It should be noted, however, that individual requirements may be significantly different, depending on the nutritional status of the subject before the physiological demand has increased.

Adolescence

The adolescent period is associated with rapid physical growth (growth spurt), sexual maturation and psychological stresses. There is an increase in energy requirements due to the rapid physical growth and increased physical activity, particularly in adolescent boys. Adolescent girls tend to store fat, which is considered as a secondary sexual characteristic of the female body. Protein requirements increase and should supply 12–15% of the total energy. Carbohydrates should supply about 50% and fat about 30–45% of the daily energy intake.

Calcium requirements increase dramatically by about 400 mg/day, due to rapid skeletal growth in boys and girls. Iron intake should be particularly monitored, as anaemia can easily occur due to the rapid increase in muscle mass and blood volume. Females are more likely to become deficient due to menstrual losses (15–30 mg/period) and the possibility of pregnancy, especially in societies where early marriage is common. An additional 10–15 mg/day of iron is recommended for females and 5–10 mg/day for males. These represent about 80% more for boys and 120% more for girls.

As to the need for vitamins in adolescence, high amounts of thiamine, riboflavine and niacin are recommended because of the high energy requirements. In most cases, because of increased energy intake, the intake of B vitamins will also increase.

Pregnancy

The foetus, placenta and maternal response to

pregnancy constitute the three factors which determine the nutritional needs. If pregnancy occurs before the age of 17 years, the additional demands of adolescence may compromise the growth of both mother and foetus. The nutritional status and therefore the nutrient reserves of the mother before gestation are an important determinant of her nutritional requirements.

The daily energy requirements of pregnancy increase by about 1.5 MJ on average per day — about 10–15% more than the previous requirement. This is due to a general increase in the basal metabolism, energy storage and energy expenditure. As anticipated from the physiological changes of body composition, most of the energy intake goes into the fat stores during the second trimester. This period is usually associated with a good appetite and rapid weight gain. During the third trimester more energy expenditure is associated with the mother's movement, due to the increased body-weight. At this stage, energy is needed to spare protein for foetal growth. However, the total energy requirements may not increase very much because at this stage most mothers tend to keep their physical activities to a minimum.

Additional protein is obviously needed during pregnancy to support the synthesis of maternal and foetal tissues, and needs increase as pregnancy proceeds. The recommended daily intake of protein during pregnancy increases by 6 g/day. Table 8.14 gives the nutritional requirements during the latter half of pregnancy as recommended by the World Health Organization (WHO). Other authorities recommend 20 g/day. During the first trimester little extra protein is required. Protein is required mainly for the growth of the uterus and the foetus, both of which are greatest during the third trimester. Protein is also required for growth of the breasts and the increased blood volume. Adverse results of protein deficiency during pregnancy are difficult to separate from the effects of energy deficiency in clinical situations. Almost all cases of limited protein intake are accompanied by limitation in availability of energy; under such conditions, decreased birth weight and greater incidence of pre-eclampsia have been reported.

Table 8.14 Nutritional allowances of the pregnant mother compared with non-pregnant requirements

	Non-pregnant	Pregnant
Energy (MJ)	9.2	10.7
Protein (g)	29	35
Calcium (mg)	200	400
Iron (mg)	14–28	14–28
Vitamin A (μg)	750	750
Vitamin D (μg)	2.5	10
Vitamin B_1 (mg)	0.9	1.0
Vitamin B_2 (mg)	1.3	1.5
Niacin (mg)	14.5	16.8
Vitamin C (mg)	30	50
Vitamin B_{12} (μg)	2.0	3.0
Folic acid (μg)	200	400

The mother's diet should supply adequate iron throughout pregnancy. During the first trimester a normal diet will give the required iron supply. A marginally adequate diet, however, will not be satisfactory during the second and third trimesters. It is therefore advised that iron therapy (30–60 mg/day) should be given to mothers in developing countries throughout pregnancy to build up maternal stores and to prevent iron-deficiency anaemia. This is essential for the extra maternal haemoglobin, foetal blood and liver stores and maternal stores of iron to meet delivery losses and to prepare for lactation. An anaemic woman is clearly less able to tolerate haemorrhage with delivery and she is more prone to develop puerperal infection.

The mother needs calcium, mainly for the growth and calcification of foetal bones. This is most active during the third trimester. During early pregnancy, most of the extra calcium requirements can be provided from the mother's bones. Milk is a good source of both calcium and protein. All mothers should be advised to take 0.5 litres of milk per day, especially during the second and third trimesters. This amount provides an extra 17 g of protein and 600 mg of calcium. The recommended daily allowance is 1000–1200 mg. As only about 30% of the dietary calcium is absorbed, this intake should provide an adequate margin of safety.

The vitamin requirements of pregnant women are only moderately increased. Therefore, a normal balanced diet is usually adequate. However, in developing countries vitamin supplements are usually given as part of antenatal care when the economic status of the family is poor. Table 8.14 indicates the marked increase in vitamin D requirements during pregnancy to increase the absorption and use of calcium and phosphorus. This supply can normally be synthesized in the body if adequate exposure of the skin to sunlight is ensured. This is important to facilitate the absorption of calcium and for active calcification in the foetal skeleton. Interest in vitamin K in pregnancy largely centres on haemorrhagic disease of the new-born due to vitamin K deficiency. Parenteral vitamin K is often administered to parturient women since the new-born has a sterile gut and bacterial synthesis of vitamin K is thus an unreliable source in the neonatal period. Ascorbic acid deficiency has not been shown to affect the course or outcome of pregnancy in humans. However, because low plasma levels of ascorbic acid have been reported to be associated with premature rupture of the membranes, as well as pre-eclampsia, questions have arisen about its possible association with these conditions. Folic acid demands increase by 100%, mainly for maternal haemopoiesis and foetoplacental growth. Folate deficiency will lead to megaloblastic anaemia, which frequently occurs in pregnancy.

Intrauterine nutrition is an important prerequisite for foetal development and attaining a good weight at birth. It should be emphasized that foetal stores of minerals and vitamins are used during the early postnatal period. In particular, good iron stores are necessary for the prevention of anaemia during infancy. For the first 6 months, the baby depends on these stores because milk is a poor source of iron. More information on the foetoplacental unit will be found in Chapter 16.

An important aspect of foetal growth during the last trimester is the rapid growth of the foetal brain. It has a maximum growth rate about the 34th week of pregnancy. Therefore, adequate nutrition at this critical stage is of vital importance for intrauterine brain development.

Lactation

The nutritional requirements of lactating mothers are similar to those during pregnancy. Milk production is the main concern. Table 8.15 shows that the quantity of milk produced increases as the baby grows.

One litre of milk has about 2.7 MJ (650 kcal) of energy. The metabolic energy cost of its production is estimated at about 1.6 MJ (400 kcal). However, the total recommended daily energy intake is 2.1 MJ (500 kcal). It is assumed that the extra energy will come from the body stores of fat deposited during pregnancy (about 4 kg). The daily energy intake of the mother will depend on the amount of milk she is producing and the state of her body stores.

The WHO recommends an additional protein intake of 17.5 g during the first 6 months and 13 g thereafter. Some other authorities recommend an additional 20 g daily. The amounts of calcium and iron required during lactation are the same as those required by the pregnant mother. Vitamin C should be increased to 50–60 mg/day. Vitamin A recommended intake should be increased by 400 µg. The vitamin B group is also marginally increased above those required for pregnancy, except folic acid, which is about 300 µg/day compared with 200 µg in non-pregnant women and 400 µg during pregnancy.

Lactating mothers should take adequate amounts of fluid, especially in hot weather, to ensure adequate milk production. Breast-feeding mothers should be discouraged from dieting to control body-weight. They should also be advised that use of oral contraceptives may suppress lactation, especially in the first 6–10 weeks after giving birth.

Assessment of nutritional status

Doctors have to assess the nutritional status of their patients for two main purposes: (i) so as to take nutritional decisions in the management of individual patients, whether a child, a mother or a surgical or a medical patient; and (ii) to assess the nutritional status of a group of people (a community) so as to describe nutritional problems and their causes and to make decisions about their prevention.

Table 8.15 The relationship between milk production and age of lactating baby

Age of baby (months)	Milk production (ml)
1	500
2	800
3	900
4−6	900−1000
7−9	1000−1200
10−12	1300−1500

ASSESSMENT OF NUTRITIONAL STATUS OF A PATIENT

Physical examination

Examination of the patient can reveal signs related to nutrition (Table 8.16). These signs should be looked for within the general examination of the patient. Musculoskeletal, abdominal, cardio-vascular and neurological signs may be super-imposed upon other signs of organic disease.

Table 8.16 Signs of nutritional deficiencies

Part of the body	Clinical signs	Nutritional deficiency
Hair	Thin, sparse, dyspigmental, easy to pluck	PEM
Eyes	Pale conjunctiva	Anaemia
	Bitot's spots	Vitamin A
	Xerosis	Vitamin A
	Keratomalacia	Vitamin A
	Corneal opacities	Vitamin A
Lips	Angular stomatitis	Riboflavine (B_2)
	Cheilosis	B_2, niacin
Tongue	Purple redness	B_2
	Inflammation (glossitis)	B_2, niacin, folic acid, B_{12}, iron, B_6
Teeth	Mottled enamel	Fluoride excess
	High prevalence of caries	Fluoride deficiency
Gums	Bleeding, swollen	Vitamin C
	Pallor	Iron
Nails	Koilonychia (spoon-shaped)	Iron
Skin	Amount of subcutaneous fat	Energy
	Oedema	PEM
	Follicular hyperkeratosis	Vitamin A, vitamin C
	Flaky paint dermatosis	Kwashiorkor
Muscle	Wasting	PEM
	Tenderness, haematoma	Vitamin C
	Bleeding in joints (pain)	Vitamin C
Skeleton	Bow legs ⎫	
	Knock knees ⎬ rickets	Vitamin D
	Beading of ribs ⎭	
Cardiovascular system	Cardiac enlargement	Thiamine (B_1)
	Tachycardia	B_1, anaemia
Abdomen	Liver enlargement	PEM
Nervous system	Apathy	Kwashiorkor
	Anxious expression	Marasmus
	Sensory loss	B_1
	Loss of tendon jerks	B_1, B_6, B_{12}
	Night blindness with impaired dark adaptation	Vitamin A

PEM, protein-energy malnutrition

Body measurements (anthropometry)
(Figs 8.7–8.10)

The most important body measurement is that of body-weight. For adults, the relation of weight to height should be noted (Table 8.17). It is found that people with similar height are expected to have ideal or recommended weights depending on the size of their skeletal configuration, hence the variation in the range according to body frame (small, medium and large). Weight between the higher value of the range and the obese weight is considered overweight. Obesity, which is the main nutritional problem in affluent societies, can be graded using what is known as the Quetelet index, derived by dividing weight in kg by the square of the height in metres (wt/ht²). Table 8.18 describes the degrees of obesity.

Anthropometry includes the assessment of body fat by measurement of skin-fold thickness and the measurement of mid-upper arm circumference for assessment of muscle mass. The sites for skin-folds and muscle circumferences are numerous and beyond the scope of this book.

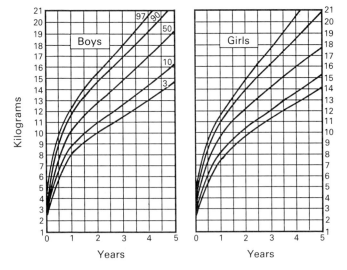

Fig. 8.7 Standard weights for boys and girls 0–5 years old. (Reproduced from Passmore, R. & Eastwood, M.A. (1986) *Human Nutrition and Dietetics.* With permission from Churchill Livingstone, Edinburgh.)

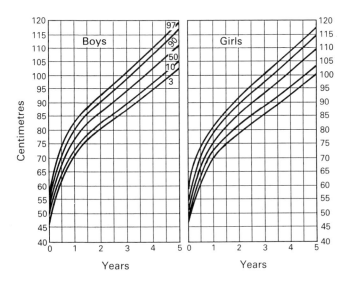

Fig. 8.8 Standard heights for boys and girls 0–5 years old. (Reproduced from Passmore, R. & Eastwood, M.A. (1986) *Human Nutrition and Dietetics.* With permission from Churchill Livingstone, Edinburgh.)

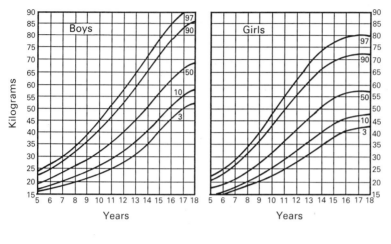

Fig. 8.9 Standard weights for boys and girls 5–18 years old. (Reproduced from Passmore, R. & Eastwood, M.A. (1986) *Human Nutrition and Dietetics.* With permission from Churchill Livingstone, Edinburgh.)

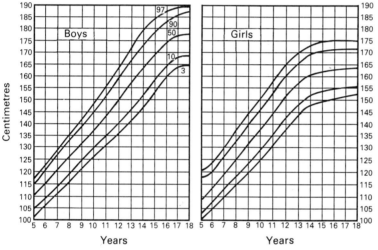

Fig. 8.10 Standard heights for boys and girls 5–18 years old. (Reproduced from Passmore, R. & Eastwood, M.A. (1986) *Human Nutrition and Dietetics.* With permission from Churchill Livingstone, Edinburgh.)

Laboratory tests

Tests relevant to nutritional assessment include haematocrit, haemoglobin, RBC and WBC counts, serum albumin, serum transferrin, delayed hypersensitivity response, iron, vitamins and minerals. The following are some important indicators of malnutrition:

Haemoglobin	<12 g/dl males ⎱ anaemia <11 g/dl females ⎰
Haematocrit	<40% for males <37% for females
Mean corpuscular volume	<80 fl, iron deficiency >99 fl, B_{12}, folate deficiency
Total lymphocytes	<1500/mm³, protein malnutrition
Blood film for WBC	Multilobular granulocytes

indicate B_{12} and folate deficiency

Serum albumin	<3.5 g/dl, PEM
Iron-binding capacity	<250 mg/dl, protein malnutrition
Serum iron	<50 µg/dl, chronic iron deficiency
Serum folic acid	<6 µg/dl folate deficiency
Serum ascorbic acid	<0.5 mg/dl vitamin C deficiency
Serum vitamin A	<0.15 mg/dl vitamin A deficiency

ASSESSMENT OF NUTRITIONAL STATUS OF THE COMMUNITY (NUTRITIONAL SURVEILLANCE)

Nutritional surveillance comes in the realm of

Table 8.17 Recommended weights (kg) for men and women[a]

| Height (m) | Weight (kg) | | | | | |
| | Men | | | Women | | |
	Average	Range	Obese	Average	Range	Obese
1.45				46.0	42–53	64
1.48				46.5	42–54	65
1.50				47.0	43–55	66
1.52				48.5	44–57	68
1.54				49.5	44–58	70
1.56				50.4	45–58	70
1.58	55.8	51–64	77	51.3	46–59	71
1.60	57.6	52–65	78	52.6	48–61	73
1.62	58.6	53–66	79	54.0	49–62	74
1.64	59.6	54–67	80	55.4	50–64	77
1.66	60.6	55–69	83	56.8	51–65	78
1.68	61.7	56–71	85	58.1	52–66	79
1.70	63.5	58–73	88	60.0	53–67	80
1.72	65.0	59–74	89	61.3	55–69	83
1.74	66.5	60–75	90	62.6	56–70	84
1.76	68.0	62–77	92	64.0	58–72	86
1.78	69.4	64–79	95	65.3	59–74	89
1.80	71.0	65–80	96			
1.82	72.6	66–82	98			
1.84	74.2	67–84	101			
1.86	75.8	69–86	103			
1.88	77.6	71–88	106			
1.90	79.3	73–90	108			
1.92	81.0	75–93	112			

a Accepted by the Fogarty Conference (USA) 1979 and the Royal College of Physicians (UK) 1983.

community health. Monitoring of the population for signs of malnutrition can best be achieved by examining a sample of a given population, especially those at risk, i.e. infants, children, adolescents and pregnant mothers. The sample must be carefully selected (see Chapter 19) so as to be reliable as representative of the population.

Table 8.18 Degrees of obesity

Non-obese	W/H^2 index
Non-obese	<25
Grade I	25–29.9
Grade II	30–40
Grade III	>40

W, weight; H, height.

Nutritional surveys are carried out using the same methods as described above for individual patient assessment, i.e. clinical examination, anthropometry and laboratory tests. Figure 8.11 shows heights and weights of children in rural Khartoum, studied in a nutritional survey.

Haematological indices are among the most valuable and relatively simple tests which can be used in nutritional surveillance. Figure 8.12 shows some haematological findings in a rural population in western Sudan.

Energy metabolism

The term metabolism means 'change'. It is used to describe enzyme-catalysed chemical reactions in which large molecules are broken into smaller ones (catabolism) and others in which smaller

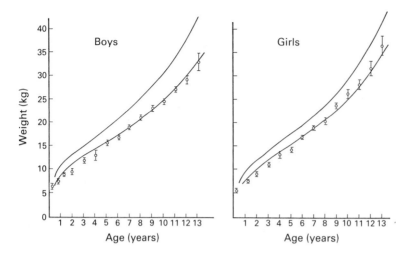

Fig. 8.11 Weights of children in rural Khartoum compared with British standards. The means are seen to be comparable to the lower 10% of the standard curve. (Adapted from Sukkar, M.Y. *et al.* (1979) *Annals of Human Biology* **6**, 147–58.)

molecules are used for the synthesis of larger and more complex ones (anabolism).

In general, catabolic or degradative processes produce energy, whereas anabolic or synthetic processes consume energy. The body requires energy for the generation of heat, for the performance of exercise and locomotion, for the synthesis of new tissues and for various cellular activities, such as maintenance of ionic gradients, processes of secretion and detoxification. To provide these energy needs, various tissues oxidize glucose (from carbohydrate), free fatty acids (FFAs) and ketone bodies (from fat) and amino acids (from proteins). These energy-containing substrates or metabolic fuels are provided as a mixture to tissues in the bloodstream and their oxidation is linked to the generation of a limited number of high-energy compounds, of which ATP is the most important. It is, however, extremely important to appreciate that ATP does not function as a simple store of chemical energy in the cell. In combination with ADP, ATP functions as an energy transfer system in the cell: the generation of ATP from ADP during the oxidation of fuels (e.g. glucose) conserves chemical energy, which is utilized in a number of processes (see Fig. 8.13).

Since man's digestive system does not provide him with a continuous supply of nutrients at a rate suitable for his requirements, his ability to survive even everyday life depends on systems for storage of metabolic fuels when they are in excess and release of these stores as required. The meta-

bolic fuels, such as glucose, FFAs and amino acids, cannot be stored as such to any great extent, due to osmotic problems. Instead, they are stored in the form of macromolecules in adipose tissue (as triacylglycerol), in liver and skeletal muscle (as glycogen) and in skeletal muscle (as proteins); however, the available metabolic fuels are related to body composition (see Fig. 8.14). In addition, metabolic fuels are found in the circulation as their constituents: carbohydrate, mainly as glucose but also (as in exercise) lactate and pyruvate; fats as FFAs and glycerol; and protein as amino acids (mainly alanine and glutamine) (see Fig. 8.14).

The biochemical pathways for the transformation of energy are outside the scope of this book. However, the following section will deal with the rate of liberation of energy in the whole body and its regulation.

THE METABOLIC RATE

The amount of energy released in the body varies with the amount of work performed and the amount of heat produced. Any excess energy intake will go into the body stores, i.e. there is no significant excretion of energy-containing compounds under physiological conditions. However, the utilization of energy as external work is not so efficient (see Chapter 14). It has been found that the rate of energy metabolism is directly related to oxygen consumption. The uptake of oxygen by the subject is used in measuring the

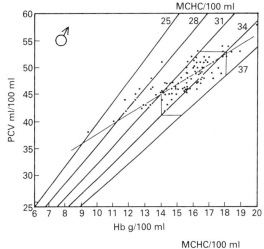

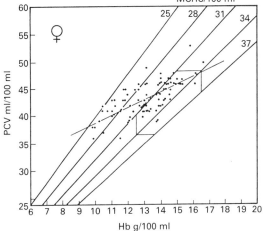

Fig. 8.12 Haematological findings in men and women of the 'Fur' tribe in western Sudan. MCHC, mean corpuscular haemoglobin concentration; PCV, packed cell volume. (Reproduced from Sukkar, M.Y. *et al.* (1976) *Afr. J. Med. Sci.* **5**, 245–54. Blackwell Scientific Publications, Oxford.)

metabolic rate, as will be described later in this section.

The production of CO_2 varies with the fuel being utilized. Thus, when glucose is the only source of energy being used at a certain time, the ratio of CO_2 produced to O_2 consumed during a given interval of time is equal to 1 (i.e. $(CO_2/O_2) = 1$). This ratio is known as the respiratory quotient (RQ). In the case of glucose oxidation, it can be appreciated from the chemical reaction:

$$C_6H_{12}O_6 + 6\ O_2 \rightarrow 6\ CO_2 + 6\ H_2O + \text{energy}$$

$$RQ\ (\text{ratio}) = \frac{6}{6} = 1.00$$

When fatty acids are used for energy production, the RQ is found to be 0.70. The RQ also goes under another name, viz. respiratory exchange ratio (R). The RQ for protein is 0.82. It appears from this description that the mixture of fuels (the metabolic mixture) being used at a given time will determine the RQ. We also know that the energy released by carbohydrates, fats and proteins is different. In the human body, 1 g of carbohydrate or protein yields 17 kJ (4 kcal) and 1 g of fat yields 39 kJ (9 kcal). The relevance of the RQ to energy metabolism is that it can be used to arrive at an energy equivalent for the O_2 utilized. Figure 8.15 gives the energy equivalent of 1 litre of O_2 utilized at different respiratory quotients, i.e. with different metabolic mixtures. In this case, the non-protein RQ is used, which means that the energy due to protein is deducted by calculation of the amount of protein used as indicated by urinary nitrogen.

The basal metabolic rate (BMR)
The rate of energy release at basal conditions is known as the BMR. A person is described to be in a basal state when he is fully rested, at least 12 hours after the last meal. Measurement of the BMR gives the minimum energy expenditure of the individual and allows comparisons to be made with normal values.

Basal metabolism represents the energy used for the vital physiological processes of the body, such as the work of the heart, respiration, brain and all the continuing cellular activities, such as synthesis of protein and the all-important active transport of ions and compounds across cell membranes.

The conditions for measuring the BMR are:
1 A good night's sleep, i.e. physical rest.
2 Last meal 12 hours before measurement.
3 The patient should be in a comfortable room temperature (no sweating or shivering or extra cardiovascular work).
4 The patient should be mentally relaxed.
A good night's sleep ensures physical and mental rest. Explaining the procedure is important so as

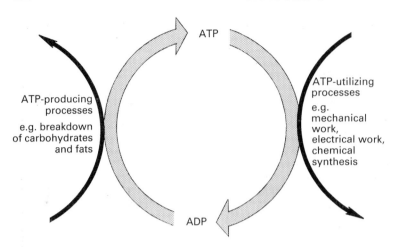

Fig. 8.13 The ATP–ADP cycle. (Redrawn from Weatheral, D.J., Ledingham, J.G.G. & Warrel, D.A. (eds) (1987) *Oxford Textbook of Medicine*, 2nd edn. With permission from the Oxford University Press, Oxford.)

not to arouse the patient's anxiety before and during measurement. A comfortable room temperature will avoid activation of mechanisms associated with temperature regulation.

The reason for having the last meal 12 hours beforehand is that food intake is associated with stimulation of secretion, gut movement and absorption processes, all of which increase energy release. But more important is the metabolic energy release in the liver stimulated by incoming food substances. The increase in energy production after food intake is known as the specific dynamic action (SDA). It is specific because each food substrate has a specific effect, e.g. protein increases the BMR by 30% of its energy content;

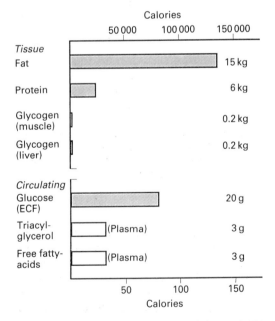

Fig. 8.14 Diagrammatic presentation of the available metabolic fuels in the normal adult. (Redrawn from Weatheral, D.J., Ledingham, J.G.G. & Warrel, D.A. (eds) (1987) *Oxford Textbook of Medicine*, 2nd edn. With permission from the Oxford University Press, Oxford.)

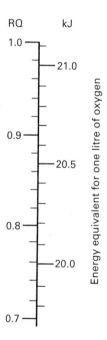

Fig. 8.15 Nomogram for determination of the energy equivalent of oxygen at different *RQs*.

carbohydrates and fats have an SDA of 6 and 4% respectively.

Factors affecting the BMR The BMR is usually expressed as kcal/m^2/hour. The unit of surface area of the body (m^2) provides a means allowing comparisons to be made between individuals of different size. The surface area can be obtained from a nomogram (Fig. 8.16) based on height and weight. Using this unified expression, it has been found that animals of different size have closely similar BMRs. This makes physiological sense because the area of the skin is closely related to the amount of heat loss. Using this expression, it was found that age and sex were factors which affect the BMR. Little infants have a higher BMR than children and adults. The BMR also falls in old age.

Females have lower BMRs than males. However, the sex difference disappears if the BMR is expressed in terms of kcal/kg lean body mass/ hour. The lean body mass is used instead of surface area. Women have more fat than men and, as fat has a lower metabolic rate than lean tissues, their BMR appears lower when this is not taken into account. Therefore, there is no real sex difference but a difference due to body composition.

Other factors include the emotional state (level of catecholamines), climate, state of the menstrual cycle, pregnancy, lactation, body temperature and level of thyroid hormones (see Chapter 10).

Measurement of the metabolic rate

Direct calorimetry This method directly measures the heat produced by the body during a given period of time. The subject sits inside a specially designed heat-insulated chamber (body calorimeter). The heat gained by the chamber is measured. This method is difficult and was used in the early days of research into the subject of energy metabolism.

Indirect calorimetry Indirect methods use the measurement of oxygen consumption for calculation of energy expenditure. A closed-circuit method employs a floating cylinder filled with

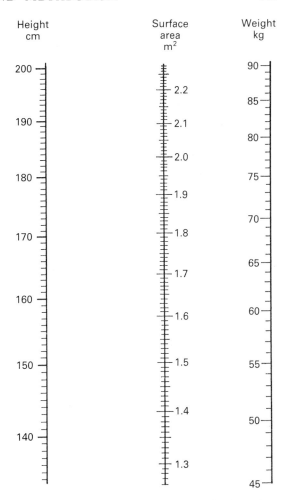

Fig. 8.16 Nomogram for determination of body surface area. A line joining the height and weight values of the subject crosses the middle scale and the value of body surface area is read in square metres.

oxygen at atmospheric pressure, from which the subject breathes. Expired air, which contains both O_2 and CO_2, is returned to the cylinder through appropriate valves (Fig. 8.17). The passage of expired air goes through a small cylinder of soda-lime, which absorbs the CO_2. Thus, when the subject continues to breathe in and out of this system, the amount of oxygen decreases gradually with each breath. This decrease is recorded graphically on a moving kymograph. Thus, a record of O_2 consumption per unit time is obtained. You

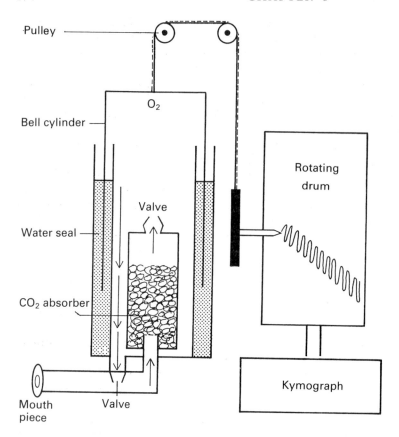

Fig. 8.17 The Benedict–Roth spirometer used for measuring O_2 consumption by the closed-circuit method.

may have the opportunity of seeing this simple method in your practical physiology course.

An open-circuit method is also used. The subject in this case breathes in atmospheric air and breathes out into a special large bag, in which all the expired air is collected over 5–10 min. The volume of expired air is then measured by a gas flowmeter. This is the Douglas bag method. A sample of the expired air is then collected from a side tube and analysed. O_2 consumption and CO_2 production can be calculated by comparison with the atmospheric air which was inspired. Other open circuits are now available which make it possible for a working man/woman to carry a small instrument into which he/she breathes out his/her expired air. The instrument measures the volume of expired air and the oxygen utilization directly. This has made it much easier to measure the metabolic rate under ambulatory and field conditions.

Metabolic homoeostasis: an integrated approach

The human body, like all biological systems, requires a regular intake of nutrients, which come from plant and animal origins. Feeding is regulated by the need for exogenous energy and other nutrient sources. The intake of food takes place at meal-times over short periods of time. Between meals, the body cells rely on endogenous stores of nutrients, which are mobilized and used according to the state of metabolism at any particular time.

Short-term changes in the metabolic state of an individual are brought about by three main variables: (i) food intake or its absence; (ii) the level of metabolic activities; and (iii) physical muscular activity. Quantitative and qualitative changes in metabolism take place rapidly with the intake of food, fasting, starvation and muscular exercise. Physiological states during periods of growth, pregnancy and lactation have a long-term influence on the state of nutrition and metabolism.

Under pathological conditions, certain changes also occur in the metabolic state of the individual, e.g. liver disease, specific nutritional deficiencies and endocrine disorders. Some diseases exert greater demand on one or more of the nutrients and thus lead to nutritional or metabolic abnormalities, e.g. loss of glucose in diabetes and albumin in renal diseases. Drug or hormone therapy, on the other hand, may also influence the state of metabolism of the patient so as to require nutritional and metabolic adjustments.

This section deals with the dynamic changes and physiological mechanisms which regulate the responses of the body to food intake and deprivation. It also addresses the needs of the body under steady physiological states and ageing.

METABOLIC RESPONSES TO FEEDING

The metabolic changes related to feeding may precede the entry of food into the mouth. The awareness of a coming meal and the sense of food in the mouth induce changes, such as delivery of saliva and gastric juice and secretion of gastro-intestinal hormones. During feeding there is a large positive fuel surplus absorbed from the gastrointestinal tract. Immediately following a meal, absorption of various nutrients results in elevated blood levels of glucose, blood lipids and amino acids. The high circulating levels of glucose stimulate insulin secretion and decrease that of glucagon. The latter is particularly evident after a carbohydrate-rich meal. Absorbed carbohydrate, mainly in the form of glucose, passes first to the liver via the portal system. The liver removes about 60–80% of carbohydrate and stores some as glycogen (via enhanced hepatic glycogenesis) for later use, and converts the rest to fatty acids and triacylglycerols. The latter two are then secreted as very-low-density lipoproteins (VLDL), whose fatty acids are subsequently stored in adipose tissue. The rest of the glucose (20–40%) enters the peripheral tissues, where it is stored as glycogen or oxidized to provide chemical energy in the form of ATP.

Meanwhile, the fat component of the meal is absorbed and then released, mainly as chylomicrons, which reach the systemic circulation via the thoracic duct to be stored in adipose tissue. In adipose tissue, triacylglycerol uptake from chylomicrons and from VLDL is stimulated by the activation of lipoprotein lipase. At the same time, the high insulin and glucose levels inhibit triacylglycerol breakdown by the hormone-sensitive lipase and enhance esterification of fatty acids. This leads to a marked decrease in the release of FFAs and consequently decreased circulating levels of FFAs.

Amino acids are not stored as such but are utilized in protein synthesis, which is increased due to the inhibition of hepatic gluconeogenesis following a meal. This effect is related to the high plasma insulin/glucagon concentration ratio. The increase in insulin concentration antagonizes the action of glucagon on the liver so that gluconeogenesis is decreased. At the same time, excess amino acids are oxidized in the liver, with the exception of branched-chain amino acids (leucine, isoleucine and valine), which tend to be broken down in extrahepatic tissues, mainly skeletal muscle.

The overall effect of the meal is therefore to promote storage of carbohydrate as glycogen in liver and muscle, storage of triacylglycerols in liver and particularly in adipose tissue, and incorporation of amino acids as protein, probably in all tissues. The metabolic changes in response to feeding are summarized in Fig. 8.18.

After the carbohydrate-rich meal the RQ will rise and approach unity, reflecting suppression of fatty acid oxidation and increased use of glucose. If the meal is rich in fat, the rate of digestion, absorption and processing of lipids to produce chylomicrons is considerably slower than the digestion of carbohydrate and absorption of glucose. A problem may arise after a meal that contains much protein and little carbohydrate, due to the fact that insulin secretion under these conditions could result in hypoglycaemia. The decreased blood glucose is due to the inhibition of glucose production and facilitation of glucose utilization. This is normally prevented by the additional secretion of glucagon in response to a high-protein meal. The absorptive phase during and after a meal blends smoothly into the post-absorptive phase.

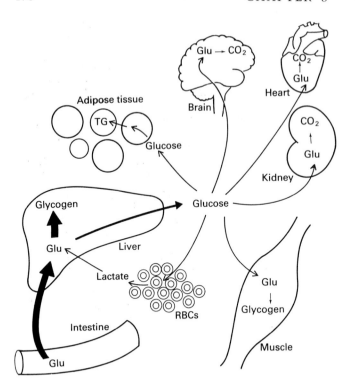

Fig. 8.18 Glucose utilization by various tissues after absorption from the small intestine. The thickness of lines indicates the flux through that route. TG, triacylglycerols.

METABOLIC RESPONSES TO FASTING AND STARVATION

For simplification, the metabolic responses to starvation can be divided, into three phases: post-absorptive phase, short-term starvation and long-term starvation. These periods are characterized by specific metabolic changes but, the transition from one to another is gradual. In practice the effects of starvation are rarely as simple as described below, since food deprivation usually occurs against a background of injury or infection, or as part of protein–energy malnutrition or famine.

Postabsorptive phase

The postabsorptive phase starts 2–3 hours after oral glucose intake. However, a mixed meal containing protein, fibre and fat delays gastric emptying and takes longer to digest and absorb from the intestine. The postabsorptive phase in this case starts 4–6 hours after the meal.

During this phase nutrients come from the body stores, i.e. the liver and adipose tissue. This process of mobilisation is triggered by the falling blood glucose. Insulin secretion is decreased and glucagon is increased. This pattern of hormone secretion leads to increased glycogenolysis and decreased glycogen synthesis (Fig. 8.19). Gluconeogenesis increases drawing upon alanine as a main source of glucose synthesis in the liver. Tissues which depend almost entirely on glucose as a source of energy continue to use blood glucose irrespective of the metabolic changes that take place during absorption or in the post-absorptive phase. It is pertinent to note that the liver is responsible for almost all of the glucose output in the post-absorptive state; most of which comes from gluconeogenesis (about 75%) and the rest from glycogenolysis.

In adipose tissue, the falling insulin and glucose concentrations decrease triacylglycerol synthesis and increase lipolysis, resulting in the enhanced release of FFAs. These are used to provide the energy needs of most tissues, such as skeletal muscle and the heart, thus sparing more glucose for other tissues. By the end of an overnight fast,

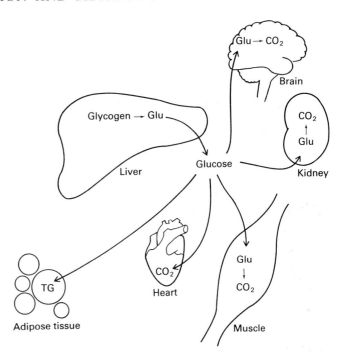

Fig. 8.19 Glucose utilization during the postabsorptive phase.

the body will be oxidizing mainly fat, as indicated by an *RQ* value in the range of 0.7–0.8, although the brain will still be using glucose almost exclusively. However, the rate of lipolysis and the decrease of insulin secretion in the postabsorptive phase is not sufficient to stimulate hepatic conversion of free fatty acids to ketone bodies.

Release of amino acids will start to occur, mainly related to a marked decrease in the synthesis of muscle proteins under the influence of low insulin levels. In skeletal muscle, amino acids will begin to provide a source of energy (although not a major one), and alanine, together with glutamine, will predominate. These amino acids are taken up in the splanchnic circulation, where they contribute to hepatic gluconeogenesis, supplementing glucose production by glycogen breakdown.

The postabsorptive phase normally extends for 6–8 hours and ends when another meal is taken, or it may extend for 16–40 hours as during fasting. The Ramadan fast lasts about 15 hours in the tropics. Therefore, it can be considered as a postabsorptive state with little or no effects attributable to starvation. As a matter of fact, the Ramadan fast enables endocrine and metabolic mechanisms to function to a fuller extent, unhampered by the introduction of meals at short intervals.

Short-term starvation
The metabolic changes that occur during periods of short-term starvation (2–7 days) are a continuation and acceleration of the processes that started during the postabsorptive period. The metabolic changes occurring in short-term starvation are summarized in Fig. 8.20.

During this period, the progressive decline in blood glucose and insulin concentrations accompanied by elevated glucagon secretion leads to increasing release of amino acids (mainly as glutamine and alanine) from muscle and increasing lipolysis in adipose tissue. Alanine, removed by the liver, lactate, produced by erythrocytes and other tissues, and glycerol, released by lipolysis, all contribute to an increasing rate of hepatic gluconeogenesis. During this period, gluconeogenesis is two to three times greater than during

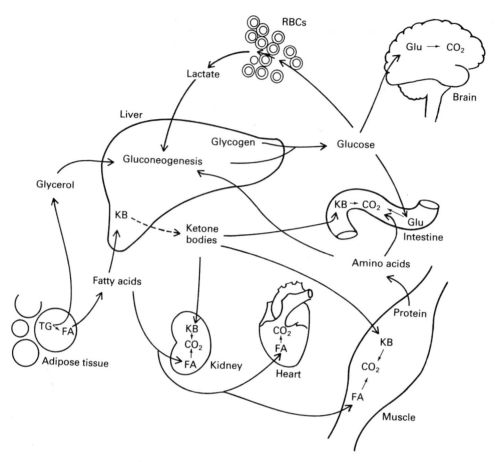

Fig. 8.20 Pattern of fuel utilization during short-term starvation.

the postabsorptive period and is responsible for virtually all of the glucose released by the liver. Estimation of glucose production from various gluconeogenic precursors during this period of starvation is presented in Table 8.19.

The increase in gluconeogenesis during short-term starvation is associated with an increase in protein degradation and thus a substantial negative nitrogen balance exists, with most of the nitrogen being excreted as urinary urea. The liver at this stage takes up increasing quantities of FFAs released from adipose tissue. An increasing proportion of FFAs, under the influence of the low insulin/glucagon concentration ratio, will be channelled into oxidation, leading to increased ketogenesis. As a result, plasma levels of ketone

bodies are increased. At this stage, ketone bodies will begin to provide a significant contribution to energy needs of various tissues, including muscle and brain. In fact, oxidation of ketone bodies may provide as much as 10–20% of the energy requirement of the brain.

Thus, the overall metabolism during short-term starvation is characterized by high rates of protein catabolism, gluconeogenesis and ketogenesis to meet the body's energy needs.

Long-term starvation
The metabolic changes that occur during long-term starvation (2–6 weeks) account for the ability of the body to survive prolonged periods without any caloric intake. These changes include

Table 8.19 Approximate estimates of gluconeogenic precursors during starvation in man

Gluconeogenic precursors	Glucose produced each day (g)		
	Duration of starvation		
	1 Day	2–4 Days	Several weeks
Lactate	40	40	40
Glycerol	20	20	20
Amino acids	60	50	15
Total glucose produced from above precursors	120	110	75
Glucose available for brain (from gluconeogenic precursors)	80	65	35
Brain energy needs (glucose equivalents)	100	100	100

Data taken from Cahill *et al.* (1966) *J. Clin. Invest.* **45**, 1751–69; Owen *et al.* (1969) *J. Clin. Invest.* **48**, 574–83; Owen *et al.* (1979) *Adv. Expt. Med. Biol.* **111**, 119–88.

decreased rate of protein breakdown and nitrogen excretion and increased dependence on triacylglycerol from adipose tissue for energy (see Fig. 8.21).

In prolonged starvation, hepatic glucose production decreases from about 200 g/day during the postabsorptive period to about 50 g/day during prolonged starvation, with about 75% of hepatic glucose being derived from non-amino acid sources (i.e. lactate, pyruvate and glycerol). At the same time, renal gluconeogenesis becomes an important source of carbohydrate, contributing about 40 g/day (derived from amino acids), with excess nitrogen being secreted as ammonia rather than urea. Consequently, nitrogen loss progressively decreases during prolonged starvation. Moreover, since the demand for glucose is decreased during this period, the rate of release of amino acids from muscle is also decreased, eventually to about one-third of that in the postabsorptive state.

The increased rate of lipolysis of adipose tissue triacylglycerol appears to be brought about mainly by the very low insulin concentration in prolonged starvation, since sympathoadrenal activity is decreased. The increased delivery of fatty acids to the liver results in an increase in the rate of ketogenesis. This results in a progressive rise in blood ketone body concentrations to a steady level around 8–9 mmol/litre. At this concentration, they provide an important fuel for the brain and, as starvation progresses, they displace glucose as the main energy source, providing 60–80% of its energy needs. The high levels of plasma ketone bodies do not usually produce ketoacidosis because under these conditions, unlike in diabetes, ketone body production is controlled. Ketoacidosis is also prevented by the production of ammonia from renal gluconeogenesis (mainly from glutamine) which increases the renal capacity to excrete H^+. The high levels of ketone bodies also appear to contribute to the decrease in gluconeogenesis associated with prolonged starvation, by inhibiting muscle proteolysis. Therefore, ketone bodies both decrease hepatic glucose production and serve as an alternative fuel for the brain. Consequently, the two major requirements for survival during prolonged starvation are met, namely, maintenance of blood glucose levels (Fig. 8.22) and conservation of body proteins. This control of protein degradation by fat oxidation is also illustrated in the

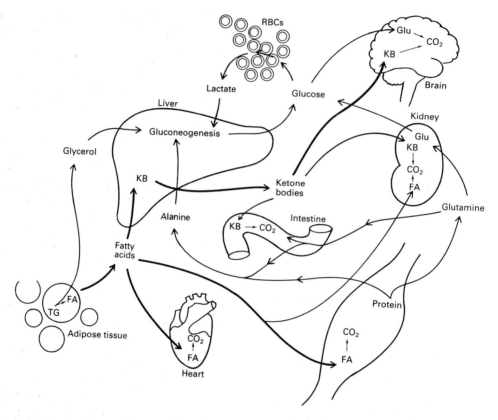

Fig. 8.21 Pattern of fuel utilization during prolonged starvation. The thickness of lines is approximately proportional to the flux through the route. FA, fatty acids; KB, ketone bodies; TG, triacylglycerols.

terminal stages of starvation. As mobilizable fat supplies eventually run out, a preterminal rise in protein catabolism starts. The ability to withstand prolonged starvation is thus dependent on the initial amount of adipose tissue.

Among other hormones, beside insulin and glucagon, that are important in the adaptation to starvation are thyroid hormones, growth hormone and cortisol. Starvation causes a dramatic decrease in plasma levels of free tri-iodothyronine (fT_3) and a corresponding increase in reverse tri-iodothyronine (rT_3) concentrations. The decrease in thyroid hormones is probably responsible for the decrease in the basal metabolic rate that usually occurs with prolonged starvation. In addition, both growth hormone and cortisol appear to contribute to the maintenance of blood glucose and FFA mobilization.

Manifestations of starvation observed in famine victims include loss of subcutaneous fat, severe muscle wasting and loss of elasticity of the skin. The skin becomes a dull colour and may show areas of pigmentation. The immune response is diminished and white cell production is decreased. The decreased body defences make starvation patients more exposed to infection, e.g. diarrhoeal diseases, amoebiasis, tuberculosis and pneumonia. The body defences are also decreased due to impairment of the immune response and decreased numbers of white blood cells. Another cause of death is shock from decreased blood volume. This may be related to hypoalbuminaemia due to diminished albumin production, which lowers the plasma colloid osmotic pressure, causing fluid to move from the blood to the interstitial space (famine oedema).

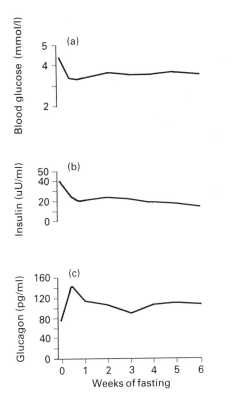

Fig. 8.22 Changes of the concentrations of: (a) plasma glucose; (b) insulin; and (c) glucagon during a period of prolonged fasting. (Based on Marliss *et al.* (1970) *J. Clin. Invest.* **49**, 2256.)

METABOLISM IN OLD AGE

Ageing is a normal process. It starts with conception and ends only with death. Different tissues age at varying rates, depending on various factors, among them nutrition. With advancing age, the senses of taste, smell, sight, hearing and touch diminish at different rates in different individuals. Various gastrointestinal changes take place with age: there is a decrease in the volume and rates of salivary, gastric, biliary and intestinal secretions and a decrease in gastrointestinal motility. The latter makes constipation a common problem in the elderly. The capacity of the lungs and the amount of blood that the heart can pump diminish, as total peripheral resistance increases with age. Ageing is also associated with decreased blood flow through the kidneys, and the number of nephrons diminishes.

The basal metabolic rate decreases with age, mainly because of the loss of a metabolically more active lean tissue mass (especially muscle) in favour of adipose tissue. Consequently, there is a decrease in energy requirements with age. In addition to the normal decline in metabolism, some 10–15% after the age of 50 years, there is almost always a decrease of physical activity, which lowers the need for energy still further. Thus, it seems advisable that most of the decrease in calories should come from a reduction in carbohydrate and fat in the diet. Elderly subjects require less energy but not less protein, vitamins and minerals. The capacity to metabolize both glucose and fats diminishes with age. Glucose tolerance appears to decrease with age. This decrease can be the result of decreased insulin secretion in response to a glucose load and/or decreased tissue responsiveness to the action of insulin. The latter may be due to an increased proportion of adipose tissue in the elderly, regardless of body-weight. Therefore, they are more susceptible to temporary hypo- or hyperglycaemia than younger subjects.

Various epidemiological studies show that the serum cholesterol level reaches a maximum between 50 and 59 years in men and between 60 and 69 years of age in women; then it falls in later years. Serum triacylglycerol levels increase with ageing and probably reflect a decreased capacity to remove dietary fat from the circulation. Restricting dietary fat, particularly saturated fat, may be helpful, but fat intake should still be about 30% of the total energy intake in the elderly.

Body protein mass decreases with age. The elderly have a poorer appetite and diminished digestive functions, which may re-establish the nitrogen balance at a lower level. This leads to a decrease in skeletal muscle mass. The decrease seems to appear more rapidly in men than in women. In healthy elderly subjects, body protein is 60–70% that of younger subjects. Stressful physical and psychological stimuli can induce a negative nitrogen balance. Infection, altered gastrointestinal function and metabolic changes caused by chronic disease can further decrease the efficiency of dietary nitrogen utilization with advancing age.

Metabolism and nutrition in disease and convalescence

An understanding of changes in the metabolic state in patients with different disease processes is useful for proper management. Nutrition of the patient, either by oral diet therapy or by parenteral nutrition, is an important part of the overall patient management. The metabolic changes which occur in tissues or organs due to disease or injury determine the nutritional needs of the patient. Diet therapy for specific diseases is outside the scope of this text and may be found in suitable reference books. The purpose of the following account is to give an overview of the general aspects of altered metabolism in two major conditions frequently encountered in hospital practice.

METABOLIC RESPONSES TO INJURY

Injury, such as major surgery or mild to severe accidental trauma, sepsis and burns, produces a continuous spectrum of metabolic changes directed towards survival. There are many variables involved that relate to nutrition and metabolism, e.g. the severity of the injury, the previous state of nutrition of the individual and the nutrients consumed following injury. The so-called metabolic response to injury is due to the powerful afferent stimuli of pain, blood loss, volume depletion and tissue damage. The responses to these stimuli include several hormonal and biochemical changes leading to the provision of readily utilizable fuels. In addition, injured individuals are subjected to variable degrees of immobility and food deprivation, which results in decreased nitrogen and caloric intakes. However, the biochemical effects of injury are not the same as those of starvation, despite some similarities (Table 8.20).

It has been found convenient to divide the metabolic response to injury into two phases: the immediate 'ebb' phase and the subsequent adaptive catabolic or 'flow' phase, in which necrobiosis and death can occur (Fig. 8.23).

The 'ebb' phase

The 'ebb' phase occurs immediately following injury and is usually relatively short in duration (12–48 hours), depending on many factors, such as the severity of the injury and the treatment given. The responses to injury may begin before the injury itself, with awareness of approaching danger activating hypothalamic mechanisms. This results in the secretion of various pituitary hormones, including adrenocorticotrophic hormone (ACTH), growth hormone (GH), prolactin and vasopressin. The increased release of ACTH stimulates the adrenal cortex to elaborate cortisol. At the same time, a sympathetic nervous reaction occurs, which results in three important effects:

1 It stimulates adrenaline release from the adrenal medulla. This hormone plays an important role in maintaining cardiovascular integrity during the 'ebb' phase, but may also produce metabolic changes if the increased secretion rate continues.

2 It stimulates glucagon and inhibits insulin secretions from the pancreas.

3 It stimulates the release of glucocorticoids from the adrenal cortex.

In addition, noradrenaline levels rise during and after injury. Excitement, pain, fear and hypovo-

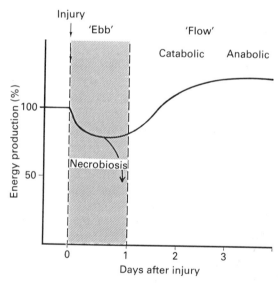

Fig. 8.23 Diagrammatic representation of the metabolic phases after injury. The time-scale is variable and the phases merge into each other. (Redrawn from Weatheral, D.J., Ledingham, J.G.G. & Warrel, D.A. (eds) (1987) *Oxford Textbook of Medicine*, 2nd edn. With permission from Oxford University Press, Oxford.)

Table 8.20 Similarities and differences between the effects of injury and starvation[a]

	Total starvation	Injury	Starvation and injury
Weight loss	++	+	+++[b]
Nitrogen loss[c]	+	++	+++
Blood glucose	Decrease	Increase	Increase
Blood alanine	Decrease	Decrease	Decrease
Blood BCAA[d]	Increase	Increase	Increase
Endocrine response			
Catecholamine	Decrease	Increase	Increase
Cortisol	Decrease	Increase	Increase
Insulin	Decrease	Decrease (early)	Decrease
Metabolic rate	Decrease	Increase	Increase
Ketonaemia	++	Variable	Variable
Water and sodium	Early loss	Retention	Retention
Potassium loss	+	++	+++

a Adapted from Smith, R. (1983) Special nutritional problems. In Weatheral, D.J., Ledingham, J.G.G. & Warrell, D.A. *Oxford Textbook of Medicine*, pp. 8–43. Oxford University Press, Oxford.
b Clearly, the weight loss resulting from injury and starvation will be more than that from either alone.
c Nitrogen loss is a reflection of the gluconeogenesis which occurs early in all these conditions; in the starving subject this falls with increasing utilization of fat.
d Branched chain amino acids—leucine, isoleucine and valine.

laemia, which accompany the injury, are potent stimuli of the sympathetic nervous system.

Stimulation of the renin–angiotensin system and alterations in serum osmolality of body fluids secondary to injury stimulate the secretion of aldosterone and antidiuretic hormone (ADH) (see Chapter 10). Aldosterone is a potent stimulator of renal sodium retention, while ADH stimulates renal tubular water reabsorption. Although the neural and hormonal mediators that result from tissue injury may stimulate aldosterone release, renin in response to hypovolaemia and afferent signals from volume receptors appear to be the major stimuli for aldosterone and ADH release respectively.

In metabolic terms the 'ebb' phase is characterized by a rapid mobilization of the glycogen and triacylglycerol fuel stores in a 'fight or flight' response, together with apparent restraint on their utilization. However, unlike the physiological 'fight or flight' response, in which the energy mobilized is freely utilized in increased physical activity, the 'ebb' phase of the response to severe injury is characterized by a lack of sensitivity to external stimuli and a decrease in metabolic rate, which is certainly not raised to the extent expected from the degree of sympathetic activity and fuel availability. In patients who do not receive hospital treatment until 30–36 hours after injuries, similar changes are still seen. Provided that early death does not occur, the 'ebb' phase merges into the 'flow' phase.

The 'flow' phase

The 'flow' phase is the catabolic phase of the response to injury. It is more prolonged than the 'ebb' phase. It is characterized by an increase in metabolic rate and breakdown of body proteins. The duration and intensity of this phase vary according to the severity of the injury. For example, in patients with long-bone fracture, the 'flow' phase peaks around 7–10 days after injury, gradually subsiding and merging into the anabolic or convalescent phase over the next 2–4 weeks, depending partly on the speed at which the patient can be mobilized.

During the 'flow' phase, injured patients have stable haemodynamic parameters. The typical metabolic changes of the 'flow' phase are:

1 Increased metabolic rate (hypermetabolism state) in association with elevated core temperature and pulse rate.

2 An increase in urinary excretion of nitrogen, mainly from muscle protein breakdown.

The latter may be of short duration and subsides as the convalescing patient is provided with adequate energy substrates and amino acids for hepatic protein synthesis. However, if this catabolic response is prolonged, the loss of body protein can pose a threat to life, through pulmonary and cardiovascular insufficiency and impaired immune function and wound-healing. It is for these reasons that intensive appropriate nutritional intervention is vital so as to reverse or at least slow down protein loss in catabolic patients.

The course of the response to sepsis or major infection is not as predictable as that described for injury, although the characteristic features of the 'ebb' and 'flow' phases of the response to injury are seen. They are more related to the severity than to the time course of the infection. Thus, the critically ill septic patient, particularly with septic shock, displays many of the features of the 'ebb' phase after injury, while the patient with more chronic but less life-threatening infection will display the hypermetabolism and catabolism typical of the 'flow' phase after injury.

In conclusion, hormonal changes act in concert to produce the metabolic changes in response to injury. They stimulate glucose production by the liver and FFA mobilization from adipose tissue. The decrease in insulin and increases in glucagon and adrenaline secretions promote hepatic glycogenolysis. Muscle glycogenolysis is stimulated, mainly via the action of adrenaline. The alterations in these hormones and increased cortisol secretion favour hepatic gluconeogenesis. Consequently, glucose production by the liver increases, leading to hyperglycaemia. At the same time, the fall in insulin and increase in adrenaline levels stimulate lipolysis in adipose tissue, thus increasing plasma FFA concentrations. The increase in cortisol output also contributes to FFA release, by enhancing the lipolytic actions of adrenaline. However, the basal levels of insulin secretion seem to have an antilipolytic effect, so that the body does not adapt by oxidizing fatty acids and ketone bodies for energy during the catabolic response to injury.

Injured patients exhibit some form of insulin resistance, as indicated by a diminished insulin response to glucose and tissue resistance to insulin action. Both effects prevent muscle tissue from utilizing glucose, and a local fuel deficit develops. To satisfy its requirements for energy, skeletal muscle oxidizes branched-chain amino acids from its own tissues. As protein is mobilized for energy, the excess ammonia that results from the oxidation is attached to pyruvate and carried back to the liver as alanine. Alanine and other amino acids stimulate glucagon secretion, which enhances hepatic gluconeogenesis and ureagenesis and decreases the blood concentrations of alanine, glycerol and lactate. Protein catabolism during the 'flow' phase also provides the amino acids used for hepatic protein synthesis needed during stress (e.g. immunoglobulins, albumin).

If the stress continues, the body maintains its sympathetic response and shows some adaptation to prolonged stress. Under these conditions, nutritional support must be started, either enterally or parenterally. The main feature of this stage is that the body uses smaller amounts of muscle protein for its energy needs and adapts to using fatty acids and ketone bodies for energy, so that the negative nitrogen balance is improved. Nutritional support is more efficiently used when the levels of adrenaline and glucocorticoids subside and no longer antagonize the actions of insulin, and thus energy and protein given at this time can be used more efficiently. Nutritional intervention should be aimed at establishing an anabolic state and restoring the tissue protein lost during the response to injury or sepsis.

As the stress of injury or sepsis is relieved by treatment or by its own natural course, the sympathetic response decreases and the parasympathetic activity increases. The patient begins to feel hungry, which is a good sign, indicating recovery.

METABOLIC RESPONSES TO INFECTION

Many of the metabolic responses to infection are similar to those described following injury (see above). Severe infection is characterized by prolonged fever, hypermetabolism and changes in fuel metabolism. The catabolic response to infection begins after the onset of fever (see Fig. 8.24). Accelerated protein degradation, increased nitrogen excretion and prolonged negative nitrogen balance occur following severe infection (e.g. typhoid fever, pneumonia and tuberculosis). These changes are also seen in patients with sepsis.

It is generally accepted that in septic patients glucose turnover is increased and that of gluconeogenesis is enhanced, despite the associated hyperglycaemia. This may be related to the dependence of inflammatory and reparative tissues on glucose, which optimizes host defence mechanisms and ensures wound repair.

The preference for fat as a metabolic fuel is more pronounced in septic than in injured patients. Therefore, fat is the major metabolic fuel in infected patients, and increased mobiliz-

ation of peripheral lipids is especially prominent during periods of inadequate nutritional support.

During starvation, hepatic uptake of FFAs is coupled with enhanced hepatic ketogenesis and the concentration of ketone bodies rises. This change does not occur in infected patients. This hypoketonaemia may be a consequence of the hyperinsulinaemia associated with septic states and may explain accelerated protein degradation in muscles of infected patients.

The hormonal responses during the hypermetabolic phase of infection are similar to those described following injury. Serum cortisol levels are increased and lose their usual circadian rhythm. Glucagon levels are increased and insulin levels are normal or even elevated. The insulin–glucagon concentration ratio remains below normal, indicating hepatic stimulation of gluconeogenesis. Levels of catecholamines, growth hormone, ADH and aldosterone are all elevated. The growth hormone level persists into convalescence, presumably to promote anabolism.

During the acute phase of infection, it is difficult to maintain a positive nitrogen balance. This is related to the patient's anorexia and the catabolic stress response which promotes tissue protein breakdown. However, the body adapts in

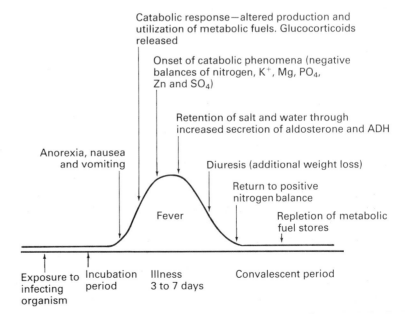

Fig. 8.24 Timing of catabolic response to infection. (Reproduced with permission from Beisel, W.R. (1976) The influence of infection or injury on nutritional requirements during adolescence. In McKigney, J.I. & Munro, H.N. (eds) *Nutrient Requirements in Adolescence.* MIT Press, Cambridge, Mass.)

Catabolic response—altered production and utilization of metabolic fuels. Glucocorticoids released

Onset of catabolic phenomena (negative balances of nitrogen, K^+, Mg, PO_4, Zn and SO_4)

Retention of salt and water through increased secretion of aldosterone and ADH

Diuresis (additional weight loss)

Return to positive nitrogen balance

Repletion of metabolic fuel stores

Anorexia, nausea and vomiting

Fever

Exposure to infecting organism

Incubation period

Illness 3 to 7 days

Convalescent period

3–4 days and it may then be possible to achieve a positive nitrogen balance, depending on the severity of the infection. Nutritional intervention should be started as early as possible. After the catabolic phase subsides, an anabolic phase follows, during which energy and protein should be provided at higher levels. The nutritional goal is to put the patient into a positive nitrogen balance.

Drugs, nutrition and metabolism

Drug metabolism may change in states of nutritional deficiency or nutritional manipulation. This is because the activity of the hepatic microsomal enzyme drug-metabolizing system (mainly the cytochrome P_{450}-dependent mixed-function oxidase system) is influenced by the dietary intake of carbohydrate, protein and fat. Thus, the ratio of protein to carbohydrate in the diet can affect the rate of drug metabolism. The half-life of both antipyrine and theophylline, for example, is markedly decreased when the diet is changed from a low protein/high carbohydrate to a high protein/low carbohydrate content. PEM may decrease the activities of drug-metabolizing enzymes, and also of substrates derived from nutrients used for the conjugation of drugs. Malabsorption caused by PEM may interfere with the pharmacological response to drugs. For example, the clearance of a known dose of chloramphenicol has been shown to be delayed in children with PEM. Mineral and vitamin deficiency may have similar effects. Thiazide diuretic-induced potassium deficiency increases the risk of cardiac arrhythmias in patients taking digitalis. Subclinical vitamin C deficiency may contribute to the increased incidence of adverse drug reactions in the elderly.

There are many ways in which drugs affect the metabolism of nutrients and some are used to do so. There are drugs that are used to decrease appetite in the treatment of obesity (e.g. fenfluramine) or to stimulate appetite in cases of prolonged convalescence or anorexia (e.g. sulphonylureas).

Carbohydrate metabolism is affected by drugs that cause either hyper- or hypoglycaemia directly or by influencing hormone secretion. Drugs causing hyperglycaemia include thiazide diuretics, which should be used with caution in the treatment of patients with impaired glucose tolerance (e.g. diabetes mellitus). Similar hyperglycaemic effects are produced by corticosteroids, oral contraceptives and anticonvulsant drugs (e.g. diphenylhydantoin, nicotinic acid). An overdose of sulphonylureas may lead to hypoglycaemia by stimulation of insulin secretion, especially when they are given with other drugs. Severe hypoglycaemia may result from ethanol ingestion by acutely starved or chronically malnourished subjects; the hypoglycaemia in the latter conditions is related to the inhibition of hepatic gluconeogenesis.

Plasma lipids are decreased or increased by a number of drugs. Some drugs cause a mild fat malabsorption (e.g. neomycin, kanamycin) or bind bile acids (e.g. cholestyramine resin used in type II hyperlipidaemia), leading to lower plasma lipids. D-Thyroxine lowers plasma cholesterol by increasing hepatic catabolism. Oral contraceptives, large doses of adrenal corticosteroids, ethanol, chlorpromazine and thiouracil are all examples of drugs which raise plasma lipids as a side-effect of their therapeutic use.

Urinary nitrogen excretion is enhanced by corticosteroids, thyroid hormone and tetracyclines. In addition, insulin and oral contraceptives decrease circulating levels of some amino acids.

Mineral and vitamin metabolism may be affected in many ways by the use of drugs. The use of corticosteroids or oral contraceptives may result in sodium retention, whereas the use of thiazide diuretics may result in potassium depletion. Prolonged use of adrenal steroids also causes osteoporosis, increasing the risk of bone fractures, especially in the elderly. The use of oral contraceptives may decrease zinc and increase copper in plasma. Impaired iodine uptake or release, leading to goitre, may be caused by sulphonylureas, phenylbutazone, cobalt and lithium.

Ethanol impairs the absorption of several vitamins (e.g. thiamine, folic acid and vitamin B_{12}). Isoniazid, used in the treatment of tuberculosis, is a pyridoxine antagonist and may also cause pellagra. Vitamin C status is impaired by the use of tetracycline, oral contraceptives, corticosteroids and high doses of aspirin.

Further reading

1 Newsholme, E.A. & Leech, A.R. (1983) The integration of metabolism during starvation, refeeding and injury. In *Biochemistry for the Medical Sciences*, 1st edn, pp. 536–61. John Wiley, Chichester.

2 Passmore, R. & Eastwood, M.A. (1986) *Davidson and Passmore's Human Nutrition and Dietetics*, Part 1: *Physiology*, 8th edn, pp. 3–168. Churchill Livingstone, Edinburgh.

9: Gastrointestinal Physiology

Objectives

On completion of the study of this section, the student should be able to:

1 Become familiar with significant structural features of the gastrointestinal tract so as to relate structure to function.

2 Differentiate between the various types of intestinal movements so as to comprehend their role in digestion and to interpret symptoms and signs related to gut motility.

3 Understand the neurological and hormonal mechanisms which control the digestive functions, in particular:

  (a) Control of salivary secretions.
  (b) Control of gastric secretions.
  (c) Control of secretions of pancreatic juice.
  (d) Control of formation and release of bile.
  (e) Control of secretion of intestinal juice.

4 Apply the above knowledge in selecting methods for assessment of the functions of the stomach, pancreas, liver and gall-bladder.

5 Understand the processes of digestion and absorption of various food substances so as to give appropriate dietary advice to patients.

6 Understand the functions of the colon in order to explain clinical problems such as diarrhoea and constipation.

Introduction

The energy which we need for warmth, movement, growth and maintenance of homoeostasis is ultimately derived from the food we eat. But food materials are mostly complex substances consisting of carbohydrates, lipids, proteins, vitamins and minerals. Before they can be absorbed into the body, carbohydrates, fats and proteins have to be broken down or modified to simpler forms. It is the function of the gastro-intestinal tract to secrete enzymes which break down the complex food materials into simple forms, a process referred to as digestion, and then to convey the end-products to the bloodstream, the process being referred to as absorption. While the food is being processed, it needs to be mixed and propelled along the tract, and this is achieved by various types of co-ordinated movement (motility). The functions of motility, secretion, digestion and absorption of food are integrated at each level of the gastrointestinal tract. To understand these functions, the reader will follow the path of food from the mouth to the anus, rather than deal with each of the major functions separately, i.e. motility, secretion and absorption throughout the tract.

Anatomical considerations

The gastrointestinal tract is simply a tube with muscle walls throughout its length; it is lined by an epithelium which is adapted for the function of the various parts of the tract. The salivary glands, the liver and the pancreas pour their secretions into the gut lumen. Basically, the tube is composed of four layers (Fig. 9.1), which are, from within outwards:

1 The mucosa, consisting of the lining epithelium, interepithelial connective tissue or lamina propria and a layer of longitudinal smooth muscle, the muscularis mucosae.

2 The submucosa, containing, in addition to connective tissue, the blood-vessels and lymphatics, a network of neurones and nerve fibres referred to as the submucous or Meissner's plexus.

3 The intestinal muscle layers consist generally of inner circular and outer longitudinal layers. This basic arrangement may be modified, e.g. the longitudinal layer is modified into the taeniae coli in the large intestine. The circular layer is modified in certain places to form sphincters. Except for the muscle in the upper third of the oesophagus and in the external anal sphincter, which is striated, the muscle in the rest of the alimentary tract is smooth. Between the two external muscle layers lies the myenteric (Auerbach's) plexus, containing a network of neurones and fibres.

4 The serosa provides an outer fibrous coat and continues on to the mesentery and the peritoneal lining of the abdominal cavity.

BLOOD SUPPLY

The gut is well supplied with blood. Blood from the stomach, most of the intestine, the pancreas and the spleen drains via the portal vein into the liver, and the combined vascular bed of these organs comprises the splanchnic circulation. The splanchnic vascular bed receives about 25–30% of the cardiac output.

Control of gastrointestinal functions

Generally, secretion and motility in the gastro-intestinal tract are controlled by nervous and

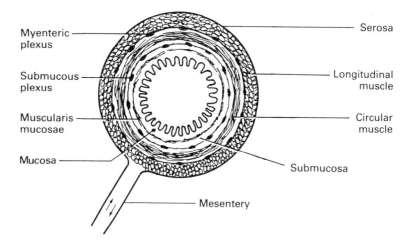

Fig. 9.1 Diagram of a transverse section through the gut to show the layers in the gut wall.

Myenteric plexus

Submucous plexus

Muscularis mucosae

Mucosa

Serosa

Longitudinal muscle

Circular muscle

Submucosa

Mesentery

hormonal mechanisms, which are closely integrated so that control may better be described as neurohormonal.

THE ENTERIC NERVOUS SYSTEM (ENS)

Mechanical and chemical changes in the gut are conveyed to the central nervous system via visceral afferent fibres travelling in the vagi and spinal nerves, while efferent fibres reach the gut in sympathetic and parasympathetic nerves.

Sympathetic fibres pass in the splanchnic nerves, which synapse in the ganglia in front of the aorta, and postganglionic fibres reach the gut. Most of the parasympathetic outflow to the gut is in the vagi, which provide fibres to the gut from the oesophagus down to the middle of the transverse colon. The remaining part of the colon, rectum and anal canal receive parasympathetic fibres in the pelvic splanchnic nerves. All the parasympathetic fibres are preganglionic; they synapse with postganglionic neurones in the submucous and myenteric plexuses, and short postganglionic fibres reach glands and smooth muscle cells of the gastrointestinal tract.

Thus, according to the classical concept of the autonomic nervous system, the neurones and nerve fibres in the wall of the gut (intrinsic neurones) are merely parts of the system consisting of sympathetic (noradrenergic) fibres and parasympathetic (cholinergic) neurones. In addition to these, the intrinsic nerve supply, consisting of the submucous and myenteric plexuses, contains many other types of neurones, some of which release serotonin (5-hydroxytryptamine (5-HT)) as transmitter (serotoninergic), some release peptides (peptidergic) and others release transmitters that are not yet known with certainty. The total number of neurones in the gut is about 100 million. These neurones have a high degree of functional independence from the central nervous system. When all the extrinsic nerves of the gut are cut, gut neurones are able to maintain the complex functions of secretion and motility on their own. The view now is that the neurones and fibres in the wall of the gut, collectively referred to as the ENS, are more than a station in the autonomic nervous system.

In general, stimulation of the extrinsic parasympathetic nerves to the gut results in increased motility, relaxation of sphincters, stimulation of secretion and vasodilatation. Stimulation of the sympathetic nerves inhibits motility, contracts sphincters and causes vasoconstriction; secretion is not necessarily inhibited, however, and on some occasions may be modestly stimulated.

On the other hand, the intrinsic nerves have important functions. Meissner's plexus subserves sensory functions, including responses to stretch and chemical composition of gut contents in contact with the epithelium. Stimulation of this submucous plexus leads to increased blood flow and increased secretions. Stimulation of Auerbach's plexus (myenteric plexus) leads to increased muscle tone in the gut, increased rate and intensity of rhythmic intestinal contraction and increased velocity of conduction of excitatory waves.

THE GUT AS AN ENDOCRINE ORGAN

The gut occupies a unique position in endocrinology. In 1902, Bayliss and Starling discovered and isolated for the first time a substance from the upper intestine which stimulated exocrine secretion of the pancreas. They called this substance secretin. They introduced the word 'hormone' (Greek for 'I arouse to activity'), which was later used to describe secretions of endocrine glands.

Endocrine cells are found scattered in the wall of the gut from the stomach to the colon in order to cater for the changing needs of secretion and motility; if collected together, their mass would be greater than any of the familiar endocrine glands. All the known gastrointestinal hormones are peptides. Many of the endocrine peptide-containing cells also contain an amine, such as serotonin, dopamine or histamine. When an amine is absent, the enzyme that forms it (amino acid decarboxylase) may be present, and this has led to the hypothesis that all gut endocrine cells possess the APUD property (amine precursor uptake and decarboxylation). It has further been suggested that gut endocrine cells are derived from the neural crest and have migrated to the gut. These cells are receptor−secretory in function, i.e. they release their secretions in response to adequate stimuli acting on their membrane receptors. They also possess neurone secretion-like or synaptic vesicle-like granules. They may more generally be designated 'paraneurones', to indicate their fundamental similarity to neurones. Many peptides and amines found in the gut endocrine cells are also present in the brain. This has led to yet another term: the brain–gut axis. What all of this seems to emphasize is the fundamental unity of the control systems of nerves and hormones.

The gut produces numerous peptides, only four of which have established status as gastrointestinal hormones. These are: secretin, gastrin, cholecystokinin-pancreozymin (CCK-PZ or CCK) and gastric-inhibitory peptide (GIP). The remaining peptides have various pharmacological effects still under investigation. Since their physiological role is not yet fully understood, they may be referred to as 'candidate' hormones.

There are three ways in which a gastrointestinal peptide may act:

1 *Endocrine*: the peptide is released into the blood, through which it reaches its target organ or cell to act (e.g. gastrin, secretin).

2 *Paracrine*: the peptide is released into the interstitial fluid and diffuses along it to act on a neighbouring target cell (e.g. somatostatin).

3 *Neurocrine*: the peptide acts as a neurotransmitter in this case, i.e. it is released from a nerve ending into a junctional cleft to act on a target cell (e.g. vasoactive intestinal peptide (VIP), gastrin-releasing peptide (GRP)).

In all cases, the target cells possess receptors which are specific for the peptide.

Mastication, saliva and deglutition

MASTICATION

Solid food is broken down by the process of mastication (chewing). Mastication also stimulates the secretion of saliva and helps to mix it with the food particles to produce a bolus which can then be swallowed. Mastication is started voluntarily but continues as an involuntary reflex. The contact of the food particles with the teeth and gums induces reflex opening of the jaws followed by closure; this produces rhythmic closure and opening at about one cycle per second. Considerable force can be applied to crush solid pieces of food; the molars can exert a force of

more than 100 kg. When the bolus is of suitable consistency and size, it will be swallowed.

SALIVARY GLANDS

Anatomical considerations

Most of the saliva in the mouth comes from three major pairs of glands, the parotid, submandibular and sublingual glands, which discharge their secretions into the mouth through ducts.

A small amount of saliva comes from small salivary glands scattered in the mucosa of the mouth and pharynx.

Salivary glands are typical exocrine glands. The basic secretory units are the acini, which are drained by intercalated ducts. These join to form intralobular or striated ducts, which lead to the interlobular ducts, which drain into the main duct opening into the mouth (Fig. 9.2).

There are two types of cells in the acini: serous cells, which contain granules and secrete electrolytes, water and the enzyme ptyalin (salivary amylase), and larger mucous cells, which secrete mucus. An acinus may be purely serous or purely mucous or may be mixed. In man, the parotid gland is purely serous, while the submandibular and sublingual glands are mixed. The cells lining the intralobular ducts deserve a special mention; their basal membrane is extensively folded and they contain a lot of mitochondria, features which give the basal part of these cells a striated appearance and account for the name of striated ducts. As evidenced by the high content of mitochondria, these cells are metabolically very active; they are, in fact, mainly responsible for active transport of electrolytes which modify the composition of the primary acinar secretion. Myoepithelial cells are found between the basement membrane and the cells lining the lumen of acini and intralobular ducts. When stimulated, these cells contract and increase salivary flow.

Nerve supply

Salivary glands receive sympathetic and parasympathetic nerves. The sympathetic nerves originate in the upper four thoracic segments of the spinal cord, synapse in the superior cervical ganglion and reach the three pairs of salivary glands in company with the arteries. The para-

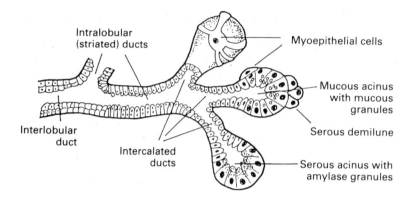

Fig. 9.2 Diagram of the acini and ducts of the submandibular salivary gland and the types of lining cells.

Labels: Intralobular (striated) ducts; Myoepithelial cells; Interlobular duct; Intercalated ducts; Mucous acinus with mucous granules; Serous demilune; Serous acinus with amylase granules.

sympathetic supply originates in two salivary nuclei in the medulla: the superior and inferior salivary nuclei. From the inferior salivary nucleus, fibres leave the medulla in the glosso-pharyngeal nerve (IX) and synapse in the otic ganglion, and postganglionic fibres reach the parotid glands. Fibres from the superior salivary nucleus leave the medulla in the facial nerve (VII), enter the chorda tympani branch and synapse in the submandibular ganglion, and post-ganglionic fibres supply both the submandibular and the sublingual glands.

Effects of nerve stimulation Stimulation of the parasympathetic (cholinergic) nerves leads to a lot of watery secretion and marked vasodilatation. If atropine is given first, the secretion is blocked but the vasodilatation is largely unaffected. This atropine-resistant vasodilatation is partly choli-nergic and partly non-cholinergic. The non-cholinergic mechanism may be mediated by vasoactive intestinal peptide (VIP). Furthermore, activation of parasympathetic nerves leads to release of the enzyme kallikrein from active gland tissue. Kallikrein acts on α-2-globulin in the interstitial fluid surrounding the gland cells to produce a vasodilator peptide called bradykinin. An injection of atropine is usually given pre-operatively in order to inhibit salivary secretion while the patient is under general anaesthesia. A side-effect of atropine-like drugs or anti-cholinergic drugs is suppression of salivary secretion, leading to a dry mouth.

Stimulation of the sympathetic nerve supply will cause vasoconstriction and usually a small amount of viscous secretion. While there is general agreement that parasympathetic path-ways operate during natural stimulation of salivary secretion, there is doubt whether the sympathetic nerves contribute under normal circumstances. If it is assumed that parasym-pathetic nerves stimulate the serous cells to produce a watery secretion and the sympathetic nerves act on the mucous cells to secrete saliva of high mucus content, it seems reasonable to expect that both divisions of autonomic nerves would be involved in natural stimulation of salivary secretion.

COMPOSITION AND FUNCTIONS OF SALIVA
The parotid glands contribute about 25% of the total saliva produced and, as they are formed of pure serous acini, they secrete watery saliva with a high content of amylase. The submandibular glands are mixed with a higher proportion of serous acini; they contribute about 70% of the saliva which contains both amylase and mucus. The sublingual glands consist of mainly mucous cells with a few serous acini and demilunes; their secretion has much mucus and little amylase and accounts for 5% of total saliva produced. Small salivary glands in the mucosa of the mouth and pharynx secrete mucus. Saliva in the mouth is a mixture of the secretions of all the glands.

About 1 litre of saliva is produced per day. It is hypotonic, with a specific gravity of about 1.003 (range 1.002–1.010) and pH that ranges from 6.2 to 7.4, increasing with the rate of secretion. Mixed saliva contains about 0.5% of solids, which are inorganic and organic in nature. The inorganic constituents include K^+, Na^+, Cl^- and HCO_3^-, the concentration of each being dependent on the rate of flow of saliva (Fig. 9.3). Potassium is found at a fairly constant concentration of 20 mmol/litre, making saliva one of the extracellular fluids with an unusually high potassium content. The concentration of calcium is 1.5 mmol/litre and phosphorus 5.5 mmol/litre. The organic constituents include mucin, the enzymes amylase, lingual lipase, kallikrein, lysozyme and carbonic anhydrase, and glycoproteins, mostly with the same specificity as the ABO blood group antigens found in red cells. About 80% of people have these antigens in their saliva and are described as secretors. Saliva also contains some disintegrating cells, from lining mucosa and glands, and micro-organisms.

The secretion of electrolytes and water in saliva occurs in two stages. There is first a primary secretion in the acini, which is secondarily modified at the level of the striated ducts. The primary acinar secretion is approximately isotonic and has an electrolyte composition similar to that of plasma. It has recently been shown that a chloride pump exists in the basolateral membrane of acinar cells, which actively transports Cl^- from the blood into the acinar lumen, to be followed by Na^+, K^+ and water. HCO_3^- is actively secreted by both the acinar and the striated duct cells. At the striated duct, Na^+ is reabsorbed, unaccompanied by water, and K^+ is secreted; this process is influenced by aldosterone. While a relatively constant amount of K^+ is secreted, irrespective of the rate of flow, less and less Na^+ is reabsorbed as the rate of flow of saliva increases. This pattern of electrolyte exchange, unaccompanied by water movement at the level of the striated ducts, results in the production of a hypotonic saliva and explains the changes in electrolyte concentrations in relation to the rate of secretion of saliva (Fig. 9.3).

The functions of saliva may be summarized as follows:

1 Saliva moistens and lubricates food and thus facilitates swallowing.

2 Saliva has a digestive function. Ptyalin is an amylase with optimum pH of 6.8. It breaks down starch to dextrins, maltotriose and maltose. Ptyalin is considered of minor physiological importance, as the food stays in the mouth for a short time. However, the action of ptyalin continues in the stomach for about half an hour and is arrested only when gastric acid penetrates the food mass. Salivary amylase may also be important for digesting away some of the food debris between the teeth. The serous salivary glands on the tongue secrete a lipase that breaks down triglycerides into monoglycerides and fatty acids. The action of lingual lipase may also continue in the stomach after food is swallowed.

3 Saliva keeps the oral mucosa constantly moist and in this way also helps the movements of the tongue and lips in speech.

4 Saliva is important for protection of the oral mucosa; its cleansing and washing effects are reinforced by the presence of the enzyme lysozyme, which is bactericidal.

5 By acting as a solvent, saliva is important for the sense of taste. Any substance must first dissolve in saliva before it can be sensed by the taste-buds.

6 Buffering action of saliva: saliva neutralizes any acids that may result from bacterial action, and swallowed saliva may help to neutralize gastric HCl in the empty stomach. The buffers in saliva include bicarbonate, phosphate and mucin. The concentration of bicarbonate is increased as the rate of salivary flow rises in response to eating.

7 Saliva serves as a vehicle of excretion for sub-

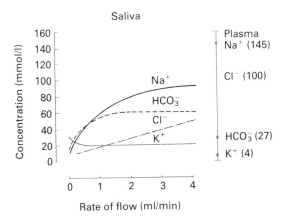

Fig. 9.3 Relationship of electrolyte concentrations to the rate of flow of saliva from the human parotid gland after parasympathetic stimulation. (After Thysen, J.H. (1954) *American Journal of Physiology* **178**, 155–9.)

stances like lead, mercury, fluorides and iodides. In parts of the world where the fluoride content of the soil and drinking water is high, fluoride may be deposited on the teeth.

If the salivary glands are congenitally absent or destroyed by disease or irradiation, xerostomia (dry mouth) results. In this condition, swallowing and speech are difficult and the patient has to take frequent sips of water. Loss of the cleansing and protective functions of saliva also leads to degeneration of the oral epithelium.

CONTROL OF SALIVARY SECRETION

While other secretions in the gut are controlled by nervous and hormonal mechanisms, salivary secretion is controlled exclusively by nervous mechanisms. Secretion of saliva is essentially mediated by nervous reflexes. The presence of food in the mouth stimulates general receptors and, especially, taste receptors. Sour-tasting substances, such as lemon drops, are particularly effective stimulants of salivary secretion. Impulses travel along afferent nerves to the salivary nuclei in the medulla and possibly to sympathetic centres. Efferent impulses travel along autonomic nerves to the salivary glands. This reflex is innate and is not acquired by learning (non-conditioned). Secretion of saliva may also occur by conditioned reflexes, resulting from seeing, smelling, hearing or even thinking about appetizing food. In this case, initial impulses arise in the parts of the brain concerned with these special sensations and stimulate the salivary centres. Pavlov, at the end of the nineteenth century, demonstrated the conditioned reflexes in dogs with salivary fistulae. The secretion of saliva in response to bell-ringing without presentation of food demonstrated the conditioned reflex. In man, mouth-watering on seeing or thinking of food provides evidence of this psychic reflex. However, in man, conditioned reflexes are less important than in dogs. Occasionally, secretion of saliva may continue even after food has been swallowed. This occurs when the food contains irritant substances or when there is nausea. The receptors for this reflex are in the stomach and upper intestine. Secretion of saliva is reduced during sleep, dehydration or severe mental stress.

DEGLUTITION (SWALLOWING)

The oesophagus

The oesophagus is a tube that connects the pharynx to the cardiac part of the stomach. It is 25–35 cm long, traversing the thorax except for the lowest 2–3 cm, which are in the abdomen. The oesophagus is lined with stratified squamous epithelium. In man, the muscle in its wall is striated in the upper third, smooth in the lower third and mixed in the middle. Both the upper and lower ends are guarded by sphincters. There is a definite anatomical band about 3 cm long at the upper end, forming the cricopharyngeal sphincter. At rest, the upper sphincter is closed at a pressure of 40–100 mmHg. At the lowest 3 cm resides the lower oesophageal sphincter (LOS). There is no definite thickened band at the lower end but the presence of the LOS can be demonstrated physiologically. If a perfused thin polythene tube connected to a manometer is gradually withdrawn from the stomach through the oesophagus, a band of pressure of about 15–35 mmHg will be recorded as the tip of the catheter traverses the region of LOS. Both sphincters are normally closed. They open only to allow a bolus to pass through.

The swallowing centre

A collection of neurones in the floor of the fourth ventricle in the medulla controls and co-ordinates the process of deglutition. Afferent impulses from the pharynx and oesophagus travel to the swallowing centre in the V, IX and X cranial nerves and efferent impulses pass down the V, IX, X and XII cranial nerves to the muscles in the tongue, pharynx, larynx and oesophagus. The vagal nuclei are particularly important as they supply the striated muscle of the pharynx and upper oesophagus as well as the smooth muscle in the oesophagus.

Swallowing

The swallowing process is considered in three stages: buccal, pharyngeal and oesophageal. The buccal stage is voluntary but both the pharyngeal and oesophageal stages are reflex.

Buccal stage Pieces of food are collected by the

tongue into a bolus. The tip of the tongue is pressed against the hard palate, thus creating a chamber between the tongue and the palate. Contraction of the mylohyoid and styloglossus muscles causes the tongue to move upward and backward. It propels the bolus towards the pharynx. The soft palate rises and the posterior pharyngeal wall moves forward as a result of the contraction of the superior constrictor muscle of the pharynx. They meet and close off the naso-pharynx. Respiration is inhibited during swal-lowing. The larynx starts to rise, signalling the end of the buccal stage. From this moment on-wards, the process of swallowing becomes involuntary. The rise of the larynx helps to push it off the path of the bolus. During the buccal stage, the lips and jaws are closed. It is very difficult to swallow when the jaws are apart. During dental operations saliva cannot be swal-lowed and collects in the mouth, from which it has to be removed by suction.

Pharyngeal stage This requires highly co-ordinated muscle contractions to propel the bolus to the upper end of the oesophagus while avoiding the entrance to the air passages at the larynx. The movements are co-ordinated by the swallowing centre in the medulla. The tongue moves further backward and pushes the bolus against the epiglottis which bends over the larynx and divides the bolus into two parts. The epiglottis is, how-ever, not essential for swallowing and can be removed without a significant effect on the process. The opening of the larynx is closed by contraction of the muscles surrounding its girdle, and the vocal folds are approximated. The bolus is propelled downwards by the contraction of the middle and inferior constrictors of the pharynx and reaches the upper end of the oesophagus. The cricopharyngeal sphincter relaxes for about 1 second. As soon as the bolus enters the oesoph-agus, the cricopharyngeal sphincter closes, the vocal folds open, the larynx drops and the epi-glottis returns to its previous position.

Oesophageal stage With efferent impulses descending in the vagi from the swallowing centre, the contraction which closes the crico-pharyngeal sphincter proceeds down the oeso-phagus as primary peristalsis which propels the bolus along. Peristalsis consists first of enlarge-ment of the lumen at the level of the bolus and ahead of it, followed after a few seconds by con-traction above the bolus, which is thus propelled down the oesophagus. This primary peristaltic wave travels down the oesophagus at about 4 cm/s for solid food. It is co-ordinated locally by the myenteric plexus. If the bolus is too big or sticky, its progression down the oesophagus is helped by development of secondary peristalsis. Secondary peristalsis is initiated by localized distension of an oesophageal segment, which sends afferent impulses in the vagi to the medullary swallowing centre. Efferent vagal impulses cause secondary peristalsis, which is also locally co-ordinated by the myenteric plexus. Thus, the main difference between primary and secondary peristalsis is in their mode of initiation: the first follows on a normal swallow, the second from localized oesophageal distension.

Vagal impulses relax the LOS ahead of the peristaltic wave. The bolus enters the stomach and the sphincter then closes rather slowly, taking 7−10 s.

Difficulty in swallowing is termed dysphagia. Swallowing may be investigated radiologically by a barium swallow. It may also be studied by oesophageal manometry, which records pressure changes in the oesophagus during swallowing.

Competence of the gastro-oesophageal junction
Except for the short abdominal segment, pressure in the oesophagus is the same as intrathoracic pressure, i.e. mostly negative, so that pressure in the stomach is always higher than in the oesoph-agus. It is therefore necessary to have a barrier at the gastro-oesophageal junction; otherwise, gastric contents will be regurgitated or refluxed (pushed back) into the oesophagus. As the lining mucosa of the oesophagus is not adapted to deal with corrosive gastric hydrochloric acid and pepsin, reflux will cause inflammation (reflux oesophagitis); the patient experiences a painful burning sensation in the chest (heartburn).

Competence of the gastro-oesophageal junction is now believed to be almost entirely due to the physiological sphincter at the lower end of the oesophagus. Mechanical factors, such as the acute

angle of entry of the oesophagus into the stomach
and the valve-like action of mucosal folds at the
cardio-oesophageal junction, seem to be un-
important. The antireflux function of the lower
oesophageal sphincter has been shown to be due
to its resting pressure (15–35 mmHg), its intra-
abdominal length, which is exposed to positive
intra-abdominal pressure, and its overall length.
The lower oesophageal sphincter is under neural
and hormonal control. Gastric overdistension
leads to increased sphincteric pressure through a
neural reflex. Between swallows, tonic vagal
cholinergic impulses maintain contraction to
keep the sphincter closed. But, during swallowing,
efferent impulses in the vagus are inhibitory, i.e.
cause the sphincter to relax, the transmitter
probably being vasoactive intestinal peptide (VIP).
The hormone gastrin, released from the stomach
by food, contracts the LOS. Secretin and CCK-PZ,
released from the upper intestine, oppose the
action of gastrin.

The stomach

ANATOMICAL CONSIDERATIONS
The stomach is a fibromuscular bag situated between the oesophagus and the small intestine. It
is approximately J-shaped with anterior and
posterior surfaces and rather sharp sides known
as the greater and lesser curvatures. The area of
the gastro-oesophageal junction is also known as
the cardia. The fundus is the dome-shaped part of
the stomach above an imaginary horizontal line
from the cardia to the greater curvature, and it
usually contains swallowed air (Fig. 9.4). An
oblique line from the angular notch to the greater
curvature separates the body of the stomach from
the antrum. The exit from the stomach is guarded
by the muscular pylorus (Greek for 'gate-keeper').
The stomach wall has three layers of muscle: an
outer longitudinal, a middle circular, which
thickens to form the pyloric sphincter, and an
inner oblique, extending from the cardia to the
greater curvature.

The innervation of the stomach consists of its
autonomic nerve supply as well as the intrinsic
nerve plexuses. Postganglionic sympathetic fibres
reach the stomach from the coeliac plexus.
Preganglionic parasympathetic fibres are supplied

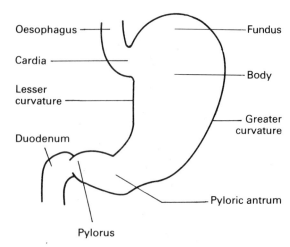

Fig. 9.4 Diagram of the stomach showing its parts.

by the right and left vagi. They synapse with the
neurones in the submucous and myenteric plexus,
and short postganglionic fibres reach the gastric
glands and smooth muscle. There are also reflex
arcs entirely within the wall of the stomach.

When the empty stomach is opened, its mucosa
is seen to be in folds or rugae. A closer look will
reveal pits, into which the gastric glands open.
The surface and pits are lined with a simple tall
columnar epithelium, which secretes a protective
layer of mucus.

Histologically, the gastric mucosa may be
divided into three major areas:
1 *The cardiac area* accounts for about 10% of the
lining mucosa. Most of the cells in the cardiac
glands secrete mucus.
2 *The main gastric area* includes the mucosa of
both the fundus and the body and accounts for
70–80% of the gastric mucosa. The pits are short
and the gastric glands are long and tubular, consisting of isthmus, neck, body or base (Fig. 9.5).
Usually four gastric glands open into a single pit.
These glands secrete all the constituents of gastric
juice. The parietal (or oxyntic) cells are most
numerous in the isthmus and neck regions but
are also seen in the body of the gland. Parietal
cells secrete hydrochloric acid. Those in the neck
region secrete the intrinsic factor as well.

Chief or peptic cells are serous cells found in
the body. They contain zymogen granules, which

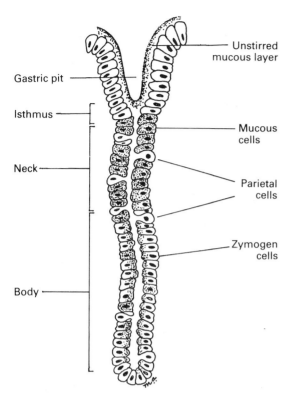

Fig. 9.5 Diagram of a main gastric gland showing the various types of cells.

are the source of the gastric pepsinogens. There are also endocrine cells, which may secrete peptides or amines, such as histamine. Mucous neck cells are found in the neck and isthmus regions; they secrete mucus and pepsinogen.

The nuclei of mucous neck cells show evidence of high mitotic activity. It is believed that either the mucous neck cell or a primitive cell similar to it acts as a mother cell for the gastric epithelium. It divides and migrates towards the surface, where it gives rise to the surface mucous cells, and also towards the base of the gland, where it gives rise to parietal cells, chief cells or endocrine cells. The gastric epithelium has great powers of repair. Every minute it replaces about half a million cells lost from the surface. The surface epithelial cells have a lifespan of a few days, while the parietal, chief and endocrine cells live for a few months. Proliferation of the gastric mucosa is stimulated by nervous and hormonal

factors; gastrin has a trophic effect on the gastric epithelium.

3 *The pyloric area* constitutes about 15% of the gastric mucosa. Most of the cells in the gastric glands are mucous cells. The endocrine cells or G cells, which secrete gastrin, are found in the pyloric glands.

FUNCTIONS OF THE STOMACH

1 The stomach stores food and regulates its passage to the small intestine. It enables the body to take large amounts of food at meal-times. In its absence, food has to be taken more frequently.

2 The stomach secretes gastric juice, which liquefies and partly digests food to produce semi-fluid chyme, which is suitable for further intestinal digestion and absorption.

3 It has a protective function. Hydrochloric acid kills ingested bacteria and thus limits their entry to the small intestine. Vomiting is a protective mechanism whereby harmful ingested material is thrown out.

4 The stomach produces the intrinsic factor, a mucoprotein necessary for the absorption of vitamin B_{12}.

5 Gastric hydrochloric acid is necessary for the absorption of iron in the upper intestine.

6 The stomach has recently been found to have an endocrine function. It produces a number of peptides, including gastrin, glucagon, somatostatin, substance P, VIP and GRP.

GASTRIC JUICE

Man secretes 2–3 litres of gastric juice per day. The main components of gastric juice are hydrochloric acid, digestive enzymes, mucus and the intrinsic factor.

Hydrochloric acid

Gastric acid is secreted by the parietal cells. These are pyramidal in shape, with the apical membrane invaginated to form an intracellular canalicular system (Fig. 9.6). Mitochondria account for about 40% of the cytoplasm, which also contains flavoproteins, cytochrome oxidase and carbonic anhydrase. When the cell is inactive, there are numerous tubulovesicles in the cytoplasm. When it is actively secreting, these vesicles seem to merge with the canaliculi, which become extensive and display microvilli.

At maximum rates of secretion, gastric juice has a pH of 0.9 and the following electrolyte composition: H^+ = 140 mmol/litre; K^+ = 15 mmol/litre; Na^+ = 10 mmol/litre; and Cl^- = 165 mmol/litre, balancing the three cations. As the pH of the plasma is 7.4, it is obvious that the parietal cell has to concentrate H^+ more than 10^6 times in order to secrete it into gastric juice. The numerous mitochondria and enzymes of the parietal cell provide it with the means to perform this task. H^+ and Cl^- ions are transported across the canalicular membrane into the canaliculi and hence the lumen of the gastric gland.

As shown in Fig. 9.6, a likely source of H^+ in the parietal cell is from the dissociation of intracellular water into H^+ and OH^-. Another source may be H^+ resulting from oxidation of substrates like glucose. As the mucosal side is always electronegative in relation to the serosal side, H^+ is not transported across the canalicular membrane against an electrical gradient, but it is transported against a huge concentration gradient.

H^+ is transported across the canalicular membrane by an ATP-driven pump, which exchanges H^+ with K^+, which is recycled inside the membrane. This pump has been called H^+, K^+ ATPase. It can be inhibited by a drug called omeprazole, which is consequently a very strong inhibitor of gastric acid secretion. OH^- will tend to accumulate in the cytoplasm, but it is neutralized by combination with H_2CO_3 to give H_2O and HCO_3^-. Carbonic acid (H_2CO_3) can be readily formed from CO_2 and H_2O under the influence of carbonic anhydrase, which is abundant in the parietal cell. HCO_3^- diffuses from the cell to the plasma and Cl^- enters via a carrier mechanism that facilitates the exchange between the two ions. Cl^- can also enter the cell in company with Na^+, which is then pumped out by Na^+, K^+ ATPase. Cl^- is transported across the canalicular membrane into the lumen in company with K^+, this being returned to the inside by H^+, K^+ ATPase and constantly recycled through the canalicular membrane.

The maximum amount of HCl that can be secreted in response to a stimulus such as pentagastrin or histamine — the maximum acid output (MAO) — is directly proportional to the number of

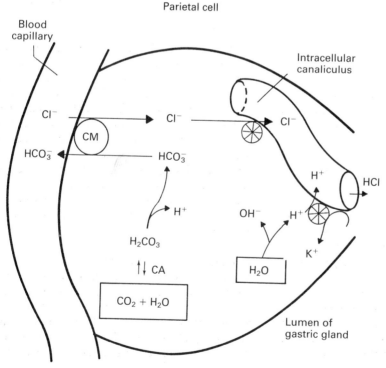

Fig. 9.6 Schematic representation of the process of production of hydrochloric acid by the parietal cell. CA, carbonic anhydrase; CM, carrier mechanism; ⊛, active transport.

parietal cells or parietal cell mass (PCM). The normal stomach has about 10^9 parietal cells, which have the capacity to secrete up to about 20 mmol H^+ per hour.

During the height of gastric acid secretion, the blood leaving the stomach has more HCO_3^- and less Cl^-, and its pH shifts slightly to the alkaline side. The excess HCO_3^- is finally excreted in urine. This may follow the ingestion of a meal, and has been described as the postprandial alkaline tide. In persistent vomiting, this situation is perpetuated, leading to metabolic alkalosis. In addition, there is dehydration and loss of Na^+ and K^+ ions.

Normally, gastric hydrochloric acid activates pepsinogen to pepsin and provides the suitable acidic medium for the action of pepsin. Its protective role in killing ingested bacteria has already been referred to. Its presence is also important for the absorption of non-haem iron in the duodenum.

Enzymes

Rennin, which clots milk, has been found in the calf stomach. It is absent from the stomach of the human infant and adult. The fundic mucosa of the human stomach secretes a lipase that hydrolyses triglycerides to monoglycerides and fatty acids. Its activity is equivalent to about one-fifth of that of pancreatic lipase. However, the main enzyme secreted into gastric juice is pepsin. The gastric glands secrete seven types of pepsinogens, which have been divided into two groups I and II. Group I pepsinogens consist of five components and are secreted by the chief cells and mucous neck cells. Group II comprises the two remaining components, which arise from these cells and other cells in the cardiac, antral and Brunner's glands.

The inactive pepsinogens are activated in the lumen by hydrochloric acid to pepsin and, once activated, they can activate more pepsinogen, i.e. autocatalysis. The pH optima of the pepsins are 1.5 to 3.5. Pepsins clot milk and break down proteins into peptones and polypeptides. They are particularly important for the breakdown of collagen and therefore the breakdown of connective tissue in meat. The digestion of protein by pepsin is not essential. If the stomach is re-moved, protein digestion can be initiated and maintained by pancreatic proteolytic enzymes.

Mucus

Mucus is an important constituent of gastric juice. It is present as a gel on the surface of the mucosa and protects the lining of the stomach against mechanical and chemical injury. By lubricating and facilitating the mixing of chyme, it protects the lining of the stomach against mechanical injury. It protects the mucosa against chemical injury by acting as a physical barrier to HCl and pepsin; it also neutralizes the acid and thus arrests the action of pepsin as well. The surface epithelial cells secrete HCO_3^-, which contributes to the protective role of mucus.

Gastric mucus is a glycoprotein with a large carbohydrate component. In about 75% of people, described as secretors, the carbohydrate part has the same antigenic specificity as the ABO antigens. It is noteworthy that people with group O are more susceptible to duodenal ulceration than other groups. The secretion of mucus is stimulated by mechanical and chemical irritation of the mucosa and is increased after either sympathetic or parasympathetic stimulation.

Intrinsic factor

Gastric parietal cells secrete a glycoprotein with a molecular weight of 60 000, which combines with ingested vitamin B_{12} (the extrinsic factor) and the complex is absorbed in the terminal part of the ileum. Neural or hormonal stimulation of gastric secretion will also increase the secretion of the intrinsic factor. The main defect in pernicious anaemia is atrophy of the gastric mucosa, including loss of parietal cells, leading to achlorhydria (absence of acid secretion) and deficiency of the intrinsic factor, which result in malabsorption of vitamin B_{12} and megaloblastic anaemia.

CONTROL OF GASTRIC SECRETION

Gastric secretion is controlled by neural and hormonal mechanisms, which are so integrated and interrelated as to warrant the use of the term neurohormonal. The control mechanisms are excitatory, followed by inhibitory means to limit the secretion. Secretion is usually considered in three phases, which correspond to the site where the stimuli to secretion start. These phases partly overlap and are also integrated:

Cephalic phase
The cephalic phase can occur by conditioned and non-conditioned reflexes. In the non-conditioned reflex, presence of food in the mouth stimulates gustatory and other receptors and afferent impulses travel to the vagal nucleus in the medulla; efferent impulses reach the stomach in the vagus nerves to cause secretion. The conditioned reflex follows psychic stimulation by seeing, smelling, hearing or thinking of appetizing food. Afferent impulses impinge on the vagal centre, which sends impulses to the gastric glands through the vagi.

Vagal impulses descending to the stomach release acetylcholine, which directly stimulates gastric glands to secrete gastric juice. Vagal impulses also release the hormone gastrin from the pyloric antrum, through a non-cholinergic mechanism via release of gastrin-releasing peptide (GRP). Thus, during the cephalic phase, the gastric glands are stimulated by the vagi both directly and indirectly through release of gastrin, which enters the blood to stimulate the glands.

Electrical stimulation of the anterior part of the hypothalamus stimulates gastric acid secretion. The cephalic phase of gastric secretion is affected by emotional factors. Chronic stress stimulates and acute stress, such as a sudden fright, reduces gastric acid secretion. All of these effects reach the stomach through the vagi, and this provides the rationale for performing surgical vagotomy to reduce gastric acid secretion as a treatment for patients with peptic ulcer.

Gastric phase
The gastric phase is mediated by nervous and hormonal mechanisms.

Distension of the stomach stimulates gastric secretion via long vagovagal reflexes and also by short intramural cholinergic reflexes. In the long reflex, both the afferent and efferent components travel in the vagi. Efferent vagal impulses act on gastric glands through acetylcholine and also release gastrin. In the short reflex, receptors that respond to stretch discharge to neurones in the

nerve plexuses of the gastric wall, and efferent fibres stimulate the gland cells by releasing acetylcholine. Gastrin may also be released by reflexes entirely within the wall of the stomach.

Presence of food in the stomach releases the hormone gastrin from the pyloric antrum. Gastrin enters the blood circulation and stimulates gastric glands.

Gastrin is secreted by flask-shaped G-cells in the pyloric antrum and upper intestine. Gastrin is released by distension and by peptides and L-amino acids resulting from the digestion of proteins. Alcohol and caffeine may also release gastrin, but carbohydrates and fats are poor releasers. Vagal excitation releases gastrin, whatever the phase of gastric secretion. Gastrin is a peptide that exists in many forms and whose occurrence in tissues and the blood has been investigated by radioimmunoassay. The physiological forms of gastrin have either 17, 34 or 14 amino acid residues, and each may occur in sulphated or non-sulphated types. The activity of gastrin resides in the four C-terminal amino acids $(Trp-Met-Asp-Phe-NH_2)$. Pentagastrin, which is widely used in gastric function tests, is a synthetic peptide with β-alanine added at the N terminus.

When given in pharmacological doses, gastrin has a wide variety of multiple effects. Gastrin is a strong stimulus of gastric acid secretion but it also stimulates the secretion of pepsin and the intrinsic factor. Gastrin also stimulates intestinal secretion, pancreatic secretion of enzymes and bicarbonate, and biliary secretion of bicarbonate and water. It contracts the LOS, stimulates gastric motility (but slows gastric emptying), stimulates intestinal motility and relaxes the ileocaecal sphincter. Gastrin has a trophic effect on the gastric mucosa. Of all these effects, the physiological ones are: stimulation of gastric acid secretion and pancreatic enzyme secretion, the trophic effect on the stomach and probably contraction of the lower oesophageal sphincter.

Gastrin is degraded and excreted by the kidneys. It is also catabolized by the small intestine and gastric fundus.

Although an islet cell in the foetal pancreas produces gastrin, it is difficult to demonstrate a cell in the pancreatic islets of adults that normally secretes gastrin. Rarely, however, an islet cell grows into a tumour that secretes gastrin, called a gastrinoma. Gastrinoma results in a syndrome of hypergastrinaemia, excessive basal and stimulated gastric acid hypersecretion, peptic ulceration and diarrhoea (Zollinger–Ellison syndrome). The tumour may more rarely occur in the stomach or duodenum.

Intestinal phase

Distension and the presence of protein digestion products in the upper intestine result in gastric acid secretion after a latent period of 1–3 hours. Since gastrin is found in the upper intestine, this phase may partly be mediated by direct release of intestinal gastrin by food. In addition, another peptide, named entero-oxyntin, may be released from the intestine by food and reaches the stomach through the blood to stimulate gastric acid secretion. Amino acids absorbed from the small intestine into the blood may also contribute to mediation of the intestinal phase.

PEPSIN SECRETION

Pepsin secretion is strongly stimulated by vagal excitation and to a lesser extent by gastrin. Acid plays an important role in pepsin secretion and its digestive action. Acid in the lumen excites a local cholinergic reflex that stimulates pepsin secretion. When acid is delivered into the upper intestine, it releases secretin, which also stimulates pepsin secretion.

HISTAMINE AND GASTRIC ACID SECRETION

Histamine is a potent stimulant of gastric acid secretion. This effect of histamine on the parietal cell is not blocked by conventional antihistamines like mepyramine maleate, i.e. the receptors involved are not H_1 receptors; they have been termed H_2 receptors. H_2 receptor blockers have been developed recently; they include cimetidine and ranitidine, both potent inhibitors of gastric acid secretion, and both are being used for the treatment of peptic ulcers. Even though the effect of histamine on gastric acid secretion has been known since 1920, it is still not certain how histamine fits into the chain of neural and hormonal stimulants of the parietal cell. Certain cells in the gastric mucosa seem to form and store

histamine. It has been proposed that the neural transmitter acetylcholine and the hormone gastrin release histamine from its cell, the released histamine diffusing to the parietal cell to act on the H_2 receptor to stimulate secretion of hydrochloric acid. This hypothesis, referred to as the mediator hypothesis, assumes that the parietal cell has no receptors for acetylcholine or gastrin, so that both are postulated to act on the parietal cell only through the mediation of histamine. More recently, work on the isolated parietal cells has shown that the parietal cell has receptors for histamine, for acetylcholine and probably for gastrin and that these receptors interact in various combinations to augment the actions of each other. It may be that gastrin and acetylcholine both release histamine from its cell and interact with it on the parietal cell.

ROLE OF INHIBITORY MECHANISMS IN THE CONTROL OF GASTRIC ACID SECRETION

In addition to excitatory mechanisms, there are inhibitory mechanisms, which limit gastric acid secretion:

1 Emotions, such as acute fear and the feeling of nausea reduce gastric acid secretion. These effects are probably mediated by reduction in the normal vagal tone to the stomach.

2 In the pyloric antrum, when the pH drops to 2.5 or less, acid secretion is reduced. Most probably acid in the antral lumen inhibits release of gastrin. This is the basis of a negative feedback loop whereby excess acid leads to a reduction in the amount of secreted acid. D-cells in the stomach and the intestine secrete somatostatin, a peptide originally isolated from the hypothalamus. Somatostatin inhibits gastric acid secretion both by directly inhibiting the parietal cell and by inhibiting release of gastrin from G-cells. It has been proposed that acid releases somatostatin from antral D-cells, which then diffuses to nearby G-cells to inhibit release of gastrin. In hypo- or achlorhydria, the negative feedback loop is interrupted; gastrin release continues unchecked, leading to hypergastrinaemia. This occurs in pernicious anaemia and may follow vagotomy. Whenever the antrum or part of it is excluded from the path of acid chyme, as may happen after a gastrojejunostomy,

hypergastrinaemia also results, which can lead to gastric acid hypersecretion and peptic ulceration.
3 The presence of acid, fat and hypertonic solutions in the upper intestine inhibits gastric acid secretion. These effects are mediated mainly by hormonal mechanisms.

(a) It has been postulated that a hormone, called bulbogastrone is released from the duodenal bulb to inhibit gastric acid secretion. Another hormone, called enterogastrone, has been postulated to inhibit both gastric acid secretion and motility when released from the upper intestine. Neither bulbogastrone nor enterogastrone has yet been obtained in pure form. A peptide with 43 amino acid residues can be released from the small intestine by carbohydrate and fats. One of its early discovered effects is inhibition of gastric acid secretion and it has thus been called gastric inhibitory peptide (GIP). However, later work has shown GIP to be a potent releaser of insulin from the β cells of pancreatic islets, and it is now believed that GIP is the physiological insulin releaser of the small intestine.

(b) Acid and digestive products of proteins and fats release the hormones secretin and cholecystokinin-pancreozymin (CCK-PZ) from the mucosa of the upper intestine into the blood, from where they stimulate pancreatic secretion. Both hormones can also act on the stomach. Secretin is a potent inhibitor of gastric acid secretion and it inhibits release of gastrin from the antrum. In small doses, CCK-PZ stimulates gastric acid secretion but in higher doses it competitively inhibits stimulation of gastric acid secretion by gastrin. Thus, secretin and CCK-PZ normally released from the intestine may contribute to intestinally initiated mechanisms that limit gastric acid secretion.

(c) Pancreatic glucagon is also secreted by A cells in the stomach and duodenum, and a substance that cross-reacts with some of the antiglucagon antibodies has been found in the rest of the small intestine and is called glucagon-like immunoreactive factor (GLI) or glicentin. In addition to its effects on carbohydrate metabolism, glucagon is a good inhibitor of gastric acid secretion in man.

(d) Other peptides inhibitory to gastric acid secretion include neurotensin, vasoactive intestinal peptide (VIP) and somatostatin.

(e) Certain types of prostaglandins inhibit gastric acid secretion in man.

TESTS OF GASTRIC SECRETORY FUNCTION
Gastric secretory function tests are usually centred on measurement of gastric acid secretion. After an overnight fast, a nasogastric tube is passed into the stomach. After the stomach is emptied, secretion is aspirated every 10 or 15 minutes. Basal acid secretion is usually collected for 1 hour. It amounts to $1-4$ mmol H^+ per hour, but it is very variable between individuals and even in the same individual at different times. Gastric acid secretion is maximally stimulated by an injection of pentagastrin (6 μg/kg) or histamine (0.04 mg/kg, accompanied by an H_1 receptor blocker), and the maximum acid output (MAO) is than determined. MAO is higher in men than in women and it tends to decline after the age of 50. It also correlates with parameters of body build, such as body-weight, lean body mass or height. The MAO is directly proportional to the parietal cell mass.

Acid secretion is usually absent (achlorhydria) in pernicious anaemia and may be reduced (hypochlorhydria) in gastric carcinoma. One-third of patients with gastric ulcer have reduced acid secretion and the rest secrete within the normal range. About one-sixth to one-third of the patients with duodenal ulcer are acid hypersecretors while the rest secrete within the normal range. Basal acid secretion in the Zollinger−Ellison syndrome (gastrinoma) is so high that it amounts to over half of the maximally stimulated output, which is also greatly increased.

After an operation of vagotomy for treatment of peptic ulcer, it is desirable to know whether the vagotomy has been complete. This is assessed by the insulin hypoglycaemia test. Insulin at 0.2 U/kg is injected intravenously to reduce glucose to about 2.78 mmol/litre (50 mg/100 ml). The hypoglycaemia stimulates the vagal centre and, if the vagi are intact, gastric acid secretion is stimulated after a latent period of about 40 minutes. Lack of an acid response indicates that the vagotomy is complete.

ABSORPTION IN THE STOMACH

A small amount of water is absorbed in the stomach. Alcohol is also partly absorbed in the stomach. Undissociated organic acids, such as acetyl salicylic acid or aspirin, may be absorbed, and generally lipid-soluble substances or drugs can be absorbed in the stomach.

GASTRIC MOTILITY

When the electrical activity of the smooth muscle in the wall of the stomach is explored by micro-electrodes, it is found that the fundus is electric-ally silent but that two types of potentials may be recorded from the rest of the stomach: slow spon-taneous waves of partial depolarization, occurring at the rate of three per minute and referred to as slow waves or basal electric rhythm (BER), and action potential spikes, superimposed on the slow waves and occurring in association with contrac-tions. BER is responsible for co-ordination of contractions and determines their frequency. It originates in a group of smooth muscle cells, located high on the greater curvature and con-sidered as the pacemaker, and sweeps along the stomach to the pylorus.

During the interdigestive period bursts of depolarizations, accompanied by peristaltic contractions, pass along the stomach. Such activity occurring in the empty stomach is referred to as the migrating motor complex (MMC). The MMC moves on along the whole length of the small intestine to reach the ileo-caecal valve after 1.5–2 hours. When the cycle reaches the ileocaecal junction, a new wave of MMC starts in the stomach. As soon as food is ingested, the activity of the MMC is terminated. The function of the MMC is to sweep remnants in the stomach and small intestine into the colon.

The motility function of the stomach after food is taken is best appreciated by recognizing two distinct motor regions: the proximal stomach, consisting of the fundus and the upper third of the gastric body, and the distal region, consisting of the lower two-thirds of the body, the antrum, the pylorus and the duodenal bulb.

The proximal stomach

As food enters the stomach, the proximal region relaxes in order to accommodate the incoming food (receptive relaxation). This is mediated through a vagovagal reflex, the efferent arm of which is inhibitory to the musculature of the upper stomach and non-cholinergic. The trans-mitter may be vasoactive intestinal peptide (VIP), but dopamine has also been suggested. Thus, the first function of the proximal stomach is receiving and storing ingested food. In addition, there are slow sustained tonic contractions in the proximal stomach that slowly push contents towards the distal region and duodenum. These contractions actually provide the pressure gradient for empty-ing of liquefied chyme from the stomach into the duodenum, which is the second major function of the proximal region.

The distal stomach

The main activity in the distal stomach is peristalsis, whereby grinding of solids and lique-faction of chyme occur and chyme is propelled to the duodenum. Peristaltic contractions are initiated by distension and mediated by a cholinergic vagovagal reflex.

The pylorus is an atypical sphincter in that it is open most of the time and closes only when a peristaltic contraction passes over it. When this happens, the stomach is closed off from the duodenum and time is afforded for grinding of the solids in chyme to reduce their size to particles that are small enough to pass through the lumen of the pylorus when it relaxes (<0.25 mm in diameter).

Gastric emptying occurs through co-ordination of the contractions of the antrum, pylorus and duodenal bulb, these being referred to as the gastroduodenal pump. As gastric contents are propelled into the distal stomach, the antrum, pylorus and proximal duodenum are relaxed and the part of the chyme which has been liquefied or is already in liquid form passes readily into the duodenum, being pushed along by the tonic contraction of the proximal stomach. The terminal antrum then contracts, further aiding the propulsion of liquid chyme. This is followed by contraction of the pylorus, which closes off the stomach and arrests emptying to allow grinding of solids in chyme. The proximal duodenum next contracts, moving contents just emptied further into the distal duodenum and jejunum. The antrum, pylorus and duodenal bulb then relax and the sequence is repeated.

CONTROL OF GASTRIC EMPTYING

The rate of gastric emptying depends mainly on the activity of the gastroduodenal pump and to a lesser extent on duodenal resistance to oncoming chyme.

Liquids are emptied rapidly from the stomach and, irrespective of their starting volume, are all emptied at the same rate. Digestible solids are emptied more slowly and indigestible solids, such as plant fibres, are emptied at very slow rates.

When chyme arrives in the duodenum, its chemical characteristics affect various duodenal receptors and the effect is then conveyed to the stomach to inhibit its motility and the activity of the gastroduodenal pump, i.e. negative feedback control.

Hypertonic chyme slows down the rate of gastric emptying. This is mediated by osmo-receptors in the duodenum and jejunum.

The arrival of acid chyme in the duodenum (pH less than 6) inhibits gastric motility and increases duodenal motility leading to increased duodenal resistance. The result is marked inhibition of gastric emptying. Excess acid in the duodenum acts on receptors which are sensitive to changes in pH.

The presence of emulsified fat and, to a lesser extent, peptides and amino acids in the duodenum slows the rate of gastric emptying.

There are duodenal receptors which respond to distension. Duodenal distension inhibits contractions of the pyloric antrum and slows gastric emptying.

All of these feedback mechanisms are conveyed from the upper intestine to the stomach by neural and hormonal means. The nervous pathways may be via a long vagovagal reflex whereby the efferent vagal impulses are inhibitory to gastric motility. Alternatively or in addition, these effects may be mediated by an inhibitory enterogastric reflex, whose centre is probably in the coeliac ganglion. However, since these effects still occur after denervation of the stomach and upper intestine, hormonal mechanisms do contribute. It has been postulated that a hormone, called enterogastrone, is released from the upper intestine by acid, fat and hypertonic solutions. It inhibits gastric motility (and acid secretion) and thus slows gastric emptying. As was mentioned above, both bulbogastrone and enterogastrone are still in crude extract form and have not been obtained in pure form yet.

The intestinal hormone secretin inhibits gastric motility, while CCK-PZ stimulates it, but both contract the pylorus, thereby increasing its resistance, and thus slow gastric emptying. Although gastrin stimulates gastric motility, it also stimulates intestinal motility and thus increases duodenal resistance, and, when the gastrin-stimulated acid reaches the duodenum, it releases secretin and CCK-PZ. Therefore, the net effect of gastrin is slowing of gastric emptying.

Emotions also affect the rate of gastric emptying. Fear prolongs while excitement speeds up gastric emptying.

After truncal vagotomy, the activity of the gastroduodenal pump is reduced, leading to slowing of gastric emptying and dilatation of the stomach. This is usually avoided by performing a pyloroplasty (enlargement of the pyloric canal) at the same time.

VOMITING

Vomiting is mainly a protective mechanism by which the stomach removes from the digestive tract harmful or potentially harmful materials. Irritation or overdistension of the stomach or duodenum is normally the strongest stimulus for vomiting to occur. Substances that induce vomiting are called emetics. Vomiting may also accompany movement, and some people seem to be specially prone to motion sickness. Vomiting may occasionally follow disgusting or shocking sights or smells.

A collection of neurones present bilaterally in the medulla oblongata is referred to as the vomiting centre. It controls and integrates the process of vomiting. The centre receives afferent impulses from the stomach and duodenum in both the vagi and the sympathetic nerves. There are also afferents from the pharynx; the receptors concerned are very potent. Mechanical stimulation of the pharynx easily induces vomiting, and this may be useful in emergency situations when a poisonous substance has accidentally been ingested. Efferent impulses travel from the vomiting centre in the V, VII, IX, X and XII cranial nerves plus the spinal nerves to the diaphragm and abdominal muscles.

Vomiting is usually preceded by nausea and retching. A deep breath is taken, the glottis is closed and the soft palate rises to close off the nasopharynx; the cricopharyngeal sphincter then opens. This is immediately followed by violent contractions of the diaphragm and abdominal muscles. In the meantime, the fundus and body of the stomach are relaxed and reverse peristalsis, usually starting at the angular notch and directs chyme towards the cardia. Sometimes antiperistalsis may even start in the lower intestine. The most important factor for expelling chyme from the stomach, however, is the squeezing of the stomach between the diaphragm and the abdominal muscles. Intragastric pressure rises sharply and propels chyme towards the cardia. The lower oesophageal sphincter relaxes and gastric contents are expelled through the oesophagus to the outside. Vomiting is normally accompanied by salivation and slowing of the heart. Skin vasoconstriction and sweating may occur.

Emetics, such as warm salt and water or mustard and water or copper and zinc sulphates, when taken by mouth, induce vomiting. They act on receptors in the stomach and duodenum which discharge to the vomiting centre. Other emetics, such as morphine, apomorphine, digitalis derivatives and tartar emetic, act centrally on an area in the floor of the fourth ventricle called the chemoreceptor trigger zone. This zone discharges to the vomiting centre. In motion sickness, initial impulses arise in the labyrinth and reach the chemoreceptor trigger zone through the cerebellum. Psychic stimuli arise in the frontal cortex of the brain and impulses pass directly to the vomiting centre.

The exocrine pancreas

When chyme is delivered from the stomach into the intestine, it is mixed with pancreatic juice, bile and intestinal secretions. The digestion of food, started in the mouth and stomach, is completed and the products of digestion are absorbed. The main source of digestive enzymes in the body is the exocrine pancreas.

ANATOMICAL CONSIDERATIONS

The greater mass of the pancreas is composed of acinar gland tissues, which produce pancreatic juice. The secretions are delivered into the duodenum via the pancreatic duct. Dispersed among these exocrine pancreatic acini are the endocrine islets of Langerhans, which secrete insulin, glucagon and somatostatin.

The fibrous capsule of the pancreas sends septa that divide the gland into lobules. Each lobule is composed of many acini. The cells lining the acini are serous cells containing zymogen granules, which are the precursors of the enzymes of pancreatic juice. The endoplasmic reticulum is extensive and contains numerous ribosomes, which form the enzyme proteins.

The duct system of the pancreas is the same as in salivary glands. Cells of intercalated ducts often extend into the central part of the acinus and are referred to in this position as centroacinar cells. The centroacinar and duct cells produce the electrolytes of pancreatic juice. In over 70% of people, the main pancreatic duct joins with the bile-duct at the ampulla of Vater (Fig. 9.7), the opening of which, into the duodenum, is surrounded by the sphincter of Oddi (the choledochoduodenal sphincter). The walls of the largest ducts contain smooth muscle and small mucous glands that open into the lumen of the ducts.

Preganglionic parasympathetic fibres arrive in the vagus to synapse with ganglion cells around the larger ducts; postganglionic fibres supply the acini. The sympathetic fibres accompany the blood-vessels and visceral afferent fibres run with sympathetic nerves.

COMPOSITION AND FUNCTIONS OF PANCREATIC JUICE

The volume of pancreatic juice secreted per day is about $1.2-1.5$ litres. Pancreatic juice is isotonic, with a specific gravity of $1.008-1.030$; it has an alkaline pH of about 8. It has 1% inorganic materials (electrolytes) and $1-2\%$ organic materials, mostly enzymes.

Electrolytes

The electrolytes are produced by centroacinar and duct cells. They include the cations Na^+, K^+ and Ca^{2+} and the anions HCO_3^- and Cl^-. The

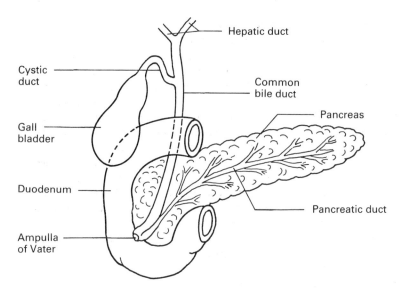

Fig. 9.7 Anatomical relationships of the pancreas, liver, gall-bladder and associated ducts with the duodenum.

greater bulk of the electrolytes is in the form of $NaHCO_3$. When pancreatic secretion is stimulated, the concentrations of Na^+ and K^+ are near to their level in the blood and remain fairly constant, but HCO_3^- and Cl^- vary with the rate of flow of secretion. HCO_3^- concentration rises and that of Cl^- drops as the rate of flow increases, such that their sum remains constant. Apparently, the primary secretion of centroacinar cells consists of $NaHCO_3$; the duct cells reabsorb HCO_3^- in exchange for Cl^-.

Details of the cellular mechanisms of pancreatic electrolytes and fluid secretion are still uncertain. The pancreas is known to contain carbonic anhydrase, which is essential for formation of HCO_3^-. Acetazolamide (Diamox) is a strong inhibitor of carbonic anhydrase and has been shown to inhibit pancreatic HCO_3^- secretion. Secretion of HCO_3^- is also inhibited by ouabain (an inhibitor of Na^+, K^+ ATPase), suggesting that secretion of Na^+ and HCO_3^- is somehow coupled.

$NaHCO_3$ in pancreatic juice makes a major contribution to the neutralization of acid chyme, along with bile and duodenal secretion, in order to create a suitable medium for the action of pancreatic enzymes.

Enzymes
The pancreas secretes enzymes that act on all the major types of foodstuffs. The protein-splitting enzymes include trypsin, chymotrypsin, elastase and carboxypeptidase. They are all secreted in an inactive precursor zymogen form, which is converted to the active enzyme in the intestinal lumen. All pancreatic enzymes are proteins. Trypsinogen is activated into trypsin by the enzyme enteropeptidase (enterokinase), secreted by duodenal mucosal cells. Once trypsin is produced, it acts similarly to activate chymotrypsinogen to chymotrypsin, proelastase to elastase and procarboxypeptidase to carboxypeptidase. Thus, the activation of trypsinogen by enteropeptidase is the key to activation of the pancreatic proteases, and enteropeptidase is considered the most important enzyme in intestinal juice. The pancreas also secretes a trypsin inhibitor, which would probably neutralize any trypsin appearing within the pancreas. In this way, activation of the proteases and consequently autodigestion of the pancreas are prevented.

Trypsin, chymotrypsin and elastase are endopeptidases, splitting proteins into shorter peptide chains. Carboxypeptidase is an exopeptidase, which splits off amino acids at the C terminus of the peptides.

Starch is split by pancreatic amylase at 1,4-glucosidic bonds to maltose, maltotriose and α-dextrins. Pancreatic lipase is the most important fat-splitting enzyme. In the presence of bile salts and colipase, which is also secreted by the pancreas, it breaks down triglycerides into fatty acids and monoglycerides. Other lipolytic enzymes are cholesterol esterase and phospholipase. Nucleic acids are broken down by pancreatic nucleases into nucleotides.

Pancreatic enzymes are synthesized inside acinar cells, by the ribosomes of the endoplasmic reticulum, from amino acids. They are later incorporated into vacuoles to appear as zymogen granules, which are secreted from the cells by exocytosis. Once discharged into the duodenum, pancreatic enzymes are adsorbed on to the glycocalyx lining the surface of the small intestine. Interestingly, a high carbohydrate content in the diet is followed by a higher proportion of pancreatic amylase, a high protein diet is followed by increased proteolytic enzymes and a high fat diet leads to increased lipase. These adaptive changes are mainly brought about by increased enzyme synthesis, probably mediated through hormonal and neural means.

In pancreatic disease, e.g. chronic pancreatitis, the amount of enzymes delivered to the duodenum is reduced. This leads to impaired digestion of fat and protein, with little effect on carbohydrate. The increased excretion of fat in the stools is termed steatorrhoea.

CONTROL OF PANCREATIC SECRETION
Like gastric secretion, pancreatic secretion is regulated by neurohormonal mechanisms and may be considered to occur in three phases: cephalic, gastric and intestinal.

Cephalic phase
The cephalic phase is mediated by the vagus nerve. In animals direct stimulation of the vagus leads to pancreatic secretion which is viscous and rich in enzymes. In man, sight or smell of appetizing food, as well as its presence in the mouth, is followed by secretion of pancreatic juice containing both enzymes and electrolytes, with the enzymes forming the greater portion. Vagal excitation during the cephalic phase releases, first, acetylcholine, which directly acts on pancreatic acinar cells, and, secondly, gastrin from the antrum of the stomach, which also stimulates pancreatic secretion.

Gastric phase
When food reaches the stomach, it activates both neural and hormonal mechanisms that lead to secretion of pancreatic juice. Gastric distension stimulates the pancreas through long vagovagal reflexes and also via local gastropancreatic reflexes. Gastrin is released from the antrum by distension acting through vagovagal and intramural reflexes and by peptides and amino acids bathing the antral lumen. Stimulation of gastric acid secretion during the cephalic and gastric phases is an important factor for stimulation of the pancreas, because the increased acid output, on reaching the intestine, releases the intestinal hormones, which are potent stimulants of pancreatic secretion (see below).

Intestinal phase
The intestinal phase is the main phase of pancreatic secretion. Gastric acid, amino acids or fatty acids, on reaching the upper intestine, stimulate secretion from the pancreas. This effect is mediated mainly hormonally. It was, in fact, the demonstration by Bayliss and Starling of the hormonal mechanism in 1902 which led to the discovery of secretin and laid the foundation of modern endocrinology. It is now established that the intestinal phase of pancreatic secretion is mainly mediated by two hormones released from the upper intestine: secretin and cholecystokinin-pancreozymin (CCK-PZ).

EFFECTS OF SECRETIN
Secretin is released into the blood from S cells in the upper intestinal mucosa, mainly by acid (pH 4 or less) and to a lesser extent by amino acids and fatty acids. Secretin is a peptide with 27 amino acid residues similar in sequence to pancreatic glucagon. Its effects are as follows:
1 Secretin acts on centroacinar and duct cells in the pancreas to stimulate secretion of water and bicarbonate.
2 It similarly acts on biliary duct cells to stimulate hepatic bile flow and bicarbonate secretion. The stimulation of fluid and electrolyte secretion by secretin in both the pancreas and the biliary system is mediated by cyclic adenosine-5'-monophosphate (cAMP).
3 Secretin also augments the action of CCK-PZ

in stimulating pancreatic enzyme secretion.

4 In the stomach, secretin inhibits gastric acid secretion and the release of gastrin, but it stimulates pepsin secretion.

5 It inhibits gastric motility, contracts the pylorus and thus slows gastric emptying.

6 Secretin relaxes the LOS.

7 Secretin inhibits intestinal motility and contracts the ileocaecal sphincter.

VASOACTIVE INTESTINAL PEPTIDE (VIP)
VIP is found in certain neurones and endocrine cells throughout the intestine; it also stimulates secretion of water and bicarbonate from the pancreas, but it is a weaker stimulus than secretin. VIP has 28 amino acid residues similar in sequence to glucagon and secretin.

EFFECTS OF CCK-PZ
Pancreatic enzyme secretion is stimulated by an upper intestinal hormone, which was originally called pancreozymin. Later work showed that the activity of pancreozymin and that of another hormone that was shown to contract the gall-bladder and which was called cholecystokinin (CCK) were found in one and the same intestinal peptide, now referred to as cholecystokinin-pancreozymin (CCK-PZ) or simply CCK. CCK-PZ is released from I cells in the upper intestine, mainly by L-amino acids and fatty acids and to a lesser extent by HCl. It is also released by bile salts. CCK-PZ is a peptide existing in many forms — with 8, 12, 33, 39 or 58 amino acid residues. The last five amino acids at the C terminus are the same as in gastrin and the physiological activity resides in the last eight C-terminal amino acids. All forms of CCK-PZ are sulphated. Its effects are as follows:

1 CCK-PZ acts on pancreatic acinar cells to stimulate secretion of enzymes, but it also augments the stimulation of water and bicarbonate secretion by secretin.

2 CCK-PZ has a tropic effect on the pancreas.

3 It contracts the gall-bladder, relaxes the choledochoduodenal sphincter and causes discharge of bile into the intestine.

4 In small doses CCK-PZ stimulates gastric acid secretion but in higher doses it inhibits gastrin-stimulated secretion.

5 Although CCK-PZ stimulates gastric motility,

it contracts the pylorus, thereby slowing gastric emptying.

6 It antagonizes the action of gastrin at the LOS and thus relaxes it.

7 CCK-PZ stimulates intestinal motility.

8 CCK-PZ has been found in the nervous system, especially the cerebral cortex, and evidence has been presented that it may be concerned with the mechanism of satiety.

Although CCK-PZ may act through cAMP, its second messenger in stimulating enzyme secretion from acinar cells seems to be mainly Ca^{2+}. Another potent enzyme stimulant is acetylcholine, which also acts through Ca^{2+}.

A peptide isolated from the intestine specifically stimulates secretion of chymotrypsinogen from the pancreas; it has been called chymodenin. Its presence raises the possibility of the presence of more enzyme-specific hormones.

During all phases of pancreatic secretion, there is close integration between neural and hormonal stimuli. Vagal stimulation potentiates the responses of the pancreas to secretin and CCK-PZ. The mutual potentiation between secretin and CCK-PZ is considered an important factor in the physiological response to food.

TESTS OF PANCREATIC FUNCTION
Duodenal juice may be aspirated for analysis but it would be a mixture of pancreatic, biliary and duodenal secretions. Recently, direct cannulation of the pancreatic duct through a duodenoscope has been achieved and pure pancreatic juice collected. The parameters determined are usually the volume of secretion and bicarbonate and amylase or trypsin outputs. The capacity of the pancreas to secrete can be assessed after an injection of secretin, followed 30 min later by an injection of CCK-PZ. The pancreas may also be challenged indirectly by introducing food, acid or amino acid solutions into the upper intestine. If the response is reduced, it may indicate either pancreatic disease or abnormal intestinal mucosa, as in coeliac disease, resulting in reduced hormone release.

Pancreatic function can also be assessed by determining faecal fat excretion over a 24-hour period. Normally, about 5 g of fat are excreted per day. In severe pancreatic juice deficiency, up to 50 g of fat may excreted per day (steatorrhoea).

The liver

ANATOMICAL CONSIDERATIONS

The liver weighs about 1.5 kg in the adult. Its basic architectural unit is the lobule (Fig. 9.8), composed of many cellular plates radiating between the central vein and the portal tracts. The hepatic plates are two cells thick. Between the cells there are bile canaliculi, which drain into biliary ducts, and between the plates are found the blood sinusoids. The hepatic sinusoids are lined by endothelial cells and macrophage cells called Kupffer cells. The sinusoids receive blood from branches of the portal vein and branches of the hepatic artery. There is a very narrow space between the endothelial lining of the sinusoids and hepatic cells called the space of Disse. This space connects directly with a terminal lymphatic, which drains into a lymphatic vessel in the portal tract. There are large pores in the endothelial lining of the sinusoids, about 1 μm in diameter. These pores are large enough to allow protein molecules to pass through with lymph. Lymph leaving the liver has a protein concentration almost as high as that of the plasma.

HEPATIC CIRCULATION (For more details on the hepatic circulation, see Chapter 6)

The liver has a dual blood supply, from the portal vein and the hepatic artery. The portal vein is formed by the confluence of the superior mesenteric and splenic veins. It therefore drains the stomach, small intestine, part of the colon, pancreas and spleen. The branches of the portal vein drain into the sinusoids. The hepatic artery arises from the coeliac axis. It supplies the bile-ducts and other structures in the portal tracts and also gives branches that open into the sinusoids. The central veins drain the lobules and collect to form the hepatic vein, which leaves the liver to join the inferior vena cava.

Since the splanchnic circulation drains into the liver, measurement of hepatic total blood flow is equivalent to measurement of flow in the splanchnic circulation. Hepatic blood flow may be measured directly by applying flowmeters to the vessels during surgical operations or indirectly by using Fick's principle. In the latter method, either of two dyes may be used, bromsulphthalein or indocyanine green, both of which are selectively removed by the liver. The dye is infused intravenously and the infusion is maintained until the arterial concentration of the dye becomes constant. This means that the liver is removing the dye at the same rate as it is being infused. Therefore, the dye concentration in the hepatic artery is assumed to be the same as in a systemic artery. A blood sample is obtained from the hepatic vein by a cardiac catheter, introduced through the cubital vein and advanced through the right atrium and inferior vena cava. After determination of the dye concentration in the

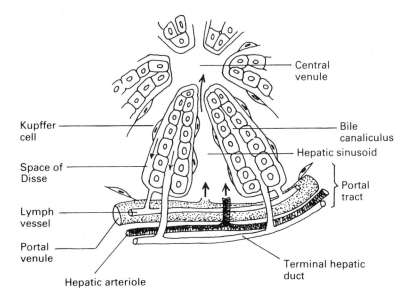

Fig. 9.8 Microscopic structure of hepatic lobule. For details, see text.

hepatic vein, hepatic blood flow is determined as follows:

$$\text{Hepatic blood flow} = \frac{\text{Rate of infusion of dye}}{C_A - C_{HV}}$$

where C_A is the arterial concentration of dye and C_{HV} is the dye concentration in the hepatic vein.

Blood flow through the portal vein is about 1 litre per minute and about 500 ml per minute flows through the hepatic artery. Total hepatic blood flow is thus about 1.5 litres per minute. This is approximately 30% of the blood volume and explains why the liver is one of the important blood reservoirs in the body.

The hepatic artery provides the liver with blood having an oxygen tension of 95 mmHg at a mean pressure of 100 mmHg. Oxygen tension of portal blood varies between 60 and 75 mmHg. Pressure in the portal vein is about 8 mmHg, while pressure in the hepatic vein, which has an oxygen tension of about 40 mmHg, averages 0 mmHg. Portal pressure in man may be measured directly during a surgical operation by introducing a needle into the portal vein, which is then connected to a manometer. In a conscious subject, a needle may be introduced through the skin into the spleen to measure pressure in the splenic pulp, which is only slightly higher than portal pressure. Alternatively, a cardiac catheter may be introduced through the cubital vein, advanced through the right atrium and the inferior vena cava into the hepatic vein, where it is further advanced until its tip cannot proceed further. In this position, it will be measuring pressure equivalent to that in the liver sinusoids and hence in the portal vein.

Since the pressure drop across the portal and hepatic veins is about 8 mmHg with 1 litre of blood flowing through each minute, resistance to flow in the liver sinusoids is normally quite low. Constriction or obliteration of the hepatic vascular channels may follow fibrosis of the liver parenchyma, occurring in liver cirrhosis or hepatic schistosomiasis, raising portal pressure to 20 or 30 mmHg (portal hypertension). In portal hypertension collateral channels to the systemic circulation enlarge and the vessels become tortuous (varices) in order to shunt blood around the partially obstructed liver sinusoids. Especially important is the development of gastro-oesophageal varices at the lower end of the oesophagus. These varices are liable to rupture and lead to vomiting of blood (haematemesis). This is one of the serious complications of portal hypertension which occurs as a result of intestinal bilharziasis and liver cirrhosis.

The liver forms about 1 ml of lymph per minute, containing about 6 g of protein/100 ml. This is equivalent to between a third and one-half of lymph formed in the whole of the body at rest.

Accumulation of fluid in the peritoneal cavity is called ascites. It occurs readily when hepatic venous pressure rises to partially obstruct drainage of blood from the liver. The amount of fluid flowing into lymph is greatly increased and part of it exudes through the liver surface into the abdominal cavity. This fluid is almost pure plasma, containing protein with a concentration of 80–90% of that in plasma. The high protein content also creates an osmotic gradient, which attracts more water. In portal hypertension, the back pressure on the gut capillaries causes oedema of the gut wall and fluid exudation into the abdominal cavity. Since the liver forms the plasma proteins, disease or partial destruction of the hepatocytes leads to hypoproteinaemia, which also contributes to ascites formation.

FUNCTIONS OF THE LIVER

The functions of the liver are so numerous that it is convenient to group them under three subheadings: vascular, secretory and metabolic functions:

1 *Vascular functions*:

(a) *Storage*. Since the liver blood flow is 30% of blood volume, it is one of the important reservoirs of blood in the body. In the case of strenuous exercise or haemorrhage, part of the hepatic blood is diverted towards the exercising muscles or the vital organs respectively.

(b) *Antigen clearance*. The liver filters blood coming from the gut. Kupffer cells in the liver sinusoids belong to the macrophage or reticuloendothelial system. They remove bacteria, debris, particulate matter or any other antigens contained in the blood arriving from the gut.

2 *Secretory functions*. The liver continuously secretes bile, which, after storage and concentration in the gall-bladder, is discharged into the duodenum. Bile is important for partial neutralization of acid chyme, for digestion and absorp-

tion of lipids and for absorption of the fat-soluble vitamins. It is also a vehicle of excretion of bile pigments, cholesterol, toxins and other substances.

3 *Metabolic functions.* The liver has a central role in the metabolism of the major types of foodstuffs as well as substances of endogenous or exogenous origin. Details about these metabolic processes may be obtained by reference to specialized texts. Only an overview is presented here:

(a) *Carbohydrate metabolism.* The liver is an important site of glycogen storage. It plays a central role in regulation of the blood glucose level. It is involved in the intermediary metabolism of carbohydrates, including the conversion of fructose and galactose to glucose and the formation of glucose from amino acids (gluconeogenesis).

(b) *Fat metabolism.* Beta oxidation of fatty acids occurs at maximal rate in the liver. The liver forms large quantities of cholesterol, lipoproteins and phospholipid and also converts carbohydrates and proteins to fat.

(c) *Protein metabolism.* The liver plays a vital role in protein metabolism. It is important for deamination and transamination of amino acids, formation of urea from ammonia and synthesis of the plasma proteins. All the plasma proteins, except the immunoglobulins, are formed in the liver.

(d) *Synthesis of blood coagulation factors.* Fibrinogen, prothrombin and factors V, VII, IX and X are all formed in the liver. Except for fibrinogen and factor V, vitamin K is essential for the hepatic synthesis of these coagulation factors.

(e) *Storage of vitamins.* The liver is an important storage site for vitamins A, D and B_{12}.

(f) *Storage of iron.* The liver cells have a protein, called apoferritin, which can combine with varying amounts of iron to form ferritin, thereby constituting a major site of iron storage in the body. The iron in ferritin is labile, being released when the serum iron level drops and combined with apoferritin when the level of iron in the blood increases.

(g) *Metabolism of hormones.* The steroid hormones cortisol, aldosterone and testosterone are conjugated and excreted by the liver.

Thyroid hormone, insulin and glucagon are also catabolized by the liver cells.

(h) *Detoxication.* The liver detoxicates many toxins and drugs, such as sulphonamides, penicillin, erythromycin, etc. The process occurs in two phases: first, oxidation, reduction or hydrolysis of the toxin or drug and, second, conjugation with glucuronic acid, glycine or sulphate to produce a water-soluble metabolite. The products are either excreted in bile or enter the blood to be removed by the kidneys.

(i) *Metabolism of alcohol.* Ethyl alcohol is oxidized by the liver, mainly by alcohol dehydrogenase, to acetaldehyde, which is further oxidized in the mitochondria of the hepatocytes. Alcoholism is associated with destruction of liver cells and fibrosis of the liver (cirrhosis).

TESTS OF LIVER FUNCTION

Liver function tests are mainly based on assessment of the synthetic, secretory or excretory functions of the liver. The synthetic functions are assessed by determination of the concentration of the plasma proteins and prothrombin time. The secretory function may be assessed by determination of bile acids in serum. The role of the liver in bilirubin metabolism may be assessed by estimation of serum bilirubin and other related tests, such as determination of urinary urobilinogen. The excretory function can be assessed after administration of the dye bromsulphthalein (BSP). BSP is conjugated with glutathione in the hepatocytes and excreted in bile. BSP is administered intravenously and its plasma level is determined 45 minutes later. Normally, not more than 5% of BSP should be retained. Excess retention of BSP is indicative of liver disease. Certain enzymes are found in high concentrations in the hepatocytes, and determination of their plasma levels has been found helpful in the diagnosis of liver disease. The serum level of alkaline phosphatase is particularly increased in liver disease associated with biliary obstruction. Serum aminotransferases are greatly increased whenever there is liver cell damage. They are also increased in other conditions in which there is tissue damage, e.g. myocardial infarction.

HEPATIC FAILURE

Hepatic failure may be acute, as in viral hepatitis or following a high dose of a hepatic toxin, such as paracetamol. It may also be the terminal stage of long-standing chronic liver disease. Failure of the liver to conjugate or excrete bilirubin leads to an elevated serum bilirubin level and jaundice. Failure of the synthetic functions of the liver may result in haemorrhage because of deficient coagulation factors and hypoproteinaemia leading to oedema. Ascites may result from a combination of factors, including haemodynamic changes and decreased plasma protein levels.

The most serious complication, however, is hepatic encephalopathy. This is a disturbance of consciousness, which may proceed to deep coma. Hepatic encephalopathy is probably due to accumulation of substances toxic to the brain in the blood, such as ammonia, mercaptans, phenols and amines derived from bacteria in the gut. Ammonia is mostly produced by intestinal bacteria and the ability of hepatocytes to detoxicate it to urea is greatly impaired. Access to the blood of other toxins derived from gut bacteria is a direct result of the failure of the liver macrophages to remove bacterial products arriving in the portal blood. Of all the neurotoxins ammonia is considered to be the most important. When ammonia combines with *glutamate* in the brain it gives rise to glutamine and leads to glutamate depletion. Since the latter is the main excitatory transmitter in the brain, its depletion could account for the neuroinhibition of hepatic encephalopathy.

The biliary system

Bile is secreted continuously by the hepatocytes. From the canaliculi, bile flows to the intralobular ductules, interlobular canals in the portal tracts and right and left hepatic ducts, which join to form the common hepatic duct. As bile is needed in the intestine only when food is ingested, it needs to be stored between meals. The gall-bladder has a capacity of 30–50 ml. It stores and concentrates bile and regulates its discharge into the duodenum. The wall of the gall-bladder consists of a layer of smooth muscle and elastic tissue; it is lined by a layer of columnar epithelial cells. The cystic duct leading from the gall-bladder

Table 9.1 Composition of human bile (mmol/l)

	Hepatic bile	Gall-bladder bile
Water	98.0%	89.0%
Total solids	2–4%	11.0%
Bile salts	26	145
Bilirubin	0.7	5.0
Cholesterol	2.6	16
Phospholipids (lecithin)	0.5	4.0
Na^+	145	130
K^+	5	12
Ca^{2+}	5	23
Cl^-	100	25
HCO_3^-	28	10
pH	8.3 ± 0.3	7.3 ± 0.3

joins the common hepatic duct to form the common bile-duct, which opens into the duodenum in company with the pancreatic duct at the ampulla of Vater (Fig. 9.7). The opening into the duodenum is guarded by the choledochoduodenal sphincter.

COMPOSITION AND FUNCTIONS OF BILE

Bile is a viscous golden-yellow or greenish fluid with a bitter taste. It is iso-osmotic with plasma and slightly alkaline. The liver produces about 5 litres of bile per day but only 500–600 ml are poured into the duodenum daily. The composition of the human bile from the liver and in the gall-bladder is given in Table 9.1. It is clear from this table that the gall-bladder not only stores bile but also concentrates it. Concentration is achieved through active absorption of Na^+, Cl^- and HCO_3^- by the lining epithelium, with associated passive water movement out of the lumen. The drop of pH of gall-bladder bile is due to the reduction of the concentration of $NaHCO_3$. The gall-bladder epithelium also secretes mucus, which increases the viscosity of bile.

The components of bile will be considered in relation to their function under three major subheadings—neutralization, fat digestion and absorption, and excretory function.

Neutralization of acid chyme

$NaHCO_3$ in bile is responsible for its alkaline reaction and participates with pancreatic and duodenal secretions in the neutralization of acid chyme delivered from the stomach.

Fat digestion and absorption
The bile salts are essential for the emulsification of fats and for their digestion and absorption. Since absorption of the fat-soluble vitamins is closely related to the absorption of fats in general, bile salts are also important for the absorption of vitamins A, D, E and K.

The bile salts are the sodium salts of the bile acids. These are steroid acids, synthesized in the liver from cholesterol by the enzyme cholesterol 7α-hydroxylase. The principal primary bile acids are cholic acid and chenodeoxycholic acid. They are conjugated in the hepatocytes with glycine and taurine, to give glycocholic and taurocholic acids respectively.

Enterohepatic circulation of bile acids About 20–30 g of bile acids are poured into the duodenum per day. Some of the bile acids are de-conjugated and dehydroxylated in the 7α position by the intestinal bacteria. Dehydroxylation results in the production of the secondary bile acids, i.e. deoxycholic and lithocholic acids. The intestine thus contains both conjugated and unconjugated primary and secondary bile acids. On reaching the terminal ileum, 90% of the bile acids are absorbed actively and reach the liver through the portal vein. In the liver cells, the bile acids are reconjugated and resecreted in bile.

In the small intestine cholic acid is absorbed faster than chenodeoxycholic acid, and conjugated bile acids are absorbed better than unconjugated bile acids. Some unconjugated bile acids are absorbed passively in the colon and also reach the liver through the portal vein. About 0.6 g of unconjugated dehydroxylated bile acids are lost in the faeces daily. These are replaced by new synthesis from cholesterol in the liver, so that the total bile acid pool is maintained constant at 2–4 g. Since the amount of bile acids poured into the duodenum each day is 20–30 g, the daily turnover of the total bile acid pool through the enterohepatic circulation must be 6–10 times.

The presence of bile acids in portal blood as a result of the enterohepatic circulation is essential for stimulating and maintaining the secretion of bile by the hepatocytes. If the enterohepatic circulation is interrupted (e.g. due to obstruction by disease or surgical removal of the terminal ileum), bile flow is markedly decreased. Synthesis of bile acids in the hepatocytes and of cholesterol in the liver and intestinal mucosa is increased. But, despite this, the output of bile acids into the duodenum is reduced. Deficiency of bile acids leads mainly to defective fat digestion and absorption and steatorrhoea (excessive fat in the stool). Absorption of fat-soluble vitamins is also reduced. Excess amounts of bile acids entering the colon may result in diarrhoea (see below).

Functions of the bile acids:
1 In the intestinal lumen the bile acids are essential for emulsification of fats. They enter into the formation of mixed micelles, needed for absorption of lipids and fat-soluble vitamins.
2 Bile acids stimulate the release of CCK-PZ from its cell in the upper intestine and thus indirectly stimulate pancreatic secretion and the discharge of bile into the intestine.
3 They have a negative feedback effect on the synthesis of cholesterol by the hepatocytes and the intestinal mucosal cells.
4 In the liver, bile acids are important for stimulating bile secretion and flow (choleresis), but they have a negative feedback effect on the synthesis of bile acids from cholesterol. Bile acids in the liver also take part in the formation of micelles, which render cholesterol soluble in bile.
5 In the colon, bile acids inhibit reabsorption of water and electrolytes and stimulate motility, and may cause diarrhoea.

Excretory function of bile
Bile is a medium for excretion of bile pigments, cholesterol and many drugs, e.g. penicillin, toxins, various inorganic substances, such as copper and zinc, and dyes, such as bromsulphthalein.

The principal bile pigment, bilirubin, is conjugated with glucuronic acid in hepatocytes, thereby rendering it water-soluble. The conjugated bilirubin is secreted by the liver cells into the bile canaliculi independent of the secretion of bile acids. The colour of bile is due to bilirubin.

About 1–2 g of cholesterol appears in bile per day. Cholesterol is water-insoluble; it is solubilized by incorporation in micelles, along with the bile acids and phospholipid. The micelles remain stable so long as the concentrations of bile acids, phospholipids and cholesterol remain within certain limits. If the relative concentration of any of the constituents alters, cholesterol may be precipitated out of solution. In people who produce bile with a high concentration of cholesterol, cholesterol gallstones may form in the gall-bladder.

CONTROL OF THE BILIARY SYSTEM
There are two aspects of control of the biliary system to consider:
1 Secretion of bile by the liver cells (choleresis).
2 Contraction of the gall-bladder with relaxation of the choledochoduodenal sphincter to pour bile into the duodenum (cholecystokinesis).

The human liver secretes bile at a pressure of about 25 cmH$_2$O. Between meals, the choledochoduodenal sphincter is normally closed, offering a resistance of about 30 cmH$_2$O. Bile secreted by the liver is thus diverted to the gall-bladder during the interdigestive periods. Pressure in the lumen of the gall-bladder varies between 0 and 16 cmH$_2$O.

Control of choleresis
Bile secretion by the hepatocytes is independent of the hydrostatic pressure in the liver sinusoids. Blood-pressure cannot therefore account for bile secretion. The main driving-force for bile secretion is the active transport of organic anions, especially bile acids, into the canaliculi, with water flowing passively along the osmotic gradient created. In the biliary ducts, bicarbonate is secreted independently of bile acid secretion and is followed passively by water. Total bile flow is thus due to two components: bile acid-dependent and bile acid-independent components.

Bile acid-dependent component There is a direct linear relationship between the rate of bile flow and bile acid secretion (Fig. 9.9). After cannulation of the bile-duct during a surgical operation in man, bile acids were introduced into the duodenum through a tube. Bile was collected from the cannula and its flow rate and bile acid output were determined. The exogenous bile acids were incorporated into the enterohepatic circulation and produced the increase in bile secretion. This bile acid-dependent component depends mainly on the integrity of the enterohepatic circulation. At least 90% of the rate of secretion of bile acids is determined by the rate of clearance of reabsorbed bile acids from the portal vein. The remaining 10% is due to synthesis of new bile acids by hepatocytes. As mentioned previously, interruption of the enterohepatic circulation results in markedly reduced choleresis.

Bile acid-independent component With reference to Fig. 9.9, the regression line does not pass through the origin. When extrapolated, it intercepts the Y axis at about 0.2 ml/min. This means that, when the rate of bile acid secretion is zero, there would still be some bile flow, i.e. the bile acid-independent component. This fraction of bile secretion is due to secretion of HCO$_3^-$, followed by water, by the biliary duct cells. It depends on active sodium transport, since it is blocked by inhibition of Na$^+$, K$^+$ ATPase. Pure stimulation of this component may actually lead to reduction in the concentration of bile acids in bile.

The bile acid-independent fraction of bile secretion is stimulated by the hormones secretin, glucagon, CCK-PZ and gastrin. They all stimulate the secretion of HCO$_3^-$ and passive H$_2$O transfer

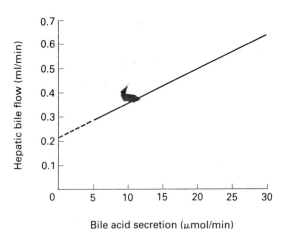

Fig. 9.9 The relationship between bile flow and bile acid secretion in man. The dotted line is extrapolated to give the bile acid-independent secretion. (From data by Sherstan, T., Nilsson, C.S., Cahlin, E., Filipson, M. & Brodin-Persson, G. (1971) *European J. Clin. Invest.*, **1**, 242–7.)

by the biliary duct cells. Vagal stimulation also stimulates bile flow. The effect is mediated mainly indirectly, through stimulation of gastric acid secretion, which leads to release of secretin and CCK-PZ. All substances that stimulate hepatic secretion of bile or choleresis are termed choleretics.

Control of the discharge of bile into the intestine
Contraction of the gall-bladder is followed almost immediately by relaxation of the choledocho-duodenal sphincter. The sphincter is kept closed by cholinergic mechanisms (which can be blocked by atropine) and relaxed by the hormone CCK-PZ.

Discharge of bile into the duodenum is regulated by nervous and hormonal mechanisms. The nervous component is mediated by the vagus nerve and follows psychic influences and food ingestion. During this cephalic phase, bile is discharged for only a brief period into the duodenum.

The main pouring of bile occurs when chyme reaches the upper intestine and the mechanism is mainly hormonal, being mediated by CCK-PZ. The presence of digestive products of fats and proteins, i.e. fatty acids and L-amino acids, releases CCK-PZ from the upper intestine into the blood. CCK-PZ contracts the gall-bladder and relaxes the choledochoduodenal sphincter, thereby discharging bile into the duodenum. Both vagal excitation and secretin augment the action of CCK-PZ on the gall-bladder. $MgSO_4$ contracts the gall-bladder and discharges bile into the intestine because it releases CCK-PZ. Generally, substances that lead to pouring of bile into the intestine by contracting the gall-bladder and relaxing the sphincter are called cholagogues.

RADIOLOGICAL TEST OF BILIARY FUNCTION
The function of the gall-bladder and biliary ducts may be assessed radiologically (cholecystography) after oral or intravenous administration of a radio-opaque substance which is excreted by the liver into the bile, e.g. tetraiodophenolphthalein or iopanoic acid. A non-functioning gall-bladder fails to show the opaque material in its cavity. Gallstones show as filling defects. The function of the gall-bladder may be assessed further by giving a cholagogue and assessing its contraction.

The cholagogues used include ingestion of a fatty meal or administration of $MgSO_4$, CCK-PZ or caerulein. Caerulein is a peptide with 10 amino acids. It has all the properties of gastrin and CCK-PZ and is used in cholecystography to contract the gall-bladder.

JAUNDICE
An increase of the bile pigment bilirubin in plasma (normal 0.5 mg/dl) to about 2.5 mg/dl or more results in yellow pigmentation of the skin, sclera and mucous membranes. If this is due to obstruction of the bile duct, it is known as obstructive jaundice. Excessive haemolysis of RBCs can also lead to jaundice, even with normal liver and biliary function. This type is known as haemolytic jaundice. It is due to formation of bilirubin in excess of the liver capacity to excrete it. In obstructive jaundice the bilirubin is conjugated and therefore gives a direct reaction with the Van der Berg test. In haemolytic jaundice most of the bilirubin is non-conjugated and bound to plasma protein. Therefore, the protein needs to be precipitated before the Van der Berg test is applied. The test is said to be indirect. This is one way to differentiate between the two types of jaundice. A third type is hepatic jaundice. In this case there is severe damage to the liver cells or intrahepatic obstruction of bile canaliculi due to liver disease. Normal amounts of bilirubin fail to be excreted and therefore accumulate and give rise to jaundice.

The small intestine
In the small intestine the digestion of food materials is completed and the products of digestion are absorbed. The small intestine is, in fact, the main site of absorption of nutrients in the body.

The small intestine extends from the pylorus to the caecum. It is arbitrarily divided into the duodenum, jejunum and ileum. The jejunum is taken as the upper two-fifths of the intestine below the duodenum. The ileum is the remaining three-fifths. The most prominent histological feature in the intestinal mucosa is the presence of the villi, small finger-like projections about 1 mm in height. In man and some animals, there are mucosal folds up to 1 cm in height and about 5 cm in length. In addition, the epithelial cells lining the villi have numerous microvilli. The mucosal folds, villi and microvilli enormously increase the intestinal surface available for absorption. The total surface area has been estimated to be about 300 square metres.

Between the bases of the villi are found the intestinal glands (crypts of Lieberkühn), which secrete intestinal juice (succus entericus). Brunner's glands are found in the submucosa of the duodenum. Their secretion reaches the lumen through ducts that penetrate the mucosa to open at the base of the crypts. Lymphoid tissue is found throughout the small intestine, but in the ileum it is aggregated into prominent nodules called Peyer's patches.

The intestinal mucosa may be divided into three layers (Fig. 9.10): the layer of epithelial cells lining the villi and crypts, the lamina propria and the muscularis mucosae. The lamina propria constitutes the connective tissue core of the villus. It contains a central arteriole that breaks up into looping capillaries that connect to venules, a single lymph vessel (lacteal), nerve

fibres, smooth muscle cells that extend along the villus and many mononuclear cells, including lymphocytes, plasma cells and macrophages. In the crypts, the lining epithelial cells include the following types: (i) undifferentiated actively dividing cells, frequently seen in mitosis; (ii) mucus-secreting goblet cells; (iii) various epithelial endocrine cells; and (iv) Paneth cells containing large eosinophilic granules. Although Paneth cells have been shown to contain lysozyme and immunoglobulins, their precise function still remains uncertain. The epithelial cells lining the villi include: (i) the digestive absorptive columnar cells with microvillous brush border called enterocytes; (ii) mucus-secreting goblet cells; and (iii) a few endocrine epithelial cells.

It has been shown that the actively proliferating cells in the crypts migrate on to the villus, taking 2–5 days. In the process these cells become differentiated into the enterocytes, and digestive enzymes appear in their brush border. The microvilli of the enterocytes are covered by a layer of glycocalyx, which is a complex of poly-

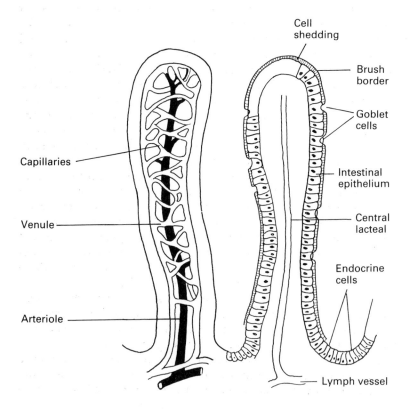

Cell shedding

Brush border

Goblet cells

Intestinal epithelium

Central lacteal

Endocrine cells

Capillaries

Venule

Arteriole

Lymph vessel

Fig. 9.10 Diagram of longitudinal section of two villi to show the histological organization of the mucosa of the small intestine (right) and the blood supply (left).

saccharide and protein. Glycocalyx mainly adsorbs pancreatic enzymes and seems to place the final products of digestion in a strategically advantageous position for absorption.

Cells are constantly shed from the tips of the villi and are replaced by upward-migrating, actively proliferating cells of the crypts. The intestinal epithelium is thus one of the most actively proliferating tissues in the body. In man 50–200 g of gastrointestinal mucosa are renewed daily. Agents that inhibit mitosis, such as ionizing radiation and cytotoxic drugs, lead to flattening of the mucosa as cells extruded fr… unreplaced.

INTESTINAL SECRETION

Brunner's glands in the duodenu… alkaline fluid that contains m… enzymes.

The intestinal crypts secrete ab… day of intestinal juice or succus e… has a pH of 6.5–7.6 and 1.6% o… organic and 1% inorganic. Intesti… turbid fluid with a fishy odour. … enzymes are found either in the b… in the cytoplasm of the enterocyt… present in only trace amounts as a… integration of desquamated muco… cells in the lumen. The enzymes th… secreted into the lumen are e… (enterokinase) and amylase.

As mentioned earlier, enteropeptide activates trypsinogen to trypsin and thereby holds the key to activation of all the pancreatic proteolytic enzymes. Enteropeptidase is, in fact, the most important enzyme secreted in intestinal juice. When it is congenitally deficient, protein digestion is severely impaired.

Succus entericus participates in the neutralization of acid chyme delivered from the stomach. Its enzymes complete the digestion of food materials.

DIGESTION IN THE SMALL INTESTINE

The polypeptides resulting from the action of the pancreatic endopeptidases (trypsin, chymotrypsin and elastase) are further attacked by the exopeptidases. These are the carboxypeptidases of pancreatic juice, which split off the terminal amino acids with the free carboxyl group, and the aminopeptidases of the succus entericus, which split off the terminal amin…

…tion of salivary, …se on starch, is … brush border of … of glucose. The …se (milk sugar) …ed, respectively, …sh border of the …own to glucose …se and galactose. All the disaccharidases are developed at the time of birth but a decrease in the activity of lactase occurs after weaning in most species. In man, regression of lactase activity occurs, but there are ethnic differences in the extent of the decline. In northern European populations, for instance, the decline is quite slow. Deficiency of a disaccharidase may be inherited or may be secondary to small-bowel disease. It leads to malabsorption of the corresponding sugar. The unabsorbed disaccharide is broken down by intestinal bacteria to lactic and other acids and this results in distension, flatulence and osmotic diarrhoea. Milk or lactose intolerance is found in a high proportion of the population in Africa and Asia and may constitute a problem in nutrition.

Deficiency of a specific disaccharidase may be investigated by giving a test dose of the disaccharide, followed by monitoring of blood glucose

the epithelial cells lining the villi include:
a) the digestive absorptive columnar cells with microvillous brush border called enterocytes.
b) mucus-secreting goblet cells
c) a few endocrine epithelial cells.

the ephitelia… the villi hav… microvilli. the… villi and mic… increase the i… available for absorption.

mucosa

level. Failure of the blood glucose level to rise is indicative of the disaccharidase deficiency. Clinically, disaccharidase deficiency is indicated by the occurrence of diarrhoea. Disaccharidase enzymes may also be estimated directly on samples of intestinal mucosa obtained by peroral biopsy.

Dietary fibre

Fibres of plant origin are mostly carbohydrates. They are not hydrolysed by pancreatic or intestinal enzymes in man. They include cellulose, hemicellulose, lignin and various gums and pectic substances. They pass unchanged to the lower small intestine and colon, where they are metabolized by bacteria, leading to an increase in osmolality and greater bulk of colonic contents. This stimulates motility, leading to reduced transit time and greater amounts of stools. Natural food, which is nearer to the diets of rural populations, has a high fibre content. In contrast, the refined food of the industrialized or urban societies is low in fibre. Epidemiological evidence has been presented to suggest that consumption of low-fibre diets is associated with constipation, diverticulosis, colonic cancer, irritable bowel syndrome, haemorrhoids, appendicitis, gallstones, hypertension, coronary heart disease and diabetes mellitus, all of which have a high prevalence in Europe and North America. A shift to more natural, moderately high-fibre diets is now advocated. As more and more people urbanize in the developing countries, they would do well to avoid repeating the experience of the developed countries.

CONTROL OF INTESTINAL SECRETION

Brunner's glands in the duodenum are stimulated mainly by hormones. Secretin can cause secretion but a hormone which specifically stimulates Brunner's gland has been postulated; it is called duocrinin and is still known only in crude extract form. Vagal stimulation can cause secretion but the role of the vagus is probably minor since a denervated duodenal segment responds to food about the same as before denervation.

Secretion of intestinal juice is very low in the interdigestive period. The most important stimulus is the presence of food in the lumen of the intestine. Distension, as well as tactile and irritating stimuli, evoke considerable secretion from intestinal glands through local neural reflexes. Watery secretion in the intestinal lumen is mainly due to active secretion of Cl^- through a specific channel activated by CAMP. Cl^- is followed by passive diffusion of Na^+ and water. The toxin of the cholera vibrio produces profuse secretion through activation of adenylate cyclase and increasing the concentration of cAMP. Indeed, death in cholera is due almost entirely to the dehydration and shock due to severe diarrhoea. If the water and electrolyte losses are promptly replaced, the patient can be saved.

Vagal stimulation is an unreliable stimulus of intestinal secretion, but parasympathomimetic drugs stimulate secretion. Sympathetic stimulation exerts an inhibitory influence on intestinal secretion. Section of the splanchnic nerves is followed by secretion, called paralytic secretion. This may also be contributed to by vasodilatation consequent on loss of sympathetic vasoconstrictor tone.

The hormones gastrin, CCK-PZ, secretin and glucagon stimulate intestinal secretion. It has been postulated that a hormone, called enterocrinin, is released from the intestinal mucosa into the blood by chyme to stimulate secretion of succus entericus; it is still known only in crude extract form. Vasoactive intestinal peptide (VIP) stimulates intestinal secretion. A tumour that produces VIP may rarely arise from a cell in the pancreatic islets; it has been called a vipoma. It causes a syndrome of severe watery diarrhoea, hypokalaemia and achlorhydria (WDHA syndrome or pancreatic cholera).

INTESTINAL ABSORPTION

Although a small amount of water is absorbed in the stomach and substantial amounts of sodium and water are absorbed in the colon along with folic acid, some amino acids and short-chain fatty acids, the main site of absorption of nutrients is the small intestine. To be absorbed, a substance traverses the brush border, cytoplasm and basolateral borders of the enterocyte. It enters either the blood, subsequently passing into the portal circulation, or the lacteals into lymph to eventually reach the systemic circulation through the thoracic duct. The mechanism of transport may be passive diffusion along an electrochemical gradient, or may be carrier-mediated. When carrier-mediated transport is along an electrical or chemical gradient, it is called facilitated diffusion; no direct energy input is required. When a carrier transports a substance against an electrical or chemical gradient, energy input is required and the process is called active transport. Active transport is fastest, followed in descending order by facilitated diffusion and passive diffusion. Some large substances may be absorbed by endocytosis.

Absorption of carbohyrates

Digestion of starch and disaccharides ends in the hexose monosaccharides glucose, galactose and fructose. Complete digestion of nucleic acids produces pentose sugars. Glucose and galactose are actively transported, fructose is transported by facilitated diffusion, while the pentoses are passively absorbed.

There is experimental evidence that shows a direct link between the absorption of glucose and that of sodium. The presence of sodium in the gut lumen increases the rate of glucose absorption, and inhibition of the sodium pump, e.g. by ouabain, reduces the rate of glucose absorption. Administration of substances that block the release of metabolic energy, such as dinitrophenol, as well as that of the glycoside phloridzin, inhibits the rate of glucose absorption. These findings lend support to the model shown in Fig. 9.11. There is a membrane carrier protein in the brush border called symport that cotransports both glucose and sodium to the inside of the cell, where they are released. Glucose is further transported by facilitated diffusion across the basolateral membrane by another carrier into the blood. The sodium is actively extruded across the basolateral membrane out to the intercellular space by the sodium pump, i.e. Na^+, K^+ ATPase. This creates a gradient for the entry of Na^+ across the luminal membrane and is the step that requires direct input of energy, so that the use of dinitrophenol or ouabain inhibits the transport of sodium to the outside of the enterocyte and consequently the absorption of glucose as well. Indeed, intestinal absorption of glucose is one of the best examples of secondary active transport, i.e. secondary to Na^+, K^+ ATPase. Phloridzin acts by competing with glucose for the carrier at the brush border. Since the carrier has more affinity for glucose than galactose, the latter is transported more slowly.

Fructose is partly converted to glucose and lactate within the enterocyte creating a steep inward concentration gradient. In addition, there is a specific membrane carrier for fructose so that it is transported by facilitated diffusion. The pentoses, being transported by passive diffusion, have the slowest rate of absorption, followed in ascending order by fructose, galactose and glucose. All of these monosaccharides enter portal blood to reach the liver for further metabolic processing.

In a mixed meal, usually all the digestible carbohydrates are absorbed. The bulk of carbohydrate absorption occurs in the upper intestine.

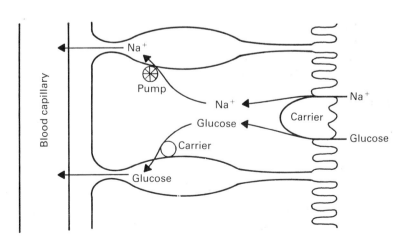

Fig. 9.11 Diagram showing the cotransport of sodium and glucose by the absorptive cells of the small intestine. Details are in the text.

Absorption of proteins

Digestion of proteins results in the liberation of amino acids and short peptides. Di- and tripeptides are, in fact, absorbed faster than their constituent amino acids. As explained previously, they are hydrolysed by brush border and cytoplasmic oligopeptidases. Amino acids are absorbed according to their optical isomerism: D-amino acids are transported by passive diffusion but L-amino acids are transported by several active transport mechanisms. The active absorption of L-amino acids, like that of glucose, is sodium-dependent; the presence of sodium accelerates the rate of amino acid transport. There seem to be separate mechanisms for the transport of basic, dicarboxylic and neutral amino acids; a fourth mechanism transports proline, hydroxyproline and glycine. As the amino acids accumulate within the enterocyte, they passively diffuse into the blood to enter the portal circulation.

Even whole proteins and polypeptides may be absorbed. In infants, protein antibodies from maternal colostrum enter the enterocyte probably by pinocytosis and exit by exocytosis into the blood. In this way passive immunity is conferred on the infant. The ability of the enterocyte to absorb whole proteins declines with age but small amounts may be absorbed in adulthood, thus providing the basis for the occurrence of food allergies. The intestine is one of the entry sites for protein antigens.

About 85–90% of ingested protein is usually absorbed and most of the absorption of protein materials occurs in the upper intestine. It is noteworthy that most of the protein in stools comes not from the diet but from cellular debris and intestinal bacteria.

The purine and pyrimidine bases resulting from the digestion of nucleic acids are absorbed by active transport.

Absorption of fats

Triglycerides, along with some monoglycerides and fatty acids, resulting from the action of lingual and gastric lipases, are delivered in crude emulsion form into the duodenum. In the upper intestine, bile salts and lecithin emulsify the fat mixture further and pancreatic lipase attacks the fat globules at their interface with water in the presence of colipase and bile salts. Colipase is secreted by the pancreas; it binds to the surface of fat droplets, displacing the emulsifying agents and allowing lipase to anchor to the droplet. The triglycerides are consequently hydrolysed to fatty acids and monoglycerides. With the presence of bile salts and monoglycerides, mixed micelles are produced. These are stable water-soluble structures 3–10 nm in diameter that can incorporate other lipid components within them. Essentially, molecules are aggregated in a micelle in such a way that water-soluble hydrophilic polar groups, i.e. hydroxyl and carboxyl groups, are facing the outer side of the micelle while fat-soluble hydrophobic hydrocarbon chains are facing the interior of the micelle. In addition to the basic building-blocks of the micelle, i.e. the bile salts and monoglycerides, long-chain fatty acids, cholesterol and fat-soluble vitamins are incorporated into the interior of the micelle. In this way, these hydrophobic compounds are made water-soluble. The mixed micelles enter the unstirred water layer (UWL) immediately adjacent to the intestinal epithelium and make contact with the brush border of the enterocyte.

The monoglycerides, long-chain fatty acids and cholesterol, being lipid-soluble, enter the enterocyte by passive diffusion, leaving the bile salts still in the lumen. The bile salt micelles diffuse back into the chyme to pick up more monoglycerides, fatty acids and cholesterol and bring them into contact with the enterocyte. Thus, the bile salts seem to perform some kind of ferrying function through the UWL that helps the absorption of monoglycerides, long-chain fatty acids, cholesterol and probably also fat-soluble vitamins. It should be remembered that bile salts are absorbed actively in the terminal ileum to enter the portal circulation and reach the liver, from which they are resecreted into bile in the upper intestine (the enterohepatic circulation).

The monoglycerides, long-chain fatty acids and cholesterol that enter the enterocyte are re-esterified to triglycerides and cholesterol esters. They are then surrounded by a layer of lipoprotein, cholesterol and phospholipid to form chylomicrons. These enter the lacteals and leave the intestine in lymph to reach the systemic circulation through the thoracic duct.

The fatty acids with more than 10–12 carbon atoms are considered long-chain and are incorporated into the chylomicrons as just explained. Short-chain fatty acids with less than 10–12 carbon atoms are directly absorbed into the enterocyte by an active sodium-dependent transport mechanism; they pass directly to the liver in portal blood.

Most fat absorption occurs in the upper intestine, although some substantial absorption takes place in the ileum. In a mixed meal about 95% of the fat is absorbed. The stools contain about 5 g of fat per day, which is equivalent to about 5% of the stool bulk, but most of this fat is derived from cellular debris and remnants of intestinal bacteria rather than from the diet. Defective assimilation of fats leads to increased excretion of fat in the stools, well in excess of 5 g/day. The stools are also pale, greasy and bulky with an offensive smell. This condition is called steatorrhoea. Steatorrhoea may result either from maldigestion (e.g. deficiency of pancreatic lipase or bile salts) or malabsorption due to diseased intestinal mucosa, e.g. in the coeliac syndrome.

Absorption of vitamins

Absorption of the fat-soluble vitamins A, D, E and K is tied to that of fat. If fat absorption is deficient, absorption of the fat-soluble vitamins will also be defective. Conditions which lead to steatorrhoea may be accompanied by symptoms of fat-soluble vitamin deficiencies.

Absorption of vitamin B_{12} (cyanocobalamin) requires the presence of a glycoprotein secreted by the parietal cells of the gastric mucosa; it has been called the intrinsic factor. It combines with vitamin B_{12} derived from food (the extrinsic factor) and the complex is absorbed in the terminal ileum. Probably the intrinsic factor stimulates endocytosis of cyanocobalamin after the complex is bound to specific receptors on the enterocyte.

It is now realized that the intestinal absorption of the water-soluble vitamins, a diverse group of chemical substances, does not mainly occur by passive diffusion. Except for pyridoxine and p-aminobenzoic acid, both of which are absorbed by passive diffusion, the rest are absorbed by carrier-mediated transport. Although there are minor species differences, ascorbic acid, biotin, inositol, nicotinic acid and thiamine are absorbed by active transport, while choline, folic acid and riboflavin are absorbed by facilitated diffusion.

Absorption of water and electrolytes

Secretions from salivary glands, stomach, bile, pancreas and intestine add up to a total of 7 litres/day. Ingested water is about 2 litres/day (may be much more in hot climates). Out of this daily input of 9 litres of water, a negligible amount is absorbed in the stomach. Net daily absorption in the small intestine is about 7.5–8 litres and in the colon about 1 litre. The remainder is excreted in the stools, 150–200 ml/day. Thus, about 98% of the water entering the lumen of the alimentary tract each day is reabsorbed. This huge internal turnover of water amounts to about 20% of the total amount of water in the human body. It is thus easy to understand that persistent vomiting or diarrhoea can readily result in dehydration and this is especially true in infants and young children.

Part of the sodium in the lumen of the small intestine can move passively in either direction across intestinal epithelium, depending upon the osmotic gradient. The other part moves across the luminal border of the small intestine and colon along a concentration gradient created by Na^+, K^+ ATPase which is located at the basolateral membrane of the epithelial cells. Such actively absorbed sodium in the small intestine facilitates the absorption of glucose, amino acids and short-chain fatty acids. In turn, the presence of glucose facilitates the absorption of sodium. This is the basis of the proved benefit of oral administration of both NaCl and glucose in diarrhoeal diseases, e.g. cholera. In the colon, active absorption of sodium is followed by passive water absorption.

For the most part, potassium moves across intestinal epithelium by diffusion. Since the gut lumen is electronegative in relation to the blood, net movement of potassium occurs into the lumen. A small component of potassium is actively secreted into the lumen as part of mucus. The net loss of potassium to the lumen of the small and large intestines explains why hypokalaemia readily develops in diarrhoea.

In the ileum and colon, chloride is actively reabsorbed in exchange for bicarbonate, which tends to make lower intestinal contents more alkaline.

MALABSORPTION SYNDROMES

There are a variety of conditions which lead to defects of digestion and absorption. Some common examples are as follows:

1 Failure of absorption of fats, carbohydrates and proteins lead to loss of weight and malnutrition.
2 Malabsorption of carbohydrates usually leads to diarrhoea of the osmotic type. The action of colonic bacteria on the unabsorbed carbohydrate results in gas formation, flatulence and abdominal distension.
3 Malabsorption of bile acids and fatty acids leads to diarrhoea with frequent loose motions.
4 Impaired fat digestion leads to a high fat content of the stool, which becomes bulky, foul-smelling and frothy.
5 Vitamins and minerals may fail to be absorbed in a variety of conditions, leading to specific deficiencies, e.g. tetany and osteoporosis in the case of Ca^{2+}, Mg^{2+} and P malabsorption; anaemia in the case of iron, folate or B_{12} malabsorption and tendency to bleed or to bruise easily when there is failure to absorb vitamin K, as in obstructive jaundice.

MOVEMENTS OF THE SMALL INTESTINE

The functions of movements of the small intestine are the mixing and churning of chyme to aid digestion and absorption and the propulsion of chyme along to the colon. Study of the electrical activity of the smooth muscle in the small intestine indicates that in the duodenum, just above the opening of bile and pancreatic ducts, there is a 'pacemaker' region from which small-bowel slow waves, also called basal electric rhythm (BER), start and move caudally. These are spontaneous waves of smooth muscle depolarization, which are responsible for co-ordination of intestinal movements. Their frequency gradually declines as they move caudally, so that the lower portions of the small intestine are less active than the upper. Spikes of action potentials associated with contractions are superimposed on the slow waves.

During the interdigestive period, the migrating motor complex (MMC), originating in the stomach, travels along the small intestine down to the ileocaecal junction. As previously explained, the MMC consists of bursts of depolarization accompanied by peristaltic contractions. The cycle of MMC may be considered as consisting of four phases: phase I is characterized by absence of contractions, phase II by irregular contractions, phase III by a burst of regular large-amplitude contractions and phase IV by return to inactivity. The function of the MMC is to propel any remnants in the stomach and small intestine during the interdigestive period into the colon.

Two other types of movement will be described below. A mixing movement (segmentation) and a propulsion movement (peristalsis).

Segmentation contractions (Fig. 9.12)
Segmentation contractions are ring-like contractions of the circular muscle layer appearing at regular intervals along a length of the small intestine. Soon they disappear and are replaced by another set of ring contractions arranged such that the parts that were contracted become relaxed and those that were relaxed become contracted. In addition to mixing the chyme with digestive secretions, the segmentation contractions seem to have a massaging influence on blood- and lymph-vessels of the intestine. Segmentation contractions persist after extrinsic denervation, but disappear after destroying the intrinsic nerve plexus in the small intestine.

Peristalsis
Peristaltic waves also occur in the oesophagus, stomach, colon and ureters. Peristalsis consists of

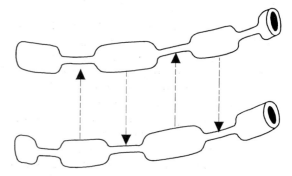

Fig. 9.12 Diagram showing segmentation contractions of the small intestine. Relaxed segments between the dotted lines have become contracted and vice versa.

a travelling wave of contraction preceded by relaxation. It persists after extrinsic denervation and is abolished after destruction of intrinsic nerves. It is controlled by an intramural myenteric reflex. The chemical transmitters involved include acetylcholine, 5-HT and substance P. Peristalsis usually travels in an oral–caudal direction. Antiperistalsis occasionally occurs in the colon and in the stomach during vomiting. A peristaltic wave travels at rates which vary between 2 and 25 cm/s.

Movements of the villi

The villi move during digestion. They do not move together, each villus seeming to move independently of the others. The villus movement consists of fast shortening and slow lengthening. The movement is executed by a short tongue of muscle that extends from the muscularis mucosa on to the villus.

Control of intestinal movements

Although movement is an inherent property of intestinal smooth muscle, it may be modified by neural and hormonal factors.

Vagal excitation and cholinergic stimulation increase motor activity while sympathetic or adrenergic stimulation inhibits movement. Vagal stimulation increases movement of the villi while sympathetic stimulation makes them pale and motionless. In addition to acetylcholine and noradrenaline, other effects on intestinal smooth muscle may be mediated by serotonin (5-HT) or peptides such as substance P, VIP and encephalin.

In man, sympathectomy does not affect intestinal movements. Vagotomy is followed by transient hypomotility; normal motility is restored after a mean time of 10 hours. Inhibition of intestinal peristalsis, largely mediated through a nervous reflex, is termed paralytic or adynamic ileus. The inhibition may be initiated by excessive irritation of the intestine during abdominal operations, by intestinal overdistension, by peritonitis or even by renal, vesical or skin irritation. The efferent arm of the inhibitory reflex is in the sympathetic nerves to the intestine.

The hormone gastrin stimulates intestinal motility and relaxes the ileocaecal sphincter. CCK-PZ also stimulates intestinal movements. Conversely, secretin inhibits intestinal motility

and contracts the ileocaecal sphincter.

There is experimental evidence indicating that chyms releases a hormone from the upper intestine that specifically stimulates movements of the villi; it has been called villikinin. The chemical nature of villikinin awaits final clarification.

The colon

The large intestine has a wider diameter than that of the small intestine. It starts with the caecum and continues as the ascending, transverse and descending colon. The sigmoid colon leads to the rectum and anal canal in the pelvis. One prominent morphological feature of the colon (Fig. 9.13) is that the outer longitudinal layer is modified to form three longitudinal bands, called teniae coli visible on the outer surface. Since the muscle bands are shorter than the length of the colon, the colonic wall is sacculated and forms haustra. The mucous membrane of the colon lacks villi and consists of simple short glands lined mostly by mucus-secreting goblet cells. There are scattered lymphoid nodules, especially prominent in the caecum and appendix. In man the colon has a length of about 150 cm and a capacity of between 0.9 and 1.8 litres.

SECRETION

Secretion in the colon is mainly mucus; no digestive enzymes are secreted by the colon. The mucus helps to lubricate faeces and neutralize any acids present; colonic contents have a pH of about 8. It also protects against irritants and provides a binding medium for faecal matter. The main stimulus to mucus secretion is the contact of the glands with colonic contents. In addition to direct stimulation of the gland cells, further secretion may occur via a reflex pathway involving the parasympathetic pelvic splanchnic nerves. Parasympathomimetic drugs, e.g. pilocarpine, stimulate colonic secretion and this stimulation can be blocked by atropine.

ABSORPTION IN THE COLON

From 0.5 to 1.5 litres of water are absorbed from the chyme delivered to the colon each day. The intestinal chyme is thus transformed into semisolid faeces with a net loss of water from the gut of 150–200 ml/day. Thus, the principal sub-

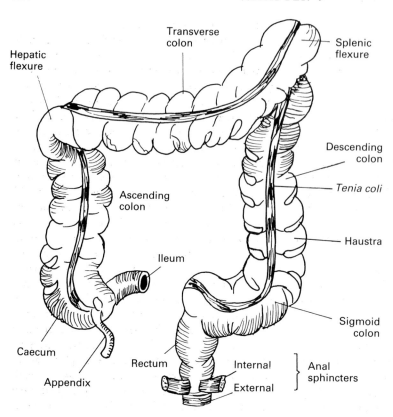

Fig. 9.13 The human colon.

stance absorbed in the colon is water. About 60 mmol of sodium are actively absorbed daily and water follows along the osmotic gradient created. Potassium is secreted into the lumen of the colon while chloride is reabsorbed in exchange for bicarbonate, which is secreted into the lumen. Persistent diarrhoea therefore results in dehydration, hypokalaemia and metabolic acidosis due to loss of HCO_3^- from the gut. Some absorption does take place in the colon. Folic acid, some amino acids and short-chain fatty acids resulting from the bacterial fermentation of carbohydrates, steroids, aspirin, anaesthetics, sedatives and tranquillizers may all be absorbed by the colonic epithelium. But carbohydrates, whole proteins and fats cannot be absorbed, so that it is not possible to adequately maintain nutrition by rectal instillation. The considerable absorptive capacity of the colon is, however, useful clinically, as it provides a convenient route of drug administration, i.e. by rectal enema, especially in children.

INTESTINAL BACTERIA

At birth, the intestine is sterile and the new-born baby passes a greenish semifluid material called meconium. This is derived from amniotic fluid swallowed by the foetus while in the uterus. Meconium consists mainly of carbohydrate substances and the green colour is due to unreduced bile pigments. Bacteria are ingested and gain access to the intestine a few days after birth and thereafter stay throughout life. It is well recognized that the colon contains a bacterial flora, but recently it has been realized that the small intestine also contains bacteria. Apparently some bacteria escape destruction by gastric hydrochloric acid and gain access to the upper intestine, from which they spread caudally, increasing in numbers as they do so.

The intestinal flora should be viewed as living in symbiosis with the host. On the whole, its effects are beneficial to the body. Vitamin K and a number of the B group of vitamins, especially folic acid, biotin, thiamine and B_{12}, are synthesized by intestinal bacteria. Vitamin B_{12} cannot be absorbed because the intrinsic factor is lacking in the colon. Evidence for absorption of vitamin K and other vitamins has been obtained in rats, which usually eat their faeces (coprophagia). It is now realized that in man only folic acid is significantly absorbed from the colon.

Bile salts are deconjugated and dehydroxylated by intestinal bacteria. Bile pigments are similarly broken down by bacteria in the intestine to produce stercobilinogen, which is excreted in faeces, the brown colour of stools being due to this pigment.

Intestinal bacteria decarboxylate some amino acids to produce amines such as histamine and tyramine. The amines skatole and indole are excreted in faeces and are responsible for its smell. The level of indole excretion in urine is used as an index of bacterial activity in the small intestine.

Urea diffuses slowly from the blood into the colon where it is broken down by bacterial urease to ammonia. Most of this ammonia is absorbed and reconverted to urea in the liver. The large intestine is a major source of ammonia in the body. Since accumulation of ammonia seems to be an important factor in causing hepatic encephalopathy, the management of hepatic failure includes administration of poorly absorbed antibiotics, e.g. neomycin, in order to sterilize the gut.

Bacterial fermentation of carbohydrates in the human colon may produce short-chain fatty acids and several gases: carbon dioxide, hydrogen sulphide, hydrogen and methane, which come out in flatus. It is important to remember, however, that normally about half of the quantity of flatus is nitrogen, mainly derived from swallowed air.

The presence of intestinal bacteria is important for the normal development of the gut and the immune system. This was shown by maintaining animals in germ-free environments. In germ-free animals, the immune system is hypoplastic. There are fewer lymphocytes and plasma cells and the concentration of immunoglobulins is greatly reduced. Furthermore, the gut in these animals becomes enlarged, with poor epithelium and villi.

MOVEMENTS OF THE COLON

The junction of the ileum with the caecum is guarded by the ileocaecal sphincter. The sphincter remains closed and opens only when an intestinal peristaltic wave reaches it, thus allowing ileal contents to pass briefly into the colon. The safety of the ileocaecal junction is further ensured by the fact that the end of the ileum containing the sphincter projects slightly into the caecum; in this way, when the pressure in the colon rises, it tends to squeeze the ileum to keep it shut, but, when ileal pressure increases, the ileal end will open. Muscle of the ileocaecal sphincter is contracted by acetylcholine, α-adrenergic stimulation and the hormone secretin. Beta-adrenergic stimulation and the hormone gastrin relax the sphincter. It is noteworthy that the effects of gastrin and secretin on the ileocaecal sphincter are opposite to their effects on the lower oesophageal sphincter. When the stomach fills, and especially when it empties into the duodenum, a fast peristaltic wave is initiated in the small intestine passing along to the ileum; the ileocaecal sphincter relaxes briefly, allowing chyme to pass through into the caecum. This is termed the gastroileal reflex, which is probably mediated by the vagus.

Movements of the colon consist of segmenting contractions, which are non-propulsive, and peristalsis, which propels colonic contents towards the rectum. Occasionally antiperistalsis may be seen, starting at about the junction of the ascending and transverse colon and travelling towards the caecum. Antiperistalsis mixes contents and probably helps absorption of water.

Segmental contractions produce localized pressure waves and are probably partly responsible for haustral contractions. They serve to mix contents and aid absorption. Colonic segmentation contractions are stimulated by cholinergic stimulation but they are not affected by gastrin. They tend to be more frequent in constipated people. Usually they are not felt but, when excessive, may lead to abdominal pain.

A special type of peristalsis occurs only in the colon; it is called mass movement or mass action contractions. These are powerful contractions advancing rapidly along the colon. They are preceded by sudden disappearance of haustral contractions, allowing rapid advance of colonic contents. Mass movements push faeces into the rectum, which is usually empty. The rectum is distended and the process of defaecation may thereby be initiated. The contractions normally occur one to three times a day and are usually not felt by the subject.

TRANSIT TIME THROUGH THE INTESTINE

If a barium meal is radiologically followed through the intestine, it will be seen to reach the caecum in about 4 hours but it will be 8–9 hours before all the meal enters the colon. The front of the meal reaches the hepatic flexure in 6 hours, the splenic flexure in 9 hours and the pelvic colon in 12 hours. Transit is greatly slowed down between the pelvic colon and the anus. After 3 days, about three-quarters of the meal will have been expelled with one-quarter still remaining in the rectum. These times may be longer for ordinary food. When small coloured beads or melon seeds are swallowed with a meal, about 70% of them are recovered from the stools in 72 hours, but complete recovery of the seeds takes more than a week. Thus, remnants from a meal may still have to be excreted in the stools 1 week after its ingestion. There is great individual variation in transit time.

DEFAECATION

The nerve supply of the distal colon and rectum is shown in Fig. 9.14. Sympathetic fibres originate in the lower thoracic and upper two lumbar segments of the spinal cord and traverse the sympathetic chain to form plexuses on the front of the aorta, from which fibres proceed to the distal colon and the rectum. Parasympathetic fibres originate from the second, third and fourth sacral segments and run in the pelvic splanchnic nerve to the rectum and internal anal sphincter of smooth muscle. The external anal sphincter consists of striated muscle and is innervated by the somatic pudendal nerve (S2, 3 and 4).

It should be noted that the pelvic splanchnic and pudendal nerves contain both efferent and afferent fibres that connect the rectum and anal sphincters with the sacral segments of the spinal cord.

Defaecation is essentially a spinal reflex, which is normally influenced by higher centres. The centre of this reflex is in the sacral segments of the spinal cord. Stimulation of the sympathetic nerves inhibits the rectum and contracts the internal sphincter but probably plays no part in the normal control of defaecation.

The rectum is normally empty and the external anal sphincter is maintained in a state of tonic contraction mediated reflexly. Impulses from muscle spindles pass in the pudendal nerve to enter the cord by the S2 dorsal root and efferent impulses pass out in the pudendal nerve to the muscle. A mass movement in the colon pushes faeces into the rectum, which is thus distended. Normally distension of the rectum by faeces is signalled to the cerebral cortex, producing the desire to defaecate. But the spinal reflex cannot be allowed to operate immediately; the subject must wait for socially convenient circumstances, like going to the toilet.

Stretch of the rectal wall is signalled to the spinal cord by afferent fibres in the pelvic splanchnic nerve, and impulses in the same nerve cause reflex contraction of the rectum and relaxation of the internal anal sphincter. This is followed by reduction in the number of tonic impulses travelling in the pudendal nerve to the external anal sphincter; the sphincter relaxes and faeces leave the rectum.

When the situation is not convenient for defaecation, the tonic contraction of the external anal sphincter is voluntarily maintained or even increased, thereby restraining operation of the spinal reflex. When proper circumstances are available, the tonic contraction of the external sphincter is lifted and the spinal reflex for evacuating the rectum operates unhindered. In infants the reflex proceeds unhindered (automatic) due to lack of cortical voluntary control.

Expulsion of faeces from the rectum is assisted by voluntary straining: a deep breath is taken and then expelled against a closed glottis. Straining greatly raises intra-abdominal pressure. In addition, contraction of the levator ani muscle in the pelvic floor helps to expel faeces by pulling the anal canal over the faecal mass. Not only the rectum but the large intestine up to the middle of the transverse colon may also be evacuated.

Usually ingestion of a meal is followed by a mass action contraction wave in the colon that pushes faeces into the rectum and produces the

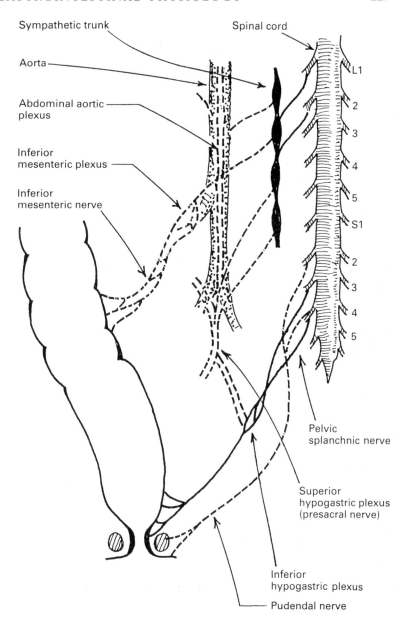

Fig. 9.14 The nerve supply of the distal colon and rectum in man. (Redrawn from Garry, R.C. (1934) *Physiological Reviews*, **14**, 107. With permission from the American Physiological Society.)

desire to defaecate. This response on the part of the colon, which may be initiated by gastric filling, gastric emptying, filling of the small intestine or even somatic movement, is better described as the gastrocolic response rather than reflex, as it is mediated by both neural and hormonal means. The hormone involved may be gastrin. As a result of the gastrocolic response,

defaecation after meals is the rule in children and some people habitually defaecate after breakfast. However, in adults, cultural factors, habit, occupation and personal circumstances alter the gastrocolic rhythm and determine the time of defaecation. It is well to remember that there is a wide range of normal bowel habit; once every 3 days as well as three times a day is normal.

Effect of spinal cord lesions on defaecation

Transection of the spinal cord is followed by a period of spinal shock, during which the rectum and sphincters are paralysed and patulous, leading to retention of faeces. If the lesion is above the sacral segments (upper motor neurone lesion), tone returns and the spinal reflex of defaecation starts to operate without interference from higher centres, after the period of spinal shock is over. The rectum is described as automatic. In this case, when it fills, it automatically empties without the patient being aware of the act.

Lesions that destroy the sacral cord or its connections with the rectum produce a lower neurone lesion. Initially, there is paralysis similar to that during spinal shock, but, after some time, tone returns and evacuation occurs at intervals; the rectum is described as autonomous in this case. Apparently, evacuation is brought about by reflex arcs entirely confined to the wall of the rectum. Whenever a person loses control over the act of defecation, he is said to have faecal incontinence.

Faeces

The amount of stools passed in 24 hours varies from 80 to 200 g. The brown colour of stools is due to stercobilinogen, derived from bacterial degradation of bile pigments. The smell of faeces is due to the amines skatole and indole and to gases produced by carbohydrate fermentation, such as hydrogen sulphide. All are produced by action of intestinal bacteria. The faecal reaction is slightly alkaline on the surface (pH 7–7.5).

About 75% of the stools is water; the rest consists of organic and inorganic solids. Organic material consists mainly of desquamated epithelial cells, bacteria, mucus and remnants of digestive enzymes. Most of the nitrogenous material, 1–2 g/day, is thus derived from non-dietary sources. Fat in the stools is also mostly non-dietary, amounting to about 5 g/day. In steatorrhoea up to 20 g may be lost/day. Inorganic material in faeces consists mainly of calcium and phosphates. Undigested plant fibres, such as fruit seeds and skin, may be seen in stools.

It is important to realize that more than half the stools is actually derived from the remnants of intestinal epithelium, digestive enzymes and intestinal bacteria and their products. The stools are thus not simply the unabsorbed food remnant. The most convincing proof of this fact is that during starvation faeces, although small in amount, are still passed. Furthermore, variations in the composition of diet have no significant effect on the composition of stools.

Further reading

1 Davenport, H.W.E. (1982) *Physiology of the Digestive Tract*, 5th edn. Year Book Medical Publishers, Chicago, London.
2 Ganong, W.F. (1991) Regulation of gastrointestinal function. In *Review of Medical Physiology*, 15th edn, pp. 448–477. Appleton & Lange, San Mateo, California.
3 Johnson, L.R. (ed.) (1985) *Gastrointestinal Physiology*, 3rd edn. C.V. Mosby Company, St Louis, Toronto, Princeton.

10: The Endocrine System

The endocrine glands

Together with the nervous system, the endocrine glands constitute the control systems in the body. They provide the necessary communication for distant control of body functions and homoeostatic regulatory mechanisms. The nerves conduct impulses to distant organs and therefore provide a fast means of communication. The endocrine glands secrete hormones directly into the bloodstream. These hormones are special chemical substances which act on specific distant organs or tissues.

NEUROENDOCRINE RELATIONSHIPS

The endocrine glands can be considered as a system which functions in close harmony with the nervous system. This is because the pituitary gland is closely connected with the hypothalamus at the base of the brain. The hypothalamus, in turn, receives information from almost all other parts of the brain directly or indirectly. (Figure

10.1 gives a diagrammatic representation of the connections of the hypothalamus.) It is therefore in a position to be influenced by incoming information, whether somatic, visceral, psychic or emotional.

The hypothalamus influences the pituitary gland through two distinct mechanisms:

1 Direct neural tracts from the hypothalamic nuclei to the posterior pituitary (neurohypophysis).

2 Secretion of releasing hormones (RH) or inhibitory hormones (IH), which are carried to the anterior pituitary (adenohypophysis) by means of the hypothalamohypophyseal portal vessels (Fig. 10.2). The RH and IH are specific for the individual hormones of the anterior pituitary gland.

The anterior pituitary gland, therefore, is closely controlled by the releasing or inhibitory hormones of the hypothalamus. Most other endocrine glands function under the control of the anterior pituitary hormones. Thus, the thyroid gland is controlled by thyroid-stimulating hormone (TSH), the adrenal cortex by adrenocorticotrophic hormone (ACTH) and the gonads (ovaries and testes) by follicle-stimulating hormone (FSH) and luteinizing hormone (LH). Growth hormone (GH) and prolactin (PRL) act on tissues rather than on specific endocrine glands.

The only two major endocrine glands not under the control of the trophic hormones of the anterior pituitary are the parathyroid glands and the islets of Langerhans of the pancreas. These are directly controlled by the chemical substrates they regulate, i.e. blood calcium and blood glucose, respectively.

The control of the neurohypophysis and the adrenal medulla deserves a special place in this description of neuroendocrine interrelations. The neurohypophysis acts as a storage organ for the granules of hormones secreted in the hypothalamic nuclei. Release of these hormones requires stimuli, which again act via the hypothalamohypophyseal tracts. The adrenal medulla, on the other hand, receives its nerve supply from the sympathetic neurones in the spinal cord (preganglionic fibres). These neurones have their connections with the hypothalamus. Therefore, whenever there is massive sympathetic activity

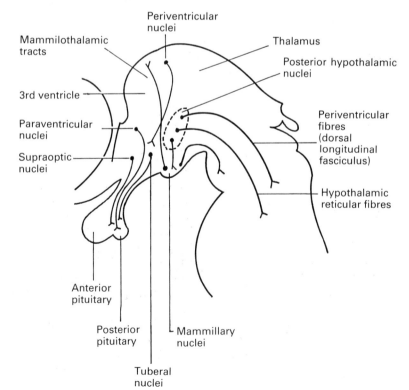

Fig. 10.1 Diagram of the hypothalamus and the pituitary gland showing major tracts and the nuclei closely related to the pituitary glands.

(fight-or-flight), the adrenal medulla is also stimulated.

The anterior pituitary gland

The anterior lobe of the pituitary gland (adenohypophysis) is the larger of the two glands. Its embryological origin is from the roof of the primitive mouth (Rathke's pouch). It consists of cells which take acid stain (40%) or basic stain (10%) or do not take any (50%). Hence the names acidophil, basophil and chromophobe cells, respectively. More recently, the cells have been classified on the basis of the hormones they secrete. Table 10.1 gives the hormones secreted by each of the cell types.

The blood supply of the anterior pituitary comes from the superior hypophyseal arteries, which are branches of the internal carotid arteries. These hypophyseal arteries supply the primary capillary plexuses of the hypothalamohypophyseal portal vessels, which travel down the pituitary stalk (Fig. 10.2). The portal vessels therefore constitute the main blood supply of the anterior pituitary. They transport the releasing hormones to the adenohypophysis. The capillaries in the anterior pituitary form sinusoids similar to those seen in the liver, thus providing close contact between the blood and the glandular cells. The hormones of the anterior pituitary are carried via the hypophyseal veins. The blood supply of the posterior pituitary gland comes from the superior and inferior hypophyseal arteries.

HORMONES OF THE ANTERIOR PITUITARY
The four trophic hormones come from basophil cells and they control three endocrine organs: the thyroid, the adrenal cortex and the gonads. The following description of the individual hormones will be given under the following subheadings: chemistry and degradation, control of secretion, actions of the hormone and abnormalities.

THYROID-STIMULATING HORMONE
(THYROTROPHIN) (TSH)

Chemistry and degradation
TSH is a glycoprotein of molecular weight 28 000 daltons and comprises two subunits, one (α) that

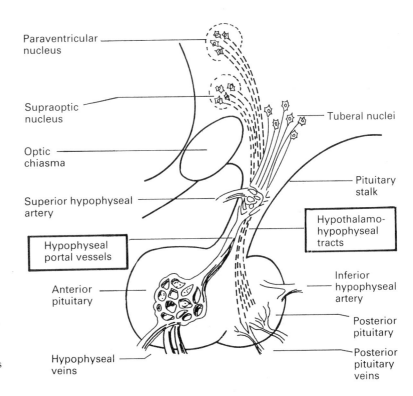

Fig. 10.2 The pituitary gland and its connections with the hypothalamus. Note the hypothalamohypophyseal portal vessels to the anterior pituitary and the hypothalamohypophyseal tracts to the posterior pituitary.

Paraventricular nucleus

Supraoptic nucleus

Optic chiasma

Superior hypophyseal artery

Hypophyseal portal vessels

Anterior pituitary

Hypophyseal veins

Tuberal nuclei

Pituitary stalk

Hypothalamo-hypophyseal tracts

Inferior hypophyseal artery

Posterior pituitary

Posterior pituitary veins

Table 10.1 Cell types of the anterior pituitary and the hormones they produce

Cell type	Stain	Hormone	Molecule
Somatotropes	Acidophilic	Growth hormone (GH)	Protein
Mammotropes	Acidophilic	Prolactin (PRL)	Protein
Corticotropes	Basophilic or chromophobe	ACTH	Polypeptide
Thyrotropes	Basophilic	TSH	Glycoprotein
Gonadotropes	Basophilic	LH and FSH	Glycoprotein

is common to LH, FSH and human placental lactogen (HPL), and one (β) that expresses TSH hormone bioactivity. It has no species specificity so that TSH extracted from ox or pig is found to be active in man. Its half-life is about 60 min. The kidney has been found to be more able to break it down than the liver.

Control of secretion

TSH is controlled by the anterior part of the median eminence of the hypothalamus, which releases thyroid-releasing hormone (TRH) (see Table 10.2):

1 TRH is controlled by the level of the hormones of the thyroid gland, i.e. thyroxine and triiodothyronine (T_4, T_3).

2 The effect of TRH on the pituitary can be blocked by T_4 and T_3 by decreasing TRH-binding sites on anterior pituitary cells. This may be the main site of the negative feedback mechanism (Fig. 10.3).

3 There seems to be a circadian rhythm (diurnal rhythm) with maximum secretion before or at midnight and minimum values at about 11 a.m. This is probably mediated by a similar TRH rhythm.

4 The effect of environmental temperature on TSH secretion has been shown in animals and in infants but there seems to be little effect in adults. However, there is a long-term effect of heat and cold on thyroid function. Thus, acclimatization to heat (when living in a hot climate) leads to slightly lower thyroid function and acclimatization to cold is associated with thyroid activity in the upper levels of normal.

Table 10.2 Hormones of the hypothalamus

Hypothalamic hormones	Molecule	Nuclei
Thyrotrophin-releasing hormone (TRH)	Tripeptide	Paraventricular and dorsomedial
Corticotrophin-releasing hormone (CRH)	Polypeptide (41)	Paraventricular
Gonadotrophin-releasing hormone (GnRH)	Decapeptide	Preoptic area and arcuate
Growth hormone-releasing hormone (GHRH)	Polypeptide (40, 44)	Arcuate
Growth hormone inhibitory hormone (somatostatin)	Peptide (14)	Ventral
Prolactin inhibitory hormone (PIH)	Dopamine	Arcuate
Prolactin-releasing factor (PRF)	?Polypeptide	Ventral

The figures in brackets indicate the number of amino acids in polypeptide hormones.

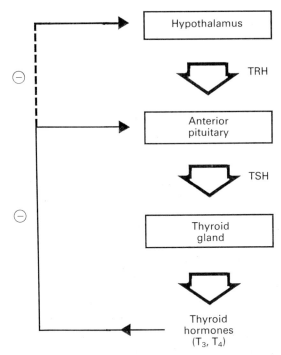

Fig. 10.3 Feedback control between the thyroid gland, the anterior pituitary and the hypothalamus. The solid arrow indicates that the main feedback is exerted on the pituitary, with a minor control mechanism on the hypothalamus indicated by the broken arrow.

Actions of TSH

Thyrotrophic hormone, as the name implies, is important for the growth and development of the thyroid gland.

The actions of TSH are mediated by cyclic AMP via receptor sites on thyroid cells. TSH accelerates the process of iodine trapping and therefore its concentration by the thyroid gland. It is also concerned with the synthesis of the thyroid hormones and their release. Thus, TSH specifically accelerates oxidative coupling of thyroid hormone precursors. It also activates the proteolytic enzymes, which release the thyroid hormones from their storage form in the thyroglobulin molecules. (For more detail see the section on thyroid hormone synthesis and release.)

Abnormalities of TSH

Primary pituitary deficiency or excess of TSH are rare. Thyroid disease, however, is common and this is usually due to diseases of the thyroid gland itself. Therefore, high or low TSH is usually secondary to thyroid disease. In hyperthyroidism and thyrotoxicosis the TSH levels are found to be low due to the negative feedback effect of the excess T_3 and T_4. In hypothyroidism the levels of TSH are high due to diminished thyroid hormone levels and therefore absence of the negative feedback.

However, in the case of endemic goitre (due to iodine deficiency) the blood levels of TSH are found to be elevated, which explains the enlargement of the thyroid gland. The function of the thyroid gland in endemic goitre is usually within the normal range.

ADRENOCORTICOTROPHIC HORMONE (CORTICOTROPHIN) (ACTH)

Chemistry and degradation

ACTH is a polypeptide of a relatively small molecular weight, 4500 daltons, consisting of a single chain of 39 amino acids. ACTH is one of a family of related peptides that are derived from a larger precursor glycoprotein, of molecular weight 31 000 daltons, called pro-opiocortin. Its half-life is short and estimated to vary between 5 and 15 minutes. The degradation of ACTH takes place in the blood. Little is known of the sites of ACTH breakdown.

Control of ACTH

1 The secretion of ACTH, like TSH, is under the control of a releasing hormone from the hypothalamus. Corticotrophin-releasing hormone (CRH) is under the control of circulating adrenal glucocorticoid hormones by means of a negative feedback mechanism. As shown in Fig. 10.4, the negative feedback works mainly on the hypothalamus but it can also work directly on the anterior pituitary.

2 ACTH is rapidly released in response to physical or emotional stress and hypoglycaemia. This response to stress is part of the general preparation to counteract injury, infections and various other stressful conditions.

3 It is also known that there is a circadian rhythm of ACTH, secretion, which results in high levels

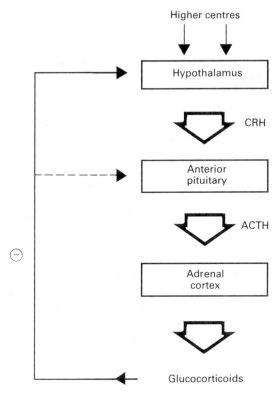

Fig. 10.4 Feedback control of ACTH. The main feedback is shown to be on the hypothalamus, with a minor control mechanism on the anterior pituitary.

of ACTH, and therefore cortical steroids, in the morning and low levels in the evening and during the night. This rhythm is related to the pattern of sleep and is reversed in workers who are on night duty.

4 Stimulation of arterial baroreceptors inhibits CRH release. This effect is mediated through the nucleus of the tractus solitarius in the medulla.

Actions of ACTH
ACTH is necessary for the healthy growth and repair of the adrenal cortex and for maintaining healthy vascularity of the gland. In cases of ACTH deficiency, the adrenal cortex atrophies and its blood-vessels disappear. Conversely, ACTH excess causes hypertrophy.

ACTH influences the adrenal cortex to stimulate the synthesis and release of cortisol and the other glucocorticoids. It increases the number of

mitochondria in the adrenal cortical cells. It acts on adenylate cyclase and a protein kinase to increase the lipid content of the adrenal cortex. Free cholesterol is also increased. ACTH decreases the ascorbic acid content of the adrenal cortex; this effect was once used as a method for biological assay of the hormone.

ACTH has a melanocyte-stimulating action which is believed to cause pigmentation of the skin in conditions where high ACTH levels are found, e.g. destruction of the adrenal cortex (Addison's disease). Its absence, on the other hand, as in panhypopituitarism, is associated with pallor of the skin.

Several other effects of ACTH have been described, e.g. stimulation of insulin secretion and mobilization of FFAs from adipose tissue. The physiological significance of these actions remains to be explained.

Abnormalities
In panhypopituitarism, there is low ACTH, which leads to hypofunction of the adrenal cortex. The levels of ACTH are found to be elevated in Addison's disease, due to destruction of the adrenal cortex and therefore absence of the inhibitory feedback effect of the glucocorticoids.

In hyperfunction of the adrenal cortex (Cushing's disease), the levels of ACTH may be high if the disease is secondary to hypersecretion by the pituitary. ACTH may, however, be low if there is primary hyperfunction of the adrenal cortex with suppression of ACTH release. ACTH levels in the blood can therefore be measured to differentiate between primary and secondary Cushing's disease.

GONADOTROPHIC HORMONES
The gonadotrophic hormones are follicle-stimulating hormone (FSH) and luteinizing hormone (LH), which control both male and female gonadal functions.

Chemistry and degradation of FSH
Human FSH is a double-chain glycoprotein of molecular weight 35 000 daltons. It is species-specific and therefore only human FSH can be used for the treatment of patients. The half-life of human FSH is about 3 hours. It is probably broken

down in the liver. FSH appears in the urine in appreciable amounts. Its urinary excretion reflects the blood levels and its production by the anterior pituitary. It is secreted in large amounts during the menopause and therefore menopausal women's urine contains high concentrations.

Control of FSH

Only small amounts of FSH can be detected during childhood. Whether the gradual increase in its secretion starts at the onset of puberty is uncertain. At puberty, the secretion of gonadotrophins is stimulated by specific gonadotrophin-releasing hormones (GnRH) under the influence of the nervous system, i.e. the stimulus is neural in origin and its initiation depends on the development of the nervous system and its maturation during adolescence.

After the establishment of puberty, the regulation of FSH depends upon the following:

1 The level of ovarian oestrogens has a negative feedback effect on pituitary FSH-releasing hormone (Fig. 10.5). The sharp fall of oestrogens in the middle of the cycle triggers a small but significant midcycle peak of FSH (Fig. 10.6).
2 In the male there is a negative feedback control with testicular testosterone from the Leydig cells.
3 Inhibin is a polypeptide, secreted by the seminiferous tubules and the Sertoli cells, which inhibits FSH secretion. It is also secreted by the granulosa cells in females.
4 FSH itself has an inhibitory effect on the hypothalamic RH, i.e. there is a short-loop feedback mechanism between the pituitary and the hypothalamus (Fig. 10.5).

Actions of FSH

1 FSH is responsible for the growth and maturation of a crop of primordial follicles in the ovaries up to their maturation towards the middle

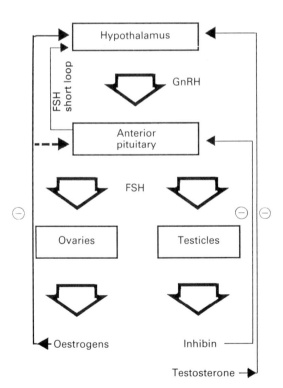

Fig. 10.5 Diagrammatic representation of the negative feedback between ovarian and testicular hormones and FSH.

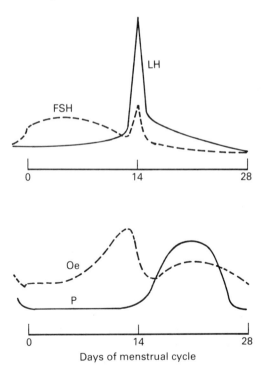

Fig. 10.6 The diagram shows the cyclic nature of secretion of the pituitary gonadotrophins and the ovarian sex hormones. FSH, follicle-stimulating hormone; LH, luteinizing hormone; Oe, oestrogens; P, progestogens.

of the menstrual cycle. The Graafian follicles produce oestrogens, which are responsible for proliferation of the endometrium during the first half of the cycle. Oestrogens are also responsible for the development of secondary sex characteristics in adolescent females.

2 Together with LH, FSH stimulates ovulation. This occurs at the time of the midcycle peaks of the gonadotrophic hormones.

3 In males, FSH is responsible for the growth and development of the seminiferous tubules and it stimulates spermatogenesis. This effect requires LH for full development and function.

Chemistry and degradation of LH

LH is a double-chain glycoprotein of low molecular weight, about 28 000 daltons, with a half-life of about 1 hour. It is filtered by renal glomeruli and appreciable amounts appear in the urine during the 3–4 days at midcycle. During the menopause, large amounts of LH can be recovered from the urine.

Control of LH secretion

Small amounts of LH are found in childhood. Like FSH, it increases at the onset of puberty, presumably triggered by a hypothalamic clock, and is secreted in a pulsatile manner (1–2 h) under the influence of GnRH.

LH secretion during the menstrual cycle is regulated as follows:

1 There is a large midcycle peak of secretion, which seems to be stimulated by a positive feedback mechanism associated with the high level of preovulation oestrogens (Fig. 10.6).

2 LH is inhibited by negative feedback associated with progesterone levels (Fig. 10.7).

3 LH secretion can act on the hypothalamus to inhibit further GnRH secretion by a short-loop negative feedback (Fig. 10.7).

4 17-Beta-oestradiol decreases the level of LH secretion but, if it is administered during the follicular phase, it increases the responsiveness of the pituitary to GnRH.

Actions of LH

LH acts synergistically with FSH in the promotion of growth and maturation of the ovarian follicles. During the middle of the cycle a sharp increase in

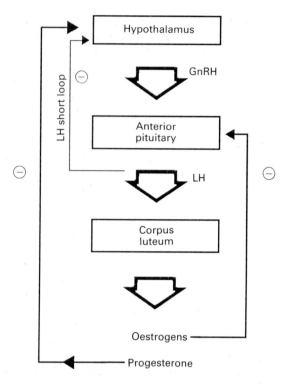

Fig. 10.7 Feedback between the ovarian hormone progesterone and hypothalamic control of LH secretion. A short-loop feedback is also shown to exist between the pituitary and the hypothalamus.

LH release (about 20-fold) is responsible for ovulation.

LH also maintains the corpus luteum during the secretion phase of the menstrual cycle. Towards the end of the cycle luteolysis occurs and results in a drop in the levels of both oestrogens and progesterone, which come from the corpus luteum. This marks the end of the menstrual cycle (Fig. 10.6).

Abnormalities of the gonadotrophic hormones

FSH and LH are affected by emotional and psychological factors. Disturbances of fertility and the menstrual cycle have been reported at times of war and stress. In some women, emotional stress may result in amenorrhoea (absence of cyclic bleeding).

In the absence of the midcycle peak of LH and FSH, there is the usual menstruation but there is no ovulation (anovular cycle). This is a known cause of infertility. Some drugs are now known to stimulate LH secretion, e.g. clomiphene, which can be used in some cases of infertility.

Failure of secretion results in delayed puberty or hypogonadism. Precocious (premature) puberty, on the other hand, can occur as a result of premature release of gonadotrophins, associated with disorders involving the posterior hypothalamus or due to tumours of the pineal body (gland). Damage to the hypothalamus can also lead to cessation of gonadotrophic hormone secretion and consequently gonadal atrophy.

GROWTH HORMONE (GH)

Growth hormone, or somatotrophin, is secreted throughout life. The name growth hormone is misleading, as it implies its secretion during the growth periods. During childhood and adolescence, it is concerned with the regulation of metabolism and attainment of the growth potential of the child. In adult life, however, it works in concert with other hormones in the short-term day-to-day regulation of energy metabolism. GH stimulates the mobilization of fat and inhibits glucose utilization. Therefore, it shifts energy metabolism towards the use of FFAs rather than glucose. It also has an anabolic effect on protein utilization in the repair process of tissues.

Chemistry and degradation of GH

GH is a peptide hormone consisting of 191 amino acids in a single chain, with two intramolecular disulphide bridges. Two forms have been described in plasma. Its molecular weight in man is 20 000 or 22 000 dalton. It is species-specific and therefore only human or synthetic GH can be used in the treatment of cases of its deficiency. Its half-life is short (20–30 min), which necessitates half-hourly serial sampling in its investigation.

Control of GH secretion

Like most of the peptide hormones, GH is stored in vesicles and secreted by the process of exocytosis in response to appropriate stimuli. The hypothalamic control is achieved by means of GH-releasing hormone (GHRH) (or somatocrinin) and GH inhibitory hormone (GHIH) (or somatostatin). The latter has also been identified in the D cells of the islets of Langerhans, where its

physiological significance is associated with insulin and glucagon regulation. The overall regulation of GH secretion is summarized in Fig. 10.8. GHRH and GHIH are involved in the control of GH secretion by the anterior pituitary. Both peptides interact with membrane receptors on the somatrophic cells of the anterior pituitary gland to effect the changes in GH release. GHRH effects are calcium-dependent and may involve adenylate cyclase. GHIH effects require changes in calcium and cyclic AMP concentration to inhibit the release of GH. In addition to the hypothalamic input to the somatotropes, feedback effects of GH and of somatomedins participate in the regulation of GH secretion.

The secretion of growth hormone takes place in discrete pulses, so that basal levels are usually very low. The pattern of its secretion is found to occur as follows:

1 GH is released under conditions of lack of glucose. Hypoglycaemia (e.g. after insulin ther-

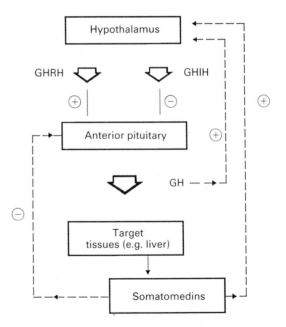

Fig. 10.8 Regulation of growth hormone by the anterior pituitary gland. Negative and positive feedback mechanisms are indicated, bearing in mind that GH has a pulsatile release pattern of short durations. GHIH, growth hormone-inhibiting hormone; GHRH, growth hormone-releasing hormone.

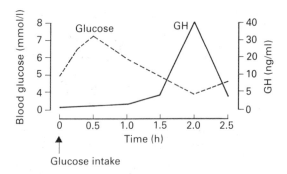

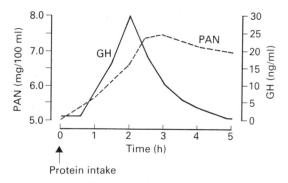

Fig. 10.9 The pattern of secretion of growth hormone during and following glucose absorption. GH (solid line) is suppressed during the absorptive phase and appears as a distinct peak burst during the postabsorptive phase of glucose intake (broken line).

Fig. 10.10 The secretion of growth hormone after protein intake. PAN = plasma amino nitrogen, indicating the absorption of amino acids. Note the secretion of GH during the absorptive phase of protein intake.

apy) is a strong stimulus, which acts by stimulating GHRH secretion. On the other hand, glucose intake and hyperglycaemia inhibit GH secretion.

2 Muscular exercise in the postabsorptive state results in GH secretion. This effect is found to be abolished if glucose is taken at the same time.

3 GH appears in the postabsorptive phase following glucose intake (Fig. 10.9). During starvation or continued fasting, GH is secreted periodically every 2–3 hours. The nature of the triggering mechanism for these peaks is not clear.

4 Intake of protein or amino acids results in the secretion of GH. This response is also inhibited by concomitant glucose intake (Fig. 10.10).

5 During sleep GH is found to be released several times. The levels of these peaks are higher and occur with greater frequency in children than in adults.

6 Stress conditions, e.g. due to trauma or emotion, are found to stimulate GH secretion.

7 Glucagon, lysine-vasopressin and L-dopa have also been found to stimulate the release of GH. The physiological significance of these responses is not fully understood.

8 On the other hand, cortisol, FFAs, GH and, of course, glucose are found to inhibit its secretion. Prolonged corticosteroid administration in children leads to stunted growth.

Actions of GH
In addition to the promotion of growth, GH has several short-term effects on the utilization of amino acids, glucose and FFAs. Its long-term effect on somatic growth depends on somatomedin C, which is secreted by the liver in response to GH. Its half-life is about 20 hours. (Other somatomedins have also been described, such as nerve growth factor and insulin-like polypeptides and relaxin.) Recently, somatomedins have been described to inhibit secretion of GH by stimulating GHIH release. They may also act directly to inhibit the effect of GHRH on the pituitary. Growth stops once the epiphyses of the long bones close. The growth-promoting action of GH also requires normal thyroid function and insulin. The sex hormones are also necessary for growth during adolescence.

Its short-term metabolic effects are as follows.

Protein metabolism GH stimulates the transport of amino acids into the cells. Their incorporation during protein synthesis is also stimulated. Generally, GH causes a positive nitrogen balance and net protein anabolism, which is known as the protein-sparing action of GH.

Carbohydrate metabolism GH inhibits peripheral utilization of glucose by muscle and adipose tissues. This is opposite to the action of insulin (insulin antagonism) and increases hepatic glucose output (diabetogenic effect). Abnormal GH secretion or prolonged administration therefore causes a type of diabetes mellitus.

Fat metabolism GH mobilizes FFAs from adipose tissue stores. This action depends on a hormone-sensitive tissue lipase. This results in an increase in FFA levels during fasting, during muscular exercise and in the postabsorptive phase of glucose intake. GH also increases the oxidation of FFAs in muscle.

In conclusion, it is clear that the effects of GH lead to a net anabolic effect on protein metabolism, shifting of energy metabolism towards the utilization of FFAs and sparing of glucose. Therefore, in adults this hormone is concerned with the short-term regulation of metabolism, while during childhood and adolescence the same metabolic effects take place, in addition to skeletal growth.

Abnormalities of GH secretion
Deficiency of GH in childhood results in retardation of somatic growth but it has no effect on mental development. The child becomes a dwarf with height much below that expected for its age (see Fig. 10.11). In some cases of dwarfism, there is normal GH secretion but a deficiency of somatomedins. In a rare type of dwarfism (Laron dwarf), there is a lack of GH receptors.

Deficiency in adult life is not fully understood. There is reason to believe that some cases of obesity may be associated with decreased GH secretion. Growth hormone deficiency in adults is one of the earliest manifestations of hypothalamic disease.

Excess of GH before or during adolescence results in an abnormal stimulation of skeletal growth and a pituitary giant (gigantism). During adult life, after skeletal maturation, the effect of hypertension results in acromegaly. This condition is characterized by enlargement of certain bones such as the jaw-bone, the frontal bone and the bones of the hands and feet. The skin of the face is thickened and the tongue is enlarged. Some internal organs, such as the liver and spleen, are also enlarged.

GH is found to be secreted continuously (i.e. it loses its periodic pattern of release) and fails to be suppressed by glucose intake. This continuous

Fig. 10.11 The girl on the left has isolated GH deficiency. The age difference with her sister is 18 months. (Reproduced with permission from Edwards, C.R.W. & McNicol, G.P. (eds) (1986) *Integrated Clinical Science*, p. 31. William Heinemann Medical Books, London.)

secretion eventually results in diabetes mellitus, which is resistant to insulin therapy.

PROLACTIN (PRL)

Chemistry
Prolactin (PRL) (or mammotrophin), also known as lactogenic hormone, is chemically similar to GH. Human PRL is a single-chain polypeptide with a molecular weight of about 22 500 daltons. Its half-life is 15 min, it is secreted in a pulsatile manner, and blood levels are higher at night and during sleep.

Control

1 PRL seems to be continuously inhibited by a hypothalamic inhibitory hormone (PIH). It is therefore stimulated by abolishing the release of PIH in the hypothalamus. PRL is at its highest at term. Immediately following labour, the hormone drops rapidly to a basal level during the first few days.

2 There are discrete bursts of release, giving rise to high plasma concentrations every time there is stimulation of the nipples during suckling. Therefore, a PRL-releasing hormone (PRH) has also been postulated to explain this phenomenon (Table 10.2). When the mother stops breast-feeding, the PRL level gradually decreases and eventually stops.

3 Other stimuli, such as muscular exercise, sleep and stress (physical or emotional), seem to work through the hypothalamus.

4 Hypothalamic TRH stimulates PRL secretion as well as the secretion of TSH.

5 PRL has a short-loop negative feedback inhibitory effect on the hypothalamus to stop its own release.

L-dopa inhibits PRL secretion. There is evidence that dopamine, which is found in the hypothalamus, is PIH. Chlorpromazine, which blocks dopamine, is found to stimulate PRL release. Bromocriptine inhibits PRL release by stimulation of dopamine receptors.

Actions of prolactin

PRL acts on the breast to produce milk. The continued secretion during pregnancy causes growth of the mammary gland tissue but does not stimulate the production of milk. This is because of the inhibitory effect of the large amounts of oestrogens and progesterone secreted by the placenta.

After delivery, PRL initiates and maintains milk production. A burst of secretion of PRL is produced during each suckling. This prepares the milk for the next feed.

PRL release during breast-feeding exerts an inhibitory effect on the ovaries and suppresses the menstrual cycle in about one-half of lactating mothers. It seems to decrease the sensitivity of the ovaries to gonadotrophins. Breast-feeding, in this case, can be regarded as a physiological

mechanism leading to birth control and spacing of pregnancies.

It is believed that PRL also acts centrally, probably on certain areas of the limbic system, to produce the characteristic maternal and nursing behaviours in the mother.

Abnormalities of prolactin secretion

Hyperprolactinaemia is found in cases with acidophil or chromophobe adenoma of the pituitary gland. This is not always associated with galactorrhoea (abnormal production of milk). On the other hand, cases of galactorrhoea have been described without elevation of blood PRL levels.

Hypersecretion of PRL occurs as a result of hypothalamic pituitary disorders, including section of the pituitary stalk and therefore interference with the production or transport of PIH. Hypersecretion is associated with secondary amenorrhoea and infertility. Treatment with bromocriptine or other dopamine agonists is used in such cases.

Deficiency can occur as an isolated pituitary PRL deficiency or as part of panhypopituitarism.

ABNORMALITIES OF THE ANTERIOR
PITUITARY

Several clinical syndromes are known to be associated with disorders of the pituitary or the hypothalamus. The predominant clinical features depend upon the position of the lesion and the hormone or group of hormones which are affected most.

Hypopituitarism

Hypopituitarism is usually due to destruction of the anterior pituitary by disease, radiation or total removal by surgery. The gonadotrophins are usually affected first. Evidence of TSH and ACTH involvement appears later. In children the predominant feature is failure to grow.

Sheehan's syndrome (Simmond's disease) is hypopituitarism due to destruction of the adenohypophysis in women, following severe postpartum haemorrhage. Loss of pubic and axillary hair is noted, together with lack of lactation. There is generalized weakness (asthenia) and loss of vigour, with hypothyroidism developing later. The menstrual cycle stops. The severity of this condition depends on the extent of pituitary destruction.

Hypothalamic hypopituitarism results from lesions in the hypothalamus and failure to produce the hypothalamic releasing or inhibitory hormones. This is secondary hypopituitarism, which will respond to administration of releasing hormones such as TRH or LRH.

Hypopituitarism in children leads to stunted growth (dwarfism). However, there are different types of dwarfs. Isolated GH deficiency will lead to dwarfism and will respond to human GH therapy (see Fig. 10.11). Some dwarfs, however, secrete GH but seem to have a faulty response. Somatomedins are peptides synthesized by the liver and probably the kidney in response to GH. The somatomedins are responsible for the growth-promoting effect on cartilage and bone. One type of dwarfism seems to have a deficiency of somatomedin production. A hypothalamic defect which results in failure of secretion of GnRH results in hypogonadism associated with obesity. This is known as adiposogenital syndrome or Fröhlich syndrome.

Panhypopituitarism is a term given to a generalized hypofunction of pituitary. In this condition, most, if not all, of the pituitary hormones are involved. Therefore, the clinical manifestations will include adrenal and thyroid deficiencies as well as dwarfism and hypogonadism.

Pituitary tumours
An adenoma of the pituitary is a benign tumour, which may be minute in size but, nevertheless, gives rise to overproduction of a particular hormone. Adenomata have been described as arising from acidophil, basophil and chromophobe cells. Therefore, the clinical manifestations of the tumours will depend on the hormone or hormones involved. Thus, acromegaly, hyperprolactinaemia and Cushing's syndrome can be produced. The tumour may also enlarge and press on neighbouring structures, such as the optic chiasma. This leads to bilateral temporal hemianopia and progressive blindness. This is particularly common with chromophobe adenoma.

The pineal body
The pineal body has recently been found to secrete

certain transmitters which influence hypothalamic and therefore pituitary functions. Melatonin, a derivative of serotonin, is released by the pineal body. This hormone suppresses the release of GnRH and causes inhibition of secretion of gonadotrophins. In young animals, this will lead to failure of sexual maturation. Melatonin is rapidly removed from the blood and metabolized in the liver.

On the other hand, the pineal body receives neural messages from parts of the central nervous system, particularly the hypothalamus. Sympathetic nerves to β-adrenergic receptors on pineal cells activate cyclic AMP and melatonin synthesis and secretion. Exposure to light influences pineal secretions and therefore the control of hypothalamic hormones.

The posterior pituitary gland
The posterior pituitary (neurohypophysis) develops from the neural tube of the embryo, i.e. it has the same embryogenic origin as the nervous system — hence the name neurohypophysis. It has direct nerve tract connections with the hypothalamus, mainly with the supraoptic nucleus and the paraventricular nuclei.

It is responsible for the storage of hormones synthesized in magnocellular neurones in the hypothalamus. Antidiuretic hormone (ADH), also known as vasopressin, is mainly produced in the supraoptic nucleus of the hypothalamus. The hormone oxytocin is mainly produced in the paraventricular nuclei. These hormones (in a larger polypeptide precursor) travel down the nerve fibres (axoplasmic flow) and remain as storage granules in the nerve endings within the neurohypophysis until they are released by specific stimuli.

ANTIDIURETIC HORMONE (ADH)
The name vasopressin has also been given to ADH because of the powerful vasoconstrictor effect of the hormone.

Chemistry and degradation
ADH (vasopressin) consists of a chain of nine amino acids (nanopeptide). Synthetic vasopressin is now available. There are slight species-specific differences in molecular structure. The half-life

of the hormone is about 5 minutes. It is degraded mainly in the liver and the kidneys.

Control of ADH release

The release of ADH, stored in the nerve endings in the posterior pituitary, is stimulated by nerve impulses, originating in the hypothalamus, which travel down the same neurones and release the hormone. The hormone is released into the blood by exocytosis of the secretory granules. Two main stimuli are known to result in its secretion.

1 *An increase in the osmotic pressure of the extracellular fluid (ECF).* This occurs in states of dehydration, due either to excessive water loss or to water deprivation. High osmotic pressure stimulates osmoreceptors in the anterior hypothalamus. These receptors are located near the supraoptic nucleus and the impulses generated in them stimulate the release of ADH. A very small change in the ECF osmotic pressure — 1% — is detected by the osmoreceptors and results in regulation of ADH (Fig. 10.12). If it is noted that the osmotic pressure of ECF is only 290 mosm per litre, then it is realized that this is a very

sensitive mechanism. The hypothalamic centres are also responsible for the sensation of thirst. This sensation, however, requires a bigger rise in the osmotic pressure.

2 *A fall in blood volume.* This occurs in conditions leading to dehydration or after blood loss. Haemorrhage initially causes a fall in the blood-pressure in the venous system. If the cardiac output is decreased, arterial blood-pressure will also fall. It is believed that there are volume receptors situated in the low-pressure system of the great veins and the right atrium. One of the earliest responses to haemorrhage is ADH release. Afferent impulses travel through the vagus nerve and are relayed in the vagal nuclei to the hypothalamus. Arterial baroreceptors may also be involved if the arterial blood-pressure falls. The response to volume receptor stimulation is more powerful than that due to osmoreceptors, i.e. hypovolaemia will stimulate ADH secretion even if ECF is hypotonic. Both mechanisms tend to release ADH, which, in turn, causes the kidney tubules to reabsorb water. The volume of urine is consequently greatly decreased and water is retained in the body so as to correct the volume and osmolality of the extracellular fluid.

3 *Other stimuli.* ADH is released by stimuli like pain, emotional stress and physical trauma. Drugs such as morphine, barbiturates and nicotine also stimulate the release of ADH. Its secretion is inhibited by alcohol — hence the diuretic and dehydrating effect of alcohol consumption.

The ADH response is an important mechanism in the concentration of urine in man and other mammals living in hot, arid environments. It is noticeable that the kidney produces a small volume of concentrated urine in the hot summer months when compared with the winter. It is important to remember this fact when dealing with patients with renal disease. A high concentration of urine leads to deposition of solids and crystallization of certain salts, such as oxalates, in the urinary tract and may lead to infection.

Actions of ADH

ADH increases the water permeability of the distal tubules and the collecting ducts of the kidney. This effect is mediated by V_2 receptors, which increase intracellular cyclic AMP. Cyclic AMP

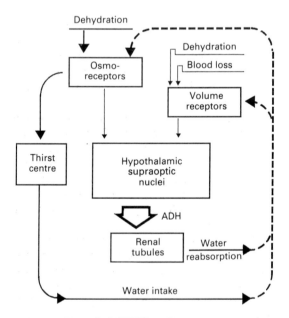

Fig. 10.12 Control of ADH by volume receptors and osmoreceptors. Dehydration is seen to work through both mechanisms.

activates a protein kinase that phosphorylates certain membrane proteins; this results in an increase in the permeability of the luminal membrane to water. And, since the interstitial fluid in the renal medulla is hypertonic, water quickly passes out of the tubules and is reabsorbed into the blood. The urine becomes more concentrated and smaller in volume.

ADH also increases the permeability of the collecting ducts to urea. This also helps the reabsorption of water.

The hormone decreases the blood flow in the renal medulla. This helps to maintain the hyperosmolality of the peritubular fluid.

In doses bigger than those normally released by the posterior pituitary, ADH causes generalized vasoconstriction and a rise in arterial blood-pressure. This vasoconstrictor effect is mediated by V_1 receptors, which apparently increase intracellular calcium ions. Acute blood loss results in vasopressin release and elevation of blood-pressure. ADH, however, is of doubtful significance in the short-term regulation of normal arterial blood-pressure. The well-known cardiovascular reflexes are more important in this respect.

Abnormalities

Lack of ADH results in the production of large volumes of dilute urine, which leads to thirst and excessive drinking (polydipsia). This condition is known as diabetes insipidus. It is usually due to a lesion in the hypothalamus. Removal of the posterior pituitary or pituitary stalk section usually produces a transient type of the disease.

ADH secretion increases after surgical operations, leading to abnormal water retention and therefore lowering of sodium concentration in the ECF (hyponatraemia). This is due to dilution of the ECF rather than a sodium deficit.

In congestive heart failure there seems to be an abnormal release of ADH, in spite of the increased blood-pressure at the right side of the heart. The volume receptor mechanism fails to suppress ADH secretion. Another explanation of water retention due to ADH is the increased half-life of the hormone in congestive heart failure. This is probably due to diminished degradation of the hormone by the liver and the kidneys.

OXYTOCIN

Chemistry and degradation

Oxytocin is also a nanopeptide hormone. Synthetic oxytocin is available for therapeutic use. Its half-life is 1–4 minutes and it is mainly degraded by the liver and kidneys. There are synthetic analogues used in clinical practice. It causes ejection of milk from the lactating breasts and contractions of the uterus during labour. Its chemical structure is similar to ADH. It is produced mainly in the paraventricular nuclei of the hypothalamus but it has been demonstrated in other parts of the central nervous system.

Control of oxytocin release

1 Oxytocin is released by stimulation of the nipples of the lactating mother. Its main action in this case is to eject milk and therefore facilitate breast-feeding. Milk ejection is a neurohumoral reflex. The afferent part of the reflex is the somatic pathway for touch sensation and the efferent part is via the hormone oxytocin. Oxytocin release may take place without stimulation of the nipples, e.g. in response to visual or auditory stimuli from the baby.

2 Distension of the uterus and stretching of the cervix stimulates oxytocin release. During childbirth this hormone facilitates contraction of the uterine muscles, which leads to the gradual expulsion of the foetal head through the birth canal (see Chapter 16).

3 During coitus, stimulation of the vagina and the genital organs may bring about oxytocin release. In the female sexual orgasm, oxytocin is released and milk may be ejected. There are also repeated contractions of the uterus and the fallopian tubes. These latter effects, however, are not necessary for helping the sperm to travel up the female genital tract.

4 Psychological and emotional factors such as fear may inhibit the release of oxytocin. Anxiety, pain and sympathoadrenal stimulation release oxytocin and alcohol inhibits its secretion.

Those stimuli that release oxytocin seem to work at the level of the hypothalamic nuclei. Impulses from the paraventricular nuclei travel in the hypothalamohypophyseal tracts to release the stored hormone from the nerve endings in the neurohypophysis directly into the bloodstream.

Actions of oxytocin

Two main physiological actions are known:

1 *Ejection of milk*. This is brought about by contraction of the myoepithelial cells of the ducts of the mammary gland. The hormone does not stimulate milk production; it only helps the ejection of the milk during suckling.

2 *Contraction of the uterus*. This is an important action during labour. The pregnant uterus becomes more sensitive to the action of oxytocin towards the end of pregnancy. It has been found that oestrogens make the uterine muscles more sensitive to oxytocin, while progesterone decreases this sensitivity. At term, there is a rapid fall in the mother's serum progesterone level and therefore increased uterine sensitivity to oxytocin. (For more detail see Chapter 16.)

Abnormalities

There is no known clinical condition related to an abnormality of oxytocin secretion. Synthetic preparations of oxytocin are widely used in obstetrics to induce labour and to promote uterine contraction during childbirth. This shortens the duration of labour and helps prevent maternal and foetal distress.

The thyroid gland

ANATOMICAL CONSIDERATIONS

The thyroid is the largest single endocrine gland. It weighs 20–25 g but it varies with age, sex and physiological condition, such as pregnancy and lactation.

The gland lies in the anterior triangle of the neck, closely applied to the trachea. It is encapsulated in deep fascia, which is attached to the cricoid cartilage. This is why the gland moves upwards on swallowing. The thyroid gland is formed of two lateral lobes joined by an isthmus. An extra lobe, the pyramidal lobe, is occasionally seen attached to the upper border of the isthmus.

The thyroid gland is the only endocrine gland that does not store its hormone within the cell but in follicular cavities surrounded by the cells.

The thyroid gland develops from the floor of the pharynx between the first and second pharyngeal pouches. Initially, a tubular elongation from the floor of the pharynx (thyroglossal duct) is formed. The lower end of the duct gives rise to the two lateral lobes of the thyroid and the isthmus. The thyroglossal duct then atrophies, but sometimes it may persist as a duct or as a series of cysts (thyroglossal cysts). The lower portion of the duct may also give rise to a third thyroid lobe (the pyramidal lobe).

Anywhere along the passage of migration of the thyroid gland, from its origin in the pharynx to its normal position in front of the trachea, thyroid tissue may be encountered. The thyroid gland may even descend further down to the mediastinum. Consequently, thyroid tissue may be found in prepharyngeal, lingual, intralingual, sublingual or even retrosternal positions.

The main histological feature is the thyroid follicle, which is a spherical structure formed of a single layer of epithelial cells, which surround a cavity filled with a colloid material. The type of epithelial cells forming the follicles varies according to the activity of the gland (Fig. 10.13). With normal activity the cells are cubical. With increased activity they become tall and columnar and microvilli can be seen projecting into the colloid. With decreased activity they become flat cells. The quantity of colloid material in the cavity of the follicle also varies with the activity of the gland. With hyperactivity the colloid material is greatly decreased such that the walls of the follicles become folded in, while with hypoactivity the follicles become distended with colloid. A few clear cells or C cells may be seen in the walls of the follicles but they have no contact with the cavity of the follicle. They are therefore also referred to as the parafollicular cells. They produce the hypocalcaemic hormone, calcitonin.

A rich network of capillaries surrounds the follicles. The nerve fibres, seen within the thyroid supply blood-vessels, control blood flow to the gland, but have no direct control on secretion.

HORMONES OF THE THYROID

The gland produces thyroxine (T_4), tri-iodothyronine (T_3) and calcitonin. Calcitonin is concerned

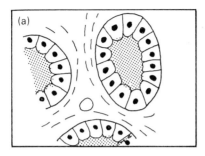

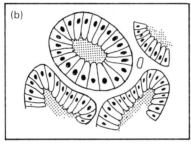

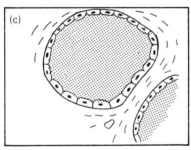

Fig. 10.13 Diagram showing: (a) the normal microstructure of the thyroid follicles; (b) smaller follicles lined with columnar epithelium in hyperthyroidism; and (c) the flattened epithelium with wide follicles full of stored colloid in hypothyroidism.

with calcium metabolism and will be considered with the parathyroid hormone and the control of plasma calcium.

Iodine forms an integral part of the thyroxine and tri-iodothyronine molecules. Iodine is taken in food and in drinking-water. A minimum daily intake of about 100–150 μg (in the form of potassium or sodium iodide) is required for normal thyroid function. The iodide is easily absorbed in the intestine to circulate in plasma in a concentration of about 0.3 μg/100 ml; 30–50% of iodide circulating in blood is taken up by the

thyroid, while the rest is lost in urine. The thyroid can therefore remove iodide from blood against a concentration gradient ranging from about 40 times in a normal gland to several hundred times in an overactive gland. This selective uptake of iodide by the thyroid gland, or iodide trapping, can be achieved only by active transport (iodide pump). Also, the follicular cell has a resting membrane potential of −50 mV. The iodide pump has to work against electrical and concentration gradients. Various other tissues can also trap iodide, particularly the salivary and mammary glands, but the thyroid contains 90–95% of body iodide.

STAGES OF SYNTHESIS OF THYROID HORMONES (Fig. 10.14)

Thyroglobulin synthesis
Thyroglobulin is a large protein synthesized by the endoplasmic reticulum of the thyroid cells and secreted into the colloid. Its molecular weight is 660 000 daltons. The importance of thyroglobulin lies in the fact that it has 140 molecules of the amino acid tyrosine, which is used for thyroid hormone synthesis. The synthesis of thyroid hormones takes place within the thyroglobulin molecules in the colloid.

Oxidation of iodide
The trapped iodide ions are oxidized into iodine in the thyroid cell by means of peroxidase enzyme. The oxidized form is capable of combining with tyrosine.

Iodination of tyrosine
The tyrosine molecules on the surface of thyroglobulin are iodinated to give monoiodotyrosine (MIT) and di-iodotyrosine (DIT). This iodination process is rapid and it occurs as soon as the thyroglobulin is synthesized and during its passage through the cell membrane into the follicles of the thyroid.

Coupling of MIT and DIT
The next stage is condensation or the coupling reactions of iodotyrosines (MIT and DIT) to give iodothyronines. This is a slow process, which

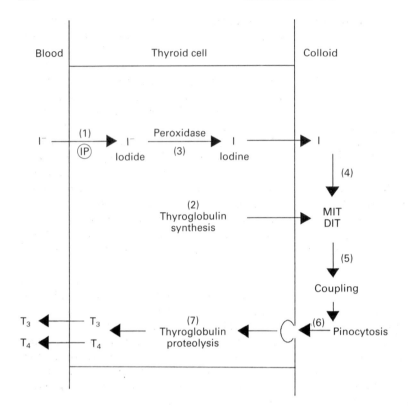

Fig. 10.14 Diagram of thyroid gland cell showing the steps of thyroid hormone synthesis: (1) iodine trapping, showing iodide pump (IP); (2) thyroglobulin synthesis; (3) oxidation of iodide to iodine; (4) iodination of tyrosine molecules within the thyroglobulin molecule to give monoiodotyrosine (MIT) and di-iodotyrosine (DIT) at the apical border of cell; (5) coupling of MIT and DIT to give T_3 and T_4, still within the thyroglobulin molecule in the colloid; (6) pinocytosis of thyroglobulin; (7) proteolysis of thyroglobulin and release of T_3 and T_4.

starts in a few minutes but continues for several days within the follicular colloid. When one MIT and one DIT molecule are coupled, tri-iodothyronine (T_3) is formed, and, when two DIT molecules are coupled, thyroxine (tetraiodothyronine, T_4) is formed. The coupling reactions take place on the surface of the thyroglobulin molecule and, in the same way as MIT and DIT, both T_4 and T_3 are attached to the thyroglobulin molecule by peptide linkages.

The active thyroid hormones, T_3 and T_4, and their inactive precursors, MIT and DIT, are stored in the colloid of the thyroid follicles. They can only be released from the colloid by proteolysis of thyroglobulin.

Release of T_3 and T_4

In the thyroid cells, the next step results in the release of T_3 and T_4 from the thyroglobulin molecule. First, thyroglobulin is taken up by the cells by pinocytosis. There are proteases present in an inactive form, but, when activated, they break the peptide linkages between the thyroglobulin molecules and the iodinated compounds. MIT, DIT, T_4 and T_3 are set free. T_3 and T_4 will leave the follicle by their concentration gradient to reach the rich capillary network around the follicles.

In the thyroid cell there is another enzyme, deiodinase, which only acts on iodinated tyrosines (MIT and DIT) after they are released from their attachment with the thyroglobulin molecule. Deiodinase has no effect on T_4 and T_3. The iodine released by deiodinase from MIT and DIT is reused for iodination of tyrosine. Very minute amounts, if any, of MIT and DIT escape deiodination and appear in the circulation. Considerable loss of MIT and DIT may occur if the enzyme deiodinase is deficient.

Figure 10.14 illustrates the various steps of thyroid hormone synthesis. Drugs such as perchlorate and thiocyanate interfere with iodide trapping. Thiouracil and related compounds inhibit iodination and coupling reactions and, as such, can be used to treat hyperthyroidism. Propylthiouracil also inhibits conversion of T_4 to T_3.

TRANSPORT AND DEGRADATION OF THYROID HORMONES

T_4 and T_3, once they reach the circulation, are transported in blood attached to plasma proteins, namely, thyroxine-binding globulin (TBG), thyroxine-binding prealbumin (TBPA) and albumin (TBA). The proportions of the hormones passed to each of these proteins depend on their affinities for the hormone (Table 10.3). TBG has a high affinity for T_4 and therefore it binds more T_4 than T_3. Albumin has a higher concentration in plasma than TBG and therefore it has a high capacity to bind T_4 and T_3 but it has less affinity. The normal plasma levels are $60-160$ nmol/litre for T_4 and $1.2-2.8$ nmol/litre for T_3. The half-lives of T_4 and T_3 are about 9 and 1.5 days respectively. Only minute amounts of the hormones are found free within plasma. The physiological actions of T_3 and T_4 and their negative feedback with TSH are due to the free fractions.

Most of T_4 is deiodinated in the liver, kidneys and other tissues to give active T_3 (3,5,3-tri-iodothyronine) and an inactive form known as reverse T_3 (r-T_3) or 3,3,5-tri-iodothyronine. Further deiodination takes place, resulting in iodotyrosines (MIT and DIT). In the pituitary and brain T_4 is deiodinated to T_3 intracellularly. The conversion of T_4 to T_3 is decreased by hypermetabolic states, such as fever, trauma, burns and malignancies. Starvation depresses T_3 formation but increases r-T_3 levels, and overfeeding increases T_3 and decreases r-T_3.

CONTROL OF THYROID FUNCTION

The function of the thyroid gland is controlled by

Table 10.3 The proportions of T_3 and T_4 bound to different plasma proteins

Binding protein	T_3 (% bound)	T_4 (% bound)
TBG	75–80	70–75
TBPA	Trace	15–20
Albumin	10–15	10–15

the hypothalamopituitary axis. The hypothalamus produces a thyrotrophin-releasing hormone (TRH) which reaches the adenohypophysis via the hypophyseal portal vessels. Under the influence of TRH, the adenohypophysis produces thyroid-stimulating hormone (TSH), which controls the function of the thyroid gland. The thyroid hormones circulating in blood regulate the relase of TSH by a negative feedback control. The outcome of this negative feedback regulation is that, as the level of thyroid hormones in the blood rises, they inhibit TSH release. Conversely, when the level of circulating thyroid hormones fall, the inhibitory effect on the hypothalamus and pituitary is removed and release of TSH is increased (see Fig. 10.3 and the section on the control of TSH).

TSH influences thyroid function in different ways. It increases the vascularity of the gland to facilitate more efficient removal of the hormones from the gland. TSH stimulates the steps involved in the synthesis and release of thyroid hormones, from the process of iodide trapping to the activation of proteases which liberate thyroid hormones from their attachment to the thyroglobulin molecule into the colloid.

ACTIONS OF THYROID HORMONES

Although the thyroid gland produces more T_4 than T_3, the latter hormone is more potent and acts much faster than the former. T_3 enters the cell more easily than T_4. The latter is also transformed to T_3 in the cytoplasm. T_3 binds avidly to receptors on the cell nucleus and promotes messenger ribonucleic acid (mRNA) and ribosomal RNA (rRNA) synthesis. Most probably, a variety of protein enzymes are synthesized in various tissues to account for the widespread actions affecting practically all tissues and organs of the body:

1 *Effects on general metabolism.* Thyroid hormones have a calorigenic action. They stimulate oxygen consumption and energy expenditure by all active tissues of the body, with some exceptions, such as the brain, pituitary and testes. Because the metabolic rate of the body is accelerated, heat production will increase. In hyperthyroidism, heat loss mechanisms are stimulated, i.e. sweating and vasodilatation of skin vessels.

2 *Effects on growth and development.* Thyroid hormones are important for normal growth and development. This is clearly seen in congenital deficiency of thyroid hormones, resulting in the thyroid dwarf (the cretin). GH can exert its full effect on skeletal growth only in the presence of adequate thyroid hormones.

3 *Effects on the nervous system.* Thyroid hormones are essential for normal development and function of the central nervous system. Mental retardation is an important feature of cretinism. There is defective synaptic development and myelination in the brain. If treatment of cretinism is not started early after birth, mental retardation will persist.

In hyperthyroidism, the activation of the central nervous system may be due to potentiation of the effects of catecholamines, which stimulate the reticular activating system.

Deficiency of thyroid hormones later in life (myxoedema) depresses mental functions. The patient becomes mentally slow and forgetful. Reflex activities are also slowed down. The relaxation phase of a stretch reflex becomes significantly prolonged. Prolongation of the ankle jerk relaxation phase is a useful clinical sign in hypothyroidism.

4 *Effects on metabolism.* Since thyroid hormones are essential for normal growth, in physiological quantities they stimulate protein synthesis, but, in large amounts, because of the accelerated metabolic rate, thyroid hormones indirectly cause protein breakdown to provide fuel for the increased energy expenditure. The patients go into negative nitrogen balance. Catabolism of muscle protein leads to muscle-wasting and removal of bone protein results in hypercalcaemia, hypercalciuria and osteoporosis.

In carbohydrate metabolism, thyroid hormones are hyperglycaemic (diabetogenic) hormones. They increase blood glucose in several ways. They promote glucose absorption in the intestine and stimulate glycogenolysis and gluconeogenesis. They also promote degradation of insulin and increase peripheral glucose utilization. The accelerated energy expenditure is partly explained by increased Na^+, K^+ ATPase activity and mitochondrial protein synthesis.

In lipid metabolism, thyroid hormones decrease serum cholesterol levels. This is due to increased formation of low-density lipoprotein (LDL) receptors. Thus, in hyperthyroidism the serum cholesterol level is low or normal while in hypothyroidism it is usually high. They also mobilize FFAs from adipose tissue.

5 *Effect on vitamin requirements.* Thyroid hormones promote conversion of β-carotene to vitamin A. They also facilitate absorption of vitamin B_{12}. Because thyroid hormones accelerate energy metabolism, there is a greater body requirement for vitamins of the B group, especially thiamine in hyperthyroid states.

6 *Water and mineral metabolism.* Thyroid hormones have a diuretic effect. There is an increase in the glomerular filtration rate (GFR) and an increase in renal excretion of sodium and calcium. Because of their catabolic effect on protein, potassium, creatine and uric acid, excretion is also increased.

7 *Effects on the cardiovascular system.* Thyroid hormones increase the chronotropic and inotropic effects of catecholamines on the heart. The number and affinity of β-adrenergic receptors are increased. Both the heart rate and the stroke volume are increased. As a result, cardiac output increases and systolic blood-pressure rises. Vasodilatation is brought about by the stimulated heat loss mechanism. This results in decreased peripheral resistance and the diastolic pressure tends to drop. Consequently, high pulse pressure is usually recorded and a collapsing pulse is detected. A high sleeping pulse is a clinical sign of hyperthyroidism, while low ECG voltage is seen in hypothyroidism. The effects on the heart can be decreased by β-adrenergic blockers, e.g. propranolol.

8 *Effect on the alimentary tract.* Thyroid hormones increase the motility of the gastrointestinal tract and improve the appetite. Patients with hyperthyroidism have very good appetites and complain of diarrhoea, while those with hypothyroidism have poor appetites and are usually constipated.

9 *Effects on muscles.* Excessive production of thyroid hormones causes considerable muscle weakness (thyrotoxic myopathy) due to several reasons:

(a) Protein catabolism and muscle-wasting are

important features of hyperthyroidism. There is also evidence that some changes take place in muscle myosin and ATPase activity and that there are decreased levels of muscle ATP.

(b) The water and electrolyte disturbance caused by abnormally high levels of thyroid hormones contributes to muscle weakness.

(c) Thyroid hormones increase the excitability of nervous tissue. One of the effects of this is the fine tremor and spontaneous twitches seen in hyperthyroidism. This continuous activity in the muscles ultimately leads to depletion of energy stores, fatigue and muscle weakness.

10 *The effect on gonads.* Thyroid hormones are essential for the normal development and function of the gonads. For instance, menstrual disturbances are seen in both hypo- and hyperthyroidism.

11 *Effect on mammary glands.* Thyroid hormones are important for maintenance of milk production, a fact sometimes made use of to increase milk production in dairy farming.

12 *Effects on bone marrow.* Thyroid hormones are important for bone marrow metabolism and for normal erythropoiesis. Hypothyroidism is characterized by anaemia due to depressed bone marrow activity, which can only be corrected by administration of thyroid hormones.

The actions of thyroid hormones can be well illustrated by considering the symptoms and signs of hypo- and hyperthyroidism.

ABNORMALITIES OF THYROID FUNCTION

All thyroid disturbances are more common in females than in males.

Goitre

Goitre refers to any enlargement of the thyroid gland, with or without disturbance of function. There are many causes of goitre:

1 *Physiological goitre.* This is a diffuse moderate enlargement due to increased demand for thyroid hormones, as in puberty, pregnancy, lactation and chronic exposure to cold.

2 *Iodine deficiency.* Dietary iodine deficiency over a long period of time gives rise to simple goitre. This usually occurs in populations living in highlands far from the sea, where rainfall washes away the iodine content of the soil. Water

and vegetables grown in these areas will be deficient in iodide. The condition is endemic; therefore, it is also referred to as 'endemic goitre'. Several members of the family are usually affected. The size of the gland can be extremely large (Fig. 10.15) but there is no disturbance of function. The condition is a public health problem, which can be prevented by addition of iodide to table salt or cooking oil. Treatment of endemic goitre is surgical removal for cosmetic reasons and to avoid pressure on underlying structures, e.g. the trachea.

3 *Hereditary defect in thyroid hormone synthesis.* This may be due to hereditary deficiency of any of the enzymes concerned with thyroid hormone synthesis. Deficiency of these enzymes may lead to iodide trapping defect, iodine organification defect, iodotyrosine coupling defect or deiodination defect. Consequently, low circulating thyroid hormone levels lead to increased production of TSH, which causes hypertrophy and hyperplasia of the gland. Since the condition is congenital or familial, the child is a cretin with a goitre (a goitrous cretin). The condition is more common when the mother has endemic goitre.

4 *Excessive intake of goitrogens*, e.g. cabbage, turnip or drugs such as sulphonamides, thiocarbamides, perchlorate and thiocyanate, can suppress thyroid hormone synthesis and lead to goitre formation.

Hypothyroidism

Deficiency of thyroid hormone production may occur as a congenital abnormality due to congenital absence of the thyroid or congenital deficiency of any of the enzymes concerned with hormone synthesis (goitrous cretin). Hypothyroidism can also occur at adolescence or later in life.

Cretinism When the deficiency is congenital, it is referred to as cretinism. The affected child is a cretin. Characteristically, the child is a mentally retarded dwarf. The features of cretins are so characteristic that they cannot be mistaken. The face is usually wrinkled and dry, the lips are thick and the tongue is large and usually protruding out of the mouth. Because of the enlargement of abdominal viscera and weak abdominal muscles,

Fig. 10.15 Patients with simple goitre in western Sudan. (Reproduced with permission from Bell, G.H., Emslie-Smith, D. & Paterson, C.R. (eds) (1980) *BDS Textbook of Physiology*, 10th edn. Churchill Livingstone, Edinburgh.)

the belly is protruding, usually with an umbilical hernia (pot-belly).

It is very important to diagnose the condition soon after birth and immediately start the treatment; otherwise the mental retardation becomes irreversible.

Myxoedema When deficiency of thyroid hormones is not congenital, the condition is referred to as myxoedema. It commonly occurs after the age of 50 but, if it occurs early in life, at adolescence, it is known as juvenile myxoedema. Myxoedema may be primary, due to damage of the thyroid, most commonly by an autoimmune process, or it may be due to surgical removal or radiation. It may be secondary to hypopituitarism. Thyroid autoimmune disease is triggered by thyroid damage or inflammation, leading to the release of appreciable amounts of thyroglobulin into the circulation. Thyroglobulin, will evoke specific antibody formation. These antibodies will reach the thyroid and destroy thyroid tissue.

The patient with myxoedema is usually overweight, with dry, thick, wrinkled skin, an ex-pressionless face, thick-lips and a large, thick tongue. The hair is dry, lustreless and easily lost from the scalp but most characteristically from the outer third of the eyebrows (Fig. 10.16).

Patients with myxoedema usually complain of intolerance to cold (due to their low metabolic rates). They are quite happy in a hot environment, while in the slightest cool breeze they cover themselves with blankets. They are mentally depressed and forgetful, forgetting recent events or even the names of their close relatives. The appetite is poor and the patient is usually constipated. The pulse rate is slow (bradycardia) and there is low blood-pressure. If the patient is still in the child-bearing age, a common complaint is menstrual disturbances. The voice becomes hoarse and speech is slow and monotonous. Myxoedematous patients are usually weak and get tired very easily. They cannot cope with their normal household activities and spend most of their time in bed. Anaemia is a common finding in these patients. The duration of the ankle jerk is prolonged and, in particular, the relaxation phase of the reflex.

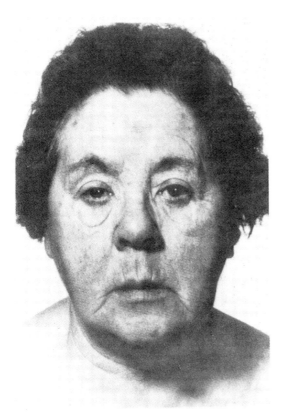

Fig. 10.16 Photograph of a patient with hypothyroidism. (Reproduced with permission from Emslie-Smith, D., Paterson, C.R., Scratchard, T. & Read, N.W. (eds) (1988) *BDS Textbook of Physiology*, 11th edn. Churchill Livingstone, Edinburgh.)

Hyperthyroidism (thyrotoxicosis)

Hyperthyroidism, which is characterized by high blood levels of thyroid hormones, can occur at an early age or it may occur later in life. The increased thyroid hormone secretion is due to abnormal immunoglobulins (IgGs) — known as thyroid-stimulating immunoglobulins (TSI). They seem to act by stimulating TSH receptors on the follicular cells. Due to the increased T_3 and T_4 output, TSH secretion is inhibited and its circulating level in thyrotoxicosis is usually low.

The patient usually complains of loss of weight in spite of good appetite and complains of weakness and tiredness after slight exertion. They are nervous, anxious and easily excited and may often quarrel with family and children. There is a fine tremor, which can be seen in the outstretched hand.

The eyes may be protruding (exophthalmos) and the upper eyelids may be retracted, with the white sclera seen above the cornea, which gives the patient a startled look (Fig. 10.17). The exophthalmos is due to accumulation of mucopolysaccharides, fat and fluid behind the eyeball. The lid retraction is caused by increased tone in the levator muscle of the eyelid, due to potentiation of sympathetic activity by thyroid hormones. The syndrome of hyperthyroidism associated with exophthalmos and enlargement of the thyroid is known as Graves' disease or exophthalmic goitre.

Patients commonly complain of palpitations. The heart rate, the stroke volume and the cardiac output are all increased. The systolic blood-pressure will rise, while the diastolic blood-pressure may fall due to vasodilatation. This may lead to high cardiac output failure. In some

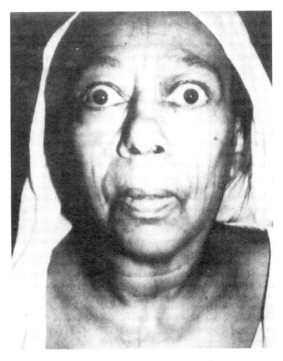

Fig. 10.17 Photograph of a patient with hyperthyroidism. (Courtesy of Professor Khalil M. Salman, Dept. of Surgery, King Abdulaziz University Hospital, Jeddah, Saudi Arabia.)

patients cardiac symptoms may be the only manifestation of thyroid disease. This should be remembered before they are referred to the cardiologist.

Thyrotoxic patients usually have a wet, warm skin; the palms are wet and warm. Menstrual disturbances are frequent and diarrhoea is a common complaint.

TESTS OF THYROID FUNCTION

A large variety of tests are used to assess thyroid function. Only some of these tests will be cited here:

1 Measurement of the basal metabolic rate (BMR) is a valuable indicator of the status of thyroid function, but, because its measurement requires highly standardized conditions, the results obtained may not be totally reliable. For this reason it has been replaced by other tests.

2 Radioactive-iodine uptake (4- or 24-hour) and thyroid scanning are two tests that give a good indication of thyroid function and may assist in the diagnosis of the pathology of the condition. An oral dose of a radioactive isotope such as ^{123}I or ^{131}I is given and the uptake by the thyroid is measured. The uptake varies from one population to another, according to dietary iodine intake. Normal standards for the indigenous population have to be worked out for the tests to be of value. High dietary iodine intake dilutes the test dose and may give misleadingly low radioactive iodine uptake values.

The percentage of a dose of radioactive iodine uptake is calculated and plotted against time. Figure 10.18 gives typical findings in cases of hyper- and hypothyroidism.

In thyroid scanning, after the administration of a dose of radioactive iodine, a graphic record of the density of isotope uptake (Fig. 10.19) is made by means of a scintiscanner. This test has the added value of detected localized sites of activity (Fig. 10.20).

3 Estimation of serum T_3, T_4 and TSH: levels of total and free forms of T_3 and T_4, together with TSH, by radioimmunoassay or other methods in plasma, give a direct assessment of thyroid function.

4 Measurement of protein-bound iodine (PBI) and of TBG is also used for assessment of thyroid

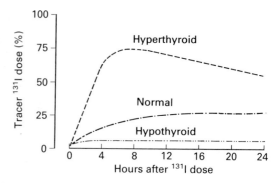

Fig. 10.18 Radioactive iodine uptake by the thyroid gland. The graph shows the percentage of the administered tracer dose taken up by the thyroid over a period of 24 hours in a normal person compared with hyper- and hypothyroidism.

function. The first is difficult and not specific and the second is influenced by a variety of drugs and physiological conditions.

The parathyroid glands and calcium homoeostasis

ANATOMICAL CONSIDERATIONS

The parathyroid glands are small, elongated nodules, usually four in number, embedded in the posterior surface of the lobes of the thyroid gland, one in relation to each pole. But they may be less or more than four and may even be found in the mediastinum. They are 6 mm in length and weigh 20–50 mg. In contrast to their rich blood supply, the glands have a poor nerve supply, from branches of the superior and recurrent laryngeal nerves, which supply the blood-vessels.

Development

The parathyroid glands develop from the endothelium of the dorsal portions of the distal ends of the third and fourth pharyngeal pouches, one pair from each pharyngeal pouch. The pair of glands arising from the third pouch descend further than those arising from the fourth pouch to form the inferior parathyroids.

Structure

In man the parathyroid glands have two main types of cells. The chief (principal) cells form the major component of the cellular structure of the

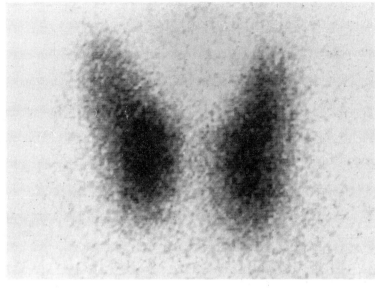

Fig. 10.19 (a): Scintiscan of a normal thyroid gland using technetium 99 m pertechnetate (iodine analogue). It shows normal size and shape (butterfly appearance) of both lobes of the gland. The distribution of the tracer within the thyroid lobes is uniform and its concentration is within the normal limit. The time taken to obtain this image was 8 minutes. (b): Scintiscan of the thyroid gland in a patient showing symmetrical diffuse enlargement of both lobes of the gland. The distribution of the tracer is even and its concentration is increased. Note that the time required to obtain this image was only 1 minute. (Courtesy of Dr Tariq Al-Baghdadi, Nuclear Medicine, King Abdulaziz University Hospital, Jeddah, Saudi Arabia.)

gland. They are small cells with poorly staining cytoplasm. They are the secretory cells of the gland. The less abundant cells are the oxyphil cells, which are seen only after puberty. They are larger in size than the chief cells and contain oxyntic granules in their cytoplasm, which have proved to be mitochondria. Although the function of these cells has not yet been clarified, their high content of mitochondria indicates that they are metabolically active and may have a secretory function as well.

PARATHYROID HORMONE

Chemistry and degradation

The glands produce parathyroid hormone (PTH), which is a polypeptide with a molecular weight of 9500 daltons that is formed within the parathyroid cells as pre-pro-PTH (115 amino acids). Pre-pro-PTH is then converted to pro-PTH, from which PTH is formed (84 amino acids). In the circulation there is further cleavage, mainly in the liver, to form two fragments (C-terminal and

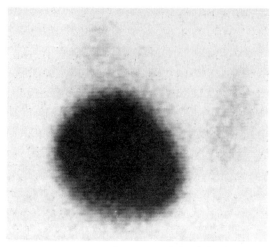

Fig. 10.20 Scintiscan of a toxic nodule of the thyroid gland. It shows an enlarged right lobe with a marked, localized, abnormally increased concentration of the tracer occupying its lower half. The upper part of the right lobe, as well as the normal-sized and shaped left lobe, are faintly visualized. The time required to obtain this image was only 1.5 minutes. (Courtesy of Dr Tariq Al-Baghdadi, Nuclear Medicine, King Abdulaziz University Hospital, Jeddah, Saudi Arabia.)

N-terminal). Biological activity is associated with the N-terminal fragment. Degradation occurs mainly in the kidney. The C-terminal fragment has a longer plasma half-life (about 60 min) than the N-terminal fragment (about 5 min). Particularly high levels of PTH are found in patients with renal failure.

Control of secretion
1 The parathyroid glands are not under anterior pituitary control nor are they under nervous control. The main mechanism known to regulate PTH release is a negative feedback with plasma ionized calcium (Ca^{2+}). Decrease in plasma Ca^{2+} level stimulates the parathyroids to release their hormone. Conversely, an increased plasma ionized calcium level decreases hormonal release.
2 Ionized magnesium has similar effects to Ca^{2+} but its physiological significance is unknown.
3 Stimulation of β-adrenergic receptors on parathyroid cells increases PTH secretion. This mechanism seems to modify the sensitivity of the gland to Ca^{2+}.
4 High levels of 1,25-dihydroxycholecalciferol

inhibit PTH secretion. It also acts as a negative feedback to regulate the activation of vitamin D by the kidneys.

Under certain physiological conditions where there is an increased demand for calcium, such as pregnancy and lactation, the parathyroid glands hypertrophy and become more vascular and more active. Their hormone production consequently increases.

Actions of parathyroid hormone
PTH acts at different sites in the body to increase the level of plasma calcium:
1 In bone it promotes calcium reabsorption. This effect is achieved by activating oesteoclasts, by formation of new oesteoclasts and by retarding the conversion of oesteoclasts into oesteoblasts. As calcium is removed from bone, phosphate is removed too. The effects on bone are mediated by an increased production of cyclic AMP.
2 In the kidney its action is manifested in several ways:
   (a) It increases phosphate excretion by decreasing tubular reabsorption.
   (b) It increases tubular reabsorption of calcium.
   (c) It converts 25-hydroxycholecalciferol into active 1,25-dihydroxycholecalciferol (the active form of vitamin D_3). This has recently been named calcitriol. It facilitates mobilization of calcium from bone and promotes intestinal calcium absorption. Calcitriol also facilitates active transport of Ca^{2+} from the osteoblasts to the extracellular fluid.

CALCITONIN
Calcitonin is the other hormone concerned with calcium homoeostasis. It is produced by C cells of the thyroid gland. It has the opposite effects to those of parathyroid hormone. Calcitonin is a polypeptide consisting of 32 amino acids with a molecular weight of 3000 daltons.

The control of calcitonin secretion is stimulated by hypercalcaemia. It is released when there is a high Ca^{2+} level in the extracellular fluid, e.g. after a meal. Its secretion, in this case, is also stimulated by the gastrointestinal hormones gastrin, CCK-PZ and glucagon. Thus, it provides a second hormone for the regulation of the serum calcium level. However, this feedback mechanism is much faster than that of PTH and therefore it provides a short-term regulatory mechanism, e.g. preventing postprandial hypercalcaemia. It must be noted, however, that in the long term it is PTH which is more important in the regulation of serum calcium.

The actions of calcitonin produce a rapid decrease in serum Ca^{2+} concentration through the following mechanisms:

1 It suppresses osteoclastic activity in bone, leading to a decrease in calcium mobilization. This effect is more marked in children, because they have a greater osteoclastic activity due to rapid bone remodelling. In adults this is a weak effect, because the amount of calcium mobilized by oesteoclasts is much smaller.

2 It increases the activity of oesteoblasts. This leads to increased mineralization of bone at the expense of extracellular fluid calcium concentrations.

3 In the long term, it decreases the number of oesteoclasts formed in the bone. The number of oesteoblasts is also decreased. Thus, its long-term effect is neutralized.

4 It has an inhibitory effect on the transport of calcium from the intracellular fluid to the extracellular fluid.

REGULATION OF PLASMA CALCIUM LEVEL

The total body calcium in an adult man amounts to about 1200 g (30 000 mmol); 99% of it is found in bone while only 1% is found in soft tissues and body fluids. The skeleton is therefore an important store of calcium to be drawn from when the plasma calcium level tends to fall. In fact, there is continuous exchange of calcium between bone and blood. The exchangeable calcium pool in bone amounts to about 4000 mg (100 mmol). The plasma calcium level ranges between 9 and 10.5 mg/dl (2.1 and 2.6 mmol/litre). About 50% is found in un-ionized complexes, mainly bound to plasma albumin. The rest is found in ionized form (1.00–1.3 mmol/litre). This range is kept reasonably constant by a delicate regulatory mechanism which involves PTH, calcitonin, vitamin D and calcitriol.

The total body calcium is maintained within the normal limits by a balance of calcium intake and excretion (see Fig. 10.21).

Calcium balance

Dietary calcium varies greatly in various communities. A rich diet provides a daily intake of approximately 1000 mg, of which about 300 mg are absorbed. Conversely, about 200 mg of cal-

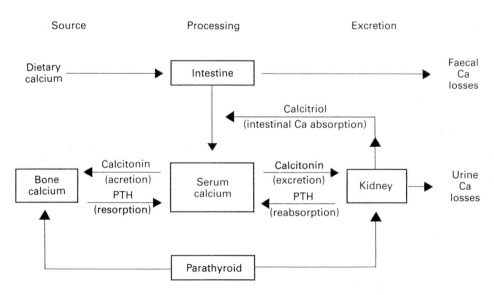

Fig. 10.21 Regulation of serum calcium level. Note that the main sites of regulation are: intestinal absorption, bone resorption and deposition, and renal tubular reabsorption.

cium are secreted into the gastrointestinal tract and about 50 mg are reabsorbed. Consequently, about 950 mg of calcium appear in faeces daily (see also Fig. 8.1). In the kidney about 10 000 mg of calcium are filtered daily but only 150 mg appear in urine; the rest is reabsorbed by the renal tubules.

It can be seen, therefore, that under normal conditions, calcium intake is accurately balanced with the calcium appearing in urine and faeces. Absorption of calcium in the gastrointestinal tract is facilitated by calcitriol, formed in the kidney under the effect of PTH. However, absorption may be impaired by phytates, fats and oxalates, which form insoluble, poorly absorbed compounds of calcium. Gastrectomy renders the intestinal contents alkaline. Since calcium is more soluble in an acid medium, its absorption from the intestine will be impaired by gastrectomy.

In the kidney PTH stimulates tubular reabsorption of calcium. Only 150 mg are excreted in the urine per day.

The Ca^{2+} levels in the serum range between 1.0 and 1.3 mmol. The level remains reasonably constant. About 500 mg of calcium are mobilized and the same amount deposited in bone every day. It appears from the above that calcium is added or taken from the plasma pool in such a way as to maintain a constant level (internal balance). Dietary intake, on the other hand, balances the calcium losses in faeces and urine (external balance). In all this, bone calcium provides a huge reserve, which is used in the short-term maintenance of Ca^{2+} levels in the blood and extracellular fluid.

DISTURBANCE OF PARATHYROID FUNCTION

Hyperparathyroidism
Excessive production of PTH due to an active tumour or hyperplasia of the parathyroids leads to a series of disturbances in various organs and tissues:
1 Mobilization of calcium from bone leads to osteoporosis; the bones become weak and may fracture following minor trauma (pathological fractures). With excessive removal of calcium, the bones become fibrous, with formation of bone cysts, a rare condition called osteitis fibrosa cystica. PTH stimulates the production of osteo-

clasts, which aggregate to form bony swellings (osteoclastomas).
2 In the kidneys, a raised plasma calcium level increases the filtered load of calcium and the tubular maximum is exceeded. The excess calcium will be excreted in the urine together with increased excretion of phosphate. The osmolar concentration of fluid in kidney tubules will rise, leading to osmotic diuresis (polyuria). Deposition of calcium phosphate crystals in the kidney ultimately leads to renal stone formation. The stones are characteristically calcium phosphate, are multiple and bilateral and tend to recur when removed. Kidney damage and renal failure are the final stages of the untreated condition.
3 Occasionally the presence of a high plasma calcium level may lead to abnormal calcification in soft tissues, such as muscles, cartilage, joint capsules and tendons.
4 Persistent hypercalcaemia leads to tiredness and mental confusion and may inevitably result in coma and death.

Diagnosis Hyperparathyroidism may be suspected from the clinical picture. The diagnosis is confirmed if there are persistent high plasma calcium levels, especially if precautions are taken when blood samples are obtained from the patient. The main precaution is to take blood without using a tourniquet. It has been shown that using a tourniquet will raise the calcium level in the sample due to stasis of blood. The syringe and container used must be calcium-free. The diagnosis can be confirmed by the steroid suppression test. Giving hydrocortisone in daily doses of 40 mg for 8–10 days will suppress the plasma calcium level elevated by other conditions but not that in hyperparathyroidism.

The development of radioimmunoassays for measuring PTH in plasma has made it possible for the direct diagnosis of dysfunction of parathyroid glands. However, measured PTH may be falsely high when inactive fragments of the hormone are detected by the assay.

Hypoparathyroidism
Hypoparathyroidism is usually secondary to surgical removal or damage of the glands. It may also be due to an autoimmune process. The clinical picture is a result of a persistently low plasma

calcium level. The presentation is that of latent or overt tetany.

Latent tetany is characterized by hypersensitivity (low threshold) of nerves and muscles, which can be demonstrated by two signs:

1 *Chvostek's sign*. Tapping the facial nerve at its emergence from the anterior border of the parotid gland in front of the ear causes contraction of the facial muscles on the ipsilateral side.

2 *Trousseau's sign*. Carpal spasm can be demonstrated by arresting blood flow to the forearm by blowing a sphygmomanometer cuff above the systolic pressure and maintaining it for a few minutes. Characteristically, there will be flexion at the wrist, thumb and metacarpophalangeal joints with extension at the interphalangeal joints.

Tetany (latent or overt) may also result from lowering of plasma Ca^{2+} due to alkalaemia, e.g. following hyperventilation or excessive vomiting. Hypocalcaemia can also be caused by a vitamin D deficiency in children or adults or in renal failure, due to a deficient production of calcitriol and consequent poor intestinal absorption of calcium.

Long-standing hypocalcaemia may lead to overt tetany with spasm in the hands and feet (carpopedal spasm), spontaneous laryngeal spasms (laryngismus stridulus), convulsions and epileptic fits. Muscle cramps and muscle weakness are quite common. The patient may also present with cataracts, sometimes associated with mental retardation.

The endocrine pancreas

Diabetes mellitus is by far the most common endocrine and metabolic disorder encountered in medical practice. Insulin, the main hormone of the pancreas, is widely used in the treatment of diabetic patients.

The endocrine part of the pancreas consists of the islets of Langerhans. These are clusters of cells scattered between the acini of the exocrine pancreas. The islets are about 150 μm in diameter. They are richly supplied with blood-vessels. The cells of the islets consist of four main types: A cells (which constitute 20% of the islets' cells), B cells (about 75%) and D cells (3–5%). F cells (<2%), whose physiological function is uncertain, constitute the remainder.

The A cells secrete the hormone glucagon. The B cells secrete insulin. The D cells secrete somatostatin. F cells release a hormone known as pancreatic polypeptide (PP), with, as yet, uncertain physiology.

INSULIN

Insulin is the main hormone concerned with regulation of carbohydrates metabolism. In particular, insulin controls the level of blood glucose, glucose storage and utilization.

Chemistry and degradation of insulin

Insulin is a polypeptide consisting of 51 amino acids. Human insulin has a molecular weight of about 5800 daltons. The amino acid sequence has been worked out for human, bovine and porcine insulin. There are minor differences between species but their biological activity is similar. Synthesized human insulin is now available for therapy.

The molecules of insulin consist of two chains: an A chain, consisting of 21 amino acids, and a B chain, consisting of 30 amino acids. The two chains are connected by two disulphide bridges. In the storage form, the molecule of proinsulin consists of a single chain (Fig. 10.22). The C chain, or connecting peptide (C peptide), consists of 31 amino acids. The C chain is removed before insulin is secreted by the B cells. The C peptide

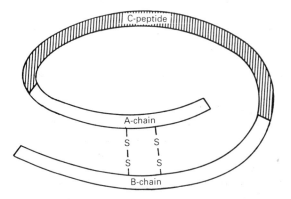

Fig. 10.22 The proinsulin molecule showing the insulin A and B chains connected by disulphide bridges. The C chain (connecting peptide) is shown as the shaded part of the proinsulin molecule. On cleavage of the C chain, equimolar quantities of insulin and C peptide are released from the B cells.

has no known function, but its blood levels indicate insulin secretion, because it is formed in equimolar quantities to insulin.

The half-life of insulin is about 5 minutes. It is carried into the pancreatic vein and goes through the portal vein to the liver before it reaches the systemic circulation. About half of it gets bound to liver cells. Almost all tissues bind and metabolize insulin. Hepatic glutathione insulin transhydrogenase cleaves the disulphide bridges between the A and B chains and this results in loss of biological activity. The kidneys also degrade significant amounts of insulin.

Regulation of insulin secretion

Insulin is found in plasma at a basal level (after an overnight fast) of about $5-10 \mu U/ml$. Overweight and obese patients have higher-than-normal fasting levels. Its rate of secretion is markedly increased immediately after food intake.

1 The main stimulant to its secretion is glucose. Insulin secretion follows very closely the changes in the blood glucose level (Fig. 10.23). It seems that glucose should be used by the B cells in order to stimulate insulin secretion. 2-Deoxyglucose (an unmetabolizable isomer of glucose) inhibits insulin secretion.

2 Amino acids in the diet are also a potent stimulus to insulin secretion. Both carbohydrate intake and protein intake stimulate the release of insulin. Together they have a synergistic action, which produces a marked increase in the amount secreted as a result of either alone.

3 The role of the autonomic nerve supply: there is evidence that stimulation of the right vagus nerve increases insulin secretion. This effect is blocked by atropine. Sympathetic stimulation, however, inhibits the release of insulin by means of an α-receptor mechanism.

4 The hormones of the gastrointestinal tract have been found to stimulate the secretion of insulin. Glucagon, gastrin, secretin, CCK-PZ and, in particular, gastric inhibitory peptide (GIP) all cause insulin release. GIP may be the main physiological gastrointestinal hormone responsible for insulin regulation after meals. It stimulates insulin secretion in very small amounts, similar to those normally found after a meal. These hormones potentiate the effect of glucose. Therefore, glucose taken by mouth results in far more insulin secretion than glucose given intravenously, although, in the latter instance, higher levels of blood glucose are attained.

5 The hormone somatostatin, secreted by D cells, inhibits the secretion of insulin. The physiological role of somatostatin, however, is not yet fully understood.

6 Insulin secretion is stimulated by a high level of ketoacids such as acetoacetic acid. This response helps to counteract further production of ketoacids.

7 Many drugs have effects on insulin secretion. Sulphonylureas (e.g. tolbutamide) stimulate insulin secretion. Newer additions, such as glibizide and gliburide, stimulate insulin secretion and promote its action at the cellular level. These drugs have proved useful in treating patients with residual pancreatic insulin-secreting capacity. Biguanides (e.g. phenformin) act by increasing glucose catabolism and decreasing its absorption in the intestine, but they have the tendency to produce lactic acidosis, which is an undesirable side-effect. Thiazide diuretics, however, inhibit insulin secretion. Diazoxide (once used for treatment of hypertension) is a potent inhibitor of insulin secretion.

In summary, the physiological control of insulin secretion depends on the blood glucose and amino acid levels. These are usually elevated after meals and they fall during the postabsorptive phase. Gastrointestinal hormones, somatostatin and autonomic nerves also play a role in the modulation of insulin release. These seem to modify the effects of the substrate levels in the blood.

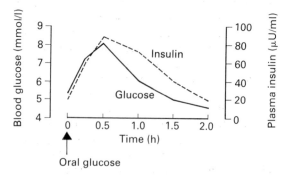

Fig. 10.23 Graph showing the normal pattern of insulin secretion stimulated by oral glucose intake.

The actions of insulin

The actions of insulin can be conveniently discussed in relation to glucose, amino acid and fat metabolism and in relation to electrolyte regulation. Specific insulin receptors have been demonstrated in many tissues, particularly in muscle, adipose tissue and liver. The binding of insulin to receptors on the cell membrane is a prerequisite to its intracellular actions.

Actions on glucose metabolism The strong hypoglycaemic action of insulin is the predominant effect, which results soon after the administration of the hormone. This hypoglycaemia is produced in several ways:

1 Insulin facilitates glucose uptake by many tissues. Glucose transport into skeletal muscle and adipose tissue cells depends upon insulin. Some tissues do not need insulin for glucose entry into their cells, e.g. the liver, brain, kidneys, red blood cells and intestinal mucosa.

2 Insulin increases glycogen synthesis in the liver and in skeletal muscle.

3 Insulin decreases the glucose output from the liver by decreasing gluconeogenesis and by increasing glycogen synthesis.

4 Insulin increases the rate of utilization of glucose by skeletal muscle and by adipose tissues.

Actions on protein metabolism Insulin is an anabolic hormone as far as protein is concerned. The following is a summary of its main actions:

1 It increases the uptake of amino acids by cells. In the liver and skeletal muscles, transport of amino acids into cells is greatly enhanced.

2 The hormone increases the rate of protein synthesis, especially in the liver and skeletal muscle. There is an increase in RNA formation and ribosomal protein synthesis by means of mRNA translation.

3 Protein catabolism is decreased. This leads to a positive nitrogen balance and decreased release of gluconeogenic amino acids.

Actions on fat metabolism Insulin promotes storage of fat and decreases its utilization by the following actions:

1 It increases synthesis of FFAs in adipose tissue cells.

2 It increases the synthesis of glycerol phosphate.

3 It promotes the formation and deposition of triacylglycerols.

4 Insulin inhibits FFA mobilization from adipose tissue and therefore lowers the levels of circulating FFAs. Hormone-sensitive lipases are inhibited.

5 Ketone body formation (ketogenesis) in the liver is decreased and the uptake of ketone bodies by skeletal muscle is increased.

Actions on electrolytes Insulin increases the active transport of sodium and potassium across the cell membrane and there is an increase in the resting membrane potential. In the treatment of diabetic acidosis, insulin causes a net shift of K^+ from the extracellular to the intracellular fluid compartment. This may lead to the development of hypokalaemia with its attendant serious complications.

On the other hand, K^+ depletion has been found to diminish insulin secretion, e.g. the thiazide diuretics may cause K^+ depletion and consequently decrease the secretion of insulin.

Abnormalities of insulin secretion

Diabetes mellitus is a syndrome due to an absolute or a relative deficiency of insulin secretion. *Type I*, or juvenile, diabetes mellitus mainly affects children and adolescents and is due to absolute or severe shortage of insulin. This makes it necessary for the patient to depend on insulin therapy — hence the name insulin-dependent diabetes mellitus (IDDM). *Type II*, or maturity onset, diabetes mellitus, however, is usually of a less severe nature. The patient may require insulin or he/she may not need it if his/her own pancreas can be stimulated by oral hypoglycaemic agents — hence the name non-insulin-dependent diabetes mellitus (NIDDM). Some patients may only require to adjust their diets so that the remaining insulin output is sufficient. The pathophysiology of NIDDM is heterogeneous. It is mainly due to decreased or delayed insulin secretion but the main defect is *insulin resistance*. In mild NIDDM the fault is a decreased number of insulin receptors; in others there is an abnormality in the receptors themselves or in the intracellular events following receptor stimulation. Post-receptor insulin resistance is believed to be the main defect in severe NIDDM. The disease

may also be secondary to other endocrine disorders, such as excess GH, glucocorticoid and
somatostatin secretions, which also cause different forms of insulin resistance.

The diabetic child fails to grow well. This is
due to lack of the anabolic action of insulin
which helps GH to achieve full utilization of the
protein and energy intake. Insulin lack results in
more catabolism of amino acids and decreased
protein synthesis.

Diabetics suffer from persistent high levels of
blood glucose (hyperglycaemia) even during the
postabsorptive and fasting state. This is mainly
due to increased basal and postprandial hepatic
glucose production, associated with a deficient
insulin response and decreased peripheral glucose
utilization. Hyperglycaemia causes several
complications. Glucose appears in the urine
(glucosuria) when the blood glucose rises above
the renal threshold and exceeds the tubular
maximum for its reabsorption. The excretion of
glucose in the urine leads to osmotic diuresis, i.e.
the glucose by its osmotic activity prevents reabsorption of water from the renal tubules. Thus,
a large volume of urine is passed. The diuresis, or
polyuria, leads to dehydration and thirst. The
diabetic patient, therefore, needs to drink frequently. It is rather distressing to wake up several
times at night to pass urine and to drink. Osmotic
diuresis also leads to loss of electrolytes in the
urine.

The metabolic abnormalities of diabetes
mellitus are the result of the inability to utilize
glucose. The fuel for energy alternatively comes
from fat and protein. Fat mobilization results in
loss of weight. The increased catabolism of FFAs
leads to ketone body formation, ketosis, due to
accumulation of acetyl-CoA (Fig. 10.24). In some
NIDDM patients the anticatabolic effects of
insulin are still sufficient to prevent ketoacidosis,
but its anabolic effects fail. The result is excessive
hyperglycaemia and *hyperosmotic coma.*

The breakdown of protein corrects energy
deficits due to loss of glucose in the urine. More
amino acids are used to make glucose in the liver
or catabolized to CO_2 and H_2O. Insulin insufficiency also results in decreased protein
synthesis. The result of these abnormalities is
muscle-wasting and further weight loss.

Electrolyte disturbances also occur in diabetic

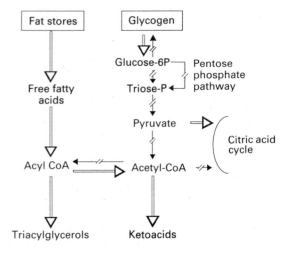

Fig. 10.24 Abnormalities of carbohydrate and fat
metabolism in diabetes mellitus. The double-line
arrows indicate increased formation of intermediates or
end-products. The crossed arrows indicate impaired
metabolic pathways.

patients. Intracellular acidosis causes K^+ to leak
into the extracellular fluid and become lost in the
urine. On administration of insulin, K^+ enters
the cells. Hypokalaemia occurs due to the presence of K^+ depletion. This is a serious effect of
insulin administration in the management of
diabetic acidosis, which may result in cardiac
failure. Therefore, careful administration of K^+
has to be considered in cases of severe acidosis, as
hyperkalaemia can be equally fatal.

Complications of diabetes mellitus Long-
standing diabetes mellitus inevitably leads to
the development of vascular, renal and other
pathologies. Table 10.4 gives a list of commonly encountered pathological changes and
complications.

GLUCAGON
Glucagon is secreted by the A cells of the islets of
Langerhans. In general, it exerts opposite effects
to those of insulin. It plays important roles in the
control of blood glucose and the regulation of fat
and protein metabolism.

Chemistry and degradation of glucagon
Glucagon is a single-chain polypeptide hormone
consisting of 29 amino acids. Its molecular weight

Table 10.4 Some common long-term effects of diabetes mellitus

Vascular pathologies	Eye changes
Atherosclerosis	Diabetic retinopathy
Ischaemic heart disease	Cataract
Cerebral stroke	*Nervous system*
Ischaemia and gangrene of the foot	Peripheral neuropathies
	Autonomic insufficiency, diarrhoea
Microangiopathies	
Diabetic nephropathy	*Infections*
Glomerular damage	Skin infections
Renal failure	Pulmonary tuberculosis
	Urinary tract infections

is about 3500 daltons. In contrast to other peptide hormones, very few species differences are seen in the structure of glucagon. Glucagon-like substances have been identified in the mucosa of the gut. Its half-life in plasma is 3–6 minutes. Appreciable amounts of the glucagon secreted by the pancreas are taken up by the liver. The portal vein usually contains concentrations about twice the glucagon concentration in the peripheral circulation. The kidney also removes and degrades glucagon. It has been observed that the serum glucagon level is increased in patients with renal disease.

Control of secretion

1 Glucagon is mainly regulated by the blood glucose level. Hypoglycaemia results in glucagon secretion and hyperglycaemia inhibits its release.
2 Ingestion of protein or amino acids stimulates secretion of glucagon, which, in turn, promotes hepatic gluconeogenesis. Oral protein intake is a more potent stimulus, probably because of a stimulating action of the gastrointestinal hormones such as CCK-PZ and gastrin. It is noteworthy that protein intake also stimulates insulin release; therefore, glucagon release will help prevent hypoglycaemia after a protein meal.
3 The hormone somatostatin, which is secreted by the D cells of the pancreatic islets, inhibits glucagon secretion as well as insulin.
4 Sympathetic nerves cause release of glucagon. This is a β-receptor response. Its secretion is inhibited by an α-receptor mechanism but, during stress, the β-receptors predominate. Vagal stimulation decreases glucagon secretion. It is possible that this vagal effect is activated by hypo-

glycaemia. Thus, hypoglycaemia not only stimulates the A cells directly but also works via the hypothalamus and relays to give vagal and sympathetic stimulation, both of which increase the release of glucagon.
5 Muscular exercise, pain, trauma, infection and other forms of physical stress increase glucagon secretion.

Actions of glucagon

The main action of glucagon is in the regulation of blood glucose. It safeguards against hypoglycaemia, while insulin prevents hyperglycaemia; however, glucagon has several other important actions:
1 It increases the blood glucose level by stimulating glycogenolysis (breakdown of glycogen) in the liver, but not in muscle. Therefore, its hyperglycaemic action depends on the presence of normal glycogen stores in the liver. Glucagon stimulates adenylate cyclase in liver cells, leading to activation of glycogen phosphorylase, which increases glycogen breakdown.
2 Protein breakdown in muscle is increased and protein synthesis in the liver is decreased. Glucagon increases gluconeogenesis by the conversion of some amino acids, lactate or pyruvate into glucose. It has a calorigenic action, which is probably due to deamination of amino acids.
3 Glucagon stimulates fat mobilization, acting via cyclic AMP. It therefore increases the level of FFAs in the blood and enhances ketoacid formation. It also decreases triacylglycerol release from the liver and suppresses triacylglycerol synthesis in the liver.

Table 10.5 gives a summary of the sites of action and effects of glucagon. The mechanism

Table 10.5 Sites of action and effects of glucagon

Site	Effect
Liver	Increased glycogenolysis
	Increased gluconeogenesis
	Increased lipolysis
	Increased ketogenesis
Adipose tissue	Increased lipolysis
Heart	Increased force of contraction
	Increased cardiac output
Pancreatic B cells	Increased insulin secretion

of action of glucagon at the cellular level is summarized in Fig. 10.25.

Abnormalities of glucagon secretion

There is evidence that glucagon may play an important role in the pathogenesis of diabetes mellitus in some patients with this disease. It has been found that diabetic patients have higher than normal levels of glucagon in spite of the hyperglycaemia. This situation seems to be similar to that due to GH secretion in acromegaly; in both cases hyperglycaemia fails to suppress the release of the hormone.

SOMATOSTATIN

Somatostatin is a cyclic peptide, consisting of 14 amino acids with a molecular weight of 1640 daltons. It is produced by the D cells of the islets of Langerhans and intestinal mucosa. It is also produced in the hypothalamus, where it is identified as the inhibitory hormone of pituitary growth hormone. It has also been found in other parts of the central nervous system.

The control of somatostatin secretion is linked to food intake. Glucose, fat and protein all seem to increase its release. This effect is probably related to a β-adrenergic receptor mechanism. On the other hand, α-receptor stimuli inhibit its secretion. Like other peptide hormones, somatostatin appears to exert its effects through the adenylate cyclase system. However, evidence indicates that somatostatin effects on calcium permeability in target cells may be important in its mechanism of action.

Its physiological role in the inhibition of GH release is established. It also inhibits both insulin and glucagon secretion, but the physiological significance of this effect is not yet clear. However, there is evidence to indicate that somatostatin may be involved in the pathophysiology of diabetes mellitus and obesity.

Somatostatin is secreted by gastrointestinal mucosa, where it has been shown to decrease HCl secretion by inhibiting gastrin release. It has also been found that somatostatin inhibits the secretion of other gut hormones such as VIP and GIP.

The overall effects of somatostatin on gastrointestinal function are to decrease nutrient absorption as well as to inhibit digestion of nutrients. Motility of the stomach and small intestine is inhibited by somatostatin, and splanchnic blood flow is decreased.

Abnormalities

Tumours which secrete somatostatin give rise to hyperglycaemia and other features of diabetes mellitus. This condition is also associated with stagnation of bile and gallstone formation due to decreased gall-bladder contraction. Gastric motility and emptying is also decreased.

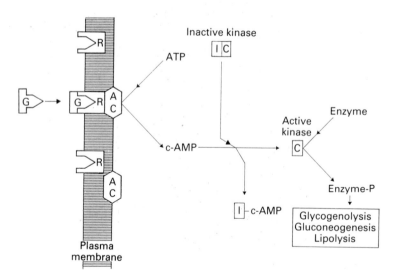

Fig. 10.25 The mechanism of glucagon action on a target cell. AC, adenylate cyclase; G, glucagon; I, inactive component; R, receptor.

The adrenal gland

The adrenal gland is composed of two distinct glands, cortex and medulla, which in lower animals are anatomically separate. In spite of their close association in mammals, they are still totally different in development, structure and function.

DEVELOPMENT

The adrenal cortex, which forms the outer part of the adrenal gland, develops from coelomic epithelium, while the medulla, which forms the central part of the gland, is of ectodermal origin, derived from the neural crest. The adrenal medulla, on the other hand, can be considered as a modified sympathetic ganglion, which receives preganglionic sympathetic fibres.

The adrenal cortex

STRUCTURE

The adrenal cortex is composed of large cells rich in cholesterol. The cells are arranged in three distinct layers (Fig. 10.26). The superficial layer lying immediately under the capsule is a thin layer of cells arranged in irregular groups and is referred to as the zona glomerulosa. It secretes the mineralocorticoids. The middle layer (zona fasciculata), which forms the bulk of the gland, is formed of long radial columns of cells. The inner layer lying next to the medulla (zona reticularis) is formed of a reticulum of branching and interconnected columns of cells. The two inner zones secrete the glucocorticoids and the adrenal sex hormones. The zona glomerulosa is responsible for regeneration of the adrenal cortical cells.

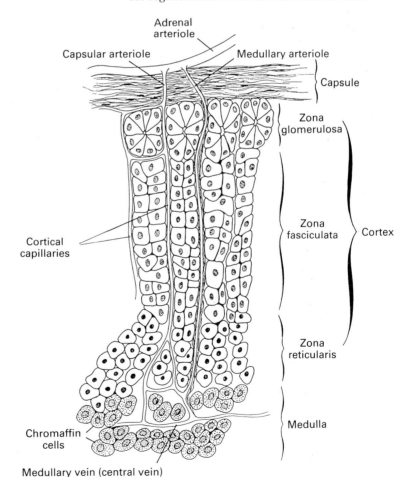

Adrenal arteriole
Capsular arteriole
Medullary arteriole
Capsule
Zona glomerulosa
Cortical capillaries
Zona fasciculata
Cortex
Zona reticularis
Medulla
Chromaffin cells
Medullary vein (central vein)

Fig. 10.26 Cell layers of the adrenal cortex. (Reproduced from Borysenko, M. & Beringer, T. (1989) *Functional Histology*, 3rd edn, with permission from Little Brown & Co., Boston.)

The blood supply of the adrenal cortex deserves special attention. It receives three arteries, which enter its capsule and form two plexuses within it. One plexus is subcapsular and the other is deeply situated in the zona reticularis. The latter plexus gives rise to venules which penetrate into the adrenal medulla and drain into the medullary sinusoids. Thus, adrenal blood rich with glucocorticoids reaches the medulla and induces the synthesis of phenylethanolamine-N-methyl transferase, the enzyme required for adrenaline synthesis. However, the adrenal medulla also receives a direct blood supply from branches of suprarenal arteries, which pass through the cortex without giving branches.

HORMONES OF THE ADRENAL CORTEX

The adrenal cortex is essential to life. Destruction of the glands is fatal unless replacement therapy is given. The adrenal cortex secretes steroid hormones while the medulla produces catechol-amines. The cortex produces three main groups of steroid hormones, categorized according to their actions. The glucocorticoids, mainly cortisol, are concerned with regulation of metabolism. The mineralocorticoids, mainly aldosterone, are concerned with electrolyte regulation. The sex hormones are mainly androgens and some oestrogens. Progesterone is found in the adrenal cortex as an intermediate product of synthesis of glucocorticoids and mineralocorticoids.

Chemistry and metabolism of corticosteroids

Adrenal cortical hormones and gonadal sex hormones are synthesized from either acetate or cholesterol and proceed via various enzymatic steps (see Fig. 10.27). The precise order in which these enzymatic actions take place is not known for certain. The molecular structure of the hormones is characterized by the cyclopentanoperhydrophenanthrene nucleus. C_{21} and C_{19} steroids contain 21 and 19 carbon atoms in their

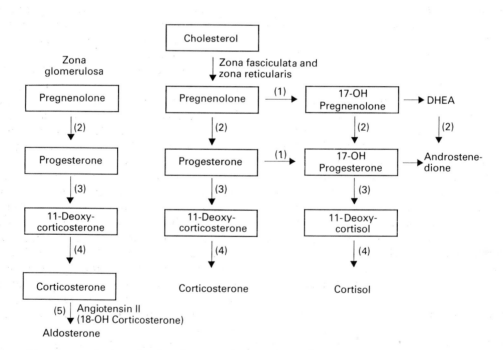

Fig. 10.27 Hormone synthesis in the adrenal cortex. The zona glomerulosa cannot convert corticosterone to cortisol. The main hormones secreted are shown in capital letters. The numbers 1–5 indicate the enzymes responsible for hormone synthesis: (1) 17α-hydroxylase (lacking in zona glomerulosa); (2) 3β-dehydrogenase; (3) 21β-hydroxylase; (4) 11β-hydroxylase; (5) corticosterone methyloxidase.

molecules, respectively. The gluco- and mineralo-corticoids are C_{21} steroids and androgens, such as dehydroepiandrosterone (DHEA), are C_{19} steroids. Figure 10.27 shows the pathways and enzymes for synthesis of the corticosteroids.

Cortisol in plasma is 75% bound to transcortin, an α-globulin, also known as corticosteroid-binding globulin (CBG). This binding makes the half-life of cortisol rather long (about 90 min) and little cortisol appears in the urine. Small amounts (15%) are loosely bound to albumin and the remaining 10%, which is metabolically active, circulates in the free form. Cortisol is catabolized mainly in the liver by reduction and conjugation with glucuronic acid. It is also converted to corti-sone, which has strong glucocorticoid activity but is physiologically insignificant because it does not reach the circulation. It is rapidly conjugated in the liver. The soluble conjugated products of glucocorticoids go into the circulation and are excreted in the urine.

Aldosterone synthesis occurs in the zona glomerulosa region of the adrenal cortex. Progesterone is hydroxylated at the C_{21} position to form 11-deoxycorticosterone (DO^{21}), which is then hydroxylated at C_{11} to produce cortico-sterone. The conversion of the C_{18} methyl group of corticosterone to an aldehyde group leads to the formation of aldosterone (see Fig. 10.27). Aldosterone circulates in plasma weakly bound to albumin and more tightly bound to transcortin and a specific binding protein, aldosterone-binding globulin. Its half-life is about 20 minutes. The principal site of catabolism is the liver, where over 90% of aldosterone is cleared from the blood during a single passage. Measurements of 'urinary aldosterone' usually refers to conjugated aldosterone rather than to aldosterone itself, which is excreted in small quantities.

DHEA and androstenedione (adrenal androgens) have a more complex metabolism. DHEA is converted to androstenedione in the circulation. Androstenedione is also a precursor of testosterone and oestrogens. It is a major source of testosterone in young females. In menopausal women, it is the main source of oestrogens. DHEA and DHEA-sulphate, secreted by the adrenal cortex, are conjugated in the liver and excreted in urine.

Control of cortisol secretion

Secretion of cortisol is controlled by the hypo-thalamus via its corticotrophin-releasing hormone (CRH) and the adenohypophysis through adrenocorticotrophic hormone (ACTH). A negative feedback mechanism between cortisol and ACTH operates via the hypothalamus (see also Fig. 10.4). There is a distinct diurnal variation in the release of cortisol, being highest in the early hours of the morning and lowest late at night (Fig. 10.28). Characteristically, the diurnal variation disappears with disturbance of adrenal cortical function, especially with hyperactivity.

ACTH and cortisol are released in response to injury and other stressful stimuli, as mentioned above.

Actions of cortisol

Like other steroid hormones, cortisol enters the cell and induces protein enzyme synthesis. The hormone binds first to a cytoplasmic receptor. The receptor-bound steroid enters the nucleus, where transcription of mRNAs is increased. This results in the synthesis of the specific enzymes responsible for the numerous actions of cortisol in various tissues:

1 It is a *hyperglycaemic* hormone. It stimulates *gluconeogenesis*, i.e. the formation of glucose from non-carbohydrate precursors. This may be a direct effect but it is mainly due to facilitation of the effect of glucagon. This is known as the permissive role of glucocorticoids. More glucose is also released from the liver.

Cortisol opposes the peripheral effects of insulin. It is one of the hormones known as insulin antagonists. The brain and the heart are not affected by the anti-insulin effect and therefore make use of the increased glucose supply. Overproduction of cortisol (Cushing's disease) leads to hyperglycaemia. The continuous stimulation of insulin production may ultimately lead to exhaustion of the B cells of the islets of Langerhans and result in frank diabetes mellitus (adrenal diabetes).

2 Protein catabolism: the gluconeogenic action of cortisol promotes protein breakdown in muscle, skin, connective tissue and bones, especially in the vertebrae and ribs. This leads to retardation of growth, muscle-wasting, thinning of the skin and subcutaneous tissues and loss of connective tissue, especially in bone.

Although cortisol has an overall protein catabolic action, in the liver it increases protein synthesis.

3 Fat metabolism: cortisol mobilizes FFAs from adipose tissue and increases ketone-body formation in the liver. Excess of cortisol leads to abnormal fat redistribution. It causes mobilization of fat from certain parts and deposition in others. Fat is deposited mainly in the trunk (truncal distribution) and removed from the limbs.

4 Effects on blood-vessels and water metabolism: cortisol makes blood-vessels more sensitive to catecholamines; it potentiates the pressor effect of adrenaline and noradrenaline on blood-vessels. It may therefore increase the effective glomerular filtration pressure, glomerular filtration rate (GFR) and water excretion. Patients with adrenal insufficiency cannot excrete a water load rapidly. There is usually a high level of ADH in these patients and a low GFR, both of which contribute to the above defect.

Glucocorticoids also increase the entry of tissue fluid into the capillaries, thus increasing plasma volume. Together with their potentiation of catecholamines, this effect is an important part of the stress responses which stimulate ACTH release. Both effects help to return the blood-pressure back to normal.

5 Cortisol has a mild mineralocorticoid action in physiological doses, but in large quantities, as in adrenocortical overactivity, this effect may be significantly high.

6 The central nervous system requires an optimum supply of cortisol for normal function. Both excess and deficiency of cortisol may result in abnormal brain function, as shown by electro-encephalogram (EEG) changes. Many patients with excess cortisol production (Cushing's syndrome) present with psychiatric manifestations and are mistakenly admitted to psychiatric hospitals.

7 Cortisol is anti-inflammatory and antiallergic. It suppresses or decreases inflammatory reactions and combats allergic reactions by inhibition of histamine release. Both of these effects are not seen with physiological amounts of the hormone but can be effectively demonstrated with therapeutic doses.

In therapeutic doses it destroys fixed lymphoid tissues and decreases formation of antibodies. Thus, it suppresses the immune response; hence its use in organ transplantation to prevent graft rejection. The patient, however, becomes more susceptible to infection.

Cortisol stimulates erythropoiesis and increases neutrophil and platelet counts, while it decreases the lymphocyte, eosinophil and basophil counts.

Control of aldosterone secretion

Aldosterone is mainly regulated via the renin–angiotensin system (see Fig. 11.10). ACTH plays a minor role in its control. The following are the main factors which stimulate aldosterone release:

1 The renin–angiotensin system: renin is an enzyme released by the juxtaglomerular cells of the kidney glomeruli and regulates aldosterone secretion through the action of angiotensin II. It acts on a plasma globulin (angiotensinogen) to convert it to angiotensin I. Converting enzymes, mainly formed in the lungs, convert angiotensin I into angiotensin II. Angiotensin II stimulates the zona glomerulosa of the adrenal cortex to release aldosterone. Several physiological mechanisms activate the renin–angiotensin system:

(a) Sodium depletion: low levels of sodium in the distal tubules result in stimulation of the juxtaglomerular apparatus to release renin.

(b) Low renal blood flow: renal ischaemia due to low blood-pressure, as in hypovolaemia due to haemorrhage or dehydration, results in the release of renin. The juxtaglomerular cells, in this case, act as baroreceptors.

(c) There is an increase in renin and therefore aldosterone secretion in the upright position.

(d) In the recumbent position, e.g. patients in bed, there is a diurnal rhythm, with a high level in the early hours of the morning.

2 A small increase in plasma K^+ (1 mmol/litre) exerts a direct stimulating effect on the adrenal cortex to release aldosterone, while depletion of body K^+ is associated with decreased secretion. This regulatory action by K^+ appears to occur at the zona glomerulosa cell and appears to act independently of the renin–angiotensin system,

although some circulating angiotensin II is required for the stimulatory action of hyperkalaemia. It is unclear whether it is the K^+ concentration within the zona glomerulosa cell or the circulating K^+ level that affects aldosterone output.

3 Several stimuli, such as trauma, surgery and anxiety, release aldosterone and cortisol from the adrenal cortex. This effect is due to secretion of ACTH from the anterior pituitary. Haemorrhage and hypovolaemia, on the other hand, do not depend on ACTH in stimulating aldosterone secretion; they act through the renin–angiotensin mechanism.

Actions of aldosterone

The main action of aldosterone is to conserve sodium and therefore maintain blood volume. In its absence, sodium is rapidly lost from the body, leading to fatal sodium depletion within a few days. The exact mechanism of its action is not fully understood but it seems to exert several intracellular effects on the renal tubular cells. Na^+ is actively absorbed in the proximal tubules of the kidneys. This is largely independent of aldosterone. The adjustment of Na^+ reabsorption takes place in the distal tubules, where the effect of aldosterone is exerted. The mechanism of action of aldosterone depends on the hormone entering the cell and attaching itself to a cytoplasmic receptor. The hormone–receptor complex migrates to the nucleus and synthesis of mRNA is stimulated. This leads to increased ribosomal protein synthesis. The way in which protein synthesis is related to Na^+ reabsorption is not fully understood. However, the action of aldosterone may be produced by one or more of the following mechanisms: (i) provision of energy for ATP synthesis; (ii) a direct increase of active transport of Na^+ (sodium pump); and (iii) a passive increase in the permeability of the cell membrane to sodium. The end result is an active transport of Na^+ from the tubular lumen to the renal interstitial fluid and consequently to the blood. The protein synthesis explains the delay of about 10–30 min before an administered dose shows its effect.

Common abnormalities of adrenocortical function

Adrenocortical overactivity (Cushing's syndrome) Excessive adrenocortical hormone secretion in about 75% of cases result from abnormal ACTH production (secondary Cushing's syndrome, bilateral adrenal hyperplasia). Excess ACTH is either produced by an adenoma of the anterior pituitary or from an extrapituitary origin such as bronchogenic carcinoma. About 25% of cases are due to a secreting tumour of the adrenal cortex (primary Cushing's syndrome). In this case the ACTH will be low, due to negative feedback by cortisol.

Cortisol and androgens are the two main hormones that are produced in excess. Aldosterone excess is less evident as it is mildly influenced by ACTH. The clinical features of Cushing's syndrome are therefore the result of excess cortisol and excess androgen production. Figure 10.29 shows a typical patient with Cushing's syndrome,

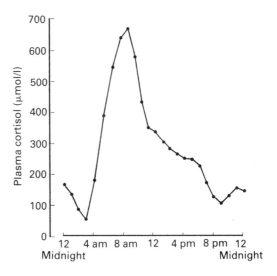

Fig. 10.28 Diurnal variation in plasma cortisol levels. It depends on rhythmic release of CRH and ACTH and gives an example of circadian rhythms. The diurnal rhythm is reversed in persons on long-term night-work. (Reproduced with permission from Emslie-Smith, D., Paterson, C.R., Scratcherd, T. & Read, N.W. (eds) (1988) *BDS Textbook of Physiology*, 11th edn. Churchill Livingstone, Edinburgh.)

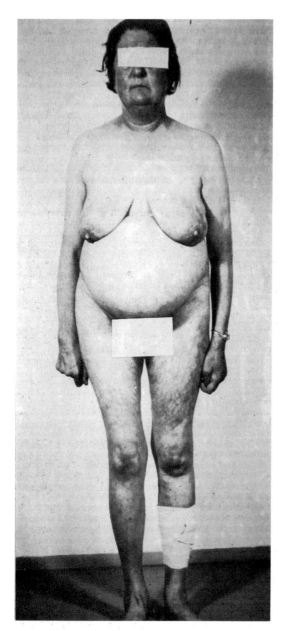

Fig. 10.29 A patient with Cushing's syndrome. (Reproduced with permission from Burke, C.W. (1987) Adrenocortical disease. In Wetheral, D.J., Ledingham, J.G.G. & Warrell, D.A. (eds) *Oxford Textbook of Medicine*, 2nd edn. Oxford University Press, Oxford.)

with central obesity, 'moon' face and thin extremities.

The excess glucocorticoid effects of cortisol result in the following:

1 Increased hepatic gluconeogenesis, leading to hyperglycaemia: this effect depends on whether or not adequate insulin is available. If there is adequate insulin, fat synthesis will increase, leading to obesity, which is a cardinal sign of Cushing's syndrome. If there is an inadequate supply of insulin, hyperglycaemia and the inhibitory effect of cortisol on peripheral glucose uptake will lead to glucosuria and ultimately to frank insulin-resistant diabetes mellitus (adrenal diabetes) with keto-acidosis.

2 The obesity of Cushing's syndrome has a characteristic distribution, which is truncal or centripetal. Fat is mainly deposited in the anterior abdominal wall, causing abdominal protrusion, in the supraclavicular region, giving the patient a hump (buffalo hump), and in the face, which becomes round (moon face). In contrast, the limbs are thin because of reduced fat content and muscle-wasting.

3 Protein depletion, due to protein catabolism in the skin and connective tissue, leads to the following:

  (a) The skin becomes thin and paper-like.

  (b) Purple striae appear, mainly in the abdomen, buttocks, thighs and shoulders, due to stretching of the skin.

  (c) The face, especially in fair-coloured patients, becomes red, as the capillaries start to show more clearly through the thin skin (plethoric face). Patients with Cushing's syndrome have a greater tendency to develop polycythaemia, which adds to the redness of the face.

  (d) Because the skin is thin and capillaries are weak, the patients easily bruise with the slightest knock and may develop petechial haemorrhages.

4 Removal of protein from bone: excess cortisol leads to removal of protein matrix from bone (osteoporosis). It is more evident in vertebral bodies, which tend to collapse, giving rise to back pain and deformity (kyphosis). Osteoporosis is also quite common in ribs, which become weak and easily fractured, commonly following a spasm of coughing (cough fracture), giving rise to the characteristic pain of fractured ribs.

5 Cardiovascular effects: patients with Cushing's syndrome often have hypertension. This is due to an increase in the peripheral resistance, due to potentiation of the catecholamines. There is also an increase in the cardiac output, due to a direct effect of glucocorticoids on the strength of cardiac muscle contraction. Another factor which leads to high blood-pressure is retention of sodium and water, as a result of the weak mineralocorticoid action of cortisol.

6 Electrolyte disturbances: there is sodium retention together with potassium loss. The wasting of muscle coupled with potassium depletion results in muscle weakness, which is a common feature of Cushing's syndrome. However, these electrolyte disturbances are not as severe as in the case of hyperaldosteronism.

7 Mental symptoms: about 60% of patients develop mental symptoms, varying from mild depression to frank psychosis, and they may present initially to the psychiatrist.

8 Anti-inflammatory responses: patients with Cushing's syndrome have poor wound-healing, e.g. after surgery, and are more liable to develop infection.

9 Features of excessive production of androgens are amenorrhoea and hirsutism in the female. The affected female may grow a beard and moustache. Acne is a common feature of excessive androgen activity. Male patients may become impotent. This is difficult to explain.

Excessive aldosterone production
1 Primary hyperaldosteronism (Conn's syndrome) is usually due to a secreting tumour affecting the zona glomerulosa cells of the adrenal cortex. Its main features are persistent hypertension, headache, hypokalaemia, alkalaemia and severe muscle weakness. The increased plasma pH leads to decreased Ca^{2+}, which may cause latent or overt tetany. Oedema is uncommon. There is increased aldosterone excretion in urine.
2 Excessive aldosterone production can also occur in disturbances of the renin–angiotensin system, as in congestive heart failure, liver cirrhosis, nephrosis and toxaemia of pregnancy. This is known as secondary hyperaldosteronism.

Adrenocortical insufficiency (Addison's disease)
This may be secondary to deficiency of ACTH production by the pituitary hypothalamic axis, or it may be primary, due to destruction of the adrenal gland by an infection such as tuberculosis or due to an autoimmune process. In the secondary type there is a high level of plasma ACTH, due to the absence of inhibitory feedback from adrenal steroids. The main features of Addison's disease are:

1 Hypovolaemia and hypotension, due to loss of sodium and water as a result of absence of aldosterone. Diminished sensitivity to catecholamines and reduced cardiac work, due to decreased glucocorticoids, also lead to lowering of blood-pressure.

2 Hypoglycaemia, due to loss of the hyperglycaemic action of cortisol and to increased sensitivity to insulin.

3 Hyperpigmentation, seen mainly in mucous membranes of the mouth, pressure points like the elbows, old scars and areas which are normally pigmented, such as the nipples, axillae and skin creases. Hyperpigmentation, due to excessive production of ACTH and melanocyte-stimulating hormone, is a feature in the primary form of the disease. Pigmentation is not seen in the secondary type because of lack of ACTH production.

4 Loss of appetite, vomiting, wasting and weakness. These symptoms are probably due to the water and electrolyte disturbances and to the hypoglycaemia.

All the above symptoms and signs may be suddenly exaggerated, leading to shock or even death, if the patient is subjected to stress or infection. The condition is referred to as adrenal or Addisonian crisis, which calls for immediate attention.

Adrenogenital syndrome Abnormal secretion of adrenal androgens, especially DHEA, occurs in adrenal ademona or in cases of congenital deficiency of an enzyme responsible for steroid hormone synthesis. The excess androgens result in precocious puberty in a male child or virilization in a female child or female adult. In a male adult, the effects are masked by normal male characteristics. Table 10.6 gives a summary of the clinical manifestations of enzyme deficiencies. These congenital deficiencies are rare; the most common of them is 21β-hydroxylase deficiency, which leads to one type of virilizing hyperplasia of the adrenal cortex.

Failure of foetal androgen synthesis or androgen excess during pregnancy causes abnormal devel-

Table 10.6 Defects of corticosteroid synthesis due to congenital enzyme deficiencies (also refer to Fig. 10.27)

Enzyme	Hormonal changes	Clinical abnormalities
17α-Hydroxylase	Failure of cortisol and sex hormone formation. Excess ACTH. High levels of corticosterone	Congenital pseudohermaphroditism in a genetic male
3β-Hydroxysteroid dehydrogenase	Gluco- and mineralocorticoid deficiency. Excess DHEA	DHEA has weak virilizing effect. Hypospadias
21β-Hydroxylase (commonest defect)	This enzyme deficiency is usually not absolute. Some gluco- and mineralocorticoids are synthesized. Excess androgens and ACTH	Virilizing hyperplasia of adrenal cortex (adrenogenital syndrome in females). Sodium loss
11β-Hydroxylase	Excess deoxycorticosterone and 11-deoxycorticosterone. Excess ACTH	Virilizing hyperplasia of adrenal cortex. Sodium retention and hypertension

opment of the external genitalia (pseudo-hermaphroditism), i.e. external genitalia different from genetic sex.

The adrenal medulla

Since the adrenal medulla represents a modified sympathetic ganglion, its function closely resembles that of the sympathetic nervous system. Its hormones, as in the case of postganglionic adrenergic neurones, are catecholamines. The only difference is that the adrenal medulla produces adrenaline from noradrenaline. The ratio of secreted adrenaline to noradrenaline is about 4 : 1. This is because the adrenal medullary cells have the enzyme system responsible for N-methylation (phenylethanolamine-N-transferase), which converts noradrenaline into adrenaline. The synthesis of this enzyme requires a high concentration of glucocorticoids, which is supplied through the arterial blood from the adrenal cortex (see Fig. 10.26).

The gland is stimulated by a nervous mechanism as part of the sympathetic nervous system. The medullary cells are supplied by preganglionic cholinergic sympathetic fibres. Stressful stimuli, such as hypoglycaemia, hypotension, haemorrhage, infection, exercise, fear, anxiety and emotion, are the main stimuli for the release of catecholamines. This is reflected in the widespread effects of catecholamines in the body. Figure 10.30 shows catecholamine synthesis.

ACTIONS OF CATECHOLAMINES

The effects of catecholamines are determined by adrenergic receptors situated in the target organs. There are two receptors for catecholamine responses: alpha (α) and beta (β). Alpha receptors mainly exhibit excitatory functions, such as contraction of smooth muscle, a notable example of which is vasoconstriction. Beta receptors can be subdivided into β_1- and β_2-receptors. $Beta_1$ receptors are found in the heart. Their stimulation accelerates the heart and increases its force of contraction (i.e. chronotropic and inotropic effects) and produces coronary dilatation. $Beta_2$ receptors are found in bronchial muscle and in other tissues. Their stimulation leads to inhibition of smooth muscles, e.g. vasodilatation and intestinal relaxation.

Noradrenaline acts almost entirely on α-receptors and has minor effects on β-receptors, except in the heart, while adrenaline acts on both α- and β-receptors — hence it may cause smooth muscle contraction in some sites and relaxation

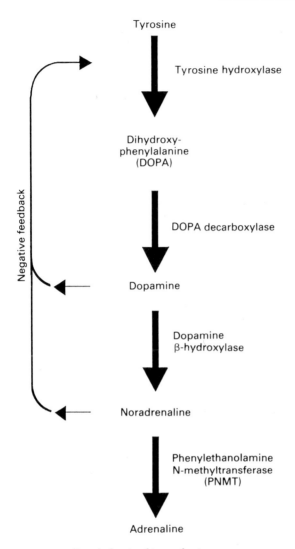

Tyrosine

Tyrosine hydroxylase

Dihydroxy-
phenylalanine
(DOPA)

DOPA decarboxylase

Dopamine

Dopamine
β-hydroxylase

Noradrenaline

Phenylethanolamine
N-methyltransferase
(PNMT)

Adrenaline

Negative feedback

Fig. 10.30 Catecholamine biosynthesis.

in others. Several drugs may be used to block selectively α- or β-receptors. For example, propranolol and practolol are β-blockers, while ergotamine, phentolamine and phenoxybenzamine are α-blockers. Atenolol is a selective β_1-blocker and is therefore used in ischaemic heart disease. A β_2-receptor agonist (salbutamol) is used in the treatment of bronchial asthma, without producing the unwanted effects of adrenaline on other receptors.

The main function of catecholamines is to prepare the body for an emergency situation –

known as the fight-or-flight reaction. Emergency and stressful situations, such as pain, hypoglycaemia, hypotension, emotional stress and muscular exercise, bring about the release of the hormones of the adrenal medulla. For this reason, the effects of catecholamines are varied and complex:

1 *Metabolic effects.* The general metabolic rate and oxygen consumption are significantly increased by catecholamines in a similar manner to that of a sympathetic stimulation. Metabolic rate may go up to 100% above normal. This is known as the calorigenic effect of catecholamines:

(a) *Carbohydrate metabolism.* Catecholamines and, in particular, adrenaline, which is about 10 times more active than noradrenaline, are glycogenolytic. They activate glycogen phosphorylase b to the active form a. In the liver, glycogen is ultimately converted to glucose via a cascade of reactions. In muscle, glycogen will only be converted to lactic acid because of very low or absent activity of glucose-6-phosphatase.

(b) *Fat metabolism.* Adrenaline stimulates lipolysis in adipose tissue, releasing FFAs and glycerol. FFAs provide an important fuel to the active muscle. Adrenaline has a direct α-receptor inhibitory action on the release of insulin by the B cells of the pancreas.

The consequences of these effects of adrenaline are to provide FFAs, which are the main fuel for muscles, and to raise the blood sugar and decrease the uptake of glucose by muscles by inhibiting the secretion of insulin. This will make glucose more available for the central nervous system, which explains its usefulness in response to hypoglycaemia.

2 *Cardiovascular effects.* Both chronotropic and inotropic effects on the heart are stimulated by catecholamines; the heart accelerates and contracts more forcibly. Most of the blood-vessels are constricted. The exceptions are vessels of the skeletal muscles, which are dilated by adrenaline through stimulation of α-receptors. Thus adrenaline increases skeletal muscle blood flow. Coronary vessels are also dilated by both adrenaline and noradrenaline.

3 *Effects on gastrointestinal and urinary tracts.* Both adrenaline and noradrenaline lead to relaxation of gut musculature and contraction of sphincters. Similar effects occur in the urinary bladder. The detrusor muscle of the bladder relaxes and the internal sphincter constricts.

4 *Effects on respiration.* Both adrenaline and noradrenaline increase pulmonary ventilation by increasing rate and depth of breathing and by broncheolar dilatation. However, if there is considerable rise of blood-pressure, inhibitory impulses from the carotid and aortic baroreceptors may lead to cessation of breathing (adrenaline apnoea).

ABNORMAL CATECHOLAMINE SECRETION

Phaeochromocytoma is a tumour of the chromaffin tissue of the adrenal medulla which secretes adrenaline and noradrenaline. The patients usually present with episodes of hypertension. When diagnosed, this type of hypertension is curable.

The gastrointestinal hormones

The gastrointestinal tract contains cells which belong to the APUD system (amine precursor uptake and decarboxylation). These cells originate from the embryonic neural crest. APUD cells are widely scattered in the body, especially in the central and peripheral nervous system. Several gastrointestinal hormones are well known for their effects on motility and secretions of the gastrointestinal tract. In the nervous system they act as neurotransmitters and parahormones.

Chemically, these hormones are polypeptides, which can be grouped into two main groups. Those in one group are similar in structure to gastrin, such as CCK-PZ, and those in the other group are similar to secretin, such as glucagon, GIP (gastric inhibitory peptide) and VIP (vasoactive intestinal peptide). In general, gastrointestinal hormones are found with different lengths of the polypeptide chain (macroheterogeneity) and they may also exhibit some minor differences in the amino acid constitution of the chain (microheterogeneity).

The secretion and effects of these hormones are discussed in Chapter 9. Abnormalities of gastrointestinal hormones are rare, the best known being excess of gastrin and VIP due to tumours (gastrinoma and vipoma, respectively). The effects of these hypersecreting tumours are also discussed in Chapter 9.

Prostaglandins, thromboxanes, prostacyclin and leucotrienes

Prostaglandins are considered as local hormones which are produced under a variety of physiological conditions in almost all tissues of the body. They are closely related to thromboxanes and leucotrienes because they are all synthesized from essential fatty acids. These local hormones have a short half-life and are rapidly inactivated in all tissues, especially the lung and the liver.

The prostaglandins received this name because they were first discovered in the seminal fluid and were thought to be produced by the prostate. Seminal prostaglandins are actually produced by the seminal vesicles. They are synthesized from unsaturated fatty acids, which are incorporated in cell membranes and can be released by the action of an enzyme (phospholipase A_2). Figure 10.31 gives an outline of the synthesis of the above local hormones. The leucotrienes shown in this diagram are aminolipids because they incorporate amino acids in their molecular structure. Thromboxane A_2 is synthesized by the platelets and prostacyclin is synthesized in blood-vessel walls.

RELEASE AND EFFECTS

Prostaglandins

Prostaglandins are found in many tissues but the regulation of their release is not certain. Their effects are as widespread as the tissues in which they are formed. They have been identified in the hypothalamus and may play a role in the regulation of the pituitary gland by hypothalamic hormones. In the stomach they inhibit gastric acid secretion and may play a role in the prevention of gastric ulceration. They have well-known effects in female reproduction. In very small doses they can induce abortion or labour. They may also play a role in the menstrual cycle by causing luteolysis towards the end of a normal cycle. In the kidney they stimulate renin secretion.

Prostaglandins appear in inflammation and may

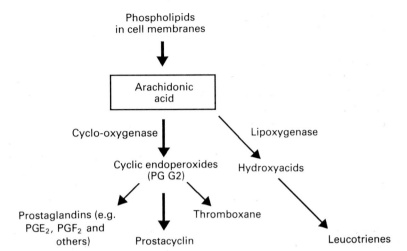

Fig. 10.31 Synthesis of prostaglandins, prostacyclin, thromboxane and leucotrienes.

play a role in the signs and symptoms of inflammation, and their effect in the hypothalamus may produce fever.

Thromboxanes

Thromboxane A_2 is synthesized by the blood platelets. Its release at the site of injury promotes blood clotting by causing vasoconstriction and platelet aggregation. Thromboxane B_2 is a breakdown product of thromboxane A_2.

Prostacyclin

Prostacyclin is produced by endothelial cells and vascular smooth muscle. It has the opposite effects to thromboxane A_2. It causes vasodilatation and inhibits platelet aggregation.

Leucotrienes

Leucotrienes are released in response to allergens acting on most cells. Their effects are those observed in allergic reactions, namely bronchoconstriction, as in asthma, and vascular permeability, as in skin and epithelial membrane allergies. They accumulate in inflamed joints, as in rheumatoid arthritis. Leucotrienes cause high concentrations of eosinophils and neutrophils in inflamed tissues.

APPLIED PHYSIOLOGY

Although knowledge of the physiology of prostaglandins is rather elementary, they have already proved their value in clinical practice. The following are some clinical applications based on knowledge of the biosynthesis and the effects of prostaglandins:

1 Cortisol and other steroids inhibit the release of arachidonic acid from cell membrane phospholipids. Thus, as anti-inflammatory agents, steroids have a rather broad inhibitory action on the synthesis of prostaglandins, thromboxane, prostacyclin and leucotrienes. Therefore, it is used in severe inflammatory conditions.

2 Aspirin in small doses (about 300 mg/day) inhibits thromboxane synthesis but not prostacyclin synthesis. It blocks cyclo-oxygenase in platelets but not in vessel walls. This results in a decrease in platelet aggregation and a decrease in the risk of intravascular clotting—hence its use in patients with coronary heart disease.

3 Indomethacin is a more selective, nonsteroidal, anti-inflammatory drug. It inhibits cyclo-oxygenase but not lipoxygenase and therefore it spares the leucotriene synthesis pathway.

4 PGE_2 is used to induce abortion so as to terminate pregnancy when a valid indication is established.

5 Pain during the menstrual period (dysmenorrhoea) has been found to be due to increased concentrations of prostaglandins in the uterus. Inhibitors of prostaglandin synthesis have been found useful in treatment of dysmenorrhoea.

6 Prostacyclin, which has a short half-life (3 min),

is used as an inhibitor of platelet aggregation in procedures such as renal dialysis and cardiopulmonary bypass operations.

7 Prostaglandins are believed to increase gastric mucosal resistance to acid. Analogues of prostaglandins may prove useful in the treatment of peptic ulcers.

Further reading

1 Emslie-Smith, D., Paterson, C.R., Scratcherd, T. & Read, N.W. (eds) (1988) The hypothalamus and the pituitary gland. In *BDS Textbook of Physiology*, 11th edn, pp. 297–309. Churchill Livingstone, Edinburgh.

2 Ganong, W.F. (1991) Endocrine functions of the pancreas and the regulation of carbohydrate metabolism. In *Review of Medical Physiology*, 15th edn, pp. 312–333. Appleton & Lange, San Mateo, California.

3 Ganong, W.F. (1991) Hormonal control of calcium metabolism. In *Review of Medical Physiology*, 15th edn, pp. 296–311. Appleton & Lange, San Mateo, California.

4 Ganong, W.F. (1991) The thyroid gland. In *Review of Medical Physiology*, 15th edn, pp. 296–311. Appleton & Lange, San Mateo, California.

5 Guyton, A.C. (1991) The andrenocortical hormones. In *Textbook of Medical Physiology*, 8th edn, pp. 842–854. Saunders, Philadelphia.

11: The Urinary System

Introduction

The kidneys play a major role in the control of the constancy of the internal environment. The blood flowing in the kidneys is first filtered (glomerular filtration) so that all the blood constituents, except blood cells and plasma proteins, go into the microtubular system. In these tubules, modi-

---

Objectives

On completion of the study of this chapter, the student should be able to:

1 Describe the structure of the nephron, which represents the functional unit of the kidney.

2 Explain the basis of glomerular filtration so as to show understanding of the role of the various factors involved.

3 Describe the mechanisms of renal tubular reabsorption and secretion in order to recognize disturbances of tubular function.

4 Explain the renal mechanisms for regulation of body water and electrolytes.

5 Describe the endocrine functions of the kidney.

---

fications of the filtrate takes place so that useful substances, including most of the filtered water, are quickly reabsorbed (tubular reabsorption) back into the blood. Unwanted substances that escape filtration are actively secreted into the tubular lumen (tubular secretion). The final concentration of electrolytes and other constituents of urine is adjusted according to the requirements of the regulation of the extracellular fluid composition. Glomerular filtration, tubular reabsorption and tubular secretion are rightly described as renal mechanisms that allow the kidney to undertake its various homoeostatic functions.

Several hormones (especially ADH and aldosterone) act on the kidney to enable it to adjust the final composition of urine in response to changes in the internal environment.

The nephron

Figure 11.1 gives the gross anatomy of the kidney, showing the cortex, medulla and renal calyces.

Nephrons are the functioning units of the kidney (Fig. 11.2). There are about 1 million nephrons in each kidney. The glomerulus (200 μm in diameter) is a tuft of capillaries covered by a fibrous tissue capsule (Bowman's capsule). The renal tubular system starts at the glomerulus. The capillaries receive their blood supply from

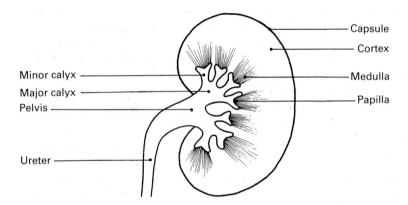

Fig. 11.1 Major structural features of the kidney.

the afferent arteriole and the blood leaves the glomerulus through the efferent arteriole.

The function of the glomerulus is filtration, which occurs across the capillary endothelial layer, the tubular epithelium and the basal membrane which separates them. The capillary endothelium has pores about 100 μm in diameter and the tubular epithelium (also called podocyte layer) has slits of 25 nm in width. The basal membrane is made of mucopolysaccharide and has no visible pores.

Functionally, this triple-filtering membrane allows particles about 8 nm in diameter or those of molecular weight about 60 000, such as plasma insulin, to pass. Due to the large number of nephrons, the total surface area of the membrane used for filtration has been estimated to be about 0.4 m² for each kidney.

The glomerular filtrate flows first into the proximal convoluted tubules. These are about 15 mm long and 5 μm in diameter. They are lined by a single layer of epithelial cells with a well-defined

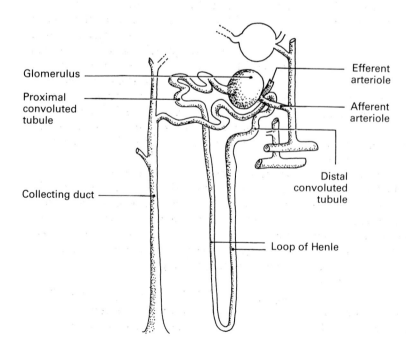

Fig. 11.2 Diagrammatic representation of the nephron.

luminal brush border. The main function of the proximal tubule is the reabsorption of useful substances, including sodium and water.

From the proximal tubules the filtrate passes into the descending limb of the loop of Henle, which varies between 2 and 14 mm in length. The loop is about 15 μm in diameter and is lined with flat epithelial cells. In man, only 15% of the nephrons have long loops descending into the renal pyramids; these are referred to as juxta-medullary nephrons, which play a major role in the urine concentration mechanism. The rest (85%) are called cortical nephrons. The ascending limb of the loop is about 12 mm in length, the more distal part of which becomes thicker as it joins the distal convoluted tubule. In this place it comes close to the afferent arteriole of the glomerulus and forms the juxtaglomerular apparatus, which plays an important role in the regulation of the renin−angiotensin−aldosterone system.

The distal convoluted tubules are about 5 mm in length and they are lined with microvilli but no brush border. The distal tubules join the collecting ducts, which are about 20 mm in length. These are lined by cuboidal epithelial cells with no brush border. The collecting ducts open at the renal papillae into the minor calyces of the renal pelvis. Each collecting duct serves about 4000 nephrons. The total length of one nephron is about 45−65 mm.

Renal circulation

The special features of the renal circulation deserve an early description. These special characteristics are essential for the nephrons to perform their various functions.

Blood reaches the kidneys through the renal arteries, which are short and come directly from the abdominal aorta. The renal arteries divide into several interlobar arteries, which give rise to the arcuate arteries (Fig. 11.3). These cross the border between the cortex and the medulla of the kidney. From the arcuate arteries many branches radiate into the renal cortex, the interlobular arteries. The afferent arterioles arise at right angles from the interlobular arteries and end in the glomeruli.

Thus, blood from the aorta has a short distance to flow until it reaches the glomeruli. The blood-pressure in the afferent arteriole is about 100 mmHg, which is about the same as the mean pressure in the aorta. This leads to a high glomerular capillary pressure, about 60 mmHg in man. Capillary pressure elsewhere in the systemic circulation is 15−30 mmHg. Thus, glomerular capillaries can only filter fluid and never allow reabsorption.

The efferent arterioles have a blood-pressure of about 18 mmHg. They also offer high resistance to blood flow and this results in a further fall of blood-pressure distal to the efferent arterioles. The efferent arterioles end in the peritubular capillary plexuses surrounding the proximal and distal convoluted tubules. In addition, some straight capillaries (vasa recta) descend into the renal medulla and drain the interstitial fluid around the loops of Henle. All these capillaries finally drain into venules, which join the interlobular veins, which eventually lead to the renal vein. The peritubular capillaries and the vasa recta have a low blood-pressure, about 10−13 mmHg, which is much lower than the colloid osmotic pressure of plasma proteins (25 mmHg). This allows only the absorption of fluid and no further filtration. It is worth noting that most of the renal blood supply goes to the renal cortex, where the glomeruli are situated; the medulla receives only 1−2% of the renal blood flow.

The kidneys receive 20−25% of the cardiac output, approximately 1−1.5 litres/min in a 70 kg man. The total renal capillary surface area is about 12 m^2, i.e. about the same as the total surface of the renal tubules. The kidney is therefore a highly vascular organ, equipped to filter blood as well as to reabsorb and secrete various constituents by means of exchange across the capillary membranes and the tubular surfaces.

Measurement of renal blood flow

The renal blood flow can be measured using the Fick principle. The blood flow to an organ can be calculated from the amount of a substance taken up by the organ and the difference in concentration of the substance between arteries and veins supplying that organ:

$$\text{Blood flow} = \frac{\text{Amount of substance taken up}}{\text{Arterial concentration} - \text{venous concentration}}$$

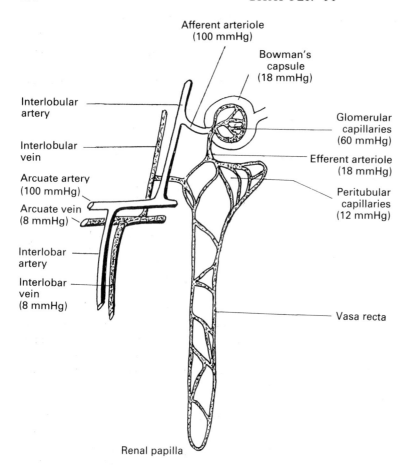

Afferent arteriole
(100 mmHg)

Bowman's
capsule
(18 mmHg)

Interlobular
artery

Interlobular
vein

Arcuate artery
(100 mmHg)

Arcuate vein
(8 mmHg)

Interlobar
artery

Interlobar
vein
(8 mmHg)

Glomerular
capillaries
(60 mmHg)

Efferent arteriole
(18 mmHg)

Peritubular
capillaries
(12 mmHg)

Vasa recta

Renal papilla

Fig. 11.3 The blood supply to the
different parts of the kidney.

A substance which is removed by the kidney can therefore be used. *Para*-amino hippuric acid (PAH) is used for measurement of the renal blood flow. PAH is filtered through the glomerulus and it is actively secreted in the renal tubules, so that blood loses 90% of PAH in one passage through the kidney.

Renal blood flow can be calculated as follows:

$$\frac{\text{Amount of PAH removed by the kidney}}{\text{Arterial concentration} - \text{venous concentration}}$$

The amount removed by the kidney can be calculated by taking a sample of urine over a known period of time — say, 15 min. Thus, urine volume × urine PAH concentration = the amount of PAH removed ($U_v \times U_{PAH}$).

There is no need to take an arterial and a venous blood sample because PAH is about 100% removed by the kidney and it is not removed by any other organ in the body. Therefore, a mixed venous blood sample taken soon after PAH administration is representative of the renal artery blood. And, since the removal by the kidney is almost complete, we can assume that the renal vein concentration of PAH is zero. Thus, the arteriovenous difference can be represented by the PAH concentration in a mixed venous blood sample:

Renal plasma flow (RPF) =

$$\frac{\text{Urine volume} \times \text{urine PAH concentration}}{\text{Plasma PAH concentration}}$$

Example:
Urine volume (U_v) = 1 ml/min

Urine PAH concentration (U_{PAH}) = 13 mg/ml
Plasma PAH concentration (P_{PAH}) = 0.02 mg/ml
Substituting in the above formula:

$$RPF = \frac{Uv \times U_{PAH}}{P_{PAH}}$$

$$= \frac{1 \times 3}{0.02} = 650 \text{ ml/min}$$

This is known as the effective renal plasma flow, i.e. blood flow through those parts of the kidney that effectively remove PAH from the bloodstream, i.e. renal tubules. This is to differentiate it from other parts of the kidney, e.g. renal capsules and perirenal fat, which receive 10% of the renal blood flow and do not transport PAH. Therefore, only 90% of the PAH is cleared by the kidney and the actual renal plasma flow is:

$$\frac{650 \times 100}{90} = 721 \text{ ml/min}$$

To calculate the renal blood flow, we use the ratio of plasma to whole blood. The packed cell volume (PCV) is about 45%. Therefore, if plasma constitutes 55% of the blood volume, blood flow can be calculated as follows:

$$\text{Blood flow} = \text{Renal plasma flow} \times \frac{100}{55}$$

$$= \frac{721 \times 100}{55} = 1309 \text{ ml/min}$$

Glomerular function

The main function of the glomeruli is filtration. This is the first of the renal mechanisms that enables the kidney to carry out its functions. It is a passive physical process and does not require energy. The glomerular capillaries are about 50 times more permeable than the capillaries in skeletal muscle. They allow substances up to 60 000 molecular weight or 8 nm in diameter to pass. Albumin, which has a molecular weight of 69 000, passes with difficulty. Only minute amounts of albumin normally appear in the glomerular filtrate. The filtrate, therefore, is described as an ultrafiltrate, because it consists of all plasma constituents of small molecular size, e.g. crystalloids, but not plasma proteins or blood cells.

THE MECHANISM OF GLOMERULAR FILTRATION
Glomerular filtration depends upon three main variables:
1 Glomerular capillary pressure (hydrostatic pressure).
2 Intracapsular pressure.
3 Plasma protein osmotic pressure.

Hydrostatic pressure of the glomerular capillaries

As indicated previously, the glomerular capillary pressure is about 60 mmHg, which is twice as much as capillary pressure in skeletal muscles. The force of this hydrostatic pressure tends to help filtration of fluid through the capillary wall, the basement membrane and the glomerular tubular epithelium into Bowman's space.

Intracapsular pressure

As Bowman's capsule restricts the size of the glomerulus, it exerts a pressure opposite to that of the capillary hydrostatic pressure. Intracapsular pressure, i.e. the pressure of fluid within Bowman's space, is usually about 18 mmHg.

Colloid osmotic pressure of plasma proteins

Plasma protein osmotic pressure acts in the opposite direction to the capillary hydrostatic pressure. It tends to retain fluid in the capillaries. Plasma colloid osmotic pressure is about 25 mmHg. Thus:

$$\text{Net filtration pressure} = 60 - (18 + 25)$$
$$= 17 \text{ mmHg}$$

This is a sizeable pressure gradient in favour of filtration. It guarantees a continuous flow of filtrate, which amounts to about 125 ml/min. This is the normal glomerular filtration rate (GFR) in a 70 kg man.

MEASUREMENT OF THE GLOMERULAR FILTRATION RATE (GFR)
A suitable substance which is filtered freely in the glomeruli is selected and its rate of excretion in urine is measured. From this the rate of filtration of fluid through the glomeruli can be calculated, as will be explained. The substance to be used should have the following characteristics:

1 It should be filtered freely in the glomeruli and neither reabsorbed nor secreted by the tubules. This ensures that all of the amount which appears in urine has come through filtration.

2 It should be stable, i.e. not metabolized or stored in the body. This ensures that the total amount administered stays in the circulation and is available for filtration in the kidney.

3 It should not bind to plasma proteins, which are not filtered at the glomerulus.

4 It should not alter the rate of filtration or glomerular function.

5 It should not be toxic.

A substance which satisfies the above characteristics is inulin, a polymer of fructose which has a molecular weight of 5200 daltons.

Procedure

1 An intravenous loading dose of inulin is given, which is followed by an intravenous drip of inulin solution to maintain the plasma level constant.

2 A timed urine sample is taken, say for 15 or 20 minutes.

3 A blood sample is collected about half-way during the collection of the urine sample.

The quantity of inulin excreted in urine (Q_u) will be the same as that which is filtered (Q_F). And, as inulin is freely filtered, its concentration in the glomerular filtrate (C_F) should be the same as that in plasma (C_p). The volume of the glomerular filtrate per minute (GFR) can therefore be calculated as follows:

$$Q_u = Q_F$$
$$Q_u = U_{in} \times U_v$$
$$Q_F = GFR \times C_p$$
$$GFR = \frac{U_{in} \times U_v}{C_p}$$

where: U_{in} = concentration of insulin in urine
U_v = urine volume/min

Example

U_{in} = 21 mg/ml
U_v = 1.5 ml/min
C_p = 0.26 mg/ml
$$GFR = \frac{21 \times 1.5}{0.26} = 127 \text{ ml/min}$$

The GFR is about 125 ml/min for a 70 kg man.

The GFR is about 7.5 litres per hour and 180 litres per day. Normally, only about 1–2 litres of urine are passed per day. This means that about 98% of the water is reabsorbed in the renal tubules.

Creatinine is a substance which is produced in the human body from skeletal muscles. It is freely filtered in the glomeruli but is secreted by the tubules to a slight extent. Using the above formula for creatinine, values of GFR closely resemble those for inulin. It is therefore the method of choice for assessing glomerular function in hospitals. Since some creatinine is secreted by the renal tubules, this makes the quantity excreted in urine (Q_u) slightly higher than that filtered (Q_F). However, some chromogenic substances in plasma give the same colour as creatinine and this overestimates the concentration of creatinine in plasma. These sources of error cancel each other and therefore the calculated GFR approximates that obtained when using inulin.

Control of the GFR

The following factors control the GFR.

CAPILLARY HYDROSTATIC PRESSURE
Capillary hydrostatic pressure is the most important factor in the control of GFR. When the hydrostatic pressure changes from as little as 75 mmHg to as high as 180 mmHg, the GFR hardly changes. This is called autoregulation of the glomerular filtration rate. Over the same pressure range, the renal blood flow also remains unchanged. It is important to maintain a GFR which is not too slow or too fast. If too slow, the glomerular filtrate will flow slowly along the renal tubules; most of its constituents will be reabsorbed and in this way the kidney will fail to excrete the unwanted waste products. On the other hand, if the flow of the filtrate is too fast, the time will not be enough to enable the tubular cells to reabsorb the vital constituents of the glomerular filtrate.

Two feedback mechanisms are believed to be responsible for the autoregulation of the GFR. These are: (i) an afferent arteriolar vasodilator feedback mechanism; and (ii) an efferent arteriolar vasoconstrictor feedback mechanism.

The renal blood-vessels constrict in response to an increased blood flow and dilate when the blood flow is reduced. This change of the vascular resistance maintains the intrarenal vascular blood-pressure almost constant in spite of changes in the systemic blood-pressure. Autoregulation is a local myogenic response of the vascular smooth muscle to the distending force of blood flow. The combination of these two feedback mechanisms is called tubuloglomerular feedback. These intrinsic regulatory mechanisms are controlled by the juxtaglomerular apparatus (Fig. 11.4). The juxtaglomerular cells secrete an enzyme called renin, which splits a decapeptide (angiotensin I) from a plasma protein (α_2-globulin) called angiotensinogen. Angiotensin I loses two amino acids in the presence of a converting enzyme to become an octapeptide, angiotensin II, which is a potent vasoconstrictor, and also causes the secretion of aldosterone from the adrenal cortex.

The juxtaglomerular cells secrete renin and therefore activate the renin–angiotensin system in response to a high rate of flow in the distal tubule or to a low sodium concentration in the distal tubular fluid. This is possible because the macula densa cells of the distal tubules come in close contact with efferent arterioles. The effect of activating the renin–angiotensin system is to lower the GFR. Angiotensin II produces constriction of the afferent arterioles and therefore lowers the hydrostatic pressure in the glomerular capillaries.

Small amounts of angiotensin II produce constriction of the efferent arterioles. This results in a rise in the glomerular capillary hydrostatic pressure and an increase in the GFR. However, as the rate of flow in the glomerular capillaries becomes slower, the rise in the concentration of plasma proteins in the capillaries tends to limit filtration. The net result is also lower GFR.

In conclusion, the intrinsic regulation of renal blood flow makes it possible for the rate of filtration in the glomeruli to be regulated by the rate of flow of the tubular fluid and the degree of concentration of the tubular fluid, so that optimum levels of filtration are achieved. If too much or too little glomerular filtrate is formed, this may result in serious changes in fluid, urea and electrolyte excretion.

CAPILLARY PERMEABILITY

The size of the pores in the glomerular filtration barrier is most important in determining the size and quantity of substances filtered. If the mem-

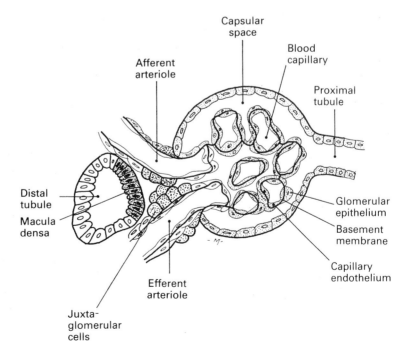

Fig. 11.4 The glomerulus and the juxtaglomerular apparatus.

branes forming this barrier (which are the capillary endothelium and the tubular epithelium) are damaged by disease, large-size molecules may pass in the filtrate and this, in turn, will influence the GFR. For example, in kidney disease, large amounts of albumin are filtered and this increases the GFR.

RENAL BLOOD FLOW

Renal blood flow is increased on lying down, during physical exercise and in pregnancy. Minor, but significant, increases of the GFR occur in these conditions.

In cases of severe haemorrhage, a reduction in systemic blood-pressure and renal blood flow may result in a serious reduction of GFR, which may lead to a complete failure of urine formation (anuria).

PLASMA ALBUMIN CONCENTRATION

The plasma albumin concentration increases in patients with dehydration, thus leading to a reduction of GFR. Patients with low plasma albumin (hypoalbuminaemia), due to loss of albumin in urine or liver disease, usually have a high GFR.

PRESSURE IN BOWMAN'S CAPSULE

The pressure in Bowman's capsule may become abnormally high in cases of intrarenal oedema or haemorrhage due to trauma to the kidney. Increased renal intracapsular pressure may also result from a tumour of the kidney. Pressure in the renal tubular system may also rise in cases of obstruction of urine flow in the renal pelvis (hydronephrosis), e.g. as a result of ureteric calculi or fibrosis due to bilharzia.

THE NUMBER OF FUNCTIONING GLOMERULI

The number of functioning glomeruli may be reduced due to disease, thus decreasing the rate of glomerular filtration. This may be due to inflammatory or parasitic disease, such as bilharzia of the urinary tract (*Schistosoma haematobium*).

Renal clearance

The term clearance is defined as the volume of plasma completely cleared of a substance by the kidneys per unit time. The concept of clearance

originated from the old observation that the function of the kidney is to clear urea and other waste products from the blood. A measure of this renal function is that volume of blood or plasma which is *completely cleared* of urea per minute. This is blood urea clearance, which is a measure of the ability of the kidney to remove urea from the blood and to deliver it in urine. In the case of PAH, clearance is used to measure renal plasma flow, as PAH is both filtered and secreted and the blood is almost completely cleared of PAH while passing once through the kidney. It is therefore a true volume in the case of PAH:

$$\text{PAH clearance} = \frac{U_{\text{PAH}} \times U_{\text{V}}}{P_{\text{PAH}}}$$

This formula can be used to calculate the clearance of any other substance which the kidney handles through filtration, reabsorption or secretion. Since substances are filtered and then reabsorbed, the value of clearance becomes a *theoretical volume*, because the blood is not completely cleared of them, e.g. clearance of urea or creatinine. However, the clearance of these substances, as measured under normal conditions, can be used as an index of normal kidney function. Measurement of their clearance in kidney disease is useful for the assessment of renal function.

The normal clearance of creatinine is about 125 ml/min. Creatinine is therefore similar to inulin in its clearance. A substance which has a clearance less than 125 ml/min is therefore reabsorbed in the renal tubules. Conversely, a substance which has a clearance more than 125 ml/min must be added to the filtrate, i.e. secreted as it passes through the renal tubules.

The clearance of urea is about 60 ml/min. This indicates that some of the urea which is filtered is reabsorbed in the tubules. The clearance of glucose is zero. All the glucose in the filtrate is reabsorbed in the tubules so that none appears in urine.

Functions of the renal tubules

The glomerular filtrate is changed both in volume and composition during its passage through the renal tubules. Useful substances such as glucose, amino acids and vitamins are reabsorbed. Most of the water is reabsorbed so that urine volume is much less than the volume of the glomerular filtrate. Sodium, potassium and chloride are partly reabsorbed, according to the requirements of the maintenance of the constancy of the composition of the internal environment (homoeostasis). On the other hand, the pH of the filtrate, which is initially similar to that of plasma (pH 7.4), becomes more acidic due to the addition of hydrogen ions by tubular cells.

The two functions of the renal tubules are therefore reabsorption and secretion.

TUBULAR REABSORPTION
Reabsorption of substances by the renal tubules is either passive or active. Passive reabsorption depends on the concentration and/or electrochemical gradient. Active reabsorption, on the other hand, is the transport of substances against concentration or electrical gradients. This requires the expenditure of energy or specialized transport systems in the tubular cells.

Glucose reabsorption
All filtered glucose is actively reabsorbed in the proximal tubules. Only minute traces may sometimes appear in urine in normal persons.

The tubules have a maximum capacity for reabsorption, called tubular maximum (Tm). For glucose it is about 300 mg/min. Usually the amount filtered per minute is much less than this. For example, if the plasma glucose is 100 mg/100 ml and the GFR is 125 ml/min, the filtered glucose load/min is only 125 mg/min. But, in cases of diabetes mellitus, the blood glucose is high and the filtered glucose load may exceed the tubular maximum, i.e. more than 300 mg of glucose pass in the filtrate per minute. As a result, the excess glucose appears in urine (glucosuria). Phloridzin blocks the source of energy for glucose reabsorption and therefore greatly reduces Tm for glucose. Thus, administration of this substance will result in glucosuria.

Figure 11.5 illustrates the tubular maximum for glucose. It shows that, when the blood glucose rises above 180 mg/100 ml, the tubular maximum is exceeded and glucose appears in urine. The level of glucose in the blood above which it appears in urine is called the renal threshold for glucose. In this case it is 180 mg/100 ml.

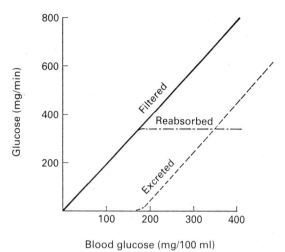

Fig. 11.5 Renal handling of glucose. The maximum amount reabsorbed per minute is known as the tubular maximum (Tm). Also note that glucose starts to appear in urine (excreted) when the blood glucose level reaches about 180 mg/100 ml.

Protein reabsorption
Small amounts of albumin (30−50 mg/day) escape in the glomerular filtrate. This is reabsorbed in the proximal convoluted tubules. Albuminuria may occur in those renal diseases which cause an increase in glomerular capillary permeability. Standing for a long time may give rise to albuminuria, probably due to increased pressure in the renal vein.

Sodium and water reabsorption
About 70−75% of the filtered water and sodium is reabsorbed in the proximal convoluted tubules. The absorption of sodium is an active process, which takes place against an electrochemical gradient. Anions and water follow Na$^+$ passively. This reabsorptive process occurs as follows.

Sodium ions enter the tubular cells passively along an electric gradient of about 52 mV. However, Na$^+$ leaves the tubular cells for the peritubular fluid against an electrical gradient (potential difference = 72 mV) by active transport. ATPase is activated by Na$^+$ and K$^+$, which results in the pumping of Na$^+$ out of the cells and K$^+$ into the cells (Fig. 11.6). Chloride ions follow Na$^+$ passively due to electrical gradient. Water also follows

Na^+ passively by osmosis. K^+ tends to leak out of the cell, due to concentration gradient.

Sodium and water are also reabsorbed in the distal tubules and the collecting ducts under the effect of the hormones aldosterone and ADH respectively. This enables the kidney to maintain homoeostasis by regulating the excretion of Na^+ and water in urine.

Urea reabsorption

Urea is highly diffusible through the renal tubules. When Na^+ and water are reabsorbed in the proximal tubules, urea concentration in the tubular lumen rises. This leads to passive diffusion of about 50% of the filtered urea from the tubules to the peritubular capillaries. When the filtrate passes down the descending limb of the loop of Henle, it loses water and regains urea from the surrounding more concentrated medullary interstitium. The fluid keeps its urea until it reaches the collecting duct. Here, as water is reabsorbed under the effect of ADH, urea concentration rises and, as a result, 40–60% of it diffuses back to the medullary interstitium and the rest is lost in urine.

Reabsorption of uric acid

Up to 90% of the filtered uric acid is reabsorbed. However, uric acid can also be actively secreted, so that its clearance can exceed that of inulin.

Reabsorption of phosphate

Phosphate is reabsorbed in the renal tubules, but it has a low Tm and therefore phosphate is always found in urine, where it acts as a main buffer system (Fig. 11.7).

Phosphate reabsorption is related to glucose reabsorption. It is possible that their transport mechanisms compete for a common source of energy. Phloridzin, which blocks glucose reabsorption, increases the reabsorption of phosphate.

Parathyroid hormone reduces the Tm for phosphate and therefore the excretion of phosphate is increased.

TUBULAR SECRETION

Tubular secretion refers to the transport of substances from peritubular blood, interstitium or tubular cells into the tubular lumen. Tubular secretion should, therefore, be viewed as a supplement to glomerular filtration since it removes from peritubular blood those substances which escaped total elimination by filtration at the glomerulus. A substance whose clearance is more than inulin must have been secreted as well as filtered.

Like reabsorption, secretion may be:
1 *Passive*, e.g. NH_3, the drugs quinidine and salicylic acid.
2 *Active—Tm-limited*, e.g. creatinine, glucuronides, PAH, penicillin, chlorothiazide and dio-

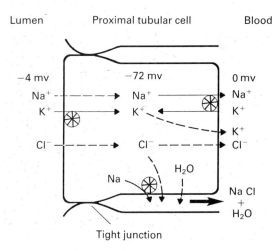

Fig. 11.6 Mechanisms of water and electrolyte reabsorption in the proximal tubules.

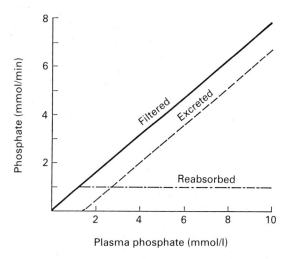

Fig. 11.7 Reabsorption of phosphate.

drast. K$^+$ and H$^+$ are actively secreted by the tubular cells into the lumen by a mechanism which does not seem to be Tm-limited.

Potassium secretion

The secretion of K$^+$ in the distal tubules is reciprocally associated with the absorption of Na$^+$, so that, when Na$^+$ reabsorption increases, K$^+$ secretion also increases. The mechanisms involved are not fully understood.

Hydrogen ion secretion

The excretion of K$^+$ in urine is inversely proportional to H$^+$ excretion. There is active secretion of H$^+$ from the renal tubules, which tends to correct acid accumulation in the plasma. The mechanism of H$^+$ secretion will be detailed in the section on renal acid−base regulation.

The excretion of water

Body fluid regulation depends largely on the role of the kidney in regulating water excretion. When there is shortage of water in the body, the kidney secretes a smaller volume of more concentrated urine. When there is an abundance of water, a larger volume of more dilute urine is secreted.

Reabsorption of water takes place in the proximal tubule, the loop of Henle, the distal tubule and the collecting ducts. In the proximal tubules water moves passively, following the active transport of Na$^+$. Micropuncture technique, which enables sampling of tubular fluid, has shown that, in the first 70% of the length of the proximal tubule, the tubular fluid is isosmotic with the plasma. The concentration of inulin at this point, however, rises to about four times that in the glomerular filtrate. This indicates that 75% of the water and other osmotically active solutes have left the proximal tubule for the peritubular blood.

In the loop of Henle about 5% of the filtrate is reabsorbed. By the time fluid reaches the distal convoluted tubules, 80% of the filtrate has been reabsorbed. Water leaves the descending limb of the loop because the osmotic pressure in the surrounding medullary interstitial fluid is much higher than that in the lumen. There is a gradual increase in osmotic pressure from normal near the renal cortex (about 300 mOsm/litre) to about

four times the normal in the interstitial fluid in the renal papillae (about 1200 mOsm/litre) (Fig. 11.8). But, as both Na$^+$ and Cl$^-$ ions also leave the ascending limb of the loop of Henle, the fluid which reaches the distal tubules is hypotonic.

Further reabsorption of water takes place in the distal tubules and the collecting ducts, where the final volume of urine is determined. This is controlled by the antidiuretic hormone (ADH) of the posterior pituitary gland.

ADH secretion is controlled by the osmotic pressure of the extracellular fluid. When there is reduction of body water (dehydration), the osmo-

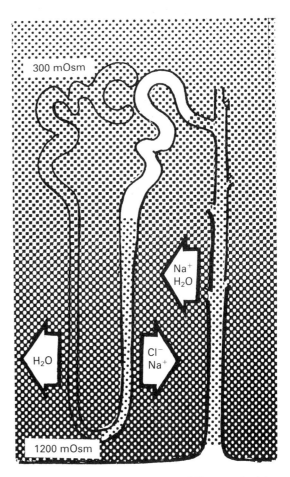

Fig. 11.8 Relative osmotic pressure differences in the various parts of the kidney.

larity of the extracellular fluid rises; this leads to ADH release and therefore more water is reabsorbed in the distal part of distal tubules and the collecting ducts. When there is a drop in the osmolarity of the extracellular fluid, e.g. after drinking large amounts of water, ADH release is inhibited. This leads to a reduction in the reabsorption of water in the distal tubules and the collecting ducts and therefore a large volume of dilute urine is formed.

ADH release is also controlled by blood volume. When there is a fall in circulating blood volume, a reflex mechanism is activated by volume receptors which results in an increase in ADH secretion. Thus, conservation of body water is achieved.

The function of the distal tubules and the collecting ducts is the final adjustment of sodium and water excretion. In the absence of ADH, the distal tubules and the collecting ducts are impermeable to water. The action of ADH is to increase the permeability of the distal tubules and the collecting ducts to water. The osmolarity of the interstitial fluid in the renal medulla is high. Therefore, water immediately passes out of the tubules and is reabsorbed into the circulation.

Mechanism of concentration of urine

The loops of Henle of the juxtamedullary nephrons play a key role in the concentration of urine, because they are responsible for the creation and maintenance of the high osmotic pressure in the interstitial fluid of the renal medulla.

The descending limb of the loop of Henle is permeable to water and this leads to loss of water to the more concentrated medullary interstitium. This leads to the gradual concentration of fluid passing down the descending limb. This fluid becomes hypertonic as it descends. Some Na^+ and Cl^- go into the descending limb, adding further to the rise in their concentration. In the thick portion of the ascending limb of the loop, Cl^- is actively pumped out of the tubule and Na^+ follows passively. This leads to the tubular fluid becoming hypotonic by the time it reaches the distal convoluted tubule.

The result of this difference in concentration of the fluid in the descending limb and the ascending limb is a similar change in the concentration of the interstitial fluid surrounding the loop. The osmotic pressure becomes progressively higher towards the tips of the renal papillae.

This process of concentration is known as the *counter-current mechanism* for the concentration of urine.

In summary, the following are the main characteristics of the counter-current mechanism:

1 Water leaves the descending limb and, as a result, the glomerular filtrate becomes hypertonic.

2 In the ascending limb, which is impermeable to water, Cl ions are actively transported and Na ions follow passively into the interstitial fluid, from where they can re-enter the descending limb.

3 Thus, there is a recirculation of solute between the ascending limb and the descending limb of the loop of Henle, which behaves as a counter-current multiplier system.

4 Fluid becomes hypotonic as it approaches the distal convoluted tubule.

5 Since the interstitial fluid in the renal medulla equilibrates with the fluid in the descending limb of the loop of Henle, its concentration increases towards the renal papillae, where it has four times the plasma osmolarity (about 1200 mOsmol/litre).

6 Active reabsorption of Na^+ from the collecting ducts and the diffusion of urea into the medullary interstitial fluid leads to further hyperosmolarity.

7 The blood supply of the medulla is poor in comparison to that of the cortex. This reduces the possibility of washing out the concentrated ions in the medullary interstitium and therefore helps to maintain the gradient of hyperosmolarity of the renal medulla.

The vasa recta are capillary loops descending from the cortex into the medulla. These loops have a similar role to that of the loops of Henle. Thus, in the descending capillaries water escapes into the interstitial fluid while Na^+ and urea enter. This leads to an increase in the osmolarity of the blood as it descends. In the ascending capillaries, water is reabsorbed and very quickly removed towards the cortex, as the capillaries rejoin the cortical circulation. Thus, the vasa recta remove any water which is reabsorbed from the distal tubules and the collecting ducts. In addition, and since they are the only blood supply to the medulla, they also transport oxygen and nutrients to it and remove carbon dioxide and waste products.

Sodium is actively reabsorbed in the distal tubules and the collecting ducts under the influence of the hormone aldosterone of the adrenal cortex. This is achieved against a high electrical gradient of about -120 mV in the tubular lumen, compared with peritubular fluid (Fig. 11.9). This leads to further concentration of the interstitial fluid in the renal medulla. Normally, about 15% of the filtrate is reabsorbed in the distal tubules by isosmotic reabsorption and a further 4% is reabsorbed in the collecting ducts.

Urine entering the renal pelvis may be hypotonic or hypertonic with respect to plasma. At the beginning of the collecting duct, urine is usually isosmotic, and the presence or absence of ADH determines whether further concentration occurs. In the absence of ADH, the absorption of Na$^+$ and other solutes is more than the absorption of water and therefore hypotonic urine is formed. In the presence of ADH, the absorption of water is accelerated and hypertonic urine is formed.

Renal electrolyte regulation

REGULATION OF SODIUM REABSORPTION
There is active reabsorption of Na$^+$ in the proxi-

mal tubules up to about 75% of the Na$^+$ in the filtrate. Further reabsorption occurs in the loop of Henle. Only about 10% reaches the distal tubules.

In the distal tubules and the collecting ducts, fine regulation of sodium reabsorption occurs under the influence of aldosterone, according to the needs of homoeostasis. In the absence of this hormone, most of the remaining sodium is lost in urine. When aldosterone is secreted in large amounts, almost all the sodium is reabsorbed.

The kidneys reabsorb quite a large amount of sodium each day. The amount which passes in the filtrate is about 625 g/day. The transport of Na$^+$ in the distal tubules is an active process, which occurs against a concentration gradient from the tubular cell to the peritubular fluid. This net movement of Na$^+$ creates a negative electrical potential inside the tubular cell, which tends to draw sodium into the cell from the tubular lumen (see Fig. 11.9).

The active Na$^+$ transport also results in pumping K$^+$ into the cell. Sodium transport, however, is greater than that of potassium. This explains the negative intracellular potential. Aldosterone acts by means of stimulating the synthesis of the mRNA responsible for the synthesis of protein carriers or enzymes necessary for the transport of sodium:

Aldosterone secretion is regulated by means of the sodium level in the extracellular fluid. Low sodium levels stimulate the secretion of renin from the juxtaglomerular cells (JG cells). Renin acts on a circulating α_2-globulin (angiotensinogen) and splits from it a decapeptide, angiotensin I (Fig. 11.10). This, in turn, becomes converted to an octapeptide, angiotensin II, under the influence of a converting enzyme, which comes mainly from the lung. Further breakdown by means of angiotensinases gives rise to angiotensin III. Both angiotensin II and angiotensin III are powerful vasoconstrictors and stimulators of the secretion of aldosterone by the adrenal cortex.

The regulation of aldosterone, therefore, enables the kidney to respond to changes in the extracellular Na$^+$ and K$^+$ concentration. This is a major mechanism for the maintenance of the constancy of electrolyte concentrations in the internal environment.

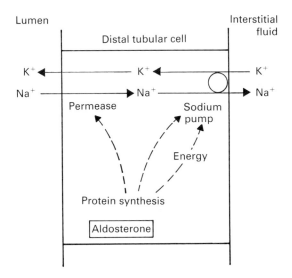

Fig. 11.9 Possible mechanisms of action of aldosterone.

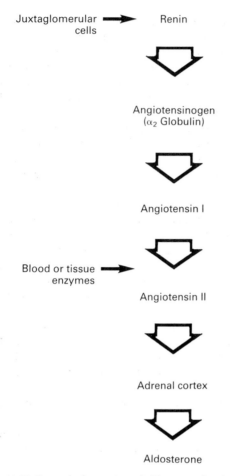

Fig. 11.10 Control of secretion of aldosterone by the renin–angiotensin system.

REGULATION OF POTASSIUM EXCRETION
In the proximal tubules and the loop of Henle, potassium moves in the same direction as sodium, i.e. both Na^+ and K^+ are reabsorbed. About 65% of K^+ is reabsorbed in the proximal tubules and 25% in the loop of Henle, which leaves about 10% or less in the distal tubules and the collecting ducts.

In the distal tubules and the collecting ducts, there is a reciprocal relationship between Na^+ and K^+. K^+ is actively transported from the peritubular fluid into the tubular cells, from where K^+ diffuses passively into the tubular lumen, due to concentration gradient. The amount of K^+ secreted is directly related to aldosterone secretion and to sodium reabsorption.

A small increase of serum K^+ directly stimulates the adrenal cortex to release aldosterone and thus K^+ secretion is enhanced. In the absence of aldosterone, K^+ secretion stops and reabsorption occurs. This shows that K^+ reabsorption is taking place all the time but it is usually masked by the normally greater amounts secreted.

Hypokalaemia (low serum K^+) usually results in alkalosis. This is because the lack of K^+ results in excess H^+ secretion instead, indicating that the same carrier system is involved.

Adequate K^+ secretion is necessary to safeguard against hyperkalaemia, which is a serious consequence of tubular damage. Hyperkalaemia can cause cardiac arrhythmia, fibrillation and death. Normally, K^+ cannot be totally reabsorbed, as in the case of sodium. A minimum of 15 mol are lost every day. Regular dietary intake should be ensured; otherwise, hypokalaemia results.

The excretion of urea
The level of blood urea is a good indicator of renal function. Normally, it is 4–6 mmol/litre (30 mg/100 ml). In renal disease, it may increase to 40 mmol/litre (200 mg/100 ml) or more.

The excretion of urea depends upon both its concentration in the blood and the GFR. These two factors determine how much urea passes in the filtrate. Normally, 40–60% of this is excreted. Major changes in the GFR greatly influence urea excretion.

Reduction of the GFR due to renal disease or to reduced renal blood flow leads to a reduction in the amount of urea excreted. Two mechanisms are responsible for this. First, the low GFR leads to less urea being filtered. Second, the low GFR results in a slow flow rate of the filtrate in the renal tubules, and more reabsorption of urea. The ultimate result is reduced excretion of urea.

When the GFR is increased, tubular flow is more rapid. This not only increases the amount of urea filtered, but also reduces the amount reabsorbed. Most of the urea filtered is therefore excreted. The significance of these observations on urea excretion is to emphasize the importance of maintaining a good GFR when renal function is decreased.

Urea clearance (normally about 60–70 ml/min) is assisted by the counter-current mechanism involving the collecting duct and the ascending limb of the loop of Henle. Urea leaves the collecting duct and enters the ascending limb because of its concentration gradient. The hypertonicity of the medullary interstitium is due to both sodium ions and urea. This is the basic urine concentrating mechanism, which ultimately enables the maximum reabsorption of water and the excretion of a concentrated urine.

Renal acid–base regulation

The pH of urine produced by the kidneys varies from 4 to 8 according to the requirements of the internal environment, where the pH should be constant at 7.4. This is achieved by the ability of the renal tubules to secrete acid and reabsorb bicarbonate.

The kidney, however, is not the only organ responsible for acid–base regulation. The lungs are important for the removal of CO_2 from the blood. Indeed, larger quantities of acid are removed by the lungs than the kidneys can handle. Blood buffers tend to minimize changes in pH, while the lungs and kidneys excrete excess acid formed during metabolism. A comprehensive account of acid–base regulation is presented in Chapter 12.

ROLE OF THE RENAL TUBULES
In the proximal tubules, reabsorption of about 85% of the filtered bicarbonate ions takes place. Further reabsorption of bicarbonate occurs in the distal tubules and the collecting ducts. Normally, almost all the bicarbonate in the filtrate is reabsorbed. This is purely a conservation process. The distal tubules and the collecting ducts also secrete hydrogen ions (less so in the distal tubules than in the collecting ducts).

A fall of urine pH to 6 can be achieved by reabsorption of bicarbonate only. Any further fall in pH requires the active secretion of H^+.

BICARBONATE

Mechanism of bicarbonate reabsorption
Bicarbonate is reabsorbed together with Na^+. Sodium in the tubular lumen is present in the form of:
1 $NaHCO_3$.
2 Disodium phosphate (Na_2PO_4).
3 NaCl and other monovalent salts.

The mechanisms of absorption of sodium and bicarbonate are shown in Fig. 11.11. Na^+ is obtained from $NaHCO_3$ in the tubular lumen. The presence of carbonic anhydrase (CA) in the brush border enhances conversion of carbonic acid into CO_2 and water. CO_2 diffuses back into the tubular cell, where it combines with water to form carbonic acid; this reaction is also enhanced by carbonic anhydrase. The dissociation of carbonic acid gives HCO_3^-, which is reabsorbed, and H^+, which is secreted actively into the tubular lumen.

Formation of 'new' bicarbonate
In the formation of 'new' bicarbonate, HCO_3^- ions, which are formed within the tubular cells (therefore the term 'new'), are added to the blood. Na^+ is obtained from disodium phosphate in the tubular lumen. The dissociation of intracellular carbonic acid gives HCO_3^-, which is reabsorbed, and H^+ is secreted into the tubular lumen. This hydrogen ion is picked up by either phosphate or ammonium buffers, rather than being used for the absorption of filtered HCO_3^- (Fig. 11.12).

All of these reactions have four features in common:
1 Na^+ and bicarbonate are reabsorbed into the peritubular capillaries.
2 Carbonic anhydrase is needed to facilitate formation of carbonic acid in the tubular cells.

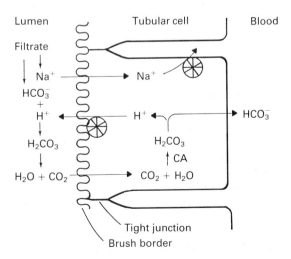

Fig. 11.11 Mechanism of absorption of bicarbonate. CA, carbonic anhydrase.

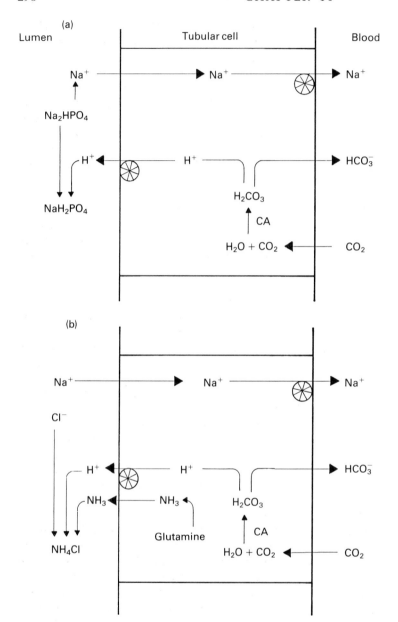

Fig. 11.12 Formation of HCO_3^- and excretion of H^+ through (a) monobasic phosphate; and (b) NH_4^+ formation. Both act as buffers which combine with H^+ in urine.

3 H^+ formed in the tubular cells is secreted into the tubular lumen.

4 H^+ secreted is accepted by urinary buffers (HCO_3^-, HPO_4^- and NH_3).

Tubular maximum for bicarbonate

It has been found that Tm for bicarbonate is about 2.9 mmol/min or 2.8 mmol/10 ml of filtrate. Figure 11.13 gives a diagrammatic representation of the tubular maximum for bicarbonate, above which it appears in urine. When appreciable amounts of bicarbonate appear in urine, this indicates the presence of alkalosis.

Under normal conditions (GFR = 125 ml/min),

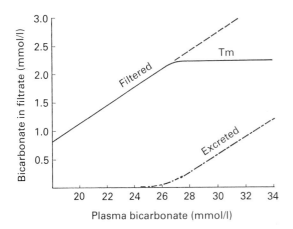

Fig. 11.13 Tubular maximum (Tm) for the reabsorption of bicarbonate.

up to 3 mmol/min of bicarbonate are filtered and almost all of that bicarbonate is reabsorbed. Any increase in filtered bicarbonate results in bicarbonate appearing in urine.

HYDROGEN IONS

Excretion of hydrogen ions

The rate of formation of H^+ by the tubular cells is directly related to the concentration of CO_2 in the extracellular fluid. The more CO_2 is formed, the more of it passes into the tubular cell and the more H^+ is secreted (Fig. 11.12).

The excretion of H^+ by tubular cells, as seen from the above account, is achieved by means of their acceptance in urine by several buffers:

1 They may combine with HCO_3^- in the filtrate, facilitating the reabsorption of filtered HCO_3^-.

2 They may be conjugated with dibasic phosphate, giving monobasic phosphate ions ($H_2PO_4^-$). Since phosphate is all filtered, its quantity is limited and therefore in acidosis it will readily be utilized as a buffer.

3 H^+ may combine with NH_3, formed by the renal tubular cells, to form NH_4^+, which combines with Cl^- to form NH_4Cl. This salt is excreted in urine. In the latter two processes, for each H^+ secreted by the tubular cells, a 'new' bicarbonate ion is added to the bloodstream.

The excretion of hydrogen ions is achieved in

exhange for Na^+ (i.e. H^+-cation exchange). There is evidence to suggest that K^+ competes with H^+ for exchange with Na^+. In hyponatraemia, more Na^+ reabsorption takes place in exchange for K^+. This leaves little K^+ to exchange with H^+, which leads to acidosis. In hypokalaemia, more K^+ is reabsorbed and therefore more H^+ is secreted, which leads to alkalosis. Hyperkalaemia, on the other hand, results in more K^+ excretion and retention of H^+, with the secretion of alkaline urine, leading to acidosis.

Secretion of ammonia

Ammonia is formed in the renal tubular cells from glutamine, which is broken down by the enzyme glutaminase to glutamic acid and ammonia. NH_3 is secreted into the lumen. H^+, secreted by the tubular cells into the lumen, combines with NH_3 to give NH_4^+. About two-thirds of the H^+ secreted by the renal tubules is excreted in this form.

NH_4^+ is an important buffer, because it forms a weak acid and therefore can take large quantities of H^+ with little change in pH. The rate of production of NH_3 is proportional to the rate of H^+ formation in the renal tubular cells. The more acidic the urine, the more NH_3 is produced. Unlike the phosphate buffer, NH_3 is produced by tubular cells and therefore the kidney can increase its NH_3 production up to 50 times when there is excessive need for excretion of H^+.

Urine acidity

Secretion of H^+ can take place until its concentration in the tubular lumen is 900 times that in the extracellular fluid. This is the maximum transport gradient for H^+ and corresponds to a urine pH of 4.5 (limiting pH). The buffers in urine determine whether this pH is reached quickly or not. If the H^+ secreted forms strong acids, urine pH falls quickly, but, if weak acids are formed, then the fall of urine pH is slower and more H^+ can be excreted.

The titratable acidity of urine can be measured by adding a strong base to urine until the pH 7.4 (normal pH of plasma) is reached. This will measure the acid excretion by the kidney. It is usually 20 to 40 meq of H^+ per day. It should be emphasized, however, that the titratable acidity does not include H^+ which is excreted as bicarbonate or in combination with ammonia.

Micturition

In the adult, passing urine is a voluntary act. However, there are underlying reflex mechanisms which deserve special description for the understanding of disturbances of micturition. The urinary bladder receives innervation from the parasympathetic outflow of the sacral nerves and also a sympathetic supply from the hypogastric nerves. These nerves are distributed to the detrusor muscle of the bladder as well as the internal sphincter. The external sphincter receives somatic nerve supply from the pudendal nerves.

ANATOMICAL AND FUNCTIONAL
CONSIDERATIONS

Urine travels down the ureters by means of peristaltic waves, which are initiated by a rise of pressure in the renal pelvis. These waves occur at the rate of one every 10 s and travel down the ureter at the rate of 3 cm/s.

The ureters enter the lower part of the bladder at the base of the trigone. They pass obliquely into the muscle coat of the bladder, forming a valve-like arrangement which makes it difficult for urine to pass back into the ureters, even when the pressure in the bladder rises.

When the bladder is empty, its internal pressure (intravesical pressure) is about zero. When up to 100 ml of urine is collected in the bladder, the pressure rises rapidly to about 10 cmH$_2$O. Additional collection of urine up to 300 ml produces no further rise in the pressure (Fig. 11.14). This is due to relaxation of the bladder. When about 350–400 ml of urine has collected, the intravesical pressure rises rapidly. This stimulates stretch receptors in the wall of the bladder and the micturition reflex is initiated. The reflex causes the bladder to contract and empty its contents. This emptying reflex has its afferent pathway in the parasympathetic sacral nerves. There are also parasympathetic efferent nerves, which come from the same sacral segments (S2, 3 and 4). These nerves trigger the contraction of the detrusor muscle of the body of the urinary bladder and relaxation of the internal sphincter.

This basic reflex arrangement can be modified in two ways:

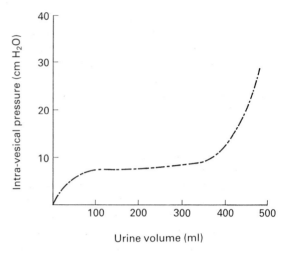

Fig. 11.14 Pressure changes in the urinary bladder.

1 Voluntary control.
2 Sympathetic modification.

Voluntary control of micturition

Reflex contraction of the detrusor muscle results in repetitive contraction of the detrusor muscle. If the time and situation are not appropriate for passing urine, there is voluntary suppression of the reflex. The reflex becomes inhibited for a few minutes up to 1 hour. Further stretch of the bladder will initiate another series of contractions that are more frequent and more powerful.

There seem to be inhibitory and facilitatory centres in the brain stem and in the cerebral cortex. Afferents to these higher centres possibly travel in the lateral spinothalamic tracts of the spinal cord. The efferents from higher centres travel deep to the lateral corticospinal tracts.

These higher centres exert their influence as follows:

1 They send impulses which keep the micturition reflex inhibited except when there is a desire to pass urine.
2 They can inhibit reflex micturition even when it has already started. This happens when the time and place are not suitable.
3 Voluntary control of the external sphincter can check emptying of the bladder.
4 Higher centres facilitate the micturition reflex and relax the external sphincter when it is convenient to empty the bladder.

Role of the sympathetic nerves in micturition
Sympathetic stimulation results in relaxation of the detrusor muscle and contraction of the internal sphincter. This occurs in condition of 'fight or flight' to prevent incontinence of urine. The sympathetic nerves have no role in the control of normal micturition.

DISTURBANCES OF MICTURITION
Damage to the spinal cord above the sacral segments results in the abolition of the voluntary control of micturition. There is no voluntary control and automatic emptying occurs every time the threshold for the reflex is reached. This occurs in patients with paraplegia. However, some paraplegic patients empty the bladder when the time is convenient and initiate the reflex by stimulating the upper thigh or the skin near the genital organs. During spinal shock, which immediately follows damage to the spinal cord, the micturition reflex is abolished. The bladder fills up and overflow incontinence occurs. In a few days to a few weeks, the shock state passes and the reflex returns.

Damage to the afferent pathway from the bladder results in loss of the reflex. The bladder tends to overfill and overflow occurs (autonomous bladder). It eventually becomes atonic.

Endocrine functions of the kidney
The kidney produces several substances some of which may not strictly be labelled as hormones. These are renin, erythropoietin, and vitamin D_3 (1,25-dihydroxycholecalciferol). These substances either act locally or are responsible for the production of other hormonal agents.

RENIN
Renin is produced by the juxtaglomerular cells (Fig. 11.4) in response to a variety of stimuli, e.g. renal ischaemia as in low blood volume, low extracellular fluid volume, low serum Na^+, high serum K^+ concentration, and others as will be explained later. Renin release is the first step in the activation of the renin–angiotensin system, which results in the production of angiotensin II. The actions of angiotensin II can be summarized as follows:
1 It produces arteriolar constriction, resulting in an increase of both systolic and diastolic blood

pressure. It is four times as potent as noradrenaline. It is not useful to use therapeutically in cases of sodium depletion because angiotensin II receptors will be already occupied with endogenously produced angiotensin II. Its physiological role in the regulation of arterial blood-pressure is mainly through the maintenance of Na^+ concentration rather than of vascular tone.
2 Angiotensin II increases the production and secretion of aldosterone by the adrenal cortex. This response tends to correct low ECF volume and blood-pressure through sodium conservation.
3 Angiotensin II modulates sympathetic activity by means of a facilitating effect on catecholamine synthesis and release.
4 Angiotensin II seems to participate in the regulation of water intake by a direct action on the hypothalamus, stimulating the sensation of thirst.

Angiotensin III is formed from angiotensin II by the action of an aminopeptidase. It has much less pressor activity than angiotensin II, but it retains all its stimulating action on aldosterone secretion. It has been suggested that both angiotensin I and angiotensin II can be converted to angiotensin III in the adrenal cortex but the physiological significance of this mechanism is still uncertain.

Regulation of renin secretion
Renin is released in conditions of low blood-pressure, hypovolaemia and Na^+ depletion:
1 Low arterial blood-pressure is detected by intrarenal baroreceptors residing in the afferent arterioles. Renin is released in response to a fall in the pressure in the afferent arterioles.
2 Low sodium and chloride ions at the region of the macula densa stimulate the release of renin from JG cells.
3 Prostaglandins stimulate renin release by direct action on JG cells. Also, the stimuli mentioned in 1 and 2 above may be mediated through the release of prostaglandins in the renal cortex.
4 Plasma concentration of K^+ and the release of renin are inversely related. This may also be associated with the rate of delivery of Na^+ and Cl^- ions at the level of the macula densa.
5 Angiotensin II inhibits the release of renin, probably through a feedback mechanism on the JG cells.
6 Vasopressin inhibits the secretion of renin.

7 Increased sympathetic discharge leads to the release of renin, both by the action of circulating catecholamines and through the sympathetic nerves to the kidney. This is a β-adrenergic receptor mechanism.

It appears that there are close interrelationships between the above mechanisms, involving hypotension, hypovolaemia and Na^+ depletion. It is possible that more than one mechanism may be working at any one time.

ERYTHROPOIETIN

It is now established that erythropoietin is produced by peritubular cells in the renal cortex and other medulla. It is released when the kidney is exposed to hypoxia. The physiological role of erythropoietin is to stimulate the bone marrow stem cells to promote erythropoiesis.

The kidneys will produce erythropoietin as long as renal tissues are not getting adequate oxygen supply, as in acclimatization to high altitude. The half-life of erythropoietin is about 5 hours. If large amounts of erythropoietin are produced, polycythaemia follows. Low levels of erythropoietin are found in severe renal damage, which accounts for the anaemia associated with kidney disease.

VITAMIN D₃

Vitamin D is synthesized in the skin or taken in food in an inactive form. It is hydroxylated in the liver at the C-25 position to give 25-hydroxycholecalciferol (calcidiol), which is the major form of vitamin D in the blood. However, this form does not seem to be active to act on the gut, kidney or bone.

Calcidiol is hydroxylated at the C-1 position by a specific kidney enzyme to give 1,25-dihydroxycholecalciferol (calcitriol). This is the active form, which acts on:

1 The gut to regulate calcium and phosphate absorption.

2 The kidney to regulate calcium and phosphate excretion.

3 Bone to regulate Ca^{2+} mobilization and exchange with serum calcium.

All of these actions are achieved together with parathyroid hormone. Low serum calcium stimulates parathyroid hormone (PTH) release, which, in turn, increases the conversion of 25-dihydroxycholecalciferol into 1,25-dihydroxycholecalciferol in the kidney. However, low serum phosphorus (P) does not require the mediation of PTH for the conversion of vitamin D into the physiologically active form. The significance of the difference in the two mechanisms lies in the fact that low Ca and low P do not usually occur at one and the same time. The stimulus is usually low Ca or low P. In case of low Ca, the released PTH mobilizes both Ca and P from the bones. PTH will enhance the loss of the excess P in urine. In case of low P, 1,25-dihydroxycholecalciferol will increase without the release of PTH. Both Ca and P will be mobilized from bone by 1,25-dihydrocholecalciferol and, in this case, excess Ca will be lost in urine due to the absence of PTH.

PROSTAGLANDINS

Prostaglandins are a group of 20-carbon unsaturated fatty acids. They are rapidly metabolized, particularly by the lungs. They have a short half-life (minutes), which makes them act as local hormones, either within the cells in which they are formed or on the organ in which they are released.

Renal prostaglandin E (PGE) is released in response to renal ischaemia due to sympathetic stimulation, catecholamine release or angiotensin II. It may be involved in the regulation of renal blood flow in these conditions.

PGE_2 and PGI_2 have a vasodilator effect and lead to increased renal blood flow on intrarenal infusion. PGA_2 and PGF_2, on the other hand, have a vasoconstrictor effect. Renal prostaglandins may also modify the action of ADH by reducing the sensitivity of renal tubular cells to the effect of ADH.

Further reading

1 Guyton, A.C. (1991) The kidneys and body fluids. In *Textbook of Medical Physiology*, 8th edn, pp. 273–329. Saunders, Philadelphia.
2 Hladky, S.B. & Rink, J.J. (1986) *Body Fluid and Kidney Physiology*. Edward Arnold, Sevenoaks, Kent.

12: Regulation of Blood pH

Objectives

On completion of the study of this section, the student should be able to:

1 Define the major clinical classifications of acid–base disturbances.

2 Describe the roles of blood buffers in order to interpret laboratory findings.

3 Explain the respiratory and renal mechanisms concerned with acid–base regulation so as to diagnose and manage the underlying causes of acid–base disturbances.

4 Identify and explain the main responses to acid–base disturbances so as to interpret clinical and laboratory findings.

5 On reviewing clinical information and laboratory data, differentiate between the various types of acidoses and alkaloses.

6 Explain the main relationships between body electrolytes and acid–base disturbances.

Introduction

Acid–base disturbances are one of the most worrying clinical problems to the physician, mainly because of the subtle nature of the physiological changes which lead to them. Furthermore, acid–base imbalance usually indicates a serious disturbance of physiological control mechanisms. It is always easy to suspect acidosis or alkalosis clinically, but it is not as simple to assess it quantitatively for the purpose of management. Most hospitals, especially in developing countries, do not have the laboratory facilities to measure blood pH or the other parameters necessary for evaluation of acid–base status. The doctor ends up relying on simple observations and clinical judgement. Under all circumstances, the underlying cause of the disturbance, whether respiratory or metabolic, should receive immediate attention.

The constancy of the pH in the blood and the extracellular fluids is one of the important characteristics of the internal environment. The normal H^+ concentration in the plasma and extracellular fluid is 0.00004 meq/litre (40 nmol/litre). The term pH is the negative log of 0.00004, which is 7.4, i.e. it is slightly alkaline. The cells of the body function within the normal range of 7.35–7.45 (36–44 nmol/litre). Minor deviations outside this range cause serious effects in almost all body functions. Death usually occurs if the pH falls below 7.0 or rises to 7.8.

Every 24 hours, metabolic processes result in the production of about 12 000 to 20 000 mmol of CO_2 (volatile acid). Other acids, such as sulphuric and phosphoric acids, are given the name non-volatile or fixed acids. It is evident that the body has to deal with these acids continuously. In clinical practice there are many conditions which lead to disturbances of the blood pH. The following section explains the major clinical abnormalities and the physiological mechanisms responsible for the maintenance of the blood pH.

Disturbances of acid–base balance

There are many clinical conditions which disturb the acid–base balance. The result is a state of *acidaemia* or *alkalaemia*, which signifies the change in pH of the blood. More commonly, the terms acidosis and alkalosis are used to signify the generalized nature of the disturbance, which affects the whole of the internal environment. Acid–base imbalance is usually a complication of diseases affecting the systems which regulate the production or excretion of H^+. The clinical classification of such disturbances follows.

METABOLIC ACIDOSIS

Metabolic acidosis results from excess production of acid, as in uncontrolled diabetes mellitus, or from failure to excrete acid, as in renal insufficiency. Metabolic acidosis can also occur if large amounts of bicarbonate are excreted, as in diarrhoea. Fluid losses from the lower intestine and colon contain appreciable amounts of bicarbonate from the pancreatic juice and the bile. This bicarbonate is normally reabsorbed as it goes down the bowel.

In metabolic acidosis, there is a decrease in the bicarbonate component of the plasma bicarbonate–carbonic acid buffer, with little or no change in the carbonic acid.

METABOLIC ALKALOSIS

Metabolic alkalosis occurs in conditions where there is excessive ingestion of bicarbonate, as in the treatment of gastric hyperacidity (heartburn) and peptic ulcer. Metabolic alkalosis also occurs as a result of excessive vomiting, such as in patients with high intestinal obstruction (e.g. due to pyloric stenosis). The same thing happens after removal of large amounts of gastric secretions through a stomach tube (gastric suction). Loss of gastric juice, which contains appreciable amounts of hydrochloric acid, leads to metabolic alkalosis due to several mechanisms:

1 *Anionic stage*. For each molecule of hydrochloric acid secreted by the gastric parietal cells, one molecule of bicarbonate is reabsorbed in the circulation (alkaline tide). In persistent vomiting, the hydrochloric acid lost from the stomach is replaced by further secretion from the parietal cells. This will lead to accumulation of bicarbonate in the blood; hence the term metabolic alkalosis.

2 *Cationic stage*. With excessive vomiting, Na^+ is lost from the body. When Na^+ depletion is excessive, the kidney will conserve Na^+ in exchange for K^+ and H^+. Consequently, H^+ is secreted by kidney tubular cells and bicarbonate is reabsorbed into the circulation.

3 *Potassium loss*. With persistent vomiting, K^+ is also lost in the vomitus and in the kidney. As extracellular K^+ falls, more K^+ will leak out of cells. This is replaced by H^+ entering into the cells. The net effect is extracellular alkalosis and intracellular acidosis.

In metabolic alkalosis, the bicarbonate fraction is high with little change in carbonic acid.

In most of the above causes of metabolic alkalosis, there is a deficit of chloride ions; hence the term *hypochloraemic* alkalosis. The anion which replaces chloride in the extracellular fluid to maintain the electrochemical balance is bicarbonate.

RESPIRATORY ACIDOSIS

Respiratory acidosis occurs in disease processes which interfere with normal respiratory functions, such as congestive heart failure, pneumonia and asthma, or in depression of the respiratory centre by drugs such as morphine and barbiturates. Therefore, respiratory acidosis is due to accumulation of CO_2 in the blood.

In respiratory acidosis, carbonic acid is high relative to bicarbonate.

RESPIRATORY ALKALOSIS

Respiratory alkalosis occurs after hyperventilation, as in hysteria, high altitude and hepatic failure. It is due to excessive washing out of CO_2 from alveolar air, which leads to a low P_{CO_2} in arterial blood.

In respiratory alkalosis, the carbonic acid fraction is low, with little change in the bicarbonate level.

Table 12.1 gives illustrative examples of the bicarbonate–carbonic acid changes which may be encountered in cases of clinical acid–base disturbances of the four types above.

In conclusion, it appears from Table 12.1 that,

Table 12.1 Examples to illustrate the bicarbonate–carbonic acid ratios in the various types of acid–base disturbances

	HCO_3/CO_2 (mmol/l plasma)	Ratio in plasma	pH
Normal	26/1.30	20/1	7.4
Metabolic disturbances			
Acidosis	13/1.30	10/1	7.1
Alkalosis	52/1.30	40/1	7.7
Respiratory disturbances			
Acidosis	26/2.60	10/1	7.1
Alkalosis	26/0.65	40/1	7.7

in metabolic acid–base disturbances, it is the bicarbonate which is disturbed, with little change in carbonic acid. In respiratory abnormalities, the main disturbance lies with the carbonic acid.

Mechanisms of acid–base regulation

Any deviation from the normal pH is counteracted by the mechanisms which tend to stabilize the pH at 7.4. There are three mechanisms which deal with the maintenance and regulation of blood pH, namely the blood buffers, the respiratory mechanisms (i.e. pulmonary ventilation) and the renal mechanisms. The following account gives the main features of these mechanisms, with the objective of integrating the knowledge already gained in studying the blood, the respiratory system and renal physiology.

THE BLOOD BUFFERS

The blood buffers are the first line of defence which prevents deviation of the blood pH. They act within minutes. Acids are produced during metabolic activity. CO_2, leading to the formation of carbonic acid, is the main acid product. Other acids, such as lactic acid and keto acids (acetoacetic acid and β-hydroxybutyric acid), are produced under conditions of metabolic stress. Lactic acid accumulates in the muscles and the extracellular fluid when inadequate supplies of O_2 are delivered to muscle, as in severe muscular exercise. The keto acids accumulate when there is increased catabolism of fat with lack of insulin, as in diabetes mellitus.

Thus, the normal metabolism is continuously producing H^+, which is taken up by the blood buffers. There are several buffer systems in the blood. The major ones are the bicarbonate–carbonic acid buffer system and haemoglobin. The first is extracellular (in the plasma) and the second is intracellular (in the red cells). Other buffers are the plasma proteins and intracellular phosphate.

The bicarbonate–carbonic acid buffer is the most important system, because it is rapidly regulated by the respiratory responses to CO_2 and H^+ via the chemoreceptor reflexes. Acidosis of whatever cause results in respiratory stimulation and washing out of CO_2. This decreases the carbonic acid level and adjusts the pH accordingly. In fact, the bicarbonate–carbonic acid buffer system determines the pH of the blood and extracellular fluid. According to the *Henderson–Hasselbalch equation*, the dissociation constant of carbonic acid (pK) and the ratio of HCO_3^-/H_2CO_3 determine the blood pH:

$$pH = pK + \log \frac{salt}{acid}$$

pK for carbonic acid = 6.1

$$\log \frac{HCO_3^-}{H_2CO_3} = \log \frac{20}{1} = 1.3$$

$$pH = 6.1 + 1.3 = 7.4$$

This buffer system is closely monitored by the reflexes of the respiratory system to maintain the bicarbonate–carbonic acid ratio at 20/1.

Haemoglobin is the second most important blood buffer for a different reason. It is present in high amounts, i.e. 14 g/100 ml. There are about 700 g of haemoglobin in the blood of an adult person. This is the reason for its high buffering capacity. The combination of haemoglobin (Hb) with H^+ sets free an equivalent amount of bicarbonate according to the following reaction:

$$H^+ + HCO_3^- + KHb \rightleftharpoons HHb + K^+ + HCO_3^-$$

Carbonic anhydrase is found inside the red cells and K^+ ions are the main intracellular cation. Therefore, this reaction can only take place in the red cells. The resulting bicarbonate ions diffuse to the plasma and they are replaced by chloride ions, which enter the cell (chloride shift). Figure 12.1 illustrates this effect during the process of

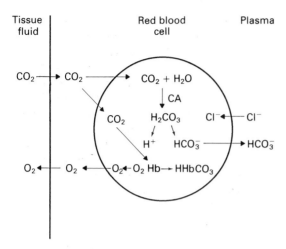

Fig. 12.1 Diagrammatic representation illustrating the combination of haemoglobin with CO_2. Note the formation of bicarbonate due to carbonic anhydrase (CA) in the red cell. Efflux of bicarbonate is associated with entry of Cl^-, the chloride shift.

CO_2 carriage in the tissues. As reduced haemoglobin is a weaker acid than oxygenated haemoglobin, its buffering capacity is greater in the tissues.

The relative importance of the blood buffers is given in Table 12.2.

THE RESPIRATORY MECHANISMS

The main advantage of the contribution of the respiratory system in acid–base regulation is its quick response. However, the correction of pH is

Table 12.2 Relative buffering capacity of the various systems in the blood

Bicarbonate–carbonic acid system	
In plasma	35%
In RBCs	18%
Total	53%
Non-bicarbonate buffers	
Haemoglobin	35%
Plasma proteins	7%
Phosphates	5%
Total	47%

incomplete. Respiration is stimulated or depressed according to the deviation of the blood pH from 7.4. The P_{CO_2} and H^+ have a potent stimulatory effect on the respiratory centre. Acidosis produces stimulation and alkalosis produces depression of respiration.

The response of the respiratory system takes place within seconds and continues for a short interval depending on the stimulus. It quickly counteracts the effect of added acid during metabolism. To illustrate this, let us consider an example where 10 mmol/litre of acid is added to the system:

$$10\ HCl + 26\ NaHCO_3 \rightleftharpoons 10\ NaCl + 10\ H_2CO_3 + 16\ NaHCO_3$$

The above reaction indicates that the system gains 10 mmol/litre carbonic acid and that the bicarbonate decreases by an equimolar amount. In the absence of the respiratory response, this is a serious deviation from normal. According to the Henderson–Hasselbalch equation:

$$pH = 6.1 + \log \frac{26 - 10}{1.3 + 10}$$

$$= 6.1 + \log \frac{16}{11.6}$$

$$= 6.1 + 0.15$$
$$= 6.25 \text{ (highly acidic)}$$

This pH is incompatible with life. It is the respiratory system which prevents this from happening. Hyperventilation will quickly decrease the P_{CO_2} and therefore the carbonic acid falls, e.g.:

$$pH = 6.1 + \log \frac{16}{1.0}$$

$$= 6.1 + 1.20$$
$$= 7.3$$

The correction of blood pH is incomplete but it is compatible with life. Furthermore, the bicarbonate level is still greatly decreased. Final correction is brought about by the kidney.

THE RENAL MECHANISMS

Food and other ingested substances are sources of non-volatile acids (i.e. other than carbonic acid). Sulphur-containing amino acids result in the production of sulphuric acid. Phosphorus from proteins and phospholipids results in phosphoric acid production. If calcium chloride is ingested, most of the calcium is not absorbed in the gut and gets excreted in the faeces; but the chloride is absorbed as hydrochloric acid. Ingestion of ammonium chloride salt gives rise to urea and hydrochloric acid. Furthermore, large amounts of keto acids accumulate in uncontrolled diabetes mellitus because of inability to complete the oxidation of these acids to carbon dioxide and water. During severe muscular exercise and in hypoxic conditions, accumulation of lactic acid can also lead to acidaemia.

The first response to the production of such acids is the immediate buffering action of plasma bicarbonate. For this reason, the plasma bicarbonate (normally 26 mmol/litre) has been given the name of alkali reserve. This explains the decreased plasma bicarbonate in cases of metabolic acidosis. The buffering effect of the plasma bicarbonate is only a temporary process. The renal function is to excrete the hydrogen ions and to recover the bicarbonate. The kidney excretes 100–150 mmol of non-volatile acid per day. This is the main reason for the acidosis which accompanies uraemic states due to renal failure. In acidosis, the renal excretion of acid may increase to 500 mmol/day. The kidney can excrete acid against a concentration gradient so that the urine pH can go down to pH 4 (compared with plasma pH 7.4). Eventually, the bicarbonate–carbonic acid ratio is returned to normal.

Three main renal mechanisms are responsible for the regulation of acid–base balance: (i) bicarbonate is reabsorbed almost completely from the filtrate; (ii) active secretion of H^+ by the renal tubules; and (iii) the synthesis of ammonia by the renal tubular cells.

The reabsorption of bicarbonate

The kidney can be viewed as adjusting bicarbonate in two ways: reabsorption of filtered bicarbonate and formation of new bicarbonate in association with H^+ secretion.

About 85% of filtered bicarbonate is reabsorbed in the proximal tubules. At the same time, water and electrolytes are reabsorbed. Because the volume of the filtrate is greatly decreased, there is little change in bicarbonate concentration or pH in the filtrate. Reabsorption of filtered bicarbonate in the proximal tubule is associated with secretion of hydrogen ions. CO_2 is formed by carbonic anhydrase in the brush order and it re-enters the tubular cell (Fig. 12.2a). The reaction results in the reabsorption of bicarbonate with equimolar secretion of H^+ in the urine.

Formation of new bicarbonate and its consequent absorption occurs during the process of H^+ secretion. In the tubular cell, carbonic anhydrase forms carbonic acid from tissue fluid, CO_2 and intracellular water:

$$CO_2 + H_2O \rightleftharpoons H^+ + HCO_3^-$$

The hydrogen ions are secreted into the tubular lumen, mainly in exchange for Na^+. The bicarbonate is reabsorbed, together with sodium. This tends to restore the bicarbonate level or to raise it above normal values in case of excess H^+ secretion by the renal tubules.

Secretion of H^+ by the renal tubules

Excess non-volatile acids can only be excreted by the kidney. This is why acidosis is one of the main features of renal failure. The mechanism for the hydrogen ion secretion is summarized in Fig. 12.2b, c.

Ammonia synthesis

Ammonia synthesis accounts for the major part of hydrogen ion secretion, especially that coming from non-volatile acids (Fig. 12.2c). The rate of synthesis of ammonia is directly proportional to the plasma pH. In this respect, the renal tubular cells are acting as chemoreceptors. Their rate of formation of NH_3 will therefore be adjusted according to the need to excrete H^+. Being a weak acid, NH_3 has a high buffering capacity. Therefore, large amounts of H^+ can be excreted in the urine in this form with little fall of urine pH.

It appears from the above account that the final adjustment of blood pH is slowly, but completely, achieved by the renal mechanisms. In particular, the kidney is the only organ which is able to excrete non-volatile acids, by virtue of its ability to synthesize ammonia and its ability to secrete H^+ actively.

Blood   Tubular cell   Lumen

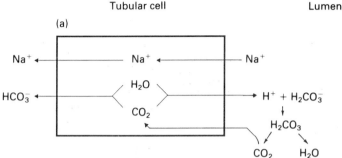

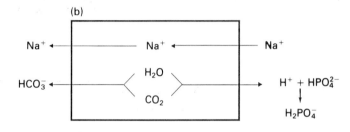

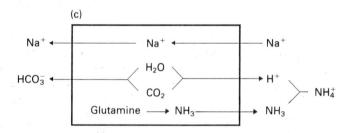

Fig. 12.2 (a) Reabsorption of bicarbonate. In the proximal tubules the filtered bicarbonate is reabsorbed in association with H^+ secretion. In this case, carbonic anhydrase present in the brush border results in CO_2 formation. Reabsorbed CO_2 is used again to form H_2CO_3 in the cell. (b) Excretion of titratable acid. Titratable acid secretion in association with monobasic phosphate in the renal tubules. (c) Excretion of H^+ with ammonia. H^+ secretion with NH_3 formed in the distal tubular cell. In this case, secretion of non-volatile acid is illustrated.

Responses to acid–base disturbances

The immediate response to any deviation of the hydrogen ion concentration in the extracellular fluid is to readjust the pH by means of the blood buffer systems. As was mentioned above, this is a temporary reaction. Adjustments have to be made by means of respiratory and renal mechanisms.

THE RESPIRATORY RESPONSE TO CHANGE
IN pH
In respiratory acid–base disturbances, the respiratory response is simply to readjust the carbonic acid fraction by increasing or decreasing ventilation and, by so doing, attempt to achieve the normal bicarbonate–carbonic acid ratio of 20/1.

In the case of metabolic acid–base disturb-

ances, however, the respiratory system cannot deal with excess non-volatile acids or get rid of excess bicarbonate. It can only readjust the carbonic acid level. Therefore, in metabolic acidosis the respiratory response of hyperventilation results in a lower carbonic acid level than normal. Conversely, in metabolic alkalosis the compensatory suppression of ventilation results in a higher P_{CO_2} and therefore a higher carbonic acid level (Fig. 12.3).

THE RENAL RESPONSE TO ACID–BASE
DISTURBANCES
The kidney helps to restore normal pH by adjusting the bicarbonate fraction and therefore works to restore equilibrium according to the

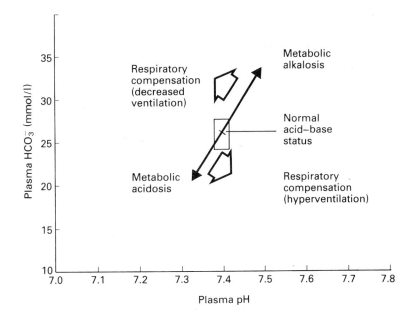

Fig. 12.3 The graph shows the relationship between plasma bicarbonate and plasma pH in metabolic acid–base disturbances. The arrows indicate the directions of change due to respiratory compensation.

Henderson–Hasselbalch relationships. HCO_3^- can only be absorbed when it reacts with H^+ in the lumen of the tubules. Therefore, alkalosis will result in decreased bicarbonate reabsorption due to lack of H^+ in the tubular lumen. Consequently, excess bicarbonate will be excreted. This decreases the bicarbonate level and helps to lower the plasma pH. In acidosis filtered bicarbonate decreases and H^+ secretion increases. The H^+ is excreted in combination with other urine buffers (such as phosphate and ammonium ions). For each H^+ secreted, there is a bicarbonate ion formed. This restores the alkali reserve and gets rid of excess H^+. These changes help to increase the plasma pH (Fig. 12.4).

Intracellular pH

In metabolic and respiratory acid–base disturbances, the cell pH remains relatively constant. The intracellular pH differs from one type of cell to another. It is generally lower than that in the extracellular fluid (e.g. 6.9 in muscle, 7.3 in renal tubule cells). It is maintained by active transport of H^+ to the extracellular fluid or the absorption of bicarbonate ions.

In respiratory abnormalities the pH of the intracellular fluid is rapidly affected by the changes in extracellular pH. This is due to the high permeability of the cell membrane to CO_2.

In metabolic abnormalities the abnormality of the extracellular fluid may not affect the intracellular pH. Change will take place rather slowly because of the relative impermeability of the cell membrane to H^+ and bicarbonate. For example, if bicarbonate is administered, it depresses respiration, leading to accumulation of CO_2. The CO_2 rapidly crosses the cell membrane, causing intracellular acidosis. As HCO_3^- cannot cross the cell membrane, there will be extracellular alkalosis but intracellular acidosis.

In alkalosis due to vomiting, loss of Na^+ and K^+ leads to exit of intracellular K^+, to be replaced by H^+ from extracellular fluids. This aggravates the extracellular alkalosis and causes intracellular acidosis.

Assessment of acid–base balance

The presence of acidosis or alkalosis can be assessed by direct measurement of blood pH, using appropriate pH metres. To determine the type of acidosis or alkalosis, the status of the bicarbonate–carbonic acid buffer system has to be measured.

Blood pH and $P\text{CO}_2$ are readily and accur-

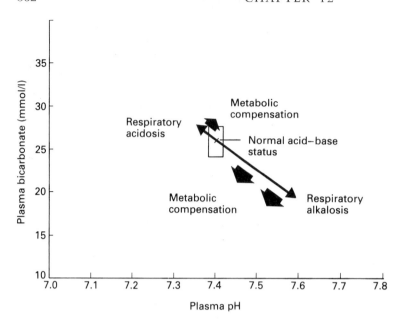

Fig. 12.4 The graph shows the relationship between plasma bicarbonate and plasma pH in the case of respiratory abnormalities of acid–base balance. The arrows indicate renal (metabolic) compensation.

ately measurable in the laboratory, using suitable equipment. The Henderson–Hasselbalch equation is used to calculate the bicarbonate concentration:

$$pH = 6.1 + \log \frac{[HCO_3^-]}{[H_2CO_3]}$$

First, P_{CO_2} is used to calculate $[H_2CO_3]$. The CO_2 dissolved in plasma depends on the partial pressure of CO_2 (P_{CO_2}) and its solubility coefficient (0.03). Therefore, $[H_2CO_3]$ can be represented in the formula as follows:

$$pH = 6.1 + \log \frac{[HCO_3^-]}{0.03 \times P_{CO_2}}$$

Thus, measurement of pH and P_{CO_2} will enable us to calculate the bicarbonate concentration. By an analysis of pH, bicarbonate, carbonic acid and P_{CO_2}, the status of acid–base balance in any clinical situation can be assessed. The following are some examples.

METABOLIC ACIDOSIS

In metabolic acidosis the bicarbonate is much reduced but the P_{CO_2} is within normal limits. If hyperventilation occurs, then P_{CO_2} will also be decreased and the pH becomes normal (compensated metabolic acidosis).

To differentiate between different types of metabolic acidosis, the *anion gap* is calculated. It is the difference between the sum of anions and the sum of cations in the extracellular fluid. It is determined by measuring Na^+ and K^+, which account for 95% of the cations, and Cl^- and bicarbonate, which account for 86% of the anions. As the total anions and cations should be equal, the anion gap is mainly accounted for by unmeasured anions, such as protein, phosphate, sulphate and organic acids.

$$\text{Anion gap} = ([Na^+] + [K^+]) - ([Cl^-] + [HCO_3^-])$$

The normal value of the anion gap is 10–16 mmol/litre, with an average of 12 mmol/litre. In metabolic acidosis the presence of acid will change the concentrations of HCO_3^- and Cl^-. The change in the anion gap indicates the type of acidosis. In metabolic acidosis due to keto acids or lactic acid, the anion gap is increased. The bicarbonate becomes used up for buffering and is replaced by fixed acids. Thus, the anion gap provides an indirect estimate of unmeasured anions such as SO_4^{2-}, HPO_4^{2-}, and other organic acids. In diarrhoea or renal acidosis, the lost bicarbonate becomes replaced by Cl^- and the anion gap remains normal. This type is known as hyperchloraemic acidosis, to differentiate it from the first type, in which the anion gap is increased and Cl^- concentration is normal, i.e. normochloraemic.

The anion gap is not increased in metabolic acidosis due to carbonic anhydrase inhibitors or ingestion of NH_4Cl. Thus, the estimation of the anion gap provides a useful laboratory test for the evaluation of patients with acidosis.

RESPIRATORY ACIDOSIS

In respiratory acidosis the P_{CO_2} is high. An attempt will be made to approximate the bicarbonate−carbonic acid ratio to normal, as CO_2 is the source of plasma bicarbonate. This will result in a slight elevation of the bicarbonate fraction, but the pH will still be low. In prolonged respiratory acidosis, renal mechanisms will retain bicarbonate further and this leads to correction of the pH but both the bicarbonate and the carbonic acid will be high, i.e. compensated respiratory acidosis.

METABOLIC ALKALOSIS

In the case of metabolic alkalosis, there is a high pH and a high bicarbonate level, with little change in the carbonic acid level. Carbonic acid may be slightly elevated (to satisfy the Henderson−Hasselbalch relationship). When the respiratory suppression takes place in response to the low H^+ concentration, the P_{CO_2} rises to high levels, bringing the pH to normal. This compensation is rather limited due to the O_2 demand, which soon stimulates respiration. Therefore, in compensated metabolic alkalosis, both the P_{CO_2} and bicarbonate are found to be elevated. This should be differentiated from compensated respiratory acidosis (discussed above) by knowledge of the clinical history and of the previous condition of the acid−base balance (Table 12.3).

RESPIRATORY ALKALOSIS

In the case of respiratory alkalosis, the P_{CO_2} is low because of the underlying hyperventilation. Due to the Henderson−Hasselbalch relationship, the bicarbonate also falls slightly. However, in chronic respiratory alkalosis, renal compensation sets in, with increased excretion of bicarbonate. Both the bicarbonate and carbonic acid will be low.

Table 12.3 Illustrative examples of plasma bicarbonate−carbonic acid values in cases of partially compensated acid−base disturbances

	HCO_3/CO_2 (mmol/l plasma)	Ratio in plasma	pH
Normal	26/1.3	20/1	7.40
Partially compensated acidosis			
Metabolic acidosis	13/0.8	16.2/1	7.31
Respiratory acidosis	46/2.6	17.7/1	7.35
Partially compensated alkalosis			
Metabolic alkalosis	39/1.5	26/1	7.51
Respiratory alkalosis	22/0.8	27.5/1	7.54

Table 12.3 gives examples of possible values obtained in compensated acid−base disturbances. Complete compensation is rather unlikely and the values given in Table 12.3 are illustrative of situations which may be encountered in clinical practice.

The electrolytes in acid−base imbalance

GENERAL CONSIDERATIONS

There are several situations where Cl^-, Na^+ and K^+ disturbances may result in abnormalities of acid−base balance. There are also situations where a primary acid−base disturbance may lead to abnormal serum electrolytes. These abnormalities may be explained by the relationship of the bicarbonate ion with Cl^- on the one hand and that of H^+ with body cations (Na^+ and K^+) on the other. Such relationships lead to disturbances in two ways: (i) exchange of similar ions across the cell membrane, i.e. between the ICF and ECF; and (ii) exchange in the renal tubules. The following account gives explanations of these relationships.

EXCHANGE ACROSS THE CELL MEMBRANE

1 Bicarbonate has a reciprocal relationship with Cl^-, e.g. the chloride shift during bicarbonate synthesis. In chloride deficiency, bicarbonate ions take its place and therefore HCO_3^- in the extracellular fluid increases.

2 H^+ and K^+ have a reciprocal relationship in the ICF, i.e. K^+ tends to leave the cells when the

intracellular H^+ rises, as in acidosis. This leads to hyperkalaemia.

EXCHANGE IN THE RENAL TUBULES

1 The secretion of H^+ is achieved in exchange for Na^+ or K^+ ions. In the case of lack of Na^+ and K^+ in the tubular fluid, H^+ excretion will be curtailed. Therefore, correction of the Na^+ deficit is one of the important measures which enable the kidney to excrete acid.

2 Reabsorption of Na^+ in the distal tubules takes place in exchange for K^+ secretion in the renal tubules, e.g. in hypersecretion of mineralocorticoids excessive K^+ will be lost in exchange for Na^+.

3 H^+ and K^+ seem to compete for the same carrier mechanism in the distal tubules. In the case of excessive K^+ in the tubular fluid, as in hyperkalaemia, H^+ secretion will be curtailed (acidosis). In hypokalaemia, H^+ secretion is enhanced (alkalosis).

Alkalosis, whether metabolic or respiratory in origin, results in increased excretion of K^+ in the urine. This is possibly due to increased entry of K^+ in the distal tubules to replace intracellular H^+. The lack of H^+ secretion will result in greater excretion of K^+, which would otherwise be reabsorbed in exchange for H^+ secretion. Hence, the association of alkalosis with hypokalaemia.

Acidosis, especially if it develops rapidly, results in decreased K^+ excretion and hyperkalaemia. This is probably due to high H^+ content of cells, including the renal tubular cells, which tends to decrease the secretion rate of K^+.

On the other hand, excessive ingestion of K^+ may lead to acidosis. The K^+ will enter the renal tubular cells and cause a net increase in secretion of potassium in the urine. This will tend to curtail H^+ excretion and therefore lead to acidosis.

EXAMPLES OF ELECTROLYTE DISTURBANCES

Diabetic acidosis

1 Diabetics with keto-acidosis are usually sodium-deficient. This is brought about in several ways:

(a) The polyuria results in excessive loss of water and sodium, as a result of shrinking of the extracellular fluid volume (dehydration).

(b) The NH_4^+ mechanism for secretion of excess H^+ fails to cope and therefore Na^+ gets excreted with fixed acid in the urine. Thus, acidosis makes the Na^+ deficit worse.

(c) On the other hand, the sodium deficiency curtails the excretion of H^+ in the distal tubules and the collecting ducts. The H^+ ions tends to stay in the extracellular fluid in the place of Na^+ and therefore the acidosis is made worse.

It appears from the above that there is a vicious circle in diabetic acidosis involving Na^+ and H^+. One of the key measures in the management of this condition is to give intravenous NaCl. The correction of the sodium deficit enables the kidney to secrete H^+. Insulin therapy also helps correction of acidosis by oxidation of keto acids. If the pH falls to 6.9, bicarbonate administration is recommended.

2 Potassium abnormalities are also observed in diabetic acidosis. Severe acidosis is usually associated with entry of H^+ and net exit of K^+ from cells. This leads to hyperkalaemia. Another cause for hyperkalaemia is a decrease in renal excretion of K^+. Potassium is believed to enter the tubular cells due to the electrochemical gradient between the interstitial fluid and the intracellular fluid. In acidosis this is curtailed by excessive entry of H^+ into the tubular cell, which leads to less K^+ secretion. However, diabetics are usually K^+-deficient, because of tissue destruction and catabolism of protein. Therefore, the hyperkalaemia does not reflect the true picture of potassium balance. On treatment with insulin and glucose, K^+ ions quickly enter into cells and may therefore precipitate hypokalaemia.

Chronic renal failure

Renal failure (uraemia) is characterized by high blood urea, associated with metabolic acidosis. This is due to failure of the renal mechanisms to excrete adequate amounts of H^+. There is also an accumulation of toxic products of metabolism, particularly those resulting from a breakdown of amino acids and nucleic acids. Uraemic subjects tend to have low intracellular potassium, probably due to toxic interference with the $Na^+ - K^+$ pump. The extracellular potassium is usually normal or high. Insulin normally augments the entry of potassium into the intracellular compartment. In uraemia there is a resistance to the actions of insulin and this probably adds to the possibility of developing hyperkalaemia.

In general, severe acidosis is associated with hyperkalaemia. A simple explanation is generalized K^+ efflux of intracellular potassium in exchange for H^+. This can be corrected by giving bicarbonate and limiting K^+ intake in the diet or in drugs containing potassium.

Patients with uraemia have a slightly expanded extracellular fluid volume, with Na^+ retention. Therefore, it is useful to restrict Na^+ intake. Renal function in these patients may become rapidly worse if they lose fluid by vomiting or diarrhoea. In this case, correcting the fluid and Na^+ loss has to be done cautiously in order not to exceed the required amount, which might result in overloading the circulation and heart failure.

In cases of renal insufficiency, hypokalaemia can be induced by vomiting or diuretic therapy. This is due to the inability of the kidney to conserve K^+ efficiently in case of excessive losses.

Abnormalities of corticosteroids
In Cushing's syndrome there is excessive secretion of cortisol, which exerts some mineralocorticoid activity. There is Na^+ reabsorption and K^+ excretion. This leads to hypokalaemia. Less K^+ will be available in the tubular cells and more H^+ will take their place, leading to excessive secretion of H^+ and reabsorption of HCO_3^-. This is probably the main explanation of the alkalosis which occurs in hyperactivity of the adrenal cortex.

In corticosteroid deficiency, Addison's disease, there is excessive loss of Na^+ in the urine, from failure of reabsorption in the distal tubules due to aldosterone deficiency. There is also potassium retention, less secretion of H^+ and acidosis.

Loss of chloride
There is an inverse relationship between chloride and bicarbonate ion concentrations in the extracellular fluid. Hypochloraemia is associated with high plasma bicarbonate levels. Alkalosis sets in but the exact mechanism is not fully understood. In hypochloraemia there is a high content of bicarbonate in the glomerular filtrate. This is known to result in increased reabsorption of bicar-

bonate. Conversely, low plasma bicarbonate in hyperchloraemia leads to less bicarbonate reabsorption, giving rise to acidosis. Restriction of salt intake is one of the methods used in the management of renal acidosis.

The effects of diuretic therapy
Diuretics that work by their inhibition of carbonic anhydrase, such as acetazolamide (Diamox), result in failure of the renal tubules to make HCO_3^-. This leads to decreased H^+ secretion and decreased HCO_3^- absorption. Therefore, acidosis sets in, with increased sodium and potassium losses.

Other diuretics (non-carbonic anhydrase inhibitors) result in the excretion of a large volume of urine. The kidney works very hard and rapid absorption of Na^+ takes place. Na^+ reabsorption is associated with H^+ and K^+ secretion. Therefore, alkalosis sets in, with associated hypokalaemia.

Summary
1 The whole system of blood pH regulation centres on a constant bicarbonate–carbonic acid ratio of 20/1. The normal levels in plasma are 24–28 mmol/litre for bicarbonate and 1.2–1.4 mmol/litre for carbonic acid. The pH of blood is predicted by the Henderson–Hasselbalch equation, using bicarbonate–carbonic acid as its buffer system.

2 The constancy of the pH is challenged all the time by constant production of acid during metabolism. The main acid is CO_2 (volatile acid), which can be excreted by the lungs. Other acids such as sulphuric acid, phosphoric acid, keto acids and lactic acid cannot be excreted by the lungs (nonvolatile or fixed).

3 There are two main buffers which safeguard the pH of the blood against rapid change:
   (a) The bicarbonate in the plasma neutralizes any excess acid production. The bicarbonate buffer is especially valuable because it converts H^+ from any source to carbonic acid, which can be excreted as CO_2 in the lungs.
   (b) Hb in the red cells neutralizes the excess H^+ associated with CO_2 carriage from tisues to the lungs.

Thus, the blood buffers are the first line of defence. They prevent rapid changes in pH. The buffers,

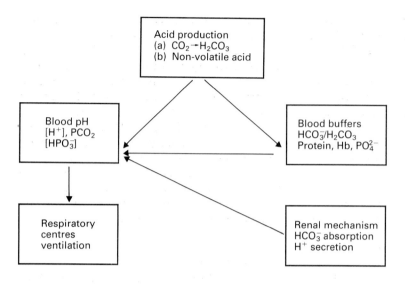

Fig. 12.5 Summary diagram of acid−base regulation.

Table 12.4 Causes and effects of acid−base disturbances

pH Disturbance	Causes	HCO_3^-	P_{CO_2}
Metabolic acidosis	Diabetic ketosis Renal failure Diarrhoea Ingestion of NH_4Cl	Low	Normal or low in respiratory compensation
Metabolic alkalosis	Ingestion of bicarbonate Excessive vomiting Excessive loss of K^+	High	Normal or high in respiratory compensation
Respiratory acidosis	Lung disease Airway obstruction Oedema of the lung Depression of respiratory centre (e.g. barbiturate poisoning)	Normal or high in renal compensation	High
Respiratory alkalosis	Hyperventilation (e.g. hysteria, high altitude, ventilators)	Normal or low in renal compensation	Low

however, have a temporary effect and cannot deal with long-term changes.

4 The respiratory system eliminates or retains CO_2 by readjustment of alveolar ventilation. Therefore, it only adjusts the carbonic acid. The respiratory response takes seconds or minutes and is incomplete in cases of bicarbonate (or metabolic) disturbances.

5 The main long-term defence is the kidney. This is achieved by reabsorption of bicarbonate and the secretion of H^+. Both mechanisms are closely linked so that, with the secretion of H^+, an equivalent amount of bicarbonate is reabsorbed. Thus, the kidney can only readjust the bicarbonate level (alkali reserve), which usually falls in acidosis as a result of the buffering

action of bicarbonate. Thus, the renal mechanisms bring about the final adjustment. This is slow but complete. It may take several days for the kidney to compensate for respiratory acidosis.

Figure 12.5 summarizes acid–base regulation.

6 Disturbances of acid–base balance are clinically classified into four types: acidosis or alkalosis, each of which may be either metabolic or respiratory in origin. Table 12.4 summarizes the main causes of these disturbances and their laboratory findings.

7 Acid–base disturbances encountered in clinical practice are usually due to failure of one of the mechanisms. The other mechanisms will attempt to compensate, e.g. in compensated metabolic acidosis, the P_{CO_2} will be low due to hyperventilation, and, in compensated metabolic alkalosis, the P_{CO_2} may be slightly elevated due to respiratory adjustments. Different types of metabolic acidosis also lead to changes in the normal value of the anion gap.

8 Acid–base disturbances are usually associated with electrolyte abnormalities. This is because of exchange between the bicarbonate and Cl^- ions on the one hand and between H^+ and both Na^+ and K^+ on the other. The exchange takes place at two sites: (i) between extracellular and intracellular fluid, e.g. the chloride shift in association with bicarbonate and the exit of K^+ from cells in exchange with H^+ in acidosis; and (ii) exchange in tubular reabsorption and secretion, e.g. Na^+ reabsorption in exchange for H^+ and K^+ secretion, H^+ secretion in exchange for Na^+ and K^+, and K^+ loss in cases of low H^+ secretion in the urine. Therefore, changes in H^+ secretion rates are reflected in electrolyte disturbances. Conversely, the availability of electrolytes in tubular fluid affects the rate of secretion of H^+.

Further reading

1 Campbell, E.J.M. (1984) Hydrogen ion (acid: base) regulation. In Campbell, E.J.M., Dickinson, C.J., Slater, J.D.H., Edwards, C.R.W. & Sikora, E.K. (eds) *Clinical Physiology*, 5th edn, pp. 218–239. Blackwell Scientific Publications, Oxford.

2 Guyton, A.C. (1991) Regulation of acid–base balance. In *Textbook of Medical Physiology*, 8th edn, pp. 330–343. Saunders, Philadelphia.

13: The Skin and Regulation of Body Temperature

Objectives

On completion of the study of this section, the student should be able to:

1 Describe the microscopic anatomy of the skin and appreciate its functional significance.

2 Acquire knowledge of the major physiological functions of the skin so as to explain abnormalities due to injury or diseases of the skin, e.g. burns, infections and loss of sensations.

3 Explain the passive and active mechanisms involved in the maintenance of thermal balance in order to interpret clinical abnormalities of temperature regulation.

4 Become acquainted with the various disorders of thermal balance so as to differentiate between them for the purpose of clinical management.

5 Become aware of the social and cultural aspects of thermal regulation in order to give appropriate advice to patients and the community.

---

ing. This may lead to disturbances of fluid and electrolyte balance.

Anatomical considerations

The skin is one of the largest organs of the body. In a 70 kg man, its total weight is about 4 kg and its surface area about 1.8 m^2. The skin has a superficial layer, the epidermis and a deep layer, the dermis. The epidermis consists of several layers of cells. That lying deepest is the basal layer of cells. Next to these are the keratinocytes, which are responsible for the continuous regeneration of the more superficial layers (Fig. 13.1). The granular cells are next. These are flat cells that have lost their nuclei and contain keratin granules. As new cells are formed, the granular cells move towards the surface and form the horny layer, which consists of dead cornified cells. It is estimated that the horny layer is entirely shed every 3 weeks.

The basal layer projects in finger-like processes (papillae) into the dermis. Thus, the epidermis is fixed to the dermis underneath. Among the basal layer cells are found the melanocytes, or pigment-

Introduction

The skin has several important functions in homoeostatic regulatory mechanisms. In addition to its protective role, the skin has a vital role in the regulation of body temperature. By adjusting the blood flow in the skin, the body can adjust the amount of heat lost to the environment. In hot climates the sweat glands secrete sweat so as to cool the skin by means of evaporation. The amount of sweat varies greatly with heat exposure. Large amounts of fluids and electrolytes can be lost through the skin during sweat-

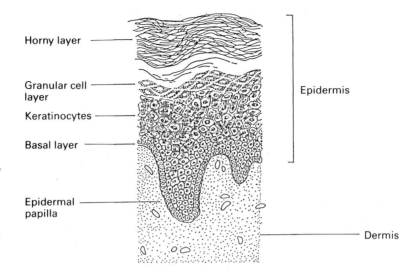

Horny layer

Granular cell layer

Keratinocytes

Basal layer

Epidermal papilla

Epidermis

Dermis

Fig. 13.1 Vertical section through the epidermis showing the layers of cells. (Reproduced from Borysenko, M. & Beringer, T. (1989) *Functional Histology*, 3rd edn. With permission from Little Brown & Co., Boston.) Students are advised to consult their histology textbook for the study of the dermis and skin appendages.

forming cells. The amount of pigment (melanin) present in the skin determines whether the skin is white, yellow, brown or black. Congenital lack of melanin formation is called albinism.

The dermis consists of dense collagen fibres, which give the skin its elasticity. Between the downward projections of the basal layer extend the papillae of the dermis, which contain the blood- and lymph vessels and nerve endings. The dermis also has extensions into the underlying subcutaneous fat. These connections fix the skin to the adipose tissue layer and prevent undue lateral displacement of the skin. When a fold of the skin is pinched, a layer of adipose tissue is picked up with it. Skin-fold thickness, as measured by a suitable instrument, is used as an indicator of the adipose tissue content of the body.

The skin has several appendages: the hair follicles, nails, sweat glands and sebaceous glands. The nail colour, texture and shape give useful clues to diagnosis of systemic disease. The hair distribution and its texture are determined by genetic and endocrine factors. The sex hormones at puberty cause hair to grow at the axillae and pubis and on the face in males. There are two types of sweat glands. The *eccrine* glands are responsible for sweating associated with the regulation of body temperature. These are present all over the body. The secretory tubules of these glands are coiled up and lie deep in the dermis.

Their ducts open on the surface of the skin. These glands are supplied by sympathetic fibres, which stimulate sweating by releasing acetylcholine. The other type of sweat glands are called *apocrine* glands, which are only found in the parts of the body with hair follicles, e.g. the axillae and pubis. Their ducts open into the hair sheath. These glands become active after puberty. Their secretion is not necessarily stimulated by heat. Decomposition of their secretion by bacteria results in the unpleasant smell of the axillae in some adults.

Blood supply of the skin

The skin has a rich supply of blood-vessels. The amounts of blood which flow in various parts of the skin may vary greatly, e.g. the face and scalp have one of the richest supplies. Wounds of the scalp and face tend to bleed more profusely.

There are two plexuses of vessels which supply the skin. Small arteries penetrate from the subcutaneous fat and form one plexus just below the dermis. Another plexus of capillaries is formed in the superficial layers of the dermis. No blood-vessels penetrate into the epidermis, which gets its nutrients through the tissue fluid. The constriction or dilatation of the arterioles determines the amount of blood flow in the skin. Normally, vasomotor tone through sympathetic α-receptors keeps the arterioles in a state of moderate balanced constriction. Decreasing the vasomotor tone will result in dilatation, while an increase will result in constriction.

There are arteriovenous shunts in the skin (glomus bodies), especially in the hands and feet. There are also venous plexuses just below the arteriolar plexuses. The veins of the skin can dilate to a larger size, e.g. on exposure to heat. This leads to accommodation of a significant amount of blood in the skin, which can be a great disadvantage when physical work is performed in hot weather. The heart has to work harder to supply the muscles with adequate oxygen and nutrients as well as maintain the increased blood flow in the skin. The heart rate increases out of proportion with the degree of muscular activity and therefore fatigue sets in much faster.

The skin blood-vessels are also regulated in response to cardiovascular reflexes (e.g. baroreceptor and chemoreceptor reflexes), which tend to regulate the cardiac output and the peripheral resistance. The changes in skin blood flow in this case are much less than those seen in regulation of body temperature.

The lymphatic drainage of the skin takes place via rich lymphatic capillaries in the dermis. These start in the tips of the papillae and collect to form larger vessels, which run into the subcutaneous fat with the larger blood-vessels. Minor or temporary blockage of superficial lymph drainage due to infection or injury causes localized lymphatic oedema around the site of infection or injury. More serious lymphatic blockage occurs in parasitic infestation with a worm, *Wucheraria bancrofti*, which causes massive swelling of distal parts, e.g. the legs and the scrotum, giving rise to the disease known as elephantiasis.

Functions of the skin

PROTECTIVE ROLE

The cornified layer of the skin is almost waterproof. It allows the body to keep its water even when the air is very dry. Conversely, it does not allow water to go in during swimming in fresh water or to go out if the body is immersed in salt water. The cornified layer also protects the underlying soft tissues from invasion by microorganisms, by chemical substances and by rough objects. When the horny layer is lost, as in cases of

burns, evaporation takes place 10–20 times faster from the underlying tissues. This is particularly serious in hot dry climates. When large areas of skin are burnt, the evaporative water loss can be enormous.

In addition to its function as a physical barrier to micro-organisms, the skin has chemical and immunological roles. The intact skin and mucous membranes have a number of secretions which are known to be germicidal, e.g. fatty acids from sebaceous glands and propionic acid produced by normal skin flora. Scattered within the stratum germinativum (Malpighi's layer) are found Langerhans cells. These are clear cells found among the keratinocytes. They seem to act as a trap for exogenous antigens which penetrate the cornified layer. They bind the antigen and migrate to the lymph nodes, where they sensitize lymphocytes. Specifically sensitized lymphocytes then migrate to the epidermis to attack the antigens trapped by the cells of Langerhans. Langerhans cells have also been found in other surface epithelium exposed to exogenous antigens such as the gastrointestinal, respiratory and genitourinary systems.

The skin pigment, melanin, which is present in the basal layer of the dermis, protects the skin against injury by ultraviolet light. The greater the amount of pigment, the less the amount of light that penetrates the epidermis. Exposure to the sun stimulates pigment formation and also leads to thickening of the horny layer. Both of these effects decrease the penetration of ultraviolet light and carcinogenic solar radiation (wavelengths less than 320 nm). It is not surprising, therefore, that individuals with congenital lack of pigment (albinos) are prone to develop skin cancer.

SENSORY FUNCTIONS

There are numerous nerve endings in the skin which serve the functions of pain, touch and hot and cold sensation. There are also specialized skin receptors for pressure, vibration, stretch and hair movement. These sensations provide the individual with information about the immediate environment. Conscious or reflex responses will take place depending upon the nature of the stimulus. In this respect, these sensations can also be considered to play a protective role.

SYNTHESIS OF VITAMIN D

Vitamin D_3 (cholecalciferol) can be synthesized in the skin from 7-dehydrocholesterol (Fig. 13.2). This indicates that this substance can also be looked upon as a hormone synthesized by the skin, which travels in the blood and acts on distant organs. As seen in the sequence of synthesis, both the liver and the kidney are essential for converting vitamin D_3 into its active form, 1,25-dihydroxycholecalciferol. The conversion of vitamin D_3 to the active form by the kidney depends on parathyroid hormone and plasma calcium levels.

REGULATION OF HEAT LOSS

The skin temperature under normal conditions is several degrees cooler than core body temperature (Table 13.1). It appears from these figures that the skin and subcutaneous tissues, shell temperature, varies with the external or ambient temperature, while the core temperature remains relatively constant. Therefore, in hot weather, the skin acts as a cooling organ for the body. Blood flowing through the skin loses heat to the skin, which, in turn, loses it to the environment. At a comfortable ambient temperature, which is usually between 23 and 30 °C, the skin blood flow is about 20 ml/min/100 g. This constitutes about 4% of the cardiac output. The regulation of heat loss through the skin is achieved in two ways:

1 *Regulation of blood flow in the skin.* This is achieved by adjusting sympathetic vasomotor tone to skin blood-vessels. In hot environments the vasomotor tone is decreased and this leads to vasodilatation and a consequent increase in blood flow. In extremely hot environments, the blood flow to the skin can increase up to 3 litres per minute (i.e. 60% of cardiac output). This means that the cardiac output has to be increased to maintain adequate flow to other organs.

In cold environments vasomotor tone is increased and this leads to decreasing the blood flow to the skin, which may fall to 2 ml/min/100 g in extremely cold climates.

2 *Sweating.* When the skin temperature or the core temperature rises, reflex sweating is stimulated. This leads to secretion of sweat on the surface of the skin. When water evaporates, it takes its latent heat of vaporization from the skin and therefore the skin temperature drops to normal or below normal levels. This is a vital mechanism for heat loss in extremely hot environments or when the body is producing large amounts of heat (as in muscular exercise, hard

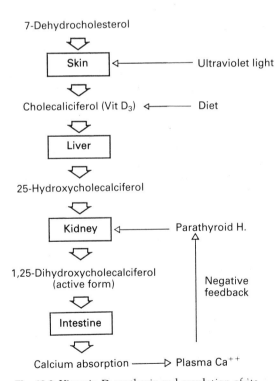

Fig. 13.2 Vitamin D synthesis and regulation of its activation by parathyroid hormone.

Table 13.1 Variations of skin and core temperatures (°C) in a subject exposed to various ambient temperatures[a]

Ambient temperature	Skin				Rectum
	Foot	Hand	Trunk	Head	
23	24.8	29.5	32.8	34.0	36.4
25	29.0	31.5	33.2	34.5	36.6
27	30.5	33.2	33.5	34.2	36.6
29	32.6	33.5	34.0	34.6	36.8
32	33.5	34.0	34.4	34.8	36.9
34	33.8	34.5	34.6	35.0	37.0

a Approximate temperatures after Dubois, F.F. (1933) *Journal of Nutrition* **15**, 477.

physical labour and fever). Failure of sweating under such conditions leads to a rapid rise of body temperature, which may lead to coma and death (heatstroke).

Thermal balance

The contributions of the skin to heat loss are under physiological control. However, the body can lose or gain heat from the environment by passive heat transfer mechanisms, i.e. radiation, convection and conduction, and also by the evaporation of water from the surface.

Homothermic, or warm-blooded, animals (e.g. mammals and birds) maintain their body temperature within a narrow range. This is achieved by regulatory mechanisms which adjust the thermal balance. Poikilothermic animals, such as fish and reptiles, are referred to as cold-blooded animals because they allow their body temperature to change with that of the environment. Camels are homothermic but they can allow their body temperature to change within a wider range than that of other mammals. This mechanism helps the camel to cool during the night and therefore not to lose water in cooling during the day. It can also allow its temperature to rise to higher levels than is compatible with normal body functions in other mammals.

Thermal balance depends on factors which cause the body to gain or lose heat. Table 13.2 gives the various factors which contribute to thermal balance. Thermal balance can be expressed by the following equation when there is no change in body temperature, as in homothermic animals (see Table 13.2 for explanation):

$$M - E \pm R \pm C = 0$$

In fevers or in hypothermia, the thermal balance can be either positive (excess of heat gain) or negative (excess of heat loss). The thermal balance equation does not come to zero:

$$M - E \pm R \pm C = S$$

where S = change in heat store due to change in body temperature, which can be either positive or negative.

Now let us consider the various factors which contribute to thermal balance. Each of the mechanisms in Table 13.2 will be explained.

HEAT LOSS

Radiation

Hot objects emit infrared heat rays (electromagnetic waves) through the air. When the body is surrounded by cooler objects, radiation will result in heat loss. If, however, the surroundings are hotter than the body, radiant heat will travel from the environment to the body. This will result in heat gain. Therefore, the contribution of radiation to heat balance can be either positive (gain) or negative (loss), i.e. $\pm R$ in the thermal balance equation. At a comfortable room temperature, about 60% of heat loss occurs by radiation.

Conduction

Heat transmission by conduction takes place between objects in direct contact with each other. Thus, clothes, chairs and beds will exchange heat with the body. Therefore, the contribution of conduction can be either positive or negative depending on the temperature of the object in contact with the skin. However, the amount of heat exchanged by conduction is very small, except when the temperature difference is high and the surface of the skin in contact is extensive, e.g. cold sponging on large areas of body surface in treatment of fevers or heatstroke. Heat gain by conduction can be increased by the use of an electric blanket and heat loss by the use of cooling beds and jackets, as in the management of heatstroke.

Conduction of heat to the air or water molecules in contact with the skin also takes place. But this loss is assisted by movement of the heated air or water in immediate contact with the skin, to be replaced by cooler air by convection. Hot air or water rises and is replaced by cooler molecules,

Table 13.2 The components of heat gain and heat loss

Heat gain	Heat loss
Metabolic heat production (M)	Evaporation (E)
Radiation (R)	Radiation
Conduction ⎫(C)	Conduction
Convection ⎭	Convection

thus creating convection currents. Similarly, conduction between the skin and fluid can take place, e.g. when the body is immersed in cool water. Convection currents also move the water in contact with the body, to be replaced by cooler water. If immersed in warm water, the body can gain heat very quickly by conduction.

Convection

It is clear from the above that convection takes place in air and water. Convection currents move the layers of air or water in contact with the body, to be replaced by other molecules with a different temperature. In a cool room, about 10% of the heat loss occurs by conduction and convection. This method of heat loss can be greatly increased by air movement, e.g. the use of fans.

Evaporation

Evaporation of water takes place from the body surface in contact with air, i.e. the lungs and the skin. About 500–600 ml of water are lost every day in this way. This is referred to as inevitable water loss, which occurs irrespective of body temperature or environmental conditions. Evaporation from the skin takes place even in cool environments, due to simple diffusion and evaporation of water from the skin. This is also known as *insensible perspiration*, to differentiate it from sweating, which is an active secretory process. Sweating greatly increases the evaporative heat loss. At comfortable temperatures, the inevitable water loss from skin and lungs accounts for about 30% of the heat loss, and this is relatively constant. Evaporation is clearly associated only with heat loss and therefore its contribution to the heat balance is always negative $(-E)$ in the thermal balance equation. The evaporation of 1 ml of water from the skin or lungs results in losing 0.58 kcal of heat.

Sweating

There are about 2.5×10^6 sweat glands in an adult person. However, the number of functioning glands varies with acclimatization. People who live and work in hot environments tend to have more active sweat glands. Persons living in cold environments and who do not perform hard physical labour tend to have a minimum number of active sweat glands. Even in these people, the number of active glands can be increased in a few days during acclimatization to heat and can therefore achieve maximum rates of sweating very quickly.

Thermal sweating is controlled by the hypothalamic heat regulating centres. Stimulation of these centres by the blood temperature leads to electrical discharge carried in the sympathetic nerves to the sweat glands. It is also possible to produce reflex sweating by stimulation of heat receptors in the skin. Quadriplegic patients can sweat reflexly on the trunk and limbs. This indicates that there are centres in the spinal cord capable of producing reflex sweating.

Sweat is a dilute fluid similar to extracellular fluid. For practical purposes, it is a dilute solution of sodium chloride. It has a specific gravity of 1.002–1.003. The concentration of NaCl in sweat varies between 50 and 100 mmol/litre. It is found that, with acclimatization, the concentration of NaCl decreases and therefore more sweat is produced without severe losses of NaCl.

HEAT GAIN

Metabolic heat gain

Under basal conditions almost all the energy produced appears as heat. If measured by O_2 consumption, this is about 1 kcal/kg/hour. As the body is composed mostly of water, its specific heat is almost 1. This means that, if there were no heat loss, the body temperature would rise by 1 °C every hour at basal conditions. It is clear, therefore, that the body must lose heat all the time, especially when the metabolic rate is high, e.g. during physical work or exercise.

Fevers cause an increase in metabolism of about 10% or more. In children this can cause a rapid rise in body temperature, leading to convulsions. Sweating during an attack of fever is a good sign and helps to cool the body.

Metabolic heat gain can be increased by means of shivering and non-shivering thermogenesis. The mechanisms responsible for body temperature regulation are described in the following section.

Mechanisms of body temperature regulation

The core temperature of the body remains relatively constant in homothermic animals in spite of the wide variations in heat production and the changes in environmental temperature. In man, we can measure the core body temperature by inserting a clinical thermometer into the rectum. Alternatively, the mouth temperature is taken as indicative of the core temperature, provided the thermometer is kept in the mouth for $1-2$ minutes to equilibrate. The body temperature has a physiological diurnal rhythm. It is highest in the afternoon and lowest in the early hours of the morning (Fig. 13.3). This is an example of circadian rhythm. The core body temperature, however, stays within a range of $1-2\,°C$ throughout the day.

The maintenance of a constant body temperature depends on regulation mechanisms, mainly mediated through the hypothalamus, where thermal regulatory centres have been identified. In the anterior hypothalamus, the preoptic nuclei are sensitive to changes in core body temperature. These centres are mainly associated with responses to heat exposure. Parts of the posterior hypothalamus are associated with the body responses to cold exposure. Sensory inputs to these regulatory centres come from two main thermal receptors:

1 *Central thermoreceptors.* Receptors sensitive to core body temperature changes have been identified in the anterior hypothalamus. They are responsible for monitoring the temperature of the blood flowing through the hypothalamus. Central thermoreceptors are also found in other parts of the central nervous system, e.g. the spinal cord.

2 *Receptors in the skin.* Nerve endings in the skin that are specialized in detection of heat or cold sensations detect changes in the temperature of the immediate environment in contact with the body.

Information from both the central and peripheral thermoreceptors is integrated in the hypothalamus and other parts of the central nervous system, so as to bring about the appropriate responses.

NEURAL MECHANISMS

Lesions in the anterior hypothalamus result in elevation of body temperature. It seems that these centres are constantly keeping the temperature of the body down, i.e. lowering the heat production. The absence of this effect results in fever and the body temperature can only be brought down by cooling the skin artificially. The barbiturate drugs have been found to bring down body temperature in cases of fever caused by central lesions.

On the other hand, lesions of the posterior hypothalamus result in a lower body temperature or hypothermia. This is particularly severe in cold environments, where the body temperature can fall to fatal levels unless artificial warming is introduced. Neural mechanisms which control heat production depend on integrated signals from both central and peripheral thermoreceptors. Impulses from the skin receptors alone or from central receptors alone do not seem to have an effect. The details of central neural mechanisms responsible for integration of thermal regulatory responses are not fully understood. However, the following responses have been amply described.

Effects of exposure to heat

The first response to heat exposure is vasodilatation of skin vessels. This is mediated by skin temperature receptors and carried in the spinothalamic tracts to the thalamus. Impulses are relayed to the hypothalamic nuclei and from there to the vasomotor centres in the medulla. Heat exposure causes an inhibition of vasomotor tone to skin vessels, which results in vasodilatation. Increased blood flow takes place and consequently heat loss by radiation increases.

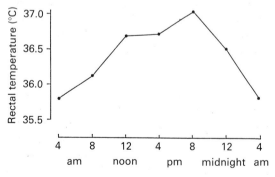

Fig. 13.3 Diurnal circadian rhythm of core body temperature.

Conduction and convection also increase, but these are responsible for smaller amounts of heat loss. Vasodilatation alone can take care of regulating thermal balance as long as the environmental temperature is 30 °C or less and as long as heat production is kept at basal or resting levels.

If the rate of heat production is high (e.g. muscular activity) or the environmental temperature is higher than 30 °C, then the core body temperature may start to rise. An increase of core body temperature of a fraction of 1 °C results in stimulation of sweating. The set point for thermal regulation is around 37 °C. It may be a fraction of a degree above or below. Figure 13.4 shows the initiation of sweating in a subject exposed to heat in relation to core body temperature. The sweat glands are controlled by the hypothalamic heat-regulating centres, which send impulses via the cholinergic sympathetic supply to the sweat glands.

Effects of exposure to cold

Vasoconstriction of the skin is the first response to cold. This results in a reduction of heat loss by radiation, convection and conduction. On further exposure, other responses take place.

Shivering Heat gain can be increased during exposure to cold by increasing the tension of the skeletal muscles at first. Then reflex shivering occurs. The mechanisms of shivering depend on the somatic nerves and cannot occur in muscles deprived of voluntary function (various types of paralysis). Both central and peripheral temperature receptors can initiate shivering. Figure 13.5 shows a diagrammatic representation of the neural mechanisms involved in thermal regulation.

The posterior hypothalamus initiates the shivering response via the midbrain and extrapyramidal nuclei, and the impulses descend in the tecto- and vestibulospinal tracts to the anterior horn cells. Impulses travel via the somatic motor nerves to produce alternate excitation in small groups of motor units in an asynchronous on-and-off fashion. Whole muscles do not contract and therefore gross movement does not occur. Shivering starts at a critical skin temperature of 27 °C in a nude person but this can be greatly lowered on wearing suitable clothes.

The rigors of fever (e.g. malaria) are due to intense cutaneous vasoconstriction with a rapid fall in skin temperature, giving the sensation of cold. Shivering can be induced by cold receptors in the skin even if the core temperature is high. The core and skin temperatures sensed by the receptors seem to be integrated at the level of the hypothalamus. Thus, we find that the thresholds for sweating and for shivering vary with various combinations of skin and core temperatures (Fig. 13.6).

Non-shivering thermogenesis In the new-born child there is a special type of fat, known as brown fat, which is found around the neck and shoulders. Brown fat disappears after the first 6 months. It is also found in animals that live in cold climates. Brown fat is richly supplied with sympathetic nerves and blood-vessels. Stimulation of these nerves releases noradrenaline,

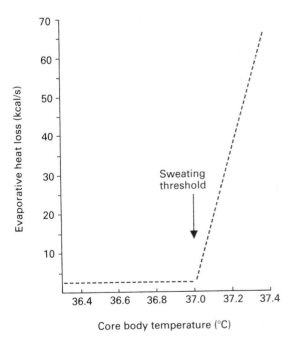

Fig. 13.4 Stimulation of sweating in a subject, showing initiation and increase in sweat rate in relation to core body temperature. (Based on data by Benzinger, T.H. (1959) *Physiological Reviews* **49**, 671.)

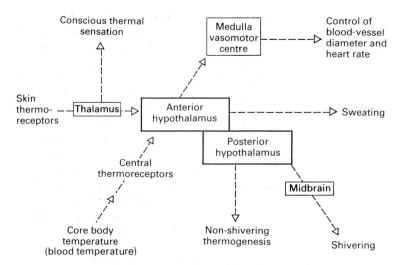

Fig. 13.5 Diagrammatic representation of the pathways of thermal regulation.

which results in increased metabolism of the brown fat cells and consequently increases heat production. Vasodilatation in brown fat helps to increase the distribution of heat throughout the body.

HORMONAL AND CHEMICAL MECHANISMS
Adrenergic stimulation results in the release of adrenaline in the circulation. This has a profound calorigenic effect. It is fast-acting but lasts for only a short time. It accounts for the sensation of heat which accompanies anxiety and stress. Sympathetic stimulation is lower during fasting and higher after feeding. In response to cold, sympathetic stimulation is responsible for the non-shivering thermogenesis described above.

Thyroid function increases slowly on prolonged exposure to cold environments. It is considered one of the mechanisms for cold adaptation, resulting in increased basal metabolic rate (BMR). In laboratory animals TSH is stimulated by exposure to cold; this mechanism is uncertain in humans.

Adaptation
Long-term exposure to hot or cold climates causes a variety of adaptive changes in thermal regulatory functions. In essence, these changes lead to more efficient maintenance of thermal balance under extremes of environmental conditions. Persons who move from one environment to another also adapt themselves by means of choice of clothes and by behavioural patterns suited to the new environment.

ACCLIMATIZATION TO HEAT
On moving from a temperate to a tropical region, an individual experiences discomfort at first but soon becomes 'used to it'. This is due to acclimatization. One of the main features of acclimatization to heat is a marked increase in the rate of sweating. This adaptive change leads to more efficient cooling, a smaller rise in body temperature, a smaller rise in pulse rate and therefore less discomfort. The stimulus for heat acclimatization is the rise in body temperature. The more frequently the body temperature rises, the greater the activity of the sweat glands.

The increase in the sweat rate is also associated with lower salt concentration in the sweat. This is due to aldosterone secretion, which acts on the sweat glands. The salt balance also affects the amount of salt in the sweat. Experiments on Sudanese and British medical students revealed that the salt intake was 13 g/day in Cambridge and only about half that amount in Khartoum. This indicates that the salt balance can be maintained on much lower salt intakes than those seen in temperate climates.

During heat acclimatization, there is an increase in blood volume. This safeguards against circulatory insufficiency on exposure to heat.

Behavioural changes also take place, e.g. avoid-

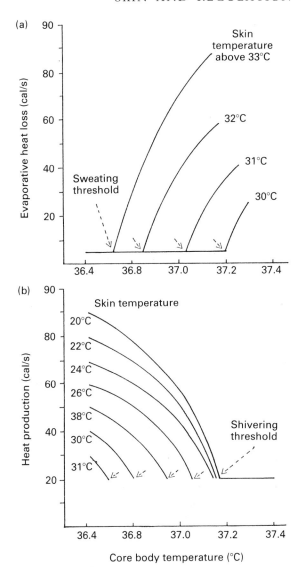

Fig. 13.6 The variation in thresholds of (a) sweating and (b) shivering, with various combinations of skin and core body temperature. (Adapted from Benzinger, T.H. (1959) *Physiological Reviews* **49**, 671.)

ing the heat by staying in the shade and by working and moving about at a slower pace. The choice of appropriate clothes plays an important part in maintaining a thermal balance.

ACCLIMATIZATION TO COLD

There is evidence that, in animals exposed to cold, thyroid gland activity increases by means of

an increase in hypothalamic thyrotrophin-releasing hormone (TRH). In man, such evidence is not available. There is evidence, however, that the metabolism increases in individuals exposed to cold. Increased metabolism is partly explained by increased sensitivity to the calorigenic effect of catecholamines. There is also an increase in corticotrophin-releasing hormone (CRH), which stimulates ACTH release and consequently the production of cortisol. Animals also increase their heat production through non-shivering thermogenesis by increasing the amount of brown fat.

Behavioural factors include avoiding cold surroundings, wearing warm clothes, consuming hot drinks and using indoor heating.

Cultural and behavioural factors in thermal regulation

HOUSING

The types of houses which form human habitats vary widely. There are the grass huts of the African tropics, the open flat tents of the nomads of Arabia, the ice igloos of the Eskimos and the concrete blocks of flats and high-rise buildings seen in large cities. The choice of building material and the design of human homes are determined by the climate, the social requirements and the economics of constructing them. The grass and mud houses of the tropics provide protection from heat by insulation. Stone and concrete housing was originally developed in temperate parts of the world, usually with fireplaces installed. More recently, with 'economic development', concrete, steel and stone have come to be used for housing in the tropics. Without the advent of air-conditioning (at high capital and energy cost), these buildings would have been uninhabitable compared with the traditional houses.

What has been said about houses also applies to the work environments. A comparison of the peaceful traditional village market-place, on the one hand, with the modern city shopping centre and industrial areas, on the other, reveals huge differences relevant to thermal regulation in man.

SOCIAL AND BEHAVIOURAL FACTORS

Climatic conditions also influence human behaviour and attitudes. In rural tropical environments, people tend to stay out of doors more often than in urbanized communities. Therefore, in the rural community there is more social interaction and more interdependence. This is reflected in local traditions and customs. Communities in the tropics meet in large groups on social occasions, such as marriage or mourning. These gatherings usually take place in the open air or in the shade of a large tree or tent erected for the occasion. Thus, climate, together with social interaction, determines many of the social and behavioural characteristics of the community. It is not surprising, therefore, that indoor living in the cities of some tropical countries, with more dependence on artificial cooling, has produced significant social and behavioural changes in the community.

The predominant ambient temperature influences individual behaviour. Heat tends to predispose to laziness. On a hot day the normal tendency is to reduce one's physical activity so that metabolic heat gain is kept to a minimum; hence the common siesta or afternoon nap. This, in turn, leads to variation in work and leisure time distribution. Studies on the behaviour of desert animals have revealed that they, too, tend to reduce their activity during the day and become more active at night. This helps to maintain thermal balance and to conserve body water. On the other hand, in temperate climates long walks during the day and long working hours are quite tolerable.

CLOTHING

The choice of suitable clothing is of direct relevance to the subject of thermal balance. The material of which clothes are made, their thickness and their colour are all determined by the climate. In tropical environments, cotton clothes are more suitable than woollens and industrial thread (e.g. nylon and polyester textiles). Wool tends to trap air in the mesh of its fibres and thus forms a thick insulating layer of air, which is a poor heat conductor. A tight-fitting garment made of wool efficiently prevents heat loss. Light cotton cloth allows air to pass through and, when it is made into a loosely fitting garment, it allows air movement between the skin and the outside by convection currents. Sweat can evaporate from the skin directly instead of wetting the clothes and evaporating from them.

Black surfaces absorb radiant heat and white surfaces reflect it. Also, radiation from white surfaces is much greater than from black ones. It follows that white clothes are most suitable for hot climates. The loosely fitting white garments made of light cotton material which are worn in many tropical countries are a remarkable example of cultural adaptation, developed long before the laws of thermodynamics were discovered.

Effect of environmental temperature on work performance

The human body has a high capacity to lose heat during muscular exercise. Even at an energy expenditure level of 20 kcal/min in an adult (about 20 times the resting level), the heat-losing systems can maintain a thermal balance. This is only possible if the environmental conditions are favourably cool, with adequate air movement for efficient evaporative heat loss. The main mechanism is the activation of sweating. Heat-acclimatized athletes can maintain thermal balance and, at the same time, achieve high muscular performance (e.g. marathon runners).

However, there are limitations to high performance in hot environments. There is the heat stress which goes with the inevitable rise in body temperature. On exposure to heat at rest, the first response is to increase the blood flow in the skin through vasodilatation. The total skin blood flow increases to about 2 litres or more. This throws an extra burden on the heart, resulting in an increase in cardiac output. In unacclimatized individuals, this is achieved by increasing the heart rate. There is evidence, however, that work in hot environments does not result in a greater cardiac output.

At the onset of exercise in the heat, there is transient cutaneous vasoconstriction. The core body temperature starts to rise; thermal cutaneous vasodilatation takes over. The vasodilatation results in pooling of blood in the periphery, which leads to decreased cardiac filling and consequently causes a decrease in stroke volume. There is no difference in cardiac output for a given exercise in hot compared with comfortable environments. However, the heart rate is increased in hot environments to maintain the higher cardiac output required for heat loss.

The haemodynamic adjustments during muscular exercise in hot environments are brought about by several reflex mechanisms. Cutaneous vasodilatation, without a concomitant increase in cardiac output, inevitably results in a fall in the arterial blood-pressure. This activates the baro-receptor reflex to increase vasomotor tone in the splanchnic circulation. Receptors in contracting skeletal muscles are also activated and they, too, stimulate the vasomotor centre and produce vaso-constriction in the viscera. Diminished discharge from baroreceptors, together with stimuli from muscle receptors, activate the cardiac sympath-etic nerves, causing an increase in the heart rate. Furthermore, there is redistribution of the blood flow, so that the liver and kidneys receive a considerably smaller share of the cardiac output. Exercise also results in haemoconcentration, due to increased formation of tissue fluid. In the presence of profuse sweating, this also leads to a reduction in blood volume.

In conclusion, it is difficult to maintain heavy muscular work in hot environments. This is partly due to the demands of an increased cu-taneous blood flow. There is also greater pooling of blood in the venous system, together with a reduction in blood volume. Consequently, there is a reduction in cardiac filling and a decrease in stroke volume, and the heart rate for the same work load in hot conditions is faster than it is in comfortable conditions. Many studies have shown that the productivity of workers falls significantly in hot working conditions (more than 24–27 °C). It is also known that accidents and injuries become more frequent in work situations where the temperature is above 24 °C or below 13 °C.

Mental work is also affected by uncomfortable environmental temperatures. It has been found that typists and wireless operators make more mistakes when working in hot or cold environ-ments. Arithmetical calculations performed at 28 °C showed fewer mistakes than either 33.5 °C or 20 °C.

Abnormalities of thermal regulation

DISTURBANCES DUE TO HEAT
Disturbances of temperature regulation can be caused by excessive exposure to hot environments at work or in cars or by direct exposure to the hot sun. Muscular work under such conditions in-creases the metabolic heat production. This results in a heat load beyond the capacity of heat-losing mechanisms. When the Muslim pilgrimage to Mecca (haji) occurs in the summer, large numbers of people are exposed to extremely hot environments by walking long distances in the open sun. Hundreds of cases of thermal, fluid and electrolyte disturbances due to heat are brought to the health units for urgent attention. This has necessitated giving appropriate medical advice and the establishment of a special service for cases suffering from various forms of heat exposure.

Exposure to high heat loads causes a variety of heat imbalance syndromes. These can be broadly classified as heatstroke and heat exhaustion.

Heatstroke (heat hyperpyrexia)
Heatstroke may be brought about by a primary failure of sweating or by an inadequate sweating response. The patient usually, but not always, has hot, dry skin. The nature of the primary failure is not fully understood. It may be explained by damage to hypothalamic sweating centres by the rapid rise in body temperature. Post-mortem studies of cases dying of heatstroke have shown evidence of central nervous system damage. It is also known that certain pyrogens (e.g. in malaria) can produce failure of sweating. This produces hyperpyrexia, associated with cutaneous vaso-constriction and shivering (rigors).

Insufficiency of the sweating response occurs in some cases of heatstroke, i.e. heat loss is inadequate for balancing the heat gain. Such in-sufficiency of sweating occurs when the skin does not have adequate numbers of active sweat glands (e.g. lack of acclimatization or damage by prickly heat rash). In some cases, insufficiency of sweating can result from fatigue of the sweat glands, e.g. after prolonged sweating in hot humid environments.

Heat exhaustion
Heat exhaustion can be caused by: (i) circulatory deficiency; (ii) dehydration; and (iii) salt de-ficiency. In most instances, there is a combination of two or more factors.

Circulatory deficiency This predisposes to heat exhaustion, which presents with weakness and fainting without a rise in body temperature or with failure of sweating. It is brought about by exposure or working in hot environments, usually in unacclimatized individuals or persons with poor physical fitness. The blood-pressure is low and the pulse may be slow (vasovagal effect). The skin is moist and cold. The fainting is caused by pooling of blood in the skeletal muscles and the skin, particularly in the lower limbs, due to failure of adjustment of the circulation.

Dehydration This is usually caused by insufficient water intake to compensate for evaporative water losses. Infants in the tropics are particularly prone to develop this condition. Normal food and salt intake replaces the salt loss but not the fluid deficit. There is a decrease in blood and ECF volume and an increase in ECF osmotic pressure. The pulse rate is high and is closely related to the degree of dehydration. Death occurs as a result of cell damage brought about by abnormally high osmotic pressure and heatstroke. It is of particular importance to be aware of this in the management of dehydrated children in the tropics. High salt or sugar intake without adequate hydration can be fatal.

Salt deficiency Salt deficiency heat exhaustion is due to excessive loss of chloride after profuse sweating, accompanied by water intake without salt replacement. Fatigue and dizziness develop, together with muscle cramps. These are painful muscle spasms of limbs and abdominal muscles. Salt depletion and heat exhaustion occur in unacclimatized individuals or workers in hot environments, e.g. boilers, bakeries, the iron industry and mining in the tropics.

The salt deficiency syndrome presents itself more often without muscle cramps. The osmotic pressure of ECF is low, which results in movement of water to the intracellular compartment. This affects muscle fibres. The kidneys continue to secrete urine. Salt conservation by the kidney is accompanied by an increase in urea reabsorption; hence there is usually a rise in blood urea.

Prevention and treatment of this condition depend on adequate salt intake.

DISTURBANCES DUE TO COLD

Hypothermia

Hypothermia is clinically defined as a drop of body temperature to 35 °C or less. Exposure to cold affects old people and babies more often than it does other age-groups. This is due to the inadequacy of their heat regulatory mechanisms, particularly the shivering response. It is generally easier to raise the body temperature than to lower it. This is due to the efficiency of shivering. Therefore, the dangers of hypothermia in the general public are much less than heatstroke when extremes of climatic conditions prevail.

The elderly and newborn babies are particularly prone to suffer from cold exposure. In the elderly, there is evidence that the sensation of cold develops less readily than in young adults. There is also an impairment of the shivering response. There may also be disturbances of vasomotor responses suggesting deterioration in autonomic functions. On the other hand, newborn babies do not have a shivering response but they can produce heat from brown fat, as described before. However, it is important to remember that, even at normal room temperatures in the tropics, a newborn or a premature infant can lose enough heat overnight to develop fatal hypothermia.

Swimmers can remain in cold water for 10–20 hours without developing hypothermia. This is achieved in two ways: (i) a well-developed layer of subcutaneous fat, which acts as an efficient insulator; and (ii) a high rate of heat production due to muscular work. However, if fatigue sets in, the swimmer becomes slower and can rapidly lose heat.

Hypothermia causes mental confusion and visual hallucinations. First there is skeletal muscle weakness. When the body temperature falls to 33 °C or below, shivering stops, and consciousness becomes affected at 32 °C. At 30 °C the patient becomes completely unconscious. When rewarmed, patients usually recover completely.

Cold injury

Freezing of the extremities (e.g. fingers and toes) occurs at temperatures appreciably below zero. This is due to the high salt content of cell water.

When freezing occurs for more than a few minutes, permanent damage to tissues occurs. This is known as frost-bite. Vasomotor control of the extremities fails because nerve conduction stops when the temperature of the limbs drops to 9 °C or less. Vasodilatation occurs with painful hyperaemia and loss of sensation.

FEVER

Fevers are characterized by an elevation of core body temperature by 1−4 °C due to infection. The increase in temperature may be fast or slow, depending on the nature of the infective organism. The fever of malaria, brucellosis and typhus (relapsing fever) is intermittent, with rapid rises of body temperature and periods of remission. Fevers caused by typhoid, enteric fevers, pneumonia and tuberculosis tend to be maintained at a continuous plateau.

During the course of an attack of fever, the temperature seems to be regulated, only it becomes maintained at a higher level than normal. It is believed that the regulatory centres are reset at a higher level. The mechanism of this resetting of the hypothalamic 'thermostat' is not understood. If the rise of temperature is rapid, a sensation of cold or chill is experienced by the patient. Actual shivering or rigors may occur. Chills and rigors are probably produced by skin cold receptors, due to vasoconstriction and cooling of the skin. On the other hand, blood coming from the periphery is relatively cooler than the temperature at which the hypothalamic 'thermostat' has been reset. It is believed that both peripheral and central thermoreceptors contribute to the sensation of cold and rigors.

The mechanism of production of the fever response to infection has been attributed to substances, called pyrogens, which act on the hypothalamus. Such substances are of two types:

1 Exogenous pyrogens are lipopolysaccharides of large molecular weight (several million). They are part of the cell membrane of pathogenic microorganisms. If injected into animals, exogenous pyrogens produce fever after a delay of up to 1 hour.

2 The endogenous pyrogen (EP), also known as interleukin 1, is a protein molecule of about 14 000 daltons. It is produced by leucocytes, macrophages and hepatic Kupffer cells in response to bacterial endotoxins and products of tissue damage. EP acts on the preoptic area of the hypothalamus to produce resetting of the thermostat at a higher level, leading to fever (Fig. 13.7). It has such a powerful effect that an injection of a few nanograms in animals produces fever in 2−3 minutes.

Prostaglandin E_1 has been found to produce fever under experimental conditions. Aspirin blocks prostaglandin synthesis but does not inhibit fever produced by its administration. It is therefore believed that the antipyretic effect of aspirin is due to its inhibitory action on prostaglandin synthesis.

Summary

EFFECTS OF EXPOSURE TO HEAT

Mechanisms which increase heat loss
1 Cutaneous vasodilatation — increased blood flow to the skin.
2 Increased heart rate — increased cardiac output.
3 Sweating.
4 Increased respiration (panting in animals).

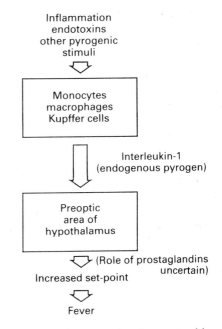

Fig. 13.7 The mechanism and pathogenesis of fever.

5 Behavioural changes — selection of suitable clothes and assuming a stretched-out, relaxed posture to increase surface area used for heat loss.

Mechanisms which decrease heat production
1 Decreased appetite, decreased specific dynamic action of food (SDA).
2 Decreased sympathetic discharge.
3 Decreased thyroid function and BMR in the long term.
4 Decreased muscular activity (behavioural adaptation).

EFFECTS OF EXPOSURE TO COLD

Mechanisms which decrease heat loss
1 Cutaneous vasoconstriction — decreased skin blood flow.
2 Assuming a curled-up position to decrease the surface area exposed to cold.

3 Hair erection (horripilation) in hairy animals.

Mechanisms which increase heat production
1 Increased appetite (increased SDA).
2 Shivering.
3 Increased voluntary muscular activity (behavioural adaptation).
4 Increased secretion of catecholamines.
5 Increased thyroid function as a slow and long-term mechanism.

Further reading

1 Emslie-Smith, D., Paterson, C.R., Scratcherd, T. & Read, N.W. (eds) (1988) Thermoregulation. In *BDS Textbook of Physiology*, 11th edn, pp. 510–533.
2 Guyton, A.C. (1991) Body temperature, temperature regulation and fever. In *Textbook of Medical Physiology*, 8th edn, pp. 797–808. Saunders, Philadelphia.

14: Exercise and Work Physiology

Introduction

Exercise physiology is concerned with the description and explanation of the changes that occur in the human body during single or repeated bouts of muscular exercise. During exercise, physiological control systems can be extended (tested) to their limit. Skeletal muscles convert chemical energy into mechanical work. To do this they use their own stores; but they have to be supplied with additional energy sources and oxygen. Co-ordination of muscle work is the function of the nervous system. The chemical and physical changes which take place during muscular exercise trigger numerous responses in almost all systems of the body, to respond to the increased demands on energy and oxygen, and to deal with the increased production of heat and waste products.

The study of muscular exercise has enabled athletes to attain better standards at different sports and made it possible to increase the efficiency of manual workers to do more work with less

Objectives

On completion of the study of this chapter, the student should be able to:

1 Understand the physiology of muscle contraction kinetics so as to differentiate between different types of muscular exercise.

2 Identify the sources of energy and control of metabolism during exercise.

3 Describe and explain the cardiovascular and respiratory adjustments which take place during muscular exercise.

4 Understand the homoeostatic mechanisms invoked during different grades of muscular work.

5 Explain the effects of training on muscles and the supporting systems.

6 Become aware of the basis and uses of various exercise tests in clinical practice.

energy cost. Exercise tests are used in clinical medicine to assess the condition of patients with cardiac, vascular or lung diseases for the purpose of diagnosis and follow-up of their treatment.

Skeletal muscle

Skeletal muscle in the human body comprises more than 40% of the total body-weight. Since it is the organ needed for work, the student requires a good knowledge of its physiology. The mechanism of skeletal muscle contraction has been dealt with in Chapter 4.

There are two types of skeletal muscles: red and white. Red muscles usually have a large blood flow, numerous mitochondria and a high concentration of oxidative enzymes. The speed of contraction and relaxation of red muscles is slow. Red muscles are therefore suitable for *aerobic work* of moderate intensity and long duration. White muscles, on the other hand, contain a highly developed anaerobic enzyme system and therefore they are more suitable to perform *anaerobic work* of high intensity and short duration. However, it was found that the colour of a muscle was not necessarily a good indication of its physiological or metabolic characteristics. A better way of distinguishing between different fibre types is by making a section of a muscle and measuring the activity of oxidative enzymes by histochemical techniques. According to the sensitivity of myosin ATPase to extremes of pH, muscle is classified into at least three types of fibre: I, IIA and IIB. However, type IIA fibres are not found in appreciable amounts in humans.

Type I fibres contain high activities of enzymes of the Krebs' cycle, of FFA oxidation and of the electron-transfer chain. They also show high content of triacylglycerols and low glycolytic capacity. Type I fibres are referred to as *slow twitch* since the time course of the maximal twitch is longer. They are also known as fatigue-resistant, a feature that depends on their capacity for FFA oxidation.

Type IIA fibres have high oxidative and glycolytic capacities with an intermediate content of triacylglycerols, whereas type IIB fibres have low oxidative and high glycolytic capacities and a low content of triacylglycerols. Type II fibres are known as *fast twitch* fibres—type IIA fibres as fast twitch—oxidative, and type IIB fast twitch—glycolytic. Therefore, red muscles contain a large number of type I fibres and white muscles a large number of type II fibres. Although it is recognized that a whole range exists between characteristically fast and slow muscle fibres, some characteristics of fast and slow twitch muscle fibres are shown in Table 14.1.

All muscle fibres receiving one motor nerve belong to one type. A motor neurone, together

Table 14.1 Characteristics of fast and slow twitch muscle fibres

---

Characteristics greater in fast twitch fibres
Myosin ATPase activity
Anaerobic glycogen and glucose metabolism
Recruitment during maximal exercise of short duration

Characteristics greater in slow twitch fibres
Activities of enzymes of the Krebs' cycle and electron transport system
Activities of enzymes for fatty acid catabolism
Number of mitochondria per muscle cell
Recruitment during submaximal exercise
Number of capillaries in the muscle

---

with all the muscle fibres it innervates, constitute one motor unit. The number of muscle fibres supplied by a single motor neurone is largely related to the complexity of the movement subserved. For example, the temporalis muscle, which performs only gross movements, has about 1000 fibres supplied by one neurone. On the other hand, the eye muscles, which carry out more complex movements, have around five muscle fibres innervated by one motor neurone. Muscle contraction is graded, since, as the strength of stimulation is progressively increased from threshold onwards, more and more tension is developed by single twitches (contractions). This is due to a gradual recruitment of more motor units. Ultimately, when all motor units are recruited, the maximum tension possible by separate twitches is reached. Further increase in muscle tension is possible only if the frequency of stimulation is increased, so that contractions add up and get fused (summation and tetanus; see Chapter 4).

TYPES OF SKELETAL MUSCLE
CONTRACTIONS
There are two types of muscle contractions:
1 *Isometric contractions*. These are also called static contractions. In this type of contraction there is no change in the length of the muscle and therefore no external work is done.
2 *Isotonic contractions*. The muscle shortens to produce a given force with constant tension. The force produced in a given movement is not really constant throughout the whole range of movement. Therefore, isotonic contractions are better referred to as *isokinetic*. During isokinetic contractions the length of the muscle may be decreased and the work is called positive (concentric) or the length may increase and the work is called negative (eccentric). The amount of work done is the product of distance times force.

The tension developed by a muscle is related to its initial length. This applies for a single fibre as well as for the whole muscle because the muscle fibres are arranged in parallel. The length at which a muscle produces maximum tension is called the resting length. Usually this is the length of the muscle in the resting state. If this length is decreased or increased, the tension developed is less. This is due to the fact that in both cases formation of cross-bridges between actin and myosin will be less efficient (see Chapter 4).

Energy sources during exercise

ATP is a high-energy compound formed inside skeletal muscle cells. In combination with ADP, ATP functions as a system to transfer chemical energy derived from the breakdown of metabolic fuels (e.g. carbohydrates and fats) under aerobic and anaerobic conditions. The concentration of ATP in muscle is only $5-7$ μmol/g fresh muscle, which will be depleted in less than 1 second during intense muscular activity and therefore requires regeneration from ADP. The ATP/ADP system couples the cellular oxidative reactions with the contractile process which is entirely dependent on it.

The basic fuels that can supply substrates for oxidation by muscle are glucose transported from the liver via the blood, locally stored glycogen and FFA removed from the blood or, to a lesser extent, derived from triacylglycerol depots within the muscle. When oxidative energy production is insufficient, muscle makes use of anaerobic metabolism to obtain energy.

The stimulation of the myofibrillar ATPase by Ca^{2+}, which initiates the process of muscular contraction, leads to a decrease in the concentrations of ATP and an increase in those of ADP, phosphate and protons; the latter changes favour ATP synthesis. This occurs when oxidative energy production is insufficient to cover the expenditure, as in the early stages of exercise, before readjustment of the blood supply to satisfy oxygen requirements. During short periods of high-intensity exercise, regeneration of ATP is essentially anaerobic, and for this there are two metabolic pathways. The first is from phosphocreatine, which, in the presence of creatine kinase, can directly rephosphorylate ADP to ATP:

Phosphocreatine + ADP ↔ ATP + creatine

The second route is from glycogen and/or glucose, with the formation of lactate.

In prolonged heavy work, oxygen supplies may also be insufficient. In such conditions glycogen is used for energy production, leading to formation of lactate. The accumulation of lactate decreases muscle activity and decreases utilization of fat to obtain energy. However, when oxygen supplies are sufficient, more fats are used for energy. The increase in circulating fat-mobilizing hormones during exercise is important for the breakdown of triacylglycerols to FFAs, which are then oxidized.

FFAs may provide up to 65% of the energy requirements during exercise, thus sparing glycogen and blood glucose. In aerobic conditions, ADP is phosphorylated to form ATP during oxidative phosphorylation by the mitochondria. The energy sources and metabolic pathways are given in Fig. 14.1 and 14.2 respectively.

The metabolic rate

The metabolic rate is the rate of energy expenditure. Energy expenditure at complete physical and mental rest is known as the basal metabolic rate (BMR). This has been dealt with in Chapter 8.

For practical purposes, energy expenditure during exercise is measured by indirect calorimetry, i.e. calculated from the rate of oxygen consumption. This is because more than 95% of the energy produced is derived from reactions of oxygen with different energy substrates. The rates of energy expenditure at different exercise or work loads are given in Chapter 8 (see Table 8.6).

Physiological responses to exercise

CARDIOVASCULAR RESPONSES

The changes that occur in the cardiovascular system during exercise are directed to increase the amount of oxygen delivered to the working skeletal muscles to meet the increased consumption. The cardiac output is increased and consequently the blood flow to the working muscles.

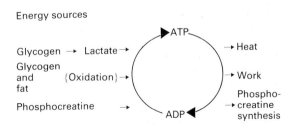

Fig. 14.1 Sources of energy for working muscles.

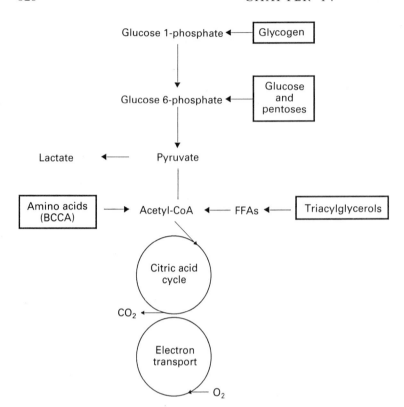

Fig. 14.2 Pathways of energy metabolism during muscular exercise.

In addition to that, the ability of the muscles to extract oxygen from the blood also increases.

Cardiac output
Cardiac output (COP) is the product of stroke volume and heart rate. At rest it is approximately equal to 5 litres/min in a young adult.

During exercise there is a significant increase in the COP. This depends on the intensity of exercise and also on the state of training of the individual. A COP of 40 litres has been reported in an athletic individual during maximal activity. Sedentary young individuals can attain a COP of 20 litres with maximal activity. The increase in COP during exercise is attained by an increase in both heart rate and stroke volume.

Heart rate
Even before the beginning of exercise there is a considerable increase in heart rate. This is due to the effect of higher centres of the brain on the sympathetic nervous system in anticipation

(expectation) of exercise. The increase in circulating catecholamines also contributes to the increase in heart rate at this stage.

When exercise starts, there is a progressive rise in heart rate. The increase in heart rate during exercise is explained by several mechanisms:
1 Receptors in joints and muscle spindles are stimulated by movement. Impulses are then transmitted from these receptors to centres in the spinal cord and then to the cardiac centres in the medulla. This results in an increase in the sympathetic discharge and a decrease in the vagal tone to the heart.
2 Catecholamines are released from the adrenal medulla due to increased sympathetic activity.
3 There is an increase in sinus rhythm due to stretch of the sinoatrial node by the increased volume of blood returning to the right atrium. The increased blood temperature also stimulates the sinoatrial node to produce a faster rhythm.
4 Muscular work results in formation of substances, such as lactic acid, which diffuse from

muscle into the blood, thus decreasing the blood pH. This stimulates the peripheral chemoreceptors, which, in turn, send impulses to the cardiac centres in the medulla.

5 Stretch receptors in the lungs also send impulses to the cardiomotor centre to increase the heart rate.

The maximum heart rate attained during exercise varies with age. Younger individuals can attain higher heart rates than older individuals. At the age of 20 years, the maximum heart rate is approximately 210/min (beats per minute). This decreases to about 170/min at the age of 60 years. The maximum heart rate at a particular age can be derived from the following equation:

$$\text{maximum heart rate} = 210 - (0.65 \times \text{age in years})$$

Stroke volume

The stroke volume is the amount of blood pumped by each ventricle per beat. At rest it is approximately 70 ml. During exercise stroke volume may increase up to twofold. Increased sympathetic discharge and catecholamines released from the adrenal medulla increase the force of contraction of the heart muscle, thus increasing the stroke volume. Another factor that causes a rise in stroke volume is the increased venous return, which increases the stretch of ventricular muscle fibres (Starling's law; see Chapter 6).

Blood circulation

Changes in the diameter of arteries and veins occur during exercise. On the arterial side, as a result of increased sympathetic discharge, there is vasoconstriction of arterioles supplying visceral structures. However, both the adrenaline secreted by the adrenal medulla and the sympathetic cholinergic nerves supplying skeletal muscle vessels cause dilatation of these vessels. An important contribution to vasodilatation in working muscle is due to the build-up of vasodilator metabolites. Therefore, skeletal muscle receives most of the cardiac output (Table 14.2). On the venous side during exercise, there is a decrease in the volume of blood contained in the veins. This is brought about by two mechanisms, namely: (i) the working muscles compress the deep

Table 14.2 Cardiac output distribution at rest and during heavy muscular exercise

	Rest	Heavy exercise
Cardiac output	5 l/min	25 l/min
Skin	5%	a
Muscle	15%	80%
Liver and gastrointestinal tract	20%	3.5%
Kidneys	20%	2.5%
Heart	4.5%	4.5%
Brain	15%	3.5%
Adipose tissue	10%	1%
Bone	3.5%	0.5%

a Skin blood flow depends on the ambient temperature and status of core body temperature regulation.

veins (muscle pump); and (ii) sympathetic venoconstriction.

Visceral blood flow From the above discussion, it is seen that the blood flow to the splanchnic area and kidneys is decreased during exercise. A significant reduction of the blood flow to these areas occurs when exercise of about 30% of the individual's maximum capacity is performed.

Coronary blood flow The coronary arteries dilate as a result of low P_{O_2}, increased P_{CO_2} and increased adenosine, lactic acid and H^+ ion concentrations. All these metabolites accumulate as a result of muscular exercise. Dilatation of the coronary arteries increases the amount of oxygen transported to the cardiac muscle, which depends almost completely on aerobic metabolism.

Skin blood flow The heat produced during exercise is lost mainly through the skin by the various physical processes (radiation, convection, evaporation). This results from dilatation of skin blood-vessels, which transfers heat from the body core to the shell.

Cerebral blood flow This is not usually affected by moderate exercise. Autoregulation of the cerebral circulation safeguards against major fluctuations due to redistribution of the cardiac output and changes in blood-pressure.

Arterial blood-pressure

Blood-pressure is the product of cardiac output and peripheral resistance. The cardiac output increases during exercise, leading to a rise in systolic blood-pressure. However, diastolic blood-pressure, which depends on peripheral resistance, remains constant. This is due to the fact that, although there is constriction of some vascular beds, there is also dilatation of other vascular beds.

Arteriovenous (AV) O_2 difference

In addition to the adjustments in cardiac output and local blood flow in the various tissues, the working muscle extracts more O_2 from the blood. In exercise the O_2 extraction rises, but rather slowly compared with the increase in the cardiac output. Maximal extraction of the order of 13–16 ml O_2/100 ml of blood has been reported. This is about three times the resting AV O_2 difference (5–6 ml O_2/100 ml).

RESPIRATORY RESPONSES

Pulmonary ventilation increases in response to exercise so that it provides the necessary gas exchange. At rest pulmonary ventilation is about 6 litres/min and the amount of oxygen consumed by all tissues is about 250 ml/min. When oxygen consumption increases to 3000 ml/min during exercise, ventilation increases to 80–100 litres/min. Athletes can attain pulmonary ventilation of 150–200 litres/min with maximal physical effort.

During physical work of low intensity, ventilation is increased mainly by an increase in tidal volume. But with heavy physical work the frequency of ventilation (respiratory rate) also increases.

The mechanisms by which pulmonary ventilation increases during exercise are not fully understood. There are only slight changes in P_{O_2}, P_{CO_2} and pH. These are not sufficient to explain the great change in pulmonary ventilation. The arterial blood pH remains constant during light and moderate exercise. During heavy exercise the pH falls and ventilation is further stimulated by the increased H^+ concentration. However, it is clear that none of the above mechanisms can account for the marked increase in pulmonary ventilation during moderate and light exercise.

Other stimuli have been suggested, such as psychic stimuli and impulses from receptors situated in muscles and joints. These stimuli are probably responsible for the rapid rise in pulmonary ventilation during the first few seconds of exercise.

Other contributing factors include a possible increase in the sensitivity of the chemoreceptors to small changes in P_{CO_2} and an increased response to the increase in core body temperature during exercise. Another theory holds that radiation of impulses from the cerebral cortex or hypothalamus may be responsible for stimulation of the respiratory centre, thus serving as a link between central neural mechanisms and working muscle. None of the theories so far proposed is completely satisfactory.

METABOLIC AND HORMONAL CHANGES

The plasma levels of most of the hormones are affected by exercise. A brief account of the response of various hormones to exercise is presented in this section:

1 *Antidiuretic hormone.* There is increased secretion of ADH in response to exercise for the purpose of conserving body water. This effect is particularly noticed when an individual performs exercise in a dehydrated state.

2 *Growth hormone (GH).* There is increased secretion of GH during exercise in the postabsorptive state. This serves the purpose of mobilizing fatty acids from adipose tissue to be used as a source of energy. GH takes about 1 hour to produce its fat-mobilizing action, so it is particularly important in prolonged exercise.

3 *Thyroid-stimulating hormone.* An increased plasma level of TSH is noticed even before exercise begins, i.e. in anticipation of exercise.

4 *Cortisol.* An increase in cortisol secretion occurs, particularly in severe exercise, and may be due to an increase in ACTH secretion. The expected stress of strenuous exercise may be the factor responsible for stimulation of cortisol secretion.

5 *Catecholamines.* A rise in plasma levels of catecholamines occurs at an exercise intensity of about 60% of maximum. The increased plasma adrenaline causes mobilization of fatty acids and also increases blood glucose level.

6 *Insulin.* Plasma insulin level decreases signi-

ficantly after exercise. Two mechanisms are involved in this, namely, decreased secretion by the pancreas and increased uptake by skeletal muscles. In view of the antilipolytic and anti-glycogenolytic actions of insulin, this decrease creates favourable conditions for more FFA mobilization and glycogenolysis.

7 *Glucagon*. There is no change in plasma concentration of glucagon in exercise that lasts up to half an hour, but its level increases three times in prolonged exercise lasting about $1\frac{1}{2}$ hours.

It can be seen from the above discussion that during prolonged exercise, glucose production is increased, as a result of increased secretion of catecholamines, glucagon and cortisol, while insulin levels fall. Following a period of exercise, muscle glycogen is restored from dietary sources under the influence of insulin or from hepatic gluconeogenesis. In skeletal muscle the importance of FFA oxidation is to spare the glycogen stores to some extent, although it does not completely replace glucose as an energy source. FFAs provide energy at a rate of about 6 kcal/min. The rest of the energy must come from glucose. During moderate exercise this presents no problem but, when work intensity increases acutely, the release of FFAs can supply only about 65% of an individual's maximum capacity. So, in very severe exercise, glycogen stores set the limit for the capacity to undertake prolonged exercise.

The contribution of the different fuel supplies to the total energy output by the muscle will vary with both the intensity and duration of the exercise and will be further influenced by the fitness of the individual, the nutritional status both before and during exercise and the level of anxiety, and even by the environment. Morphological differences between individuals in their muscle fibre make-up may also affect their use of the different fuels available during standard exercise.

Steady-state exercise

When a subject starts to exercise at a particular workload, there is a slow adjustment of the cardiovascular and respiratory systems, which are responsible for the transport of oxygen to the working muscles. This is why oxygen consumption rises gradually during the first 2–3 minutes of exercise. There is a similar response of the heart rate, cardiac output and pulmonary ventilation. Then a state is reached when all these parameters remain constant up to the end of exercise. This is called steady-state exercise. During the steady state the oxygen requirements of the tissues equal the oxygen uptake.

This explains the phenomenon of second wind, in which respiration becomes much easier and the athlete can continue in the performance of the exercise without increased breathlessness. The rate of lactic acid removal equals the rate of its production and O_2 consumption is matched by its supply. This is due to adequate cardiovascular and respiratory adjustments.

At the end of exercise, oxygen uptake declines gradually until it reaches the resting level. More oxygen is required to repay what is called the *oxygen debt*, which is defined as the oxygen uptake during the recovery period after exercise in excess of the oxygen uptake at rest (Fig. 14.3). This excess oxygen is mainly used for removal of lactic acid, to restore ATP and myoglobin oxygenation in skeletal muscle and to satisfy the increased metabolic demands of the heart and respiratory muscles during the recovery period. The term *oxygen deficit* is defined as the difference between the theoretical oxygen requirement of a given physical activity and the oxygen actually used during exercise.

Effects of training

Certain changes occur in the functions of various organs when exercise is repeated over a period of time. The magnitude of the change depends on many factors, the most important being the intensity and frequency of exercise. Age and heredity may also play a role. The nature of the change depends on the type of exercise, the muscles used and the previous training of the individual.

To attain a training level, the intensity of exercise must be at least 60% of the maximum oxygen consumption (maximum exercise capacity) or 60% of the maximum heart rate.

Changes that are produced by training disappear after some time if the person stops training. The primary effects of training occur in skeletal muscle.

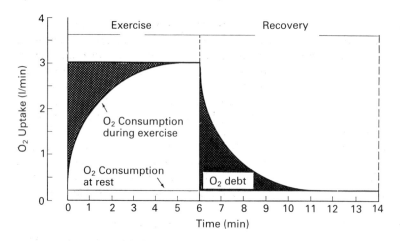

Fig. 14.3 Oxygen consumption during exercise and recovery, showing the oxygen debt.

SKELETAL MUSCLE

Muscle strength and muscle mass
Both the muscle mass and the force produced by the muscle increase as a result of training. The increase in muscle mass particularly occurs after heavy overload training (e.g. weight-lifting). There is an increase in the amount of contractile proteins (actin and myosin). The number of muscle fibres is not affected but the individual fibres increase in size. More protein synthesis and less protein breakdown occur. Amino acid uptake by muscle cells is enhanced, probably due to increased permeability of muscle cells.

The central nervous system seems to adapt by increased excitation and decreased inhibition of alpha motor neurones. As a result of this, a nerve impulse causes more actin–myosin cross-bridges after training. Muscle connective tissue, tendons and ligaments become stronger with training.

Mitochondria
Size and number of mitochondria in skeletal muscle cells increase. Also, there is an increase in the concentration of enzymes needed for utilization of fuel substances to obtain energy. Therefore, there is an increased capacity of the muscle cells to oxidize carbohydrates and lipids. Because of this greater oxidative capacity of the cell, more fat can be used. This is why we usually have a lower respiratory quotient (RQ) with training.

Energy stores
There is up to a 100% increase in the glycogen storage capacity. Phosphocreatine stores are also increased. A high carbohydrate diet enhances the storage of glycogen in muscle. The amount of glycogen is an important factor in endurance sports, e.g. long-distance running. It is also found that the activity of the enzyme systems required for oxidative metabolism are also increased. This results in about a 45% increase in the rate of oxidation.

Myoglobin content
Myoglobin stores oxygen in a manner similar to that of haemoglobin inside RBC. Training increases the myoglobin content of skeletal muscle.

Blood flow
There is an increased number of capillaries in muscle tissue, leading to increased blood flow, and therefore more oxygen is brought to the muscle cells.

There is also an increase in arteriovenous oxygen difference, which means more extraction of oxygen by muscle cells and lower lactate concentration in muscle and in blood at a given workload. This indicates that the muscles depend more on aerobic work.

THE CARDIOVASCULAR SYSTEM
With training, no change occurs in the cardiac output at any given submaximal workload. How-

ever, there is usually a marked decrease in heart rate attained at a given workload. Since the cardiac output is the product of heart rate and stroke volume, it follows that the stroke volume must be increased proportionally. Bradycardia in trained or athletic individuals is observed at rest. Resting heart rates as low as 40 beats/min have been reported. Training has been shown to increase the ventricular volume and wall thickness and this explains the rise in stroke volume. The bradycardia has also been explained by a decreased drive from trained muscles to the cardiac and vasomotor centres.

The muscles can extract more oxygen and function aerobically. Therefore, a low concentration of anaerobic metabolites is produced. This leads to a decrease in the sympathetic activity and an increase in vagal activity to the heart. Animal experiments have shown an increase in coronary vascular beds, but no evidence for this exists in man.

The mean resting blood-pressure has been shown to fall by about 10 mmHg during submaximal exercise. The blood-pressure of a trained individual increases slowly and it comes back to resting levels more quickly than in a sedentary untrained individual.

Since an athlete can attain a very high cardiac output, the diminished rise in blood-pressure indicates that there is a decrease in peripheral resistance. This can be explained by the increase in the number of capillaries and decreased sympathetic drive.

PULMONARY FUNCTION

In response to training there is a slight increase in the depth of breathing and therefore a decrease in respiratory rate during exercise.

Adaptation to exercise in hot environments

The comfortable environmental temperature for a nude person at rest is $28-30\,°C$. At this temperature the body core temperature is maintained at $37\,°C$ and the skin temperature is about $33\,°C$. During exercise, heat production is increased up to 20 times the resting values. Maximal work in a hot environment increases the core temperature up to $40\,°C$.

It takes a few days to 2 weeks for a person to get acclimatized to work in a hot and humid environment. The most important change that occurs is the ability of the person to produce more sweat with a lower salt content so that it evaporates rapidly, causing better cooling. The increase in sweating rate may be explained by the fact that more impulses are sent from the hypothalamus to the sweat glands and a larger number of sweat glands becomes active. Also, sweating begins at a lower body temperature than in unacclimatized individuals. Furthermore, physical training may increase the sensitivity of the sweat glands to signals from the brain.

When a person is acclimatized to work in a hot environment, the subject can perform a submaximal workload with a lower heart rate and a smaller increase in body temperature than before acclimatization. The lower heart rate may partly be explained by the more efficient maintenance of body temperature. However, neither cardiac output nor oxygen consumption is lowered as a result of acclimatization to heat. The plasma volume has been shown to increase by about 5%, which may also contribute to the greater sweating observed after acclimatization.

Exercise testing

In this section, a brief description of the various methods used for the assessment of physical fitness is presented. Special emphasis is given to the use of exercise testing for clinical purposes.

Two types of exercise tests are in current use: (i) tests of maximal capacity; and (ii) submaximal tests.

TESTS OF MAXIMAL CAPACITY

The aim of tests of maximal capacity is to determine the maximum ability of the individual to perform muscular work. This can be measured directly or indirectly. In the direct method, the maximum oxygen uptake (max. V_{O_2}) is measured. Max. V_{O_2} is defined as the highest oxygen uptake an individual can attain, breathing air at sea-level. When estimated under standardized conditions, max. V_{O_2} measures the capacity of the cardiovascular system to transport oxygen to the working muscles and the ability of the muscles to consume oxygen. Max. V_{O_2} is a very reproducible characteristic of the individual.

The actual procedure consists of the subject performing exercise on a treadmill or a bicycle ergometer. Usually the subject is made to work at a low workload for a few minutes and then the load is progressively increased until the subject feels exhausted and can no longer continue the exercise. Both heart rate and oxygen consumption are continuously measured. Instruments are available which can reliably record these parameters continuously. Oxygen consumption is measured in litres/min. Athletic individuals attain max. VO_2 of 5 litres/min. Max. VO_2 is taken at the point when there is no further increase in oxygen uptake with further increase in work time. It usually correlates with the maximum heart rate (Fig. 14.4).

There are some limitations for the use of max. VO_2 testing. It requires high co-operation and motivation on the part of the subject and it is not safe in patients with heart disease. Also, it is time-consuming. Because of these limitations, max. VO_2 is not suitable for population studies.

SUBMAXIMAL TESTS

There are many types of submaximal tests. Those described below are the ones that are most frequently used.

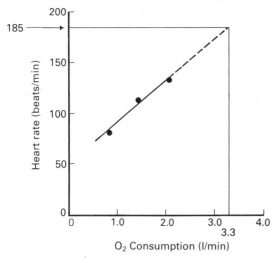

Fig. 14.4 The relationship between oxygen consumption and the heart rate at three workloads. The extrapolated broken line is extended up to a heart rate of 185 per minute and the oxygen consumption 3.3 litres/min is taken as the max. VO_2.

Tests in which both heart rate and oxygen consumption are measured

Tests measuring both heart rate and oxygen consumption are based upon the following principles:
1 The relationship between heart rate and oxygen consumption is linear throughout all workloads, including the maximum levels of work.
2 All subjects within the same age-group show a similar maximum heart rate.

The subject performs progressive workloads but is not taken to the max. VO_2. Max. VO_2 is predicted from the relationship between heart rate and VO_2 by extrapolation to the maximum heart rate for his age (see Fig. 14.4).

Tests in which only heart rate is measured
Max. VO_2 is predicted mathematically from heart rates at submaximal workloads by using regression equations relating heart rate to oxygen consumption.

Tests in which heart rate is measured at a standard oxygen consumption
The heart rate at an oxygen consumption of 1.0 litre/min or 1.5 litres/min is recorded and used as the index of physical fitness. The lower the heart rate, the better the physical fitness. The advantage of this test is that it avoids extrapolation.

Performance-scoring tests
In performance-scoring tests the score of physical fitness is the heart rate at a certain submaximal workload, i.e. the amount of work done at a specified heart rate is the index of physical fitness.

In other tests, the heart rate obtained after exercise (during recovery) is used to obtain the physical fitness score. The Harvard step test has been used extensively to assess manual and factory workers. It consists of the subject stepping up and down a stool until he is exhausted. The score is then calculated as follows:

$$score = \frac{duration\ of\ exercise}{2\ (P_1 + P_2 + P_3)}$$

where P_1, P_2 and P_3 are the pulse rates at: $1-1.5$, $2-2.5$ and $4-4.5$ min respectively. The higher the score, the more physically fit the individual is considered.

Tests based on distance run in a measured time
A test introduced by Cooper in 1968 has been used extensively to assess the physical fitness of soldiers and physical education students. The index of fitness is the distance that the individual can run in 12 min. The score is considered to be very poor when the distance run is less than 1 mile and excellent when it is more than 1.75 miles (2.28 km).

THE USE OF EXERCISE TESTS IN CLINICAL PRACTICE

It appears from the above that exercise can test the cardiorespiratory capacity to transport oxygen to the working muscles. This fact is made use of in disease conditions related to the respiratory and cardiovascular systems.

Cardiovascular disease

Exercise testing is used to confirm diagnosis and assess the severity of ischaemic heart disease (ECG stress test). ECG changes are related to the level of exercise (walking at different speeds and inclinations on a treadmill). Particular notice is given to various arrhythmias and inversion of the T wave. The Bruce exercise test protocol is used for this purpose. In addition, the test is used to screen the population in order to identify those who are at risk of developing ischaemic heart disease or sudden death.

Exercise testing is also used to assess the severity of arterial disease in the leg. A treadmill walking test will identify the level of exercise that brings claudication (leg pain due to ischaemia).

Pulmonary disease

Patients with restrictive lung disease show a marked decrease in work capacity. Some patients complain of breathlessness during performance of heavy work in their daily life. To establish a diagnosis, an exercise test is carried out and symptoms of bronchospasm are looked for.

Other patients complain of breathlessness that persists after heavy work. To confirm that, FEV_1 or peak expiratory flow is measured after 6–20 min of exercise. A summary of the uses of exercise testing is given in Table 14.3.

Table 14.3 Some uses of exercise tests

Cardiovascular system
Prediction of heart disease and surveys of cardiorespiratory fitness
Assessment of ischaemic heart disease for diagnosis and follow-up
Follow-up on drug therapy, e.g. in hypertensive patients
In rehabilitation, to restore patients' confidence after myocardial infarction
Assessment of peripheral vascular disease, e.g. intermittent claudication

Respiratory system
Assessment of exercise tolerance
Investigation of dyspnoea or hypoxia
Investigation of alveolar ventilation

Muscles
Assessment of muscle disorders

Further reading

1 Åstrand, O. & Rodahl, K. (eds) (1986) *Textbook of Work Physiology*, 3rd edn. McGraw-Hill, New York.
2 Guyton, A.C. (1991) Sports physiology. In *Textbook of Medical Physiology*, 8th edn, pp. 940–950. Saunders, Philadelphia.
3 Noble, B.J. (1986) *Physiology of Exercise and Sport*. Mosby College Pub, St Louis.

15: Physiology of Human Reproduction

> **Objectives**
>
> Upon completion of the study of this chapter, the student should:
>
> **1** Understand the neural and hormonal mechanisms concerned with the control of various reproductive functions.
>
> **2** Describe and explain the changes which occur at puberty and adolescence to be able to diagnose defects of sexual maturation.
>
> **3** Describe and explain the cyclic changes in the human ovary and uterus so as to recognize abnormalities of the menstrual cycle.
>
> **4** Understand the neural and hormonal mechanisms which control the sexual act so as to be able to recognize deviations from the normal.
>
> **5** Become aware of some common abnormalities of reproductive functions so as to pursue further study of their underlying pathology and management.

Hormonal control of reproduction

The hormonal control of reproduction in males and females is determined by the interaction of three glands: the hypothalamus, the anterior pituitary and the gonads — the ovaries in females and the testes in males.

The gonads secrete the sex steroid hormones — oestrogen and progesterone in females, and testosterone in males. These are the main hormones responsible for the sexual features and characteristics of males and females. The interrelationship between the hypothalamic—pituitary—gonadal system can be summarized as follows.

The hypothalamus secretes a decapeptide hormone, known as gonadotrophin-releasing hormone (GnRH). GnRH reaches the anterior pituitary along the hypothalamic pituitary portal vessels. It acts on the anterior pituitary to increase the synthesis, storage and secretion of the gonadotrophins: luteinizing hormone (LH) and follicle-stimulating hormone (FSH). The gonadotrophins are normally secreted in a pulsatile manner every 60–120 min. Both FSH and LH act directly on the gonads to stimulate sex steroid hormone secretion (see Chapter 10). It is also important to note that higher cerebral centres have important influences on normal development and functions of the hypothalamic—pituitary axis. Therefore, factors such as nervous stress and social or mental disorders can affect normal maturity and functions of the reproductive organs.

The differences between females and males do not only lie in the type of sex steroid hormones secreted, but also in the pattern of secretion of hypothalamic—pituitary gonadotrophins and the relationship between the gonads and the hypothalamic—pituitary system.

During the female's reproductive life, gonadotrophins are secreted in a cyclic pattern and they regulate the menstrual cycle. In males gonadotrophins have no such pattern.

Another important feature is the effect of sex hormones on the anterior pituitary gonadotrophin secretion, known as the feedback mechanism, which controls pituitary gonadotrophin hormone secretion. This mechanism is slightly different in females and males.

In females, high oestrogen levels suppress the hypothalamic–pituitary unit by inhibiting FSH secretion (negative feedback) (Fig. 15.1), while having little effect on LH secretion except at midcycle, when the high level of oestrogen induces the LH surge (positive feedback). Also, progesterone feedback at midcycle enhances the LH surge (positive feedback). In the luteal phase, the high level of progesterone acts with oestrogen to inhibit gonadotrophin secretion (negative feedback).

In males, testosterone from the Leydig cells of the testes has a negative feedback on pituitary gonadotrophins, mainly on LH and, to a lesser extent, on FSH. It seems also that other factors, e.g. the hormone inhibin produced in the testes, are responsible for the feedback regulation of the pituitary FSH hormone (Fig. 15.2). The roles of

these hormones in the control of reproductive function is discussed later in this chapter.

Puberty
Puberty is the period of maturation which follows childhood. At puberty many physiological changes take place. The most important are acceleration of somatic growth, development of secondary sexual characteristics and attainment of reproductive capacity. They are mainly due to increased secretion of the sex steroid hormones, testosterone in males and oestrogens in females. The changes at puberty occur over a period of 2–4 years, usually between ages 11 and 15 in boys and between 9 and 13 in girls (Fig. 15.3). Adolescence is another term which refers to the period between childhood and maturity; it usually extends between the ages of 10 and 19.

The exact mechanisms which initiate gonadal function at puberty are not known. But during childhood it seems that the pituitary is very sensitive to the normally detectable very low levels of sex steroid hormones, testosterone (in males) and oestrogens (in females). In addition, there is inhibition of hypothalamic gonadotrophin-releasing hormone (GnRH) secretion.

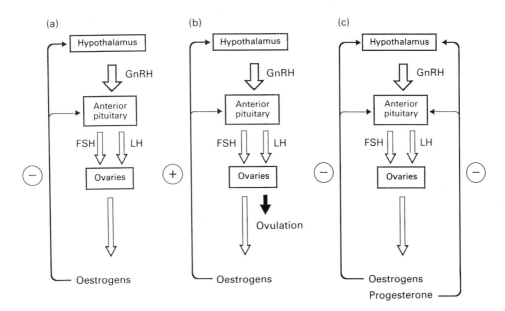

Fig. 15.1 Feedback effects of oestrogens and progesterone on the hypothalamic–pituitary axis. (a) Negative feedback in the proliferative phase. (b) Positive feedback in the preovulatory phase. (c) Negative feedback in the luteal phase.

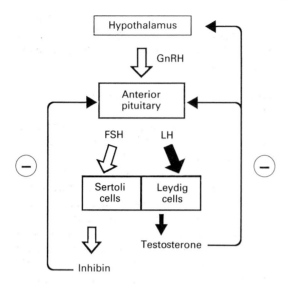

Fig. 15.2 Feedback by testicular hormones testosterone and inhibin to the hypothalamus and the anterior pituitary.

With the onset of puberty the hypothalamic–pituitary system starts to mature; it seems to become less sensitive to the inhibitory effect of gonadal steroids. The pulsatile secretions of GnRH start to appear and consequently gonadotrophins are secreted by the anterior pituitary, which, in turn, stimulate secretion of the sex steroids by the gonads. The mechanism of hypothalamic maturation is not fully understood, but it is also related to higher centres in the nervous system, which control the pulsatile secretion of the hypothalamic GnRH. Many factors affect the onset of puberty, including psychological, social, nutritional and environmental factors.

The various changes, somatic and sexual, which occur during puberty are responsible for the manifestations of puberty. It is important to recognize these features in order to be able to diagnose abnormalities. The following are the main features of normal puberty.

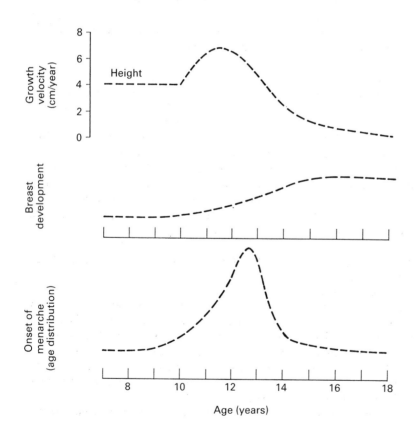

Fig. 15.3 Diagram showing timing of some of the main changes of puberty in females.

THE ADOLESCENT GROWTH SPURT

In childhood the growth curve normally runs steeply during the first 2 years, then gradually levels off. Around the age of puberty, it suddenly becomes steep again. This acceleration of growth is termed the adolescent or puberty spurt. In girls the peak is around 12 years of age; in boys it is about 2 years later. It does not begin at the same time in all parts of the body, e.g. the trunk reaches its maximal growth rate about 6 months after the legs.

DEVELOPMENT OF SECONDARY
SEX CHARACTERISTICS

1 *Distribution of fat (body contour)*. This mainly occurs in females. The body shape changes by deposition of subcutaneous fat around certain areas, such as beneath the breasts, around the hips and thighs and in the gluteal region, and the body assumes a more rounded outline.

2 *The breasts*. The development of breasts is usually described in five stages. The first stage (infantile stage) persists from infancy until puberty, when further development begins and continues for 2–5 years (average is 4 years) before the fifth stage (adult breast) is completed.

3 *Pubic hair distribution*. Pubic hair distribution has characteristic features. It has a horizontal top in females but in males it extends upward towards the umbilicus. Its growth can also be described in five stages, usually completed between 12 and 17 years of age. The axillary hair appears more or less at the same time as pubic hair.

4 *Cutaneous gland development*. Axillary and pubic apocrine glands begin to function at approximately the same time that pubic and axillary hair appear, causing a characteristic change in body odour. Also, the sebaceous gland secretion becomes thicker and predisposes to blockage and development of acne. Sweat glands of the general body become more active at about the same time.

In males there are other secondary sex characteristics which occur during puberty. The voice becomes low-pitched and deeper due to growth of the larynx and elongation of the vocal cords. Facial hair appears and the hair-line recedes at the anterolateral aspects of the forehead, with a general increase of body hair. There is also an increase in skeletal growth, broadening of the shoulders and more muscle development. Behavioural changes also occur. Boys become more active and more aggressive and develop an interest in the opposite sex.

DEVELOPMENT OF THE REPRODUCTIVE
ORGANS

There is genital growth in both sexes. The genitalia increase in size and start to show functional manifestations.

In females the vulva takes an adult appearance, with growth of both labia majora and labia minora. The vagina becomes rugose and takes a mature histological appearance, as the thickness of its lining epithelium increases. The vaginal discharge (made of vascular transudation, cervical discharge and desquamated epithelial lining) becomes acidic (pH 4.5), as it contains lactic acid. The acid comes from the breakdown of glycogen present in the vaginal epithelium by Döderlein's bacilli, which invade the vagina at puberty and become normal inhabitants. The increased thickness of the vaginal epithelium and the low pH of its secretions are protective mechanisms against ascending infection.

The uterus

In childhood the cervix is about twice the length of the uterine body. During puberty the body of the uterus grows so that it increases to about twice the length of the cervix. Its muscle fibres also increase in bulk. The cervical glands secrete a mucoid secretion. The lining of the uterus, the endometrium, proliferates under the effect of oestrogens and subsequently responds to progesterone. These cycles are responsible for the onset of menstruation (menarche).

ATTAINMENT OF REPRODUCTIVE CAPACITY

Menarche and ovulation

The word menarche is defined as the age at which the first menstruation takes place. This usually occurs when the hypothalamic–pituitary–ovarian axis matures sufficiently for the ovary to secrete some oestrogen so as to induce endometrial development followed by some bleeding.

However, menarche does not mean fertility,

since it can take 1−2 years before oogenesis and ovulation occur. During this time, cycles are often anovulatory (see ovulation and regulation of menstruation).

Spermatogenesis
Spermatogenesis means the formation of the male germ cells or mature sperm. As with oogenesis in females, spermatogenesis in males is a sign of reproductive maturity. It takes about 74 days to form a mature sperm from a primitive germ cell (Fig. 15.4).

The first phase The primitive germ cells (spermatogonia) replicate by mitotic division and produce several generations of primary spermatocytes.

The second phase The primary spermatocytes undergo a meiotic or reduction division so that each will form two haploid secondary spermatocytes (having half the number of chromosomes). A second maturation division occurs, resulting in four haploid spermatids arising from every primary spermatocyte. It has been estimated that over 500 spermatids are formed from one spermatogonium.

The third phase Further development of the spermatids changes them to spermatozoa. Further

maturation of the spermatozoa (which are closely surrounded by Sertoli cell cytoplasm) takes place before they are released into the lumen of the seminiferous tubules.

The composition of fluid in the tubular lumen is different from that of plasma. It has high concentrations of K^+, glutamic acid, aspartic acid and inositol, as well as high levels of androgens and oestrogens. On the other hand, it has low levels of glucose and protein. This is explained by the presence of a blood−testes barrier, which prevents free exchange. It is believed that the barrier is due to tight junctions between the Sertoli cells near the basal lamina of the seminiferous tubules. This barrier may serve the purpose of preventing harmful substances (e.g. drugs) in the blood from reaching the lumen. Conversely, it prevents products formed during cell division, which may be antigenic, from reaching the bloodstream and causing an autoimmune reaction.

Endocrine control of spermatogenesis The role of pituitary luteinizing hormone (LH) and follicle-stimulating hormone (FSH) in males is as important as in females. LH acts mainly on the Leydig cells of the testes to stimulate the production of testosterone, the main androgen responsible for the development and maintenance of secondary sexual characteristics and normal sexual activity in males. The target of FSH is mainly the Sertoli cells of the seminiferous tubules of the testes. In response to both testosterone and FSH, the Sertoli cells produce a high-affinity androgen-binding protein (ABP), which maintains the high androgen content within the seminiferous tubules, an essential factor for normal spermatogenesis. The Sertoli cells also secrete inhibin, which controls FSH release by negative feedback.

The menstrual cycle
The menstrual cycle is a term used to describe the monthly cycle of physiological events which prepare the female for pregnancy. The cycle ends in menstruation if pregnancy does not occur. The duration of an average normal menstrual cycle is 28 days from the onset of one menses (period) to the next, with a range of 21−35 days.

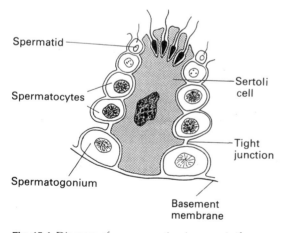

Fig. 15.4 Diagram of a cross-section in a seminiferous tubule showing the phases of sperm maturation. Note the Sertoli cells in relation to the developing spermatocytes.

THE OVARIAN CYCLE

Each month a few primordial follicles start to grow in response to the rising level of pituitary follicle-stimulating hormone (FSH) (Fig. 15.5). Most will undergo atresia while usually only one 'leading' follicle will continue development to be able to respond to LH and thereby progress to ovulation. The functional and morphological changes which occur in the ovaries and ovarian follicles during the menstrual cycle may be divided into three stages:
1 Follicular development.
2 The ovulatory process.
3 The luteal phase.

Follicular development

Follicular development involves changes in the three components of the primordial follicles: the oocyte, the granulosa cells and the theca cell layers.

The granulosa cells proliferate to form several layers, with fluid spaces appearing between cells.

These spaces eventually join to form a single fluid-filled space or antrum. The granulosa cells respond to FSH by synthesizing oestrogen hormone.

The oocyte enlarges and becomes surrounded by a zona pellucida, a mucopolysaccharide layer and some layers of granulosa cells, known as the cumulus oophorus.

The theca cells differentiate into a well-vascularized theca interna and less vascularized theca externa. The theca cells respond to LH by synthesizing androgens, which normally pass to the granulosa cells to be transformed into progesterone.

During the follicular phase, the rising level of oestrogens exerts a negative feedback on FSH secretion, causing it to decline. The further increase in oestrogen level towards midcycle, together with the increased sensitivity of the pituitary to GnRH, exerts a positive feedback on LH and FSH secretion, causing a dramatic midcycle peak of LH, which begins the second phase (or ovulation).

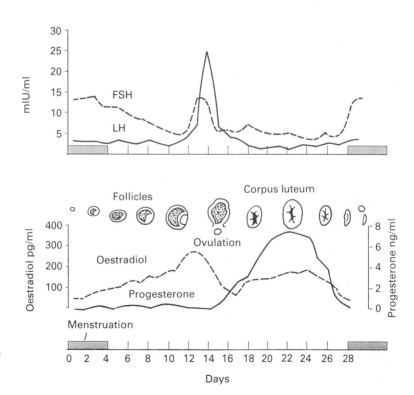

Fig. 15.5 Schematic representation of the ovulatory cycle showing the pituitary hormones and ovarian hormone secretion levels during the phases of the ovarian cycle.

The ovulatory process

The ovulatory process usually occurs about 24–36 hours following the LH peak, between the 12th and 15th days of a 28-day cycle.

The luteal phase

After ovulation the follicle becomes the corpus luteum and secretes both oestrogen and pro- gesterone. As the levels of these two hormones increase, they inhibit the secretion of gonado- trophins by a negative feedback. If no pregnancy takes place, the corpus luteum gradually reaches the end of its life and oestrogen and progesterone levels decline. This will remove the negative feedback effect from the pituitary, allowing the premenstrual rise of FSH to stimulate the matu- ration of more follicles for the next cycle.

THE UTERINE CYCLE

The most important target for oestrogen and progesterone is the uterus. In response to the ovarian hormones, both myometrial and endo- metrial tissues undergo important changes in anticipation of pregnancy.

The myometrium

Oestrogen increases the uterine blood flow. Oestrogen increases and progesterone decreases myometrial activity. This effect is important, as it ensures minimum uterine muscular activity at midcycle and in the early luteal phase, in case of early implantation.

Endometrial changes

Within 48 hours after the period of menstruation, the surface of the endometrium is covered by epithelial outgrowth from the remnants of glands in the basal part of the endometrium. Subsequent changes in the endometrium, in response to the ovarian hormones, can usually be described under three phases: proliferative phase, secretory phase and menstrual phase.

The proliferative phase This phase starts from the end of menstruation and lasts until the time of ovulation. The glands first appear tubular, lined by low columnar cells with centrally placed nuclei. Gradually, they become long and increas- ingly convoluted. The stromal cells increase in number and become more oedematous. The blood supply of the endometrium also grows. The basal layer is supplied by straight arteries and the superficial part is supplied by the spiral arteries.

During this period the endometrium grows from approximately 0.5 mm to 5 mm in height, a 10-fold increase. This is due to new tissue growth and to stromal expansion.

The secretory phase This phase occupies the time from ovulation until menstruation. During this phase, progesterone is the dominant hormone and corresponds to the luteal phase of the ovary. Under the influence of progesterone, endometrial growth ceases and functional changes begin to prepare the tissues to accept the embryo. The glands commence secretory activity, with sub- nuclear vacuoles appearing. They discharge their contents into the lumen of the glands. The se- cretion consists mainly of glycogen, sugars, amino acids, mucus and enzymes, such as alkaline phos- phatase. The arteries become more prominent and increasingly coiled, as they continue to grow while the endometrium height remains relatively static.

Further changes in the late luteal phase depend on whether implantation has taken place or not:
1 *Implantation.* If pregnancy occurs, the early embryo secretes human chorionic gonadotrophin (hCG), a hormone which maintains the corpus luteum, and therefore the secretion of oestrogen and progesterone continues. Further changes in the endometrium will mainly involve the stromal cells, which differentiate into three layers: a superficial cellular compact layer; the stratum compactum, an oedematous, less cellular layer underneath, known as the stratum spongiosum, and a basal layer of endometrium surrounding the straight arteries, known as stratum basalis. This last layer does not show much change.
2 *Non-implantation.* In the absence of chorionic gonadotrophin support, the corpus luteum de- clines and consequently both oestrogen and pro- gesterone levels fall. This results in a reduction in endometrial tissue height and more coiling of the spiral arteries, with secondary stasis, ischaemia and eventual blanching of the endometrium. Local release of prostaglandins induces waves of rhythmic contractions in the spiral arteries.

Menstruation Interstitial haemorrhages occur as a result of the breakdown of superficial arteries and capillaries. Eventually, the non-viable tissues are extruded into the uterine cavity and contribute to the menstrual flow. This process continues until all the layers, except the deep layer of the endometrium, are shed.

Eventually, the menstrual flow stops, due to vasoconstriction of the spiral arteries and the formation of thrombin–platelet plugs. Resumption of oestrogen secretion by the new crop of developing follicles induces healing and new tissue growth.

Menstrual blood does not normally coagulate and does not contain fibrin, probably as a result of proteolytic and fibrinolytic enzymes secreted by the damaged endometrial cells.

The sex hormones

The sex hormones are the hormones which are directly responsible for sexual differentiation and function in males and females. They are mainly produced in the gonads, i.e. the testes in males and the ovaries in females, but the adrenal glands also produce minimal amounts of sex steroid hormones in both sexes, which might have an important function before puberty and in females after the menopause.

The steroid hormones are clearly related to each other. Their main precursor is cholesterol, and they are easily converted from one to the other in the appropriate tissues, which possess the required enzymes (Fig. 15.6).

Steroid hormones are secreted in the blood and transported to the target organs bound to serum proteins. At the target organ cells, specific receptor proteins take up the steroid hormones, because their affinity to the hormones is usually much higher than that of the carrier protein. They act by entering the cell, where they become bound to a cytoplasmic protein receptor. The steroid–protein complex becomes attached to the nucleus and stimulates RNA synthesis, which directs protein synthesis and therefore regulates cell functions.

OESTROGENS

The main natural oestrogens found in the circulation are: 17β-oestradiol, oestriol and oestrone.

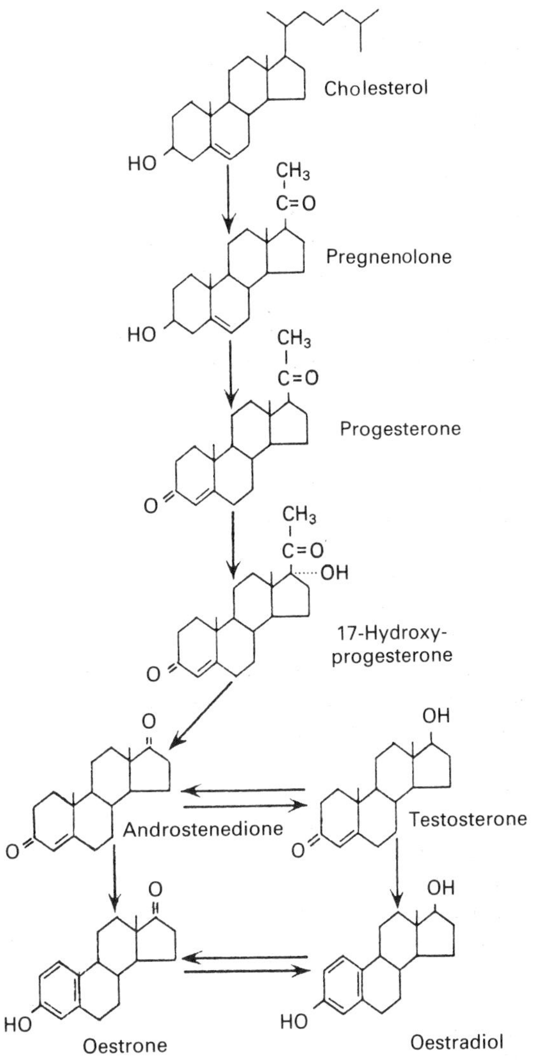

Fig. 15.6 Biosynthesis of ovarian steroids.

Oestradiol is the most active compound. It is bound to plasma proteins (60% to albumin and 37% to globulin). Oestrogens are C_{18} steroids, mainly produced by the granulosa cells in the ovaries. In pregnancy, the placenta produces a large amount of oestrogen, mainly oestriol. Oestrogens are mainly metabolized in the liver by conjugation with glucuronic acid, and most of the conjugate (65%) is excreted in the urine.

Actions of oestrogens

Ovarian oestrogens, which begin to be secreted at puberty, are responsible for the changes that take place at that period. The target tissues include the genitalia, the uterus, the breasts and certain parts of the subcutaneous fat layers. In the breast, they promote the growth of the duct system and slight pigmentation of the areola. General growth is promoted by their anabolic action. However, they cause the epiphyses to close and therefore limit skeletal growth.

During the menstrual cycle oestrogens are responsible for proliferation of the endometrium. Oestrogens also induce the growth of uterine muscle and increase the blood flow in the uterus. The myometrium becomes more sensitive to the effect of oxytocin and its electrical activity becomes more frequent.

Cyclic oestrogen secretion results in regulation of gonadotrophic hormone release, as previously described (see Fig. 15.1). This effect has made it possible to use oestrogens in combination with progesterone as oral contraceptives for inhibition of hypothalamic and pituitary hormones and thus suppression of ovulation. Progesterone seems to potentiate the inhibitory effect of oestrogens on LH secretion.

Undesirable or adverse effects of oestrogens are seen with artificially administered oestrogens (e.g. oral contraceptives). Therefore, oestrogen administration should only be given under medical supervision. The following are common examples of these unwanted effects:

1 Water and salt retention.

2 Effects on blood clotting factors. Oestrogens increase the risk of thrombosis and thromboembolism through changes in platelet adhesiveness and increased synthesis of clotting factors.

3 Biochemical disturbances. Oestrogens increase the level of circulating plasma globulins, which bind iron, cortisol and thyroxine, leading to raised plasma levels of these substances. Like other steroids, it also disturbs carbohydrate metabolism and may result in a decreased tolerance to glucose.

PROGESTERONE

Natural progesterone is mainly synthesized in the corpus luteum by the theca lutein cells of the follicle. During pregnancy the placenta produces a large amount of progesterone. Progesterone is a C_{21} steroid. It is also found in secreting tissues as an intermediary product in hormone synthesis. In the plasma it is bound to serum proteins (50% to albumin and 45% to transcortin). Progesterone is metabolized in the liver, converted to pregnanediol and conjugated to glucuronic acid, which eventually appears in the urine, where it can be measured in clinical investigation of pregnancy and the menstrual cycle.

Actions

The main action of progesterone is to induce secretory changes in the oestrogen-primed endometrium during the luteal phase of the menstrual cycle. These changes are necessary for implantation. Secretory changes also take place in the fallopian tubes. These changes provide the necessary nutrition for the fertilized ovum.

Progesterone exerts an antioestrogen effect on the endometrium by decreasing the oestrogen receptors and by increasing the rate of conversion of 17β-oestradiol to oestrogens with weaker physiological effects. Progesterone also exerts antioestrogen effects on the myometrium, making it less excitable and less sensitive to oxytocin. This is reflected in less electrical activity and a higher resting membrane potential.

In breast development progesterone exerts its effect on the alveolar tissue and lobules, which grow in size during adolescence and subsequently during pregnancy.

During pregnancy progesterone plays an important role in the maintenance of pregnancy, as it inhibits uterine muscle activity (relaxing effect). Also, together with oestrogens, it prepares the mammary glands for lactation. It exerts a thermogenic effect, which is partially responsible for the raised BMR of pregnancy. It also seems to stimulate the respiratory centre, causing hyperventilation and thus lowering arterial P_{CO_2}.

TESTOSTERONE

Testosterone is mainly synthesized in the testes by the Leydig cells, which account for about 20% of the testicular mass in adults. The adrenal glands also produce some testosterone, as well as other androgens (see Chapter 10). In females some testosterone is produced by the ovarian stroma cells. Testosterone is a C_{19} steroid, which is transported in the plasma bound to albumin (40%) and to a β-globulin (40%) and to other proteins (17%). It acts by binding to tissues which contain 5α-reductase and is converted to dihydrotestosterone (DHT). Most of the physiological effects of testosterone are due to DHT. It is metabolized in the liver and conjugated to glucuronic acid and sulphate. Excretion takes place in the urine and bile.

Actions

Testosterone and other androgens develop and maintain the male primary and secondary sex characteristics, already described in the section on puberty. Together with FSH, testosterone is required for spermatogenesis. Androgens also have a protein anabolic effect, which is evident during the growth spurt of puberty, and, like oestrogen, testosterone causes the epiphyses to close. Normal levels of testosterone secretion are also necessary for sex drive in the male.

The menopause

Menopause is a term which describes the last period or the final cessation of menstruation. It usually occurs in the late 40s to early 50s. However, women's reproductive abilities usually decline over 2−3 years before the menopause. This period is known as the climacteric.

PHYSIOLOGY OF THE MENOPAUSE

At birth the ovary contains all the germ cells that have the potential to mature into ova during a woman's reproductive life. After puberty, during each menstrual cycle, a number of follicles attempt to mature but only one reaches maturity and becomes shed at ovulation; the others degenerate to form atritic follicles. Gradually, as the woman approaches the menopause, the ovaries become depleted of oocytes and become increasingly small, until they consist mainly of fibrous tissue with no follicles.

As a consequence of ovarian ageing at the menopause, circulating oestrogens fall to very low levels. As a result of the diminished negative feedback effect of ovarian oestrogens and progesterone on the hypothalamus and the anterior pituitary, large amounts of FSH and LH are produced (up to 10 times the levels in premenopausal women). These hormones appear in blood and in urine. Measurement of gonadotrophins can be used to confirm the onset of the climacteric or menopause (Fig. 15.7).

The fall in oestrogen production causes various local and systemic effects, known as the symptoms and signs of the menopause. When severe, these manifestations are considered abnormal and require treatment. The important changes are described below.

SYMPTOMS AND SIGNS OF THE MENOPAUSE

Vasomotor symptoms

Vasomotor symptoms present as a sensation of warmth, which occurs on the trunk and spreads to the face. These are known as hot flushes and are usually accompanied by sweating and palpitations; they occur mainly at night. The exact cause is unknown, but it is suggested that a hypothalamic disturbance leads to the vasomotor responses associated with a burst of LH secretion. The attacks of hot flushes can be stopped by oestrogen therapy. Similar attacks have been reported in men and women after removal of the gonads and continue after removal of the pituitary. Therefore, the hypothalamic origin of the hot flushes, as well as the LH bursts, has been postulated.

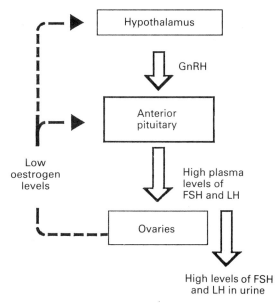

Fig. 15.7 The pituitary−ovarian axis at the menopause.

Emotional and psychological problems
The subject's attitude towards the menopause, as well as that of her husband and family, determines the extent and severity of emotional and psychological symptoms.

Skeleton
The lack of oestrogens results in increased bone resorption, associated with increased calcium and hydroxyproline excretion. Collagen is lost from other sites of the body as well as bone, namely skin and joints. It is the loss of collagen and elasticity that contributes to the overall ageing appearance. Loss of bone, particularly from the vertebrae, causes a reduction in height, with forward bending and kyphosis in extreme cases. There is an increased incidence of fractures, particularly of the neck of the femur and the distal end of the radius. Ligaments lose their elasticity, and loss of collagen leads to degenerative changes in the joints, leading to osteoarthritis.

Cardiovascular problems
Atherosclerotic changes gradually occur, due to a progressive increase in the plasma concentration of cholesterol, triacylglycerols and low- and very-low-density lipoproteins, so that the mortality rate from ischaemic heart disease comes to equal that of men.

Changes in the oestrogen target organs in the urogenital tract
The oestrogen target organs in the urogenital tract are the uterus and its appendages, the cervix, the vagina, the vulva, the urethra and the bladder. The uterus shrinks as the myometrium becomes converted to fibrous tissues. The cervix and vaginal epithelium undergo atrophic changes, resulting in thinning and dryness. The pH rises to about 7.2, compared with 4.5 in the productive years, because of absence of glycogen for Döderlein's bacilli to convert to lactic acid. This produces susceptibility to infection and atrophic vaginitis. The latter may be a cause of postmenopausal bleeding, which requires urgent investigation.

The external genitalia decrease in size and there is loss of pubic hair. The transitional epithelium of the bladder trigone undergoes atrophy. The urethra, which arises from the same ectoderm as the vaginal vestibule, also undergoes some atrophic changes.

Coitus

Sexual intercourse, or coitus, is the means of propagation of the species in all mammals. In man, unlike other species, the act of coitus can take place at any time during the menstrual cycle, whether or not reproduction is possible. The stages of physiological response to sexual stimulation are similar in males and females, namely: excitement, plateau, orgasm and resolution.

In females the excitement phase involves increased vasodilatation and congestion of the clitoris and labia minora and majora and increased vaginal exudate. These changes increase during the plateau phase of the sex act, when further lubrication occurs from the secretions of Bartholin glands.

In males penile erection occurs, which involves lengthening and increased rigidity of the penis. It is essentially a vascular phenomenon. It is accomplished by dilatation of the arteries and constriction of the veins in the penile corpora, leading to engorgement under pressure and hardening of the penis. The process is controlled by humoral and nervous mechanisms which are not fully understood. The neural control is through parasympathetic fibres in the nervi erigentes of the pelvic splanchnic nerves. It has been suggested that vasoactive intestinal peptide (VIP) may be the neurotransmitter involved. The flaccid penile state appears to be due to an intrinsic tone in the afferent arteriolar sphincters under sympathetic α-adrenergic control. Releasing the sympathetic tone results in erection. Also, nervous and psychological influences can affect the process of erection.

The orgasmic phase is short and associated with uterine and vaginal contractions in the female and ejaculation in the male.

There are three phases in the process of ejaculation:
1 *Seminal emission.* This is controlled by centres in the upper lumbar spine via sympathetic nervous supply, leading to contraction of the smooth muscles of the seminal vesicles, ejaculatory duct and prostate, to express seminal fluid into the prostatic urethra.
2 *Formation of pressure chamber.* As pressure

increases, it forces the fluid into the bulbar urethra.

3 *Expulsion of semen from the urethra.* This occurs by rhythmic contractions of the bulbospongiosus and bulbocavernosus muscles, which are under somatic control by the pudendal nerves. Ejaculation is essentially a spinal reflex but higher centres are also involved. Therefore, emotional and psychological factors can affect the threshold of ejaculation, which explains delayed or premature ejaculation.

SEMEN

Sixty per cent of the seminal fluid comes from the seminal vesicles. The rest consists of secretions of the prostate, the vas deferens and the mucous glands of the urethra. About 3–5 ml are ejaculated each time. The pH of semen is alkaline (pH 7.5), unlike that of the vagina. The vas deferens secretes fibrinogen, which causes the semen to coagulate, and the prostatic secretion contains fibrinotrypsin, which causes the clot to dissolve in about 20 minutes. Male fertility depends on a sperm count of about 100 million/ml, on sperm motility and on the absence of a high percentage of abnormal forms. A count of less than 20 million sperm/ml usually indicates infertility.

Fertilization

Fertilization is the process of penetration of the oocyte by the spermatozoon. It marks the beginning of the formation of the embryo. Fertilization normally occurs soon after ovulation, for the ovum remains viable for only about 8 hours.

After deposition of sperms in the upper vagina, they penetrate the mucus of the cervical canal into the uterus and reach the outer third of the fallopian tube, where fertilization usually takes place. The cervical canal acts as a reservoir for sperm for about 24–48 hours. So fertilization can occur for as long as 48 hours after intercourse. Sperms reach the fallopian tubes 5–10 minutes after ejaculation. They proceed by flagellar motion at about 5–6 cm/min, which is not enough to explain their rapid ascent. It is believed that uterine contractions help their progress. Of about half a billion sperm deposited, only 1000–3000 reach the fallopian tubes.

The spermatozoa have to undergo a number of physiological changes during their course in the female genital tract before they are able to penetrate the oocyte.

CAPACITATION

Sperms are not capable of fertilization immediately after ejaculation. They develop this capacity after a few hours in the female genital tract. This process is known as capacitation. It is probably due to changes in the plasma membranes induced by enzymes located in the acrosome region of the sperm. These changes usually occur as the sperms traverse the cervical canal. Capacitation under artificial conditions was one of the problems of *in vitro* fertilization.

ACROSOME REACTION

Sperms undergo structural changes, which end by loss of the acrosomal cap (the two layers of the outer acrosomal membranes), leaving the inner acrosomal membrane and the nucleus exposed. These changes usually occur in the ampulla of the fallopian tube, where the oocyte is found soon after ovulation.

OOCYTE PENETRATION

The sperm has to penetrate through several layers of the oocyte, with the aid of the enzymes hyaluronidase and protease. The outermost layer penetrated is the corona radiata (Fig. 15.8). The sperm becomes attached to the second layer or the zona pellucida. Dispersion of the corona radiata and the zona pellucida is achieved by means of

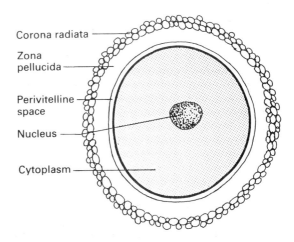

Fig. 15.8 The oocyte and its surrounding layers.

acrosomal enzymes. The sperm reaches the perivitelline space between the zona pellucida and
oocyte. Eventually, only one sperm penetrates,
activating the oocyte and blocking further sperm
penetration. This is believed to be due to the
release of a substance from cortical granules
surrounding the ovum. This substance permeates
in the perivitelline space and the zona pellucida,
and it inactivates the remaining sperms.

When the oocyte is activated, it undergoes
meiotic division and discards 23 chromosomes
as the second polar body, having lost nuclear
material as the first polar body in early ovum
formation (Fig. 15.9). These chromosomes form
the pronucleus from the ovum which fuses with
the pronucleus from the sperm, and the cell is
ready to divide with 46 chromosomes. Mitotic
division starts immediately and results in the
formation of the blastocyst.

Infertility can result from damage to the tubes,
e.g. infection, causing their blockage and/or loss
of their function in providing the ideal environment for fertilization and early development of
the blastocyst.

Abnormalities of human reproduction

Knowing the mechanisms, the signs and the
symptoms of normal changes in reproductive life
is important in order to diagnose and treat various
pathological states. Some of the abnormalities of
human reproduction have already been mentioned. The following are some examples of reproductive disorders.

ABNORMALITIES OF PUBERTY

Delayed puberty could occur due to disturbance
of gonadal hormone production, because of familial delay of maturation, disease states affecting
the hypothalamic–pituitary–ovarian axis or
congenital abnormalities.

CHROMOSOMAL ABNORMALITIES

Chromosomal abnormalities result in impairment of development of the reproductive system
and hormonal abnormalities. During the formation of the ovum or sperm, the pair of sex
chromosomes may fail to separate (non-
disjunction) during meiosis. In the case of ova,
the result is an ovum with either two X chromosomes or none. If such ova are fertilized, the

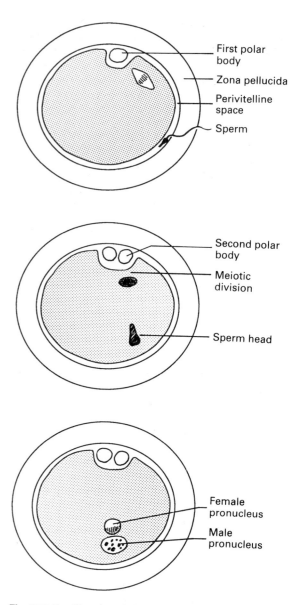

Fig. 15.9 Fertilization of the human ovum.

result is one of the following possible combinations: XO, YO, XXY or XXX. In the XO combination the individual has female internal and
external genitalia and suffers from what is known
as *ovarian dysgenesis* or Turner's syndrome. The
ovaries are rudimentary or absent. Such patients
fail to develop at puberty and they are short and

sterile. A foetus with a YO combination is usually aborted and is probably incompatible with life. The XXY combination is the most common of the sex chromosome abnormalities. The individual develops normal male genitalia. There is adequate testosterone secretion at puberty but the seminiferous tubules are not normal. This is known as Klinefelter's syndrome or *seminiferous tubular dysgenesis*. Individuals with this condition have a high incidence of mental retardation. When XXX and XYY combinations occur, the first is referred to as the superfemale and is not associated with any particular abnormalities and the second is usually a tall male with severe acne and possibly some aggressive tendencies.

INFERTILITY

Successful pregnancy requires fertilization of a mature ovum by a sperm and subsequent implantation in a prepared endometrium. This can be disturbed at several stages; male factors are responsible for infertility as much as female factors. In males, abnormal sperm counts or azoospermia can be a permanent cause of infertility. Hyperprolactinaemia has been associated with infertility in men and women. Prolactin appears to be inhibitory to the effects of LH on the ovaries and of FSH on the testes. Tubal patency and normal function are essential for fertilization. However, the commonest cause of infertility in women is failure of ovulation. This should be the first matter to investigate when there are no genital tract abnormalities in the female. Presence of progesterone in the secretory phase of the cycle indicates ovulation and its absence indicates anovular cycles.

ABNORMALITIES OF THE MENSTRUAL CYCLE

Primary *amenorrhoea*, or failure to menstruate at puberty, is usually due to failure of sexual maturation. Secondary amenorrhoea, or cessation of normal cycles, is most commonly due to pregnancy. In the absence of pregnancy, it may be brought about by an abnormality of the hypothalamic–pituitary axis, ovarian disease or some systemic disease. *Dysmenorrhoea* is the occurrence of pain during menstruation. This is mainly explained by prostaglandins accumulating in the uterus, and can be relieved by inhibitors of prostaglandin synthesis. The term *oligomenorrhoea* refers to little blood flow during menstruation and *menorrhagia* refers to much bleeding with each menstrual period. *Metrorrhagia* refers to abnormal bleeding which occurs between periods. This is often due to ovarian or uterine disease.

ABNORMALITIES OF THE TESTES

Undescended testes occur in about 10% of newborn boys; but by the end of the first year most of these complete their descent into the scrotum. When the testes remain in the abdominal cavity, this is called *cryptorchidism*, which may be treated with gonadotrophins or surgically corrected. If left in the abdomen until puberty, spermatogenesis does not occur due to the higher abdominal temperature; but testosterone is secreted. There is also the possibility of development of malignant testicular tumours.

Male hypogonadism occurs because of chromosomal abnormalities, failure of the hypothalamic–pituitary axis or testicular damage. If it occurs before puberty, the individual usually grows to be tall and thin. This is explained by the observation that their epiphyses remain open longer than normal, due to the absence of testosterone. The genitalia are small, there are no male secondary sex characteristics, and the pubic hair is of a female configuration. Such individuals are sterile and the condition is referred to as *eunuchoidism*.

When male hypogonadism occurs after puberty, there is loss of spermatogenesis and gradual loss of libido. Secondary sex characteristics are not lost as they can be maintained by adrenocortical androgens. The individual eventually becomes impotent and sterile, and may develop symptoms similar to those encountered in menopausal women.

Further reading

1 Guyton, A.C. (1991) Female physiology before pregnancy and the female hormones. In *Textbook of Medical Physiology*, 8th edn, pp. 899–914. Saunders, Philadelphia.
2 Guyton, A.C. (1991) Reproductive and hormonal functions in the male. In *Textbook of Medical Physiology*, 8th edn, pp. 885–898. Saunders, Philadelphia.

16: Pregnancy and Perinatal Physiology

Early physiological changes

The fertilized ovum undergoes a process of rapid cellular multiplication as it travels along the fallopian tube. Eventually, when it reaches the uterine cavity, the blastocyst is formed. The blastocyst acquires its nutrition from the surface secretions of the endometrium. Inadequate supplies of nutrients result in failure of the completion of implantation. The syncytiotrophoblast layer erodes the endometrium by the action of proteolytic enzymes and implantation is completed in 4–5 days after fertilization (7–8 days after ovulation). The chorion and, later, the placenta have important endocrine functions.

Pregnancy is suspected when an otherwise regular period is missed by the mother. As early as the first week after the time of the missed period, i.e. 5–6 weeks after the last period, a hormone known as human chorionic gonadotrophin can be detected in the maternal urine.

Objectives

On completion of the study of this section, the student should be able to:

1 Understand the functions of the placenta and appreciate the physiological integration of the foetoplacental unit.

2 Describe the maternal physiological responses to pregnancy in order to deal with the special needs of antenatal care.

3 Understand the physiology of normal childbirth so as to recognize obstetric abnormalities.

4 Describe the control and maintenance of lactation and its advantages so as to encourage breast-feeding.

5 Explain the special functional features of the new-born so as to give appropriate postnatal care to the neonate.

Human chorionic gonadotrophin (hCG) detection is the basis of pregnancy tests. It may be detected in the urine 14 days after conception. A positive pregnancy test is obtained when hCG is detected by an immunological technique using specific antibodies against hCG. Earlier biological methods using rabbits or toads, which were injected with maternal urine, a positive test giving rise to ovulation in these animals, are no longer used.

In the early stages of pregnancy, the embryo develops rapidly with little growth in size. About the 12th week it is only about 5 cm in length but all the organ systems have started to differentiate. It is during these first 12 weeks that the effects of radiation and teratogenic drugs are most likely to produce gross foetal deformities. During the first 12 weeks, the mother may have nausea and vomiting, known as early morning sickness. The reason for this is not known. It is possible that cell degradation products, due to the invasion of the endometrium, may have a role in this. Another possible reason may be the high levels of hCG and sex hormones produced by the placenta. In some cases, early morning sickness may develop into severe vomiting, which seriously affects the health of the mother and requires medical attention—a condition known as hyperemesis gravidarum.

The placenta

The placenta has two main functions. The first is an endocrine function and the second is the transfer of substances between maternal and foetal blood. Several hormones are produced by the placenta (Fig. 16.1).

ENDOCRINE FUNCTIONS

Human chorionic gonadotrophin (hCG)

Human chorionic gonadotrophin is a glycoprotein hormone with a molecular weight of about 3000 daltons and similar molecular structure to that of pituitary luteinizing hormone (LH). Its main function is to maintain the corpus luteum of pregnancy beyond the period of the menstrual cycle. It stimulates the corpus luteum to secrete large amounts of progesterone and oestrogens. The latter is essential for maintenance of the endometrium and suppression of menstruation. Peak levels of hCG are reached by about the 10th week. Afterwards, it declines and is maintained at a lower level until the end of pregnancy. hCG also causes the foetal testes to produce small amounts of testosterone, which stimulate the

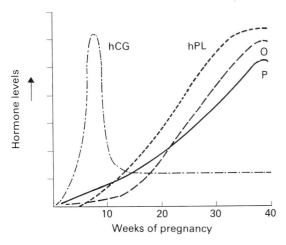

Fig. 16.1 Hormones of the placenta: levels of secretion in the maternal blood throughout pregnancy. hCG, human chorionic gonadotrophin; hPL, human placental lactogen; O, oestrogens; P = progesterone.

development of male sex organs and the descent of the testicles into the scrotum.

Progesterone

Progesterone is produced by the corpus luteum in early pregnancy and in much greater amounts by the placenta. From about the 8th week onwards, the production of progesterone is taken over by the placenta. It reaches a peak value at the end of pregnancy, which reaches levels about 10 times those of early pregnancy.

Progesterone increases the endometrial nutrient stores in early pregnancy. This provides the developing embryo with an important source of nourishment. During pregnancy, progesterone prevents uterine contractions by decreasing the contractility of the myometrium. Together with oestrogens, progesterone prepares the breasts for lactation. The effect of progesterone is to develop the duct system of the mammary glands. Placental progesterone has an important role as a precursor for foetal steroid synthesis. Thus, the progesterone of the placenta is converted to dehydroepiandrosterone (DHEA) by the foetal adrenal cortex. The latter is again converted by placental enzymes to oestriol. Thus, the production of placental oestrogens depends on the foetus and the placenta (foetoplacental unit). Figure 16.2 shows the inter-dependence of foetal and placental steroid synthesis.

Oestrogens

The placenta secretes large amounts of oestrogens, mainly in the form of oestriol, in contrast to the ovaries, which produce mainly oestradiol, which is more potent. As mentioned above, the precursors for placental oestrogens come from the foetal adrenal. Therefore, the maintenance of placental oestrogens depends on a living foetus. Intrauterine foetal death causes a sharp fall in maternal plasma oestrogen levels.

Placental oestrogens have several functions. They promote the growth and enlargement of the uterus to accommodate the developing embryo and foetus. Oestrogens also promote the growth of the glandular tissues of the breast. By the end of pregnancy, both the duct and the glandular systems of the breast are developed and ready to produce milk. The external genitalia also enlarge

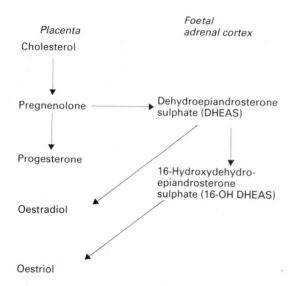

Placenta
Cholesterol

Foetal
adrenal cortex

Pregnenolone ⟶ Dehydroepiandrosterone
sulphate (DHEAS)

Progesterone

16-Hydroxydehydro-
epiandrosterone
sulphate (16-OH DHEAS)

Oestradiol

Oestriol

Fig. 16.2 Foetoplacental hormone synthesis. Placental progesterone is also used by the foetal adrenal cortex to synthesize glucocorticoids.

and the pelvic ligaments and joints relax. These changes help the passage of the foetus through the pelvis during labour.

Human placental lactogen (hPL)

Human placental lactogen, also known as human chorionic somatomammotrophin (hCS), is a protein hormone with a molecular weight of 3800. It is secreted by the placenta and is first detected in the maternal circulation around the 5th week of pregnancy. Its concentration increases in the maternal blood as pregnancy progresses. It seems to have an effect on the development of the breasts. It also has metabolic actions similar to those of pituitary growth hormone (GH), which normally decreases during pregnancy. hPL is protein-anabolic and favours mobilization of fat from maternal stores, while inhibiting glucose utilization in maternal tissues. This action contributes to the decreased sensitivity to insulin encountered during pregnancy and helps to spare glucose for the foetus. The foetus uses glucose as its main source of energy.

TRANSFER FUNCTIONS

Oxygen and carbon dioxide diffuse freely across the placental membrane lining the villi of the placenta. The intervillous spaces receive oxygen-ated maternal blood from the spiral arteries of the uterus. Maternal venous blood drains back by means of the placental sinuses, which drain into the many tributaries of the uterine veins. Deoxygenated foetal blood comes to the villi by means of branches of the umbilical arteries. At term, the foetus requires about 20 ml of oxygen per minute. As the arteriovenous oxygen difference in the uterine vessels is about 7 ml/dl and the uterine blood flow is about 500 ml/min or more, then 35 ml oxygen/min or more can be supplied to the uterus at term. The extra oxygen is used by the uterine and placental tissues.

The foetal arterial P_{O_2} is about 40 mmHg. The foetal venous blood P_{O_2} is about 20–30 mmHg. The P_{O_2} in the placental sinuses is about 50 mmHg. Thus, there is only 20–30 mmHg P_{O_2} difference between the foetal and maternal blood. Several factors favour the diffusion of O_2 across the placental membrane:

1 The surface area of the villi is huge. In a full-size placenta, which weighs about 500 g, the total surface area for exchange is about 14 m^2.

2 Foetal haemoglobin has a higher affinity for O_2 than adult haemoglobin. It can carry up to 30% more oxygen per gram.

3 The foetal haemoglobin concentration is about 50% higher than that in adults.

4 The foetal blood has a higher P_{CO_2} (48 mmHg) than maternal blood. CO_2 diffuses quickly into the maternal blood and shifts the O_2 dissociation curve to the left (Bohr effect), helping the release of oxygen from the maternal blood. At the same time, foetal blood becomes more alkaline, therefore shifting the curve to the right and favouring more O_2 uptake. Therefore, the Bohr effect works both ways in favour of oxygenation of foetal blood more than it does in adult lung.

Blood flow to the placenta is decreased during uterine contractions, which squeeze the spiral uterine arteries. Muscular exercise also decreases uterine blood flow due to redistribution of blood. Smoking mothers have various levels of carboxy-haemoglobin in both maternal and foetal blood. This tends to lower the O_2-carrying capacity of the blood in both mother and foetus, leading to hypoxia in the foetus. Heavy smoking is associated with smaller birth weights and greater risk of perinatal death.

The placenta allows a variety of other substances to cross. Generally, small-molecular-weight substances cross the placental membrane more readily. Electrolytes and lipid-soluble substances cross more easily than water-soluble substances. Foetal blood glucose is 20–30% lower than maternal blood glucose concentration. Therefore, glucose diffuses to the foetal blood down a concentration gradient by facilitated diffusion, using a specific carrier. More glucose tends to diffuse to the foetus when the mother has poorly controlled diabetes mellitus. The baby tends to become obese and overgrown. Many drugs, including anaesthetics and analgesics, cross the placental membrane and may affect the foetus. Protein molecules cannot normally pass across the placenta. However, immunoglobulins and some maternal enzymes are able to cross, possibly by pinocytosis.

Minerals are most probably transported by carrier-mediated transport mechanisms. In the case of iron, maternal transferrin does not cross the placental barrier but it interacts with placental binding sites. Iron is then transported into the foetus in bound form. There is a marked increase in the efficiency of maternal intestinal absorption of iron and calcium during pregnancy and there are active transport mechanisms for these substances. Even if the mother is anaemic, the foetus usually has normal haemoglobin levels; but foetal iron stores may be deficient. Therefore, the possibility of developing iron-deficiency anaemia during infancy is greater.

Amino acids are actively transported across the placental membrane and usually the foetal concentrations of free amino acids are much higher than those in the mother. This favours the process of protein synthesis, especially in the rapid growth stage during the second and third trimesters. The excretion of urea, uric acid and creatinine takes place by diffusion due to higher concentrations in foetal blood.

Maternal responses to pregnancy

The mother's systems become prepared for the increased workload and nutritional requirements of the foetus. Almost every system is involved so that there is a wide range of physiological changes during pregnancy.

BODY FLUIDS

There is marked water retention – nearly 2.5 litres by the middle of pregnancy, of which about 1.5 litres are in the plasma. During this period, growth of the uterus and mammary glands, amniotic fluid and increased blood volume account for most of the water and electrolytes retained. During the latter half of pregnancy, the tissue fluid and blood account for about 6 litres of retained water. This may be due to the oestrogens secreted by the placenta. There is also increased renin production, which leads to an increase in aldosterone release and sodium retention.

THE BLOOD

As mentioned above, blood volume increases. This increase is due to an increase in plasma volume as well as in red blood cell production. The increase in blood volume is rapid during the 3rd month and levels off during the last 3 months of pregnancy at about 1.5 litres. Although there is an increase in total red cell mass, the packed cell volume (PCV) is decreased, due to haemodilution, and may normally be as low as 30%. The haemoglobin concentration is similarly lower than normal (11–12 g/dl). However, the cell haemoglobin should be normal unless iron deficiency occurs. These observations emphasize the importance of haemoglobin measurement in antenatal care and the need for adequate iron supplies during pregnancy.

The plasma proteins show an increase in the β-globulin and fibrinogen. Although the total amount of albumin is increased, plasma albumin is found to be lower in concentration, which leads to a fall in plasma osmotic pressure. These changes raise the erythrocyte sedimentation rate (ESR) to high levels during pregnancy, which, at other times, would be considered abnormal.

THE CIRCULATORY SYSTEM

Cardiovascular changes prepare the mother for the increased demands of pregnancy. About 500 ml of blood/min pass to the uterus. Early in pregnancy the cardiac output increases to meet this demand. It is estimated that, by the end of the first trimester, the cardiac output increases by about one-third. This is achieved by an increase in both the heart rate and the stroke volume. One of the important changes during pregnancy is the increase in renal blood flow. This increases by just less than half a litre in early pregnancy and leads to a high glomerular filtration rate (GFR).

The blood flow in the skin is increased. This is useful in the maintenance of body temperature. The increased metabolic heat production requires vasodilatation and sweating to take place.

THE URINARY SYSTEM
There is a rise in GFR due to the high renal blood flow. It increases from a normal level of 110 ml/min to about 150 ml/min in the first 3 months and remains high throughout pregnancy. Renal glucosuria may occur in some mothers if the tubular maximum (Tm) for glucose is exceeded due to the increased GFR. The clearance of urea and uric acid increases and appreciable amounts of amino acids may also appear in the urine. Water-soluble vitamins may also be lost in the urine. In particular, folic acid is lost in appreciable amounts. The possibility of developing megaloblastic anaemia in pregnancy is increased because of higher demands and the renal losses of folic acid.

CHANGES IN BODY COMPOSITION
The increase in cell mass due to the growth of the uterus, the breasts and the uterine contents is associated with a net gain in protein. By the end of pregnancy, the net gain in protein amounts to about 1 kg. Most of this protein is used for uterine and foetal growth during the third trimester. This is associated with water and electrolyte gain, as part of the tissue fluids and intracellular water. Figure 16.3 shows the rate of change in body-weight during pregnancy broken down into the major components.

By the end of pregnancy, the pregnant mother stores about 4 kg of fat, mainly in subcutaneous adipose tissues. The larger part of this fat is stored during the middle trimester. It provides an energy surplus to be used mainly during the period of lactation. Repeated pregnancies are one of the common factors predisposing to obesity.

The pregnant mother gains about 750 mg of iron by the end of pregnancy. The iron is used mainly for the mother's haemoglobin during the first and second trimesters and for foetal blood and iron stores during the third trimester. The net gain in calcium during pregnancy is about

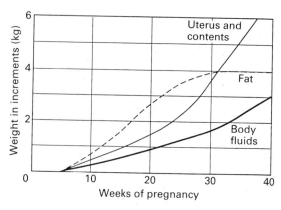

Fig. 16.3 The major components of weight gain during pregnancy. Note the rates of increase of the different components.

30 g. This is used mainly for the growth and calcification of the foetal skeleton during the third trimester.

Nutritional requirements during pregnancy
The total weight gain during pregnancy is about 12 kg. The changes in body composition during pregnancy (Fig. 16.3) serve as a useful guide for the nutritional needs of pregnant mothers. The timing of these changes also gives an indication of the important nutritional needs during the various stages of pregnancy. For more detail on nutritional requirements, see Chapter 8.

METABOLIC CHANGES DURING PREGNANCY
The metabolic responses during pregnancy can be explained by the changing needs of the pregnant mother and the foetus. The hormonal changes during pregnancy are responsible for these metabolic responses.

Carbohydrate metabolism
Carbohydrate metabolism is characterized by increased glucose utilization during the first few weeks of pregnancy, with increased glycogen storage. This is due to hyperinsulinaemia, probably mediated by high levels of oestrogen and progesterone. In the latter half of pregnancy, hPL, prolactin, cortisol and glucagon are responsible for the appearance of insulin resistance. Latent diabetes may become overt, giving rise to decreased glucose tolerance and glucosuria.

Protein and fat metabolism

Protein metabolism is mainly anabolic. There is a high protein intake, which leads to short-lived increases in plasma amino acid levels. Placental hormones play an important anabolic role. However, protein catabolism is brought about by cortisol and glucagon, which enhance the utilization of amino acids in gluconeogenesis during the latter half of pregnancy.

The above discussion indicates that there are prominent metabolic differences between early and late gestation. Maternal glucose removal from the circulation is increased in early gestation but progressively declines to subnormal levels by term. Basal plasma insulin concentrations are normal early in gestation but rise in late gestation, associated with an increased insulin response to glucose and insulin resistance. Fat storage is greatest in early gestation but almost completely stops in late gestation. All of these observations indicate enhanced maternal ability to store energy in early gestation. We call this the *anabolic phase* of pregnancy; the foetus is small at this stage and requires relatively small amounts of nutrients. However, in the later part of gestation, when foetal growth is much more rapid, there is increased nutrient flow across the placenta to the foetus, i.e. the *catabolic phase*. The insulin resistance experienced by the mother in late gestation helps to maintain glucose concentrations at a higher level than normal and decreases maternal glucose utilization. Therefore, glucose is spared for the foetus. The catabolic phase of pregnancy occurs mainly during the third trimester. During this stage the mother uses an increased amount of fat. In the fed state, the fat is derived from the diet. In the fasting state, increased amounts of fat are mobilized from adipose tissue stores, leading to a state of accelerated starvation.

MATERNAL METABOLISM IN THE FASTING STATE

The fasting state is of clinical relevance because pregnant mothers may go without food for prolonged periods of time in a variety of situations (e.g. hyperemesis gravidarum, prolonged labour). The effects of pregnancy on metabolism in the fasting state are mainly the result of continuous removal of glucose and amino acids from the maternal to the foetal circulation. The energy requirements of the developing foetus are generally believed to be met entirely by the consumption of glucose. Glucose is also utilized for protein, fatty acid and glycogen synthesis. The overall glucose uptake of the foetus at term is 20 mg/min (i.e. 6 mg/min/kg as compared with 2–3 mg/min/kg in adults). The glucose level of foetal blood is usually 20–30% below the maternal level, so that diffusion favours the net movement of glucose from mother to foetus. As maternal insulin does not cross the placenta, foetal glucose utilization is independent of maternal insulin availability. Foetal insulin is also known to play an important role in foetal growth.

In conclusion, the fasting state is characterized by maternal hypoglycaemia, hyperketonaemia and hypoalaninaemia. The sequence of events is initiated by foetal glucose utilization, which leads to a decrease in maternal plasma glucose. As a consequence, plasma insulin levels fall, leading to increased starvation ketosis. At the same time, amino acid levels (particularly alanine) are decreased in the maternal blood, due to active transport to the foetus. Maternal hypoglycaemia, initiated by foetal glucose removal, is thus the result of enhanced foetal utilization of gluconeogenic precursors. The net effect on the mother is an accelerated response to starvation.

In maternal ketonaemia, ketone bodies accumulate in the amniotic fluid and are probably available to the foetus, since the enzymes necessary for oxidation of ketone bodies are present in the foetus. In contrast to ketone bodies, free fatty acids are not transferred to the foetus. The rate of utilization of ketone bodies is primarily dependent on their availability to the foetus. It is possible that, in maternal hypoglycaemia, ketone bodies produced in the maternal liver may serve as an alternative fuel to glucose in the foetus.

Labour (parturition)

The duration of normal pregnancy is about 40 weeks from the first day of the last menstrual period (38 weeks from conception). Irregular, but usually painless, uterine contractions occur throughout pregnancy. The onset of labour is marked by regular uterine contractions, widely spaced at first and gradually increasing in frequency and strength. During labour, the uterine smooth muscles characteristically contract and retract (shorten). This phenomenon is important as it ensures gradual descent of the baby as the capacity of the uterine cavity decreases. The initiation of labour is probably due to several factors. Maternal oestrogen increases and progesterone decreases near term. These hormonal changes increase the sensitivity of the uterus to oxytocin. Oxytocin receptors on the myometrium and decidua cells increase progressively towards term (Fig. 16.4). Oxytocin stimulates the decidua to produce one of the prostaglandins ($PGF_{2\alpha}$). It is found in high concentration in the amniotic fluid and the placenta at term. It has strong stimulating action on the uterus. However, there is no detectable rise in maternal serum levels of either oxytocin or prostaglandins, but once labour starts their levels increase. Their action in maintaining effective uterine contractions is confirmed, but their role in the initiation of labour is uncertain.

When the time of delivery is near, the cervix, already softened by the action of oestrogens, is gradually taken up and shortened, a process known as cervical effacement. Once regular contractions are established, the process of cervical dilatation begins, until the cervix is fully dilated. This is called the first stage of labour.

The second stage begins after full dilatation of the cervix. The advancing head of the foetus causes stretch of the cervix and genital tract. This is a strong stimulus for the release of oxytocin, which acts directly on the myometrium, and the contractions become stronger and more frequent. Oxytocin also acts on the decidua to produce prostaglandins, which enhance the effects of oxytocin on uterine contractions. Drugs which inhibit prostaglandin synthesis delay the progress of labour. During the second stage, the foetal head is pushed slowly through the cervix and vagina until it is born. The strong uterine contractions result in poor circulation to the placenta during this stage; therefore, rapid progress is vital for a safe delivery. When the head is delivered, it is easily followed by the shoulders and the rest of the body. This marks the end of the second stage.

The third stage of labour starts after complete delivery of the foetus and ends with complete delivery of the placenta and membranes. Uterine contractions and retractions separate the placenta. Some bleeding normally occurs from the raw surface of the decidua, which helps in completion of placental separation. Further contractions of the uterus and maternal efforts complete delivery of the placenta and membranes. Bleeding is prevented by contraction of uterine muscle and blood clotting, which plug small vessels in the placental bed.

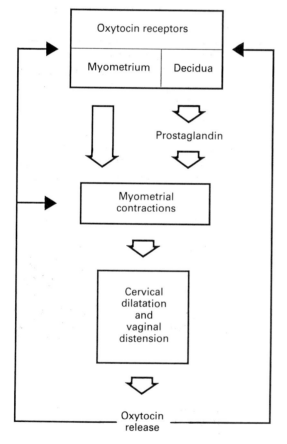

Fig. 16.4 The roles of oxytocin and prostaglandins in parturition.

Physiology of lactation

At the onset of puberty, the mammary gland starts to develop and the breast becomes prominent, due to deposition of fat more than glandular tissue. Oestrogen stimulates proliferation of the duct system and progesterone promotes the development of glandular lobules. The presence of normal pituitary function is essential for the effects of the sex hormones. In particular GH, ACTH (and therefore cortisol), TSH (and therefore T_3 and T_4) and insulin are required (Fig. 16.5).

Pituitary prolactin increases gradually throughout pregnancy. The placenta secretes human chorionic somatomammotrophin (hCS), which is also known as human placental lactogen (hPL). These hormones, together with oestrogen and progesterone, are important for the development of the lobules and alveolar tissue of the breast in preparation for lactation. However, the main function of prolactin is milk production. During pregnancy, very little milk is produced, because of the high levels of placental sex hormones, which inhibit the actions of the lactogenic hormones.

After delivery, the initial breast secretion is known as colostrum. It has a high protein content but almost no fat. It contains antibodies (IgA), produced by lymphocytes in the mammary gland. The antibodies help to protect the new-born against infections, in particular, those affecting the gastrointestinal tract. Therefore, the baby should be allowed to suckle as soon as possible after delivery. Milk secretion in appreciable amounts usually takes $2-3$ days to start, under the effect of pituitary prolactin. Oxytocin released during suckling stimulates the milk ejection reflex. The rapid decrese of oestrogens and progesterone allows the production of milk to proceed uninhibited. Similarly, after an abortion, milk production may also start. Suckling is important for continuation of prolactin secretion and also for oxytocin release during breast-feeding. Therefore, breast-feeding should be started on the first day and carried on regularly thereafter, so as to stimulate and maintain milk production.

During breast-feeding, menstrual cycles usually stop or may become irregular. Even if menstruation occurs during lactation, the cycles occur without ovulation (anovular cycles). This is due to an inhibitory effect of prolactin on GnRH release. Prolactin also decreases the effects of LH and FSH on the ovaries. Therefore, prolonged breast-feeding is recommended not only for the welfare of the baby and mother, but also for spacing of pregnancies.

Human milk is now recognized as the food most suitable for infant nutrition (see Chapter 8). Mother's milk also contains defensive cells (such as T and B lymphocytes, macrophages and polymorphonuclear WBCs) and antibacterial and antiviral substances (such as IgA, lysozymes and complement components).

THE METABOLIC RESPONSE TO LACTATION
During lactation, the energy requirements of the mother increase by about 25% above the normal energy needs, so as to provide for milk production

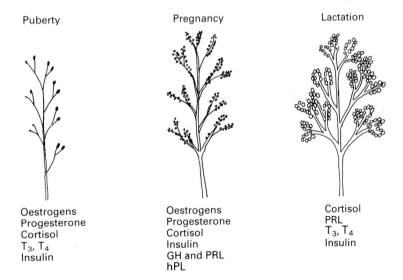

Puberty	Pregnancy	Lactation
Oestrogens	Oestrogens	Cortisol
Progesterone	Progesterone	PRL
Cortisol	Cortisol	T_3, T_4
T_3, T_4	Insulin	Insulin
Insulin	GH and PRL	
	hPL	

Fig. 16.5 The hormones acting on the mammary gland during various physiological conditions.

(see Chapter 8). The metabolic changes which occur in the mammary gland at the onset of lactation are probably larger than those in any other organ during normal development. For each volume of milk secreted, about 400–500 volumes of blood are circulated through the gland. When no milk is being secreted, metabolism is very low, but at peak lactation each cell is producing in each second $4–6 \times 10^8$ molecules of carbohydrate, fat and protein. Prolactin is the key hormone controlling milk production. The synthesis of milk protein (casein and α-lactalbumin) and fat is regulated mainly by prolactin and facilitated by insulin and cortisol.

A few millilitres of colostrum are usually produced each day throughout pregnancy. Colostrum contains similar concentrations of protein and lactose to milk but almost no fat.

At the peak of lactation, about 1.5 litres of milk may be formed each day. With that amount of lactation, large quantities of metabolic fuels are drained from the mother. For instance, the daily cost of milk production on the maternal side includes 50 g of fat, 100 g of lactose (mainly from maternal glucose) and 25 g of protein. On average, the energy cost of lactation is about 3 MJ/day. Also, some 1–3 g of calcium are used each day; therefore, lactating mothers are advised to take adequate quantities of milk and vitamin D to avoid a negative calcium balance.

Perinatal physiology

A scientific group of the World Health Organization (WHO) defined a mature infant at birth as one who shows functional competence, providing a reasonable safety margin in extrauterine life. This means that a foetus born before completion of 9 months of gestation could be functionally mature. Birth is viewed as an event in a continuing process of development, which marks the beginning of interactions between the new-born and the environment (Fig. 16.6). It is noteworthy that the highest velocity of brain growth occurs in the perinatal period.

During gestation, the foetal organs mature at different rates. An effective circulation is established with placentation, but the lungs and kidneys are not needed for survival until birth. The liver develops gradually and, until birth,

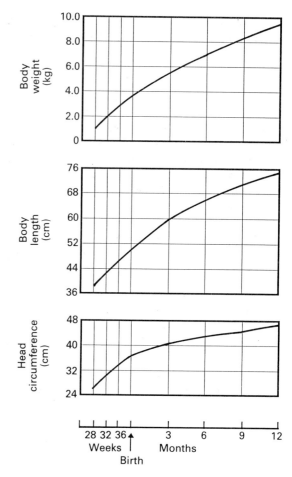

Fig. 16.6 Body weight, crown–feet length and head circumference during the perinatal period and the first year of life. Note the continuity of skeletal growth at birth while the growth of the head is much faster during the last few weeks of pregnancy. (Based on Gairdner, D. & Pearson, T. (1971) Growth charts. *Arch. Dis. Childh.* **46**, 783.)

some liver functions are still not fully developed. The nervous system and the liver continue to mature after delivery. Kidney function during gestation is required for formation of amniotic fluid. The foetus shows a wide variety of neurological and muscular activities, including breathing movements, which become increasingly vigorous and varied with age. These movements seem to be necessary for the development and maturation of the foetal lungs. For these activities, the circulatory, metabolic, endocrine and nervous support systems must be provided. Three important systems are selected for detailed discussion below.

THE FOETAL CIRCULATION

The foetal circulation differs from that of an adult by: (i) low pulmonary blood flow; (ii) high placental flow; and (iii) shunting of blood from the right atrium to the left atrium through the foramen ovale, and from the pulmonary artery to the aorta via the ductus arteriosus. These shunts allow the blood to bypass the foetal lungs (Fig. 16.7).

The placenta receives about 55% of the foetal cardiac output, through the umbilical arteries. Oxygenated blood comes back from the placenta through the umbilical vein, which joins the portal vein, but it bypasses the liver and goes to the inferior vena cava via the ductus venosus. On reaching the right atrium, most of this blood flow passes directly through the foramen ovale to the left atrium. Minimal mixing occurs with the stream of deoxygenated blood coming through the superior vena cava from the head and neck. This stream passes mainly into the right ventricle. The mixing is kept to a minimum by the position of the left atrium, which lies further posteriorly and to the right than in adults. Also, the crista dividens (the superior margin of the foramen ovale) lies over the entry of the inferior vena cava. The crista dividens directs approximately half of the oxygenated blood from the inferior vena cava directly into the left atrium and the rest goes into the right atrium.

From the left ventricle oxygenated blood is pumped into the aorta, where it is joined by the flow from the main pulmonary artery through the ductus arteriosus. The right ventricle pumps about two-thirds of the combined cardiac output. The ductus arteriosus shunts about 90% of the right ventricular output into the aorta and the rest goes to the foetal lungs. It is noteworthy that the ductus arteriosus, which brings mixed blood, joins the aorta beyond the exits of the coronary and carotid arteries, thus permitting a higher O_2 delivery to the heart and the brain. The aortic blood flow is distributed to the various parts of the foetal body, but the main bulk goes to the placenta through the umbilical arteries.

Changes at birth
Several cardiocirculatory changes take place at birth. First, there is a marked drop in the pulmonary vascular resistance. This is mainly due to the opening up of the pulmonary arterioles, no longer compressed by solid lung tissue, and to vasodilatation due to the increased P_{O_2}. The pulmonary artery pressure decreases from 60 to 35 mmHg during the first 8 hours after birth and falls to 25 mmHg by the end of 48 hours. Second, the systemic peripheral resistance increases about twofold, due to clamping of the umbilical arteries and cessation of the larger flow through the placenta. This increases the pressure in the aorta and consequently in the left ventricle and left atrium. These changes result in a reversal of the pressure difference between the right and left atria and bring about the closure of the valve of the foramen ovale. For several weeks, the foramen ovale is kept closed by these haemodynamic forces alone. A rise in right atrial pressure can

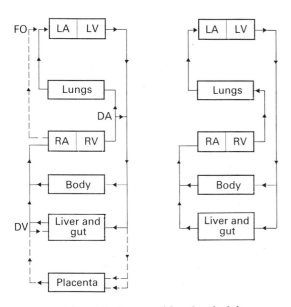

Fig. 16.7 Schematic diagram of foetal and adult circulatory systems. The broken lines indicate the channels which become obliterated after birth. LA and LV denote the left atrium and left ventricle. RA and RV denote the right atrium and right ventricle. FO denotes the foramen ovale, DA the ductus arteriosus and DV the ductus venosus.

reopen it and re-establish a right-to-left shunt. This occurs in hyaline membrane disease (HMD), leading to respiratory distress syndrome (RDS) of the new-born. The foramen ovale remains potentially open, but anatomical fusion occurs in 30% of normal adults.

Closure of the ductus arteriosus is brought about by contraction of the smooth muscle in the vessel wall. Due to the fall in the pulmonary vascular resistance described above, flow in the ductus arteriosus is reversed at birth. There is also an increase in systemic P_{O_2}, resulting from the establishment of normal ventilation and the closure of the foramen ovale. The P_{O_2} in the new-born reaches about 50 mmHg above that in the foetus. It is believed that the constriction of the ductus arteriosus is in response to high P_{O_2}. This closure is sufficient to stop the blood flow within a few hours after birth, which is known as physiological closure. Permanent or anatomical closure takes $1-4$ months, by infolding of the endothelium and disruption of subintimal layers, followed by fibrosis.

RESPIRATION

In mammals, foetal breathing movements are present from early gestation. In the human foetus, irregular breathing movements have been detected as early as the 11th week of pregnancy. They become increasingly regular after the 5th month, and by 36 weeks of gestation they acquire regular rhythm. However, these breathing movements do not seem to respond to changes in P_{O_2} and P_{CO_2}, as in postnatal life. This leads to the hypothesis that the respiratory centre must be under some form of inhibition. The foetal breathing movements are essential for normal lung development and, if interfered with as a result of harmful changes in the intrauterine environment, the normal maturation of the foetal lungs may be compromised. Changes which decrease or may even stop the breathing movements include hypoglycaemia, foetal infection, cigarette-smoking during pregnancy and administration of pethidine and barbiturates to the pregnant mother.

During foetal breathing activity, lung liquid flows up the trachea, with continuous small outward spurts of fluid synchronous with foetal breathing activity. Human lung liquid appears in the second trimester but large amounts are not produced until much later in pregnancy. The liquid keeps the future air space expanded. It is not amniotic fluid and has a completely different composition. Some of this lung fluid flows out into the liquor amnii and some of it may be swallowed. At birth, up to 35 ml of fluid drains from the mouth during normal vaginal delivery. The rest is absorbed by the pulmonary lymphatics and capillaries. Foetal adrenaline and noradrenaline, which are very high during delivery, stimulate the absorption of lung fluid. The lungs are rapidly cleared of fluid and within 60 minutes a normal functional residual capacity is established in the new-born infant.

About 90% of the epithelial lining of foetal alveoli at birth is made up of type I cells of flat epithelium. Other cells, more round and lying mostly beneath the surface, are type II cells, which produce surfactants. Early aeration of the alveoli is important for a layer of surfactant to spread and provide the stability of alveoli, preventing them from total collapse during expiration. Surfactants are complex lipoproteins which lower surface tension at the air–liquid interface. The lining of alveoli with surfactants, when the lung is inflated, keeps the pressure constant within all alveoli, irrespective of their diameter. This decreases the work of breathing during inspiration and keeps alveolar patency during expiration.

Surfactants are chemically complex substances synthesized in the alveolar epithelium. In the human lung, surfactants first appear at about 20 weeks of gestation and reach a peak at about $30-34$ weeks. Detection of the increase in surfactants concentration in the amniotic fluid indicate lung maturity and means that at birth the infant should not develop hyaline membrane disease. Administration of glucocorticoids (e.g. dexamethasone) or thyroxine to expecting mothers in premature labour helps to stimulate surfactant production.

Aeration of the lungs takes place with the first breath within 60 s of clamping the cord. Various physical stimuli, individually and in combination, can initiate respiration. These include lowering the skin temperature and physical tactile stimuli.

Hypoxia, as a result of clamping the cord, is an important factor in initiating the onset of breathing. The response to hypoxia and, particularly, to raised carbon dioxide tension in arterial blood seems to be chemoreceptor-mediated. It is abolished in experimental animals by cutting the carotid sinus nerve. At birth, the chemoreceptors become much more sensitive to small changes in arterial blood gas tensions and are important in the postnatal control of respiration. Also, the prenatal inhibition of the respiratory centre is somehow removed. Sensory inputs to the central nervous system are also involved in the onset of breathing and, if they are surgically interrupted in animals, the electrical activity of the respiratory centre is greatly reduced.

It seems, therefore, that there is not one single event that initiates the onset of respiration, or one single controlling mechanism. The enormous sensory input at birth, the change in chemoreceptor sensitivity, the hypoxia resulting from cord clamping and the increased activity of the central nervous system — all these factors combine to stimulate the new-born gasp which aerates the lungs.

THE KIDNEYS

Fluid, electrolytes and acid–base balance can be easily disturbed in the new-born because of several considerations. First, considering the weight of the newborn, the fluid turnover — intake and excretion — is seven times as much as in the adult. Second, the kidney of the foetus at birth can only concentrate urine up to 1.5 times the osmotic pressure of extracellular fluid, while the adult kidney can concentrate up to four times. The GFR rapidly increases from 20 to 60 ml/min during the first month compared with 120 ml/min in the adult. Renal tubular function is not fully developed, so that sodium reabsorption and hydrogen ion excretion are poor at birth and develop rapidly during the first month. Therefore, during the first few weeks there is inability to conserve sodium and to excrete acid. Thirdly, the metabolic rate of the new-born is twice that of the adult. Consequently, twice as much acid is produced by metabolic processes relative to body-weight. All these factors indicate that fluid and electrolyte loss can rapidly take place in new-born infants and that there is a tendency to develop acidosis.

The premature new-born kidneys are even less developed and therefore medical care should include close monitoring of fluid and electrolyte balance, as dehydration or even overhydration can easily occur.

TEMPERATURE REGULATION

The set point for thermal regulation is slightly lower in neonates. Thermal regulatory mechanisms are not fully developed. However, the neonate can maintain a constant body temperature within a narrow range of ambient temperature. Extremes of heat or cold can rapidly result in drastic changes in core body temperature. Room temperatures less than 20 °C can rapidly cause hypothermia. It is possible that maturation of hypothalamic thermoregulatory centres and availability of nutrients for peripheral thermogenesis may be among the underlying causes. Other contributory factors include the large surface area in the new-born relative to body mass and limited insulation by a thin layer of subcutaneous fat.

Mechanisms of heat production are mainly by increasing non-shivering thermogenesis. This is achieved by adrenergic stimulation of hydrolysis of triacylglycerols in brown fat to release heat energy. Brown fat is located at the root of the neck and between the scapulae, along the aorta, in the mediastinum and around the kidneys. It may constitute up to 6% of foetal body-weight. Non-shivering thermogenesis is associated with increased oxygen consumption. Therefore, hypoxic infants have a weaker response to cold. Chronic exposure to cold environments can result in energy loss and therefore poor weight gain early in infancy. Neonates only shiver when the environmental temperature falls to 15°C or less. The premature neonate is particularly less capable of maintaining body temperature (Fig. 16.8)

Heat conservation mechanisms, on the other hand, are well developed. Infants curl up and decrease their body surface exposure. Vasoconstriction is well developed in the neonate but its effectiveness may be decreased by the thin layer of subcutaneous fat.

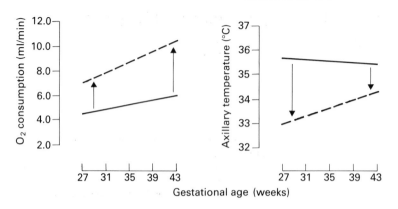

Fig. 16.8 Oxygen consumption and body temperature changes in premature neonates. The solid lines are responses in a thermoneutral environment and the broken lines are responses to a cool environment at 25–36 °C. Note that at low gestational age the premature neonate fails to raise O_2 consumption and to maintain body temperature. (Adapted from Heim, T., *et al.* (1981) Homeothermy and its metabolic cost. In *Scientific Foundations of Paediatrics*, 2nd edn, p. 105. William Heinemann Medical Books, London.)

Heat loss occurs, as in adults, through conduction, convection, radiation and evaporation. Conduction is insignificant, unless the baby is laid, without insulation, on a surface which is too cold or too warm. Convective heat loss is small and is due to air currents surrounding the skin. Radiation heat loss depends on the surroundings, either the room temperature or the walls of an incubator. If an infant is naked, e.g. for the sake of observation in the ward, heat loss by radiation is much greater in an open cot than in an incubator. Evaporative heat loss depends on the presence of liquid on the surface, e.g. amniotic fluid immediately after birth. Air movement also increases the rate of evaporative heat loss. In premature infants, the skin is more permeable to water and therefore more evaporative heat loss takes place.

If the new-born is cold, its O_2 consumption increases and surfactant formation and its efficiency is decreased. Cooling can rapidly lead to hypothermia, associated with failure of heat regulatory mechanisms, metabolic acidosis, hypoglycaemia and hypoxia.

NEONATAL JAUNDICE

During intrauterine life, foetal bilirubin crosses the placenta and is excreted by the liver of the mother. For the first few days, the foetal liver functions poorly, so that conjugation of bilirubin with glucuronic acid does not proceed satisfactorily. This leads to accumulation of bilirubin, giving rise to physiological hyperbilirubinaemia and jaundice. Fortunately, this is usually mild; serum bilirubin reaches up to 5 mg/100 ml (normal 1 mg/100 ml) and there is a mild yellow coloration of the sclera and mucous membranes.

A pathological cause of neonatal jaundice is rhesus incompatibility. The foetus is Rh-positive and the mother is Rh-negative but has had previous exposure to Rh antigens (Rh-positive red blood cells), either from a previous pregnancy or from a blood transfusion. The mother's serum will then have Rh antibodies. Maternal antibodies may cross the placenta and when they reach the foetal circulation, they start to destroy foetal RBCs. This creates a high bilirubin concentration in foetal blood. Fortunately, this condition can now be treated before the child is born.

Further reading

1 Emslie-Smith, D., Paterson, C.R., Scratcherd, T. & Read, N.W. (1988) Pregnancy. In *BDS Textbook of Physiology*, 11th edn, pp. 336–350. Churchill Livingstone, Edinburgh.

2 Guyton, A.C. (1991) Foetal and neonatal physiology. In *Textbook of Medical Physiology*, 8th edn, pp. 929–938. Saunders, Philadelphia.

17: The Central Nervous System (Clinical Neurophysiology)

Objectives

On completion of the study of this section, the student should be able to:

1 Describe the structural and functional organization of the central nervous system.

2 Understand the sensory functions of the central nervous system, so as to apply tests for sensory abnormalities.

3 Explain the mechanisms of the sensation of pain, in order to use appropriate measures in its treatment and management.

4 Comprehend the mechanisms of reflexes, so as to explain their roles in the control of motor functions.

5 Understand the voluntary and involuntary motor functions of the central nervous system, so as to test for motor disorders resulting from lesions of the motor pathways.

6 Understand the roles of the cerebellum and vestibular apparatus in the regulation of posture and equilibrium.

7 Understand the higher functions of the brain, so as to explain disturbances of memory, learning, speech and other processes.

8 Explain the relationships and functions of the hypothalamus with emotions, visceral functions and neuroendocrine integration.

Introduction

The vast complexity of body functions controlled by the nervous system is unique. These functions range from the somatic motor and sensory functions to highly integrated functions such as perception, memory, learning and emotions.

The central nervous system (CNS) consists of the brain and spinal cord. The peripheral nervous system, on the other hand, consists of 43 pairs of nerves which enter and leave the CNS. Afferent (sensory) fibres carry nerve impulses from the periphery to the CNS and efferent (motor) fibres carry impulses from the CNS to effector organs, i.e. muscles and glands.

The nervous system responds to changes in the external and internal environments. Receptors sensitive to various stimuli (e.g. light, sound, touch, blood pressure, muscle tension, etc.) are excited. Nerve impulses carried along afferent fibres to the CNS are processed and integrated before appearing in the efferent fibres to produce appropriate responses in the effector organs (e.g. muscle contraction, glandular secretion). The pathway the impulses follow from receptor to effector is called the *reflex arc* and the action that results is called a *reflex action*. Reflex arcs vary in complexity. They are simple at the level of the spinal cord, but get more complex at the level of the brain stem. The most complex reflex arcs are in the cerebral cortex.

Cells of the nervous system

NEURONES

The neurone is the functional unit of the nervous system. The human brain has 10^{11} neurones that vary in size and shape, but they all consist of a cell body (soma), one or more dendrites and only one axon, with synaptic terminals (Fig. 17.1).

The soma contains the nucleus. Its cytoplasm contains granules which stain with basic dyes known as Nissl substance. These are the source of the neurones' protein. Dendrites are tapered branches from the cell body that make contact with other neurones. Both soma and dendrites receive information from synaptic connections with other neurones. They are thus often considered sensory.

The single axon varies in length (from a few hundred micrometres to more than a metre) and also in diameter. It arises from the part of the soma known as the axon hillock, which is the most excitable part of the neurone. They have a myelin sheath and their ends branch to form terminals which make contact with other neurones at synapses.

A synapse consists of: (i) a presynaptic terminal, which contains the synaptic vesicles enclosing a neurotransmitter; (ii) a synaptic cleft, which is a small space (20–30 nm wide) between neurones that the transmitter has to cross to arrive at the postsynaptic neurone; and (iii) the postsynaptic neurone, which contains the receptors specific to the particular neurotransmitter.

It is important to note that, by the 6th month of age, the maximum number of neurones one will ever have is attained, because mature neurones do not multiply. Therefore dead neurones will not be replaced by new ones. In fact, there is progressive loss of neurones from birth onwards, due to the normal ageing process.

NEUROGLIA

Neuroglia (or supporting cells) are the matrix of the CNS which maintains its structure. There are three types of neuroglia (Fig. 17.2). *Astrocytes* are large and function also to maintain a constant K^+ environment around neurones. This is necessary for the normal functioning of the nerve cell.

Oligodendroglia form and maintain myelin sheaths around axons similar to the Schwann cells in the periphery. *Microglia* are the macrophages of the CNS which phagocytose tissue debris thus helping the process of tissue repair.

Electrical signals in the nervous system

Threshold stimuli result in action potentials or local potential changes in neuronal membranes. These are changes in the resting membrane potential produced by chemical transmitters at synapses or by specific stimuli at receptors.

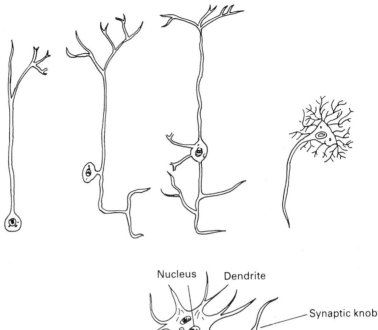

Fig. 17.1 Diagram showing various types of neurones. Below is a motor neurone. (Reproduced with modifications from Bray, J.J., Cragg, P.A., Macknight, A.D.C., Mills, R.G. & Taylor, D.W. (eds) (1989), *Lecture Notes in Physiology*, 2nd edn. Blackwell Scientific Publications, Oxford.)

RESTING MEMBRANE POTENTIAL (RMP)

The intracellular fluid has more K^+ and protein and less Na^+ and Cl^- than the extracellular fluid. This unequal distribution of ions between the two sides of the cell membrane causes a difference in electrical potential across it, called the RMP. Resting membrane potentials are maintained by an active energy-consuming process, the Na^+,K^+ pump.

The electrical potential across the membrane due to any ion can be calculated by means of the Nernst equation, e.g.:

Electrical potential for potassium, E_K

$$= 61 \log \frac{(\text{K outside cell})}{(\text{K inside cell})}$$

$$= 61 \log \frac{K_o}{K_i}$$

The RMP can be calculated by means of the Goldman equation, which expresses the contribution of the diffusion of all ions (i.e. K^+, Na^+, Cl^-) to the RMP. Accordingly, the magnitude of the RMP for nerve is about -70 mV, i.e. the electrical potential on the inner side of the mem-

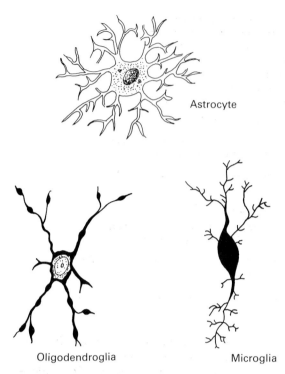

Astrocyte

Oligodendroglia   Microglia

Fig. 17.2 Three types of neuroglia. (Reproduced with modifications from Bray, J.J., Cragg, P.A., Macknight, A.D.C., Mills, R.G. & Taylor, D.W. (eds) (1989), *Lecture Notes in Physiology*, 2nd edn. Blackwell Scientific Publications, Oxford.)

brane is about 70 mV negative compared with the electrical potential on its outer side.

The RMP difference described above is changed by a net movement of charge across the membrane when its permeability changes to various ions. A decrease in the magnitude of the potential difference, with the inside becoming more positive, is called depolarization. This is caused by the movement of positive charge into (or negative charge out of) the cells. Similarly, an increase in magnitude of the potential difference, with the inside becoming more negative, is called hyperpolarization. This is caused by movement of positive charge out of (or negative charge into) the cell. When a membrane is hyperpolarized, it becomes less excitable, because it becomes more difficult to depolarize. Hypokalaemia, a decrease in the level of potassium in body fluids, causes hyperpolarization of muscle and nerve, due to a decrease

of positive charge inside the cell. Indeed, muscles show periods of paralysis in a disease called familial periodic paralysis, due to hypokalaemia.

LOCAL POTENTIALS

Local (electrotonic) potentials arise when the movement of charge responsible for the change in potential is limited to a particular site on the membrane of the neurone. This occurs when the movement of charge is small, and it usually occurs at points of discontinuity, e.g. environment-sensory receptor (i.e. receptor potentials), CNS synapses (excitatory and inhibitory postsynaptic potentials) and nerve–muscle junctions (end-plate potential).

In local potentials, the change in potential is greatest at the particular site and falls off exponentially with distance. Thus, they are typically restricted to 1 or 2 mm from their point of origin in the cell body or dendrite.

ACTION POTENTIALS

When the movement of charge is large (such as the influx of Na$^+$ at a sensory nerve terminal), the membrane of the neurone is depolarized to a point called the threshold, beyond which an all-or-nothing response, called an action potential, occurs.

Recording of an action potential

A cathode ray oscilloscope is used to record action potentials, using intracellular microelectrodes. The record of an action potential (see Chapter 4, Fig. 4.2) consists of a stimulous artefact, an isopotential latent period, which represents the time for the impulse to travel from the stimulating electrode to the recording electrodes and a spike (i.e. action potential), during which the polarity of the membrane potential temporarily reverses (i.e. the cell interior becomes positive with respect to the interstitial fluid) before returning to its original value.

If the duration of the latent period and the distance between the stimulating and recording electrodes are known, the nerve conduction velocity can be calculated. Measurements of nerve conduction velocity are useful in the diagnosis of peripheral nerve disorders, e.g. carpal tunnel syndrome.

Table 17.1 Important differences in the characteristics of local and action potentials

Action potential	Local potential (electronic potential)
Site: at axons	At dendrites and soma (postsynaptic potentials) At sensory receptors (receptor potential) At motor end plate (end plate potential)
Permeability: voltage dependent	Permeability: variable
All-or-none law: response to threshold is complete depolarization	Graded to the strength of the stimulus: non-threshold partial depolarization
Refractory period: summation impossible	No refractory period: summation possible
Propagated potential: not localized	Not propagated: localized
Amplitude: little change with distance	Decremental: maximum amplitude at point of origin, falls rapidly with distance
Conduction: potential change over long distance (range in mm or metre) (active spread)	Conduction over short distance (in mm) (passive spread)

Table 17.1 shows the important differences in the characteristics of local and action potentials.

Types and properties of nerve fibres

Nerve fibres are classified into A, B and C fibres (Table 17.2). A fibres have the thickest covering of myelin, followed by B fibres and, finally, C fibres, which are sometimes also referred to as unmyelinated fibres. The more myelin covering the nerve has, the more its thickness, conduction velocity, metabolic activity, excitability (i.e. having lower threshold) and capability of repetitive firing. Type A fibres are subdivided into Aα, Aβ, Aγ and Aδ. Examples of A fibres are spinocerebellar tract fibres and somatic motor fibres. Preganglionic autonomic fibres belong to class B.

Examples of C fibres are slow pain fibres and postganglionic autonomic motor fibres.

Larger myelinated fibres, because of their higher metabolic activity, are more susceptible to cold, compression and anoxia than smaller fibres. Therefore, these agents block conduction in myelinated fibres before unmyelinated ones. Local anaesthetic drugs, on the other hand, preferentially block unmyelinated fibres; hence their use to block pain sensation from the skin.

The compound action potential

If activity from a whole nerve (with different types of fibres in it) is recorded, we get the summated action potentials of all fibres in the nerve. This is called the compound action potential (AP). If we give stimuli of increasing intensity to the nerve, the compound AP grows in amplitude with increased stimulation strength until it reaches a maximum, when all the fibres are excited. This is because various fibres in the nerve have different thresholds, making the compound AP graded and not obeying the all-or-none law. Its absolute refractory period is the absolute refractory period of the largest fibres. The compound AP may display a number of peaks, each of which represents fibres having similar conduction velocities (Fig. 17.3). The separation of peaks

Table 17.2 Characteristics of nerve fibres

	Type of nerve fibre		
	A	B	C
Fibre diameter (μm)	1–20	<4	0.3–1.5
Conduction velocity (m/s)	5–120	3–15	0.6–2.5
Spike duration of AP (ms)	0.3–0.5	1.2	2.0

AP, action potential.

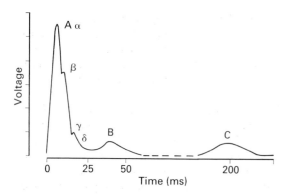

Fig. 17.3 Record of a compound action potential. The letters refer to the type of nerve fibre.

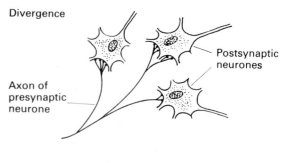

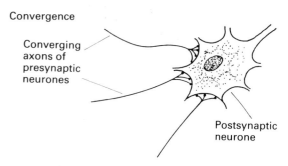

Fig. 17.4 Convergence and divergence.

increases as the distance between the stimulating and recording electrodes is increased. Type A fibres usually dominate the record, not necessarily because they are more numerous but because they have the lowest threshold and highest conduction velocity.

NEURONAL SYNAPSES

The synapse is the site of contact between the axon terminal of one neurone (presynaptic neurone) and another neurone (postsynaptic neurone), across which transmission of nerve impulses occurs.

The axon of the presynaptic neurone is called the presynaptic fibre. This divides into many branches towards its terminal end, which end in small knobs or swellings called synaptic knobs. Synaptic knobs are applied to a dendrite, to a soma or to the initial segment of the postsynaptic axon, forming axodendritic, axosomatic and axo-axonic synapses respectively.

Axons from many neurones may converge on a single neurone (Fig. 17.4). In fact, many postsynaptic neurones are covered with thousands of synaptic knobs, which are derived from hundreds of presynaptic neurones. This is known as anatomical *convergence*. Some of the converging presynaptic inputs are excitatory while others are inhibitory to the postsynaptic neurone. The process of convergence enables many neurones to act on one postsynaptic neurone. The level of excitability of the latter at any instant depends upon the number of active synapses and the re-

lative number of excitatory versus inhibitory ones. The axon of a single neurone may also terminate on many neurones across synapses. This is the opposite of convergence and is called *divergence*. It enables one neurone to influence the excitability of many other neurones (Fig. 17.4). A space called the synaptic cleft, containing extracellular fluid, separates the synaptic knob from the postsynaptic neurone. The presynaptic neurone contains vesicles, which are filled with a chemical transmitter, and is rich in mitochondria and the enzymes necessary for synthesizing the neurotransmitter.

Synaptic transmission

When an action potential arrives at the synaptic knob (Fig. 17.5), it increases the permeability of the presynaptic neurone membrane, leading to influx of Ca^{2+} ions into the neurone, through opening of *voltage-gated* Ca^{2+} channels. This causes release of the neurotransmitter by exocytosis. The amount of transmitter released is proportional to the Ca^{2+} influx. The neurotrans-

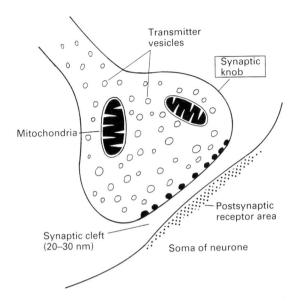

Fig. 17.5 Physiological anatomy of a synapse.

mitter diffuses across the cleft and binds to receptor molecules on the postsynaptic membrane. The neurotransmitter–receptor complex then produces changes in the permeability of the postsynaptic membrane, by opening channels, called *chemical-gated* channels, for specific ions. These ions are then allowed to pass across the membrane, causing a small local change in membrane potential (i.e. synaptic potential). There are separate Na^+, K^+ and Cl^- channels in the postsynaptic membrane.

A transmitter that opens the channels for Na^+ increases movement of these ions to the interior of the cell and so excites the postsynaptic neurone. Such a transmitter is an *excitatory* one. On the other hand, a transmitter that opens the channels either to K^+ or to Cl^- or to both, causes K^+ ions to move out and Cl^- to move inside the cell, leading to loss of positive charge from inside the cell. As a result, the neurone is inhibited, due to hyperpolarization. Such a transmitter is *inhibitory*. It is believed that, with very few exceptions, each neurone releases only one type of transmitter, and it releases this same transmitter at all of its separate terminals.

Electrical events during neuronal excitation

The resting membrane potential in most neurones is somewhat less than that found in large peripheral nerve fibres or in muscles, i.e. -65 mV. This makes the neurone more excitable.

Excitatory postsynaptic potential (EPSP) If the presynaptic neurone is one that secretes an excitatory neurotransmitter, it will increase the permeability of the postsynaptic neurones to all ions, but the movement of Na^+ predominates, due to the large electrochemical gradient across the membrane. The net effect is a slight depolarization, which increases the neurone's excitability and brings it closer to threshold (because part of the negativity of RMP is neutralized). This change in voltage above RMP is called the excitatory postsynaptic potential (EPSP).

It is important to note that the duration of the EPSP is much longer than that of an action potential. Also, note that an EPSP has to exceed a certain threshold before it triggers an action potential in the postsynaptic neurone. A single EPSP can never increase the postsynaptic neuronal potential to that level. Instead, the effects of several EPSPs must add up or summate to trigger an action potential.

There are two types of summation (Fig. 17.6):
1 *Temporal summation.* This occurs when successive postsynaptic potentials of an individual synaptic knob develop rapidly enough to summate.
2 *Spatial summation.* This occurs when postsynaptic potentials from several knobs over the surface of a neurone produce summation.

Excitability of postsynaptic neurones is thus raised and, when the threshold value of excitation is reached, an action potential is initiated at the axon hillock, this being the most excitable part of the neurone due to the existence of a much larger number of Na^+ channels in it than in the soma or dendrites.

Electrical events in neuronal inhibition

Inhibitory postsynaptic potential (IPSP) or postsynaptic inhibition When the presynaptic neurone is one that secretes an inhibitory neurotransmitter, it will increase the permeability of

Spatial summation

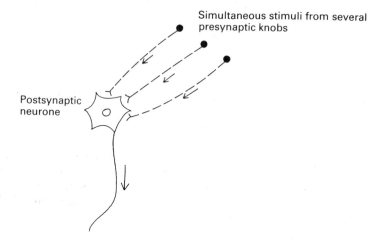

Simultaneous stimuli from several presynaptic knobs

Postsynaptic neurone

Temporal summation

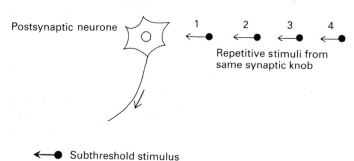

Postsynaptic neurone

Repetitive stimuli from same synaptic knob

← ● Subthreshold stimulus

Fig. 17.6 Above: Spatial summation. Simultaneous stimuli from several presynaptic fibres summate to excite the postsynaptic neurone. Below: Temporal summation. Repetitive stimuli from one source summate to excite the postsynaptic neurone.

the postsynaptic membrane to K^+ and Cl^-. K^+ efflux decreases positive charge inside, and Cl^- influx increases negative charge inside. The net effect is an increase in the degree of intracellular negativity—a state of hyperpolarization called the inhibitory postsynaptic potential (IPSP). This will inhibit the neurone because the membrane potential is now further away from the threshold for excitation. This type of inhibition is also known as postsynaptic inhibition.

IPSPs can also undergo temporal and spatial summation.

Presynaptic inhibition This is another type of inhibition, in which inhibitory synaptic knobs lie directly on the termination of the presynaptic

excitatory fibre (Fig. 17.7). The inhibitory synaptic knobs release a transmitter which inhibits the release of an excitatory transmitter from the presynaptic fibre when an action potential arrives there. The transmitter released at the inhibitory knob is thought to be gamma-aminobutyric acid (GABA).

Presynaptic inhibition takes a longer time to develop than postsynaptic inhibition but when it occurs, it persists longer.

Properties of synapses

One-way conduction At a synapse impulses always pass from the pre- to the postsynaptic neurone and never in the reverse direction.

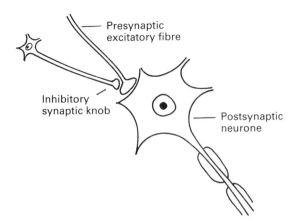

- Presynaptic excitatory fibre
- Inhibitory synaptic knob
- Postsynaptic neurone

Fig. 17.7 Presynaptic inhibition. The inhibitory synaptic knob secretes a transmitter that depresses the voltage of an action potential at the synaptic membrane of the presynaptic excitatory fibre. The transmitter released from the latter is therefore reduced in amount and the degree of excitation of the postsynaptic neurone is decreased or inhibited.

Synaptic fatigue A decrease in the number of discharges from the postsynaptic neurones occurs when the presynaptic neurone is repetitively stimulated at a rapid rate. This is mainly due to exhaustion of neurotransmitter stores in synaptic knobs and partly due to progressive inactivation of postsynaptic receptors.

Synaptic fatigue can serve as a protective mechanism. A good example is cessation of an epileptic fit (due to synaptic fatigue), which would otherwise have been fatal.

Post-tetanic facilitation Repetitive stimulation of an excitatory synapse for a period of time leads to a few moments of synaptic fatigue, followed by a period of increased excitability of the post-synaptic neurone, which might extend from a few seconds to several hours. This is called post-tetanic facilitation and is thought to be due to accumulation of Ca^{2+} in the presynaptic terminals as a result of the repetitive stimulation. The excess Ca^{2+} would eventually lead to increased release of transmitter from knobs.

Post-tetanic facilitation is considered to be one of the mechanisms for short-term memory, in which information is stored as a continuous electrical activity.

Synaptic delay This is the minimum time required for transmission across one synapse (about 0.5 ms), most of which is spent in the release of transmitter.

Uniqueness of central synapses Although synapses are found both in the CNS and in the autonomic ganglia of the peripheral nervous system, the property of inhibition is unique to central synapses.

Blood pH Alkalosis greatly enhances neuronal excitability and synaptic transmission while acidosis decreases synaptic transmission. A rise of blood pH to 7.8 causes convulsions and a fall to 7.0 or below results in coma.

Hypoxia Synaptic transmission depends on adequate oxygenation. If the blood supply to the brain stops, a subject becomes unconscious within about 5 seconds and may suffer permanent damage to the brain.

Drugs Some drugs increase the excitability of neurones and enhance synaptic transmission. Caffeine, theophylline and theobromine (found in tea, coffee and cocoa) act by decreasing the threshold for excitation of neurones. Strychnine, on the other hand, acts by inhibiting the action of inhibitory chemical transmitters.

Other drugs decrease the excitability of neurones and thereby decrease synaptic transmission. Anaesthetics act by making neuronal membranes less responsive to excitatory agents.

Functional anatomy of the central nervous system

The CNS comprises the brain lying within the skull and the spinal cord lying within the vertebral column.

SPINAL CORD

The spinal cord consists of segments, each of which has a pair of nerve roots, on each side (Fig. 17.8). The dorsal roots carry impulses from peripheral receptors into the spinal cord, while the ventral roots carry impulses to the periphery (i.e. muscles). Grey matter forms the core of the

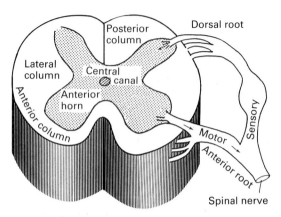

Fig. 17.8 Transverse section of the spinal cord.

spinal cord and appears like the letter 'H' in cross-section. It contains the cell bodies of neurones. White matter surrounds the grey matter. It is made up of ascending (sensory) and descending (motor) tracts.

The spinal cord functions include:

1 Transmission of sensory (afferent) impulses coming from peripheral receptors to the brain and of motor (efferent) impulses from the brain to motor neurones, which supply effector organs (i.e. muscles and glands).

2 Serving as a centre for some reflexes, some of which are the basis of movement and posture, e.g. stretch reflex.

BRAIN

The brain can be divided into four subdivisions (Fig. 17.9):

1 Brain stem.
2 Diencephalon.
3 Cerebellum.
4 Cerebrum.

Brain stem The brain stem consists of the medulla, pons and midbrain.

The medulla forms the upper extension of the spinal cord. It contains motor and sensory nuclei of the throat, mouth and neck, and nuclei for respiratory and cardiovascular control centres. The medulla also contains the nuclei of cranial nerves IX, X, XI and XII.

The pons is continuous with the medulla and also contains control centres for the respiratory and cardiovascular systems. It also contains nuclei of some sensory and motor nerves (i.e. nuclei of cranial nerves V, VI, VII and VIII).

The midbrain is continuous with the pons below and the diencephalon above. It contains nuclei of cranial nerves III and IV which mediate pupillary reflexes and eye movement.

The recticular formation of the brain stem extends from the medulla through the pons to the midbrain. It is composed mainly of ascending and descending tracts and some nuclei. The reticular formation plays an important role in the control of *muscle tone* and in *arousal* or *alerting mechanisms*.

Diencephalon The diencephalon is composed of the two thalami laterally and the hypothalamus ventrally. Thalamic nuclei are functionally divided into several groups. The most important of these are: one group that relays all types of sensation to the sensory cortex except olfaction. Another group relays signals from the cerebellum and basal ganglia to the motor cortex. The third group controls the general level of activity of the whole cerebral cortex and is therefore responsible for the level of consciousness. The hypothalamus is the higher autonomic centre (e.g. control of blood pressure, heart rate and body temperature). It also secretes hormones that control the release of other hormones from the pituitary gland. Being part of the limbic system, the hypothalamus plays a role in of generation of emotions. There are also centres for control of appetite and water intake.

Cerebellum The cerebellum lies in the posterior aspect of the brain stem and is connected to it by three thick bundles of white matter (cerebellar peduncles), which carry impulses in and out of the cerebellum. It consists of two lateral hemispheres and a central part called the vermis. All parts have the same histological structure — a thin superficial layer of grey matter or cerebellar cortex lies on the outside, enclosing an area of white matter. This consists of the fibres that enter and leave the cerebellar cortex. Masses of grey matter, the cerebellar nuclei, are found within the white matter.

(a)

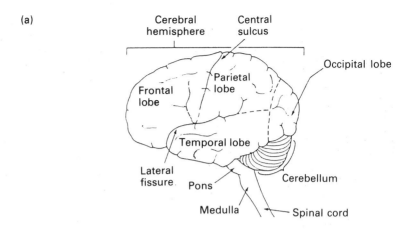

(b)

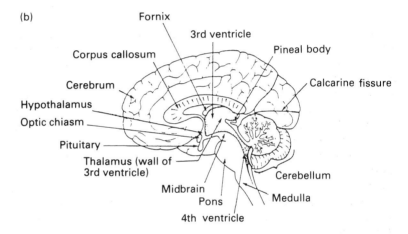

Fig. 17.9 The human brain. (a) Lateral view. (b) View at a midsagittal section. (Reproduced from Bray, J.J., Cragg, P.A., Macknight, A.D.C., Mills, R.G. & Taylor, D.W. (eds) (1989) *Lecture Notes in Physiology*, 2nd edn. Blackwell Scientific Publications, Oxford.)

The cerebellum is concerned with:

1 Control of rate, range and direction of movement.

2 Control of muscle tone.

3 Control of equilibrium and posture.

Cerebrum The cerebrum consists of the right and left cerebral hemispheres, connected in the midline by the corpus callosum. The superficial layer of each hemisphere is composed of grey matter — the cerebral cortex. This encloses a larger area of white matter inside which lie a number of nuclei, known as basal ganglia. These play a very important role in the planning and control of movement.

The cerebral cortex has elevations (gyri) and depressions (sulci), which greatly increase its surface area (Fig. 17.9).

Each hemisphere is divided into four lobes. The temporal lobe lies on the lateral surface of the hemisphere below the lateral fissure (deep groove) that separates it from the frontal and parietal lobes above. The temporal lobe contains the *primary auditory area*, which is the centre of hearing. The central sulcus forms the borderline between the frontal lobe anteriorly and the parietal lobe posteriorly. The precentral gyrus lies immediately in front of the central sulcus, and this forms the *primary motor area*, which controls muscular movements. The postcentral gyrus lies immediately behind the central sulcus and forms the *primary sensory area*, which perceives sensory information arriving from skin, muscles, joints and the viscera.

The occipital lobe lies most posterior in the

cerebral hemisphere. In it is located the *primary visual cortex*, the centre for vision.

The parietal, temporal and occipital lobes meet in the angular gyrus. Just in front of this gyrus is an area of cortex called *Wernicke's area*. This area, often called the general interpretative area, knowing area or gnostic area, plays a crucial role in higher functions of the brain, such as thinking, speech and language. On the under-surface of the medial occipital and temporal lobes is a special area which plays the important role of recognition of faces. Bilateral lesions of this area produce *prosopagnosia*, i.e. inability to recognize people by faces.

The white matter of the cerebral hemisphere constitutes three types of fibres:

1 *Association fibres*. These connect gyri in the same hemisphere.

2 *Projection fibres*. These form the ascending and descending tracts, which carry impulses to and from the hemispheres.

3 *Commissural fibres*. These connect gyri of one hemisphere with corresponding gyri of the other hemisphere. The most important commissure is the *corpus callosum*.

It is of interest to note that splitting of the corpus callosum causes each hemisphere to behave independently of the other—a condition known as split brain. A patient with split brain will identify an object placed on his hand while his eyes are closed but will be unable to identify it as the same object when it is transferred to the other hand.

The sensory system

The sensory system provides us with information about our environment, both external and internal. All information comes to us through our sense organs, which contain structures called receptors. Receptors are therefore detectors and are also transducers that convert the various forms of energy (light, sound, chemical, mechanical, etc.) they detect into action potentials. When these action potentials reach the brain and enter consciousness, we experience a sensation, e.g. vision, hearing, pain, touch, cold, etc. Every sensation is associated with a feeling that it is pleasant or unpleasant. This emotional component of a sensation is called its affect.

Anatomically, receptors are specialized structures present at the peripheral terminations of afferent fibres. From the receptors to the brain are pathways made up of chains of neurones. These pathways, e.g. visual pathway, pain pathway, etc., make up the sensory system.

SENSORY RECEPTORS

Classification of receptors

1 *Telereceptors (distance receptors)*. These are receptors that detect events at a distance, i.e. rods and cones for light and hair cells of the cochlea for sound.

2 *Exteroceptors*. These are concerned with information about the immediate external environment. They are found in the skin, e.g. touch receptors, temperature receptors.

3 *Interoceptors*. These are concerned with information about the internal environment, e.g. chemoreceptors, osmoreceptors.

4 *Proprioceptors*. These are concerned with information about the position of the body in space. They are found in muscles, tendons and joints.

5 *Nociceptors*. These are pain receptors. They are called nociceptors (from the Latin *nocere* to injure) because all painful stimuli arise due to injury or damage to tissues.

Properties of receptors

Receptors have the properties of adequate stimulus, excitability and adaptation.

Adequate stimulus Each type of receptor is most sensitive to a specific form of energy, which is called its adequate stimulus, and is almost non-responsive to the normal intensities of other forms of energy; e.g. light is the adequate stimulus for the rods and cones of the eyes but they do not respond to heat or cold (Fig. 17.10).

Pain receptors are not stimulated by a blunt object touching the skin, but they discharge as soon as the blunt object is pushed with enough force to damage tissues.

The sensation perceived as a result of stimulation of a receptor is called the *modality* of sensation. Thus, cold, warmth, touch and pain are different modalities of sensation.

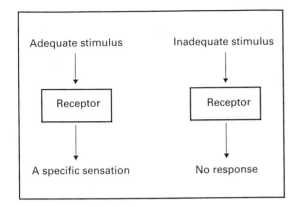

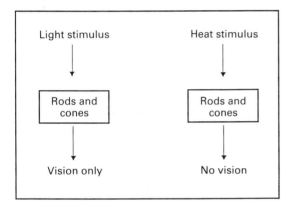

Fig. 17.10 The concept of the adequate stimulus.

Excitability When adequately stimulated, a receptor generates a non-propagated depolarizing potential, known as the generator or receptor potential. If it is of sufficient magnitude, the receptor potential triggers an action potential, which is propagated along the sensory nerve connected to the receptor.

The means of exciting a receptor varies with the different types of receptors, e.g. mechanical deformation, application of a chemical, changing the temperature or electromagnetic radiation, which somehow changes the characteristics of the receptor membrane. All these means ultimately alter membrane permeability and lead to free diffusion of ions across it. Influx of Na^+ ions leads to the generation of receptor potentials.

Adaptation When a stimulus of constant strength is continuously applied to a receptor, the frequency of action potentials in the afferent nerve fibre leading from it gradually declines. This phenomenon is called adaptation, and the rate at which it occurs varies in different types of receptors. Some adapt slowly and are known as tonic receptors. These continue to transmit impulses to the brain without decline in frequency in spite of continuous stimulation, and are therefore useful in detecting changes in the status of the body, which the brain needs to be constantly informed about (Fig. 17.11). Examples of tonic receptors are:

1 Muscle spindle, for information about posture.
2 Joint receptors, for information about the position of different parts of the body.
3 Baroreceptors, for information about arterial blood-pressure.

Pain receptors do not adapt at all. This is useful because pain serves as a protective mechanism for the body.

Another group of receptors adapt rapidly. These are known as phasic receptors because, although they show a rapid decline in frequency with maintained stimulation, they react strongly while a change in the intensity of the stimulus is taking place. They are therefore useful in transmitting information about rapid changes but are of no use in transmitting information about constant events, e.g. touch receptors.

Temperature receptors occupy a position between the above two groups. Thus, they are rated as moderately adapting receptors.

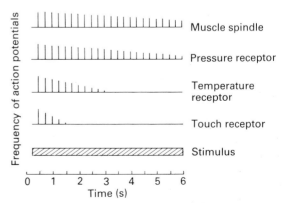

Fig. 17.11 Diagram showing a comparison between the rates of adaptation of receptors.

CODING OF SENSORY INFORMATION

Most sensory information aims at causing an awareness of the state of the body and its surroundings. Coding of sensory information enables higher centres to discriminate modality, locality and intensity of different stimuli. Although all sensory information is presented to the CNS in the form of action potentials, coding enables the above characteristics of the stimulus to be detected and perceived.

Modality

Modality discrimination is made possible by the following:

1 There is a specific stimulus for each receptor — the adequate stimulus.

2 A receptor or its nerve pathway when stimulated gives one type of sensation regardless of the method of stimulation. This is called the *law of specific nerve energies*. For example, stimulation of a pain receptor or the nerve pathway leading from it will cause a subject to perceive pain whether the stimulus was heat, electric shock or crushing of the tissues.

3 A specific pathway exists for each modality of sensation, i.e. pain pathway, touch pathway, etc.

4 Each nerve fibre in the pathway ends at a specific point in the CNS. The modality perceived is determined by that area in the CNS when the stimulus arrives there, i.e. a subject perceives touch or temperature because their responsive fibres end at a specific touch or temperature area in the brain.

Localization

Locality discrimination is explained by the *law of projection*. Concious perception of a particular sensation is always projected to the locality of the receptor, irrespective of where the sensory pathway (leading to the receptor) was stimulated.

This explains the *phantom limb*, a phenomenon in which pressure on the nerves in the stump of an amputated limb makes the patient feel the pain as though it is coming from the absent part of the limb. This is because the sensation is projected to the location of the receptors which were removed with the amputated part.

Intensity

When the intensity of a stimulus acting on a receptor is increased (see Fig. 17.12) two things happen. The number of receptors activated increases and the frequency of APs generated in the afferent fibre increases. Both these changes are interpreted by the brain as an increase in intensity.

THE SOMATIC SENSATIONS

The somatic sensations are those sensations carried from the various parts of the body by somatic nerves.

Classification of somatic sensations

Somatic sensations can be classified into three types:

1 Mechanoreceptive senses, stimulated by mechanical displacement of body tissues.

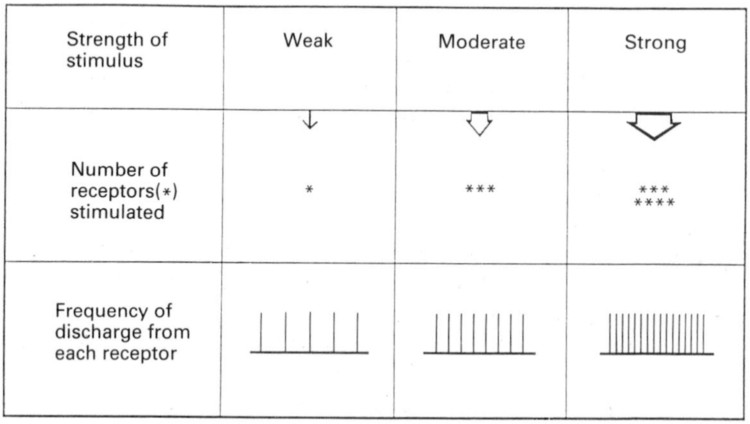

Strength of stimulus	Weak	Moderate	Strong
Number of receptors(*) stimulated	↓ *	▽ ***	▽ *** ****
Frequency of discharge from each receptor	‖ ‖ ‖ ‖	‖‖ ‖‖ ‖	‖‖‖‖‖‖‖

Fig. 17.12 Effect of increasing stimulus intensity on receptor excitation.

2 Thermoreceptive senses, stimulated by heat and cold.

3 Pain sense, stimulated by any factor that leads to tissue damage.

Mechanoreceptive senses include touch, pressure, vibration and tickle senses (which are frequently called the tactile senses) and the position sense.

Cutaneous receptors A large number of tactile receptors have been described. These include free nerve endings, encapsulated nerve endings and expanded-tip tactile receptors, the distribution of which vary in different regions of the body. Free nerve endings are widely distributed in the skin and have been shown to respond to nearly all sensory modalities. The expanded-tip tactile receptors include *Merkel's discs* and *Ruffini endings*. These are slowly adapting touch receptors, which, due to this property, enable determination of continuous touch stimuli on the skin. *Meissner's corpuscles* and *Pacinian corpuscles* are encapsulated, rapidly adapting receptors. The former are the receptors for fine touch, i.e. recognition of locality, form and texture of objects touched, while the latter are particularly important for detection of extremely rapidly changing mechanical events, e.g. vibration sense. Another important tactile receptor is the tactile hair and its basal nerve fibre, collectively called the hair end organ. This receptor detects mainly movements of objects on the surface of the body. Pacinian corpuscles and Ruffini endings are also described among receptors found in deeper tissues and joint capsules. There they subserve the sensation of deep pressure and position.

In spite of the large number of types of mechanoreceptors, they transmit their impulses via three types of fibres. Receptors such as Meissner's and Pacinian corpuscles, Merkel's discs, hair end receptors and Ruffini endings transmit impulses via quick-conducting type Aβ fibres. The less specialized free nerve endings transmit via the slower fibres, type Aδ and type C.

Thermoreceptive sensations are discriminated by two types of receptors that detect heat changes, i.e. cold and warmth receptors, and others that detect pain resulting from extremes of tempera-

ture. The receptors are free nerve endings that respond to absolute temperatures. Cold receptors respond to skin temperatures from 10 °C to 40 °C and transmit their impulses via type Aδ and probably C fibres. Warmth receptors respond from 30 °C to 45 °C and transmit impulses via type C fibres. At temperatures below 10 °C and above 45 °C, tissue damage begins to occur and the sensation experienced is that of pain, due to stimulation of two subtypes of pain receptors, namely cold-pain and heat-pain receptors respectively.

The receptors for pain are all free nerve endings, which are distributed in almost every tissue of the body. They transmit their impulses in two types of fibres, Aδ and C fibres. The physiological significance of these two types will be discussed later.

To conclude, it can be seen that receptors are classified according to the kind of energy to which they are most specific. Thus, mechanoreceptors respond to mechanical energy, thermoreceptors to heat and pain receptors to chemicals released as a result of tissue damage. In some cases, specificity can be associated with a particular structure (e.g. Pacinian corpuscle), but more often structures (or receptors) which appear histologically similar show different specificities. Thus free nerve endings that signal itch and tickle sensation are different from those free nerve endings which detect temperature.

Neural basis of sensation

Having described the characteristics and properties of receptors, we can consider the construction of a typical sensory pathway. Figure 17.13 shows a typical sensory pathway from a receptor in the head region and one from the body to the brain. Lesions can occur at any level in that pathway.

Somatosensory tracts

After entering the spinal cord, sensory fibres follow one of two ascending tracts within the spinal cord:

1 *The dorsal (posterior) columns or lemniscal tracts*. These subserve the following:

(a) Touch sensations characterized by sharp

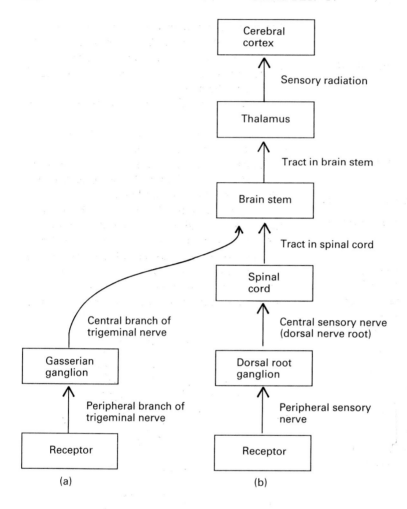

Fig. 17.13 Schematic diagram of (a) A typical sensory pathway from a receptor in the head region. (b) A typical sensory pathway from a receptor in the trunk region.

localization and high intensity discrimination (i.e. fine touch).

(b) Rapidly repetitive sensations such as the sense of vibration.

(c) Position sensations.

(d) Pressure sensations characterized by high intensity discrimination (i.e. fine pressure).

2 *The anterolateral tracts (spinothalamic tracts).* These subserve the following:

(a) Pain.

(b) Thermal sensations, including both warm and cold sensations.

(c) Touch sensations characterized by crude localization and poor intensity discrimination.

(d) Pressure sensations characterized by poor intensity discrimination.

(e) Tickle and itch sensations.

(f) Sexual sensations.

Dorsal column pathway The fibres destined for the dorsal column pathway are mainly large myelinated type Aβ fibres. These are actually the axons of first-order neurones (Fig. 17.14).

The paths of the three orders of neurones are as follows:

1 *First-order neurones.* Afferent fibres enter the spinal cord via the medial portion of the dorsal root. They enter the ipsilateral dorsal column and ascend upwards as the gracile and cuneate tracts to end, respectively, in the gracile and cuneate nuclei of the medulla.

2 *Second-order neurones.* These are located in

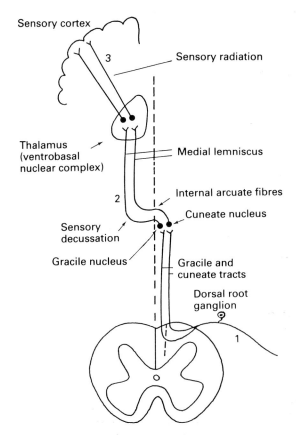

Fig. 17.14 The dorsal column pathway. (1), (2) and (3) refer to first-, second- and third-order neurones.

the gracile and cuneate nuclei and give rise to axons, which cross to the opposite side, forming the sensory decussation. They then pass upwards as the medial lemniscus, which traverses the midbrain to end in the ventrobasal nuclei of the thalamus.

In its pathway through the brain stem, the medial lemniscus is joined by additional fibres from nuclei of the trigeminal nerve. These fibres subserve the same sensory functions for the head that the dorsal column fibres subserve for the body.

3 *Third-order neurones.* The ventrobasal nuclear complex of the thalamus gives rise to axons, which project in the sensory radiation to the somatosensory cortex.

Anterolateral pathway (spinothalamic pathway)
Fibres contributing to this pathway are small, myelinated type Aδ and unmyelinated type C fibres, which enter the spinal cord via the lateral division of the dorsal roots.

The paths of the three orders of neurones are as follows:

1 *First-order neurones.* Afferent fibres, after entering the spinal cord, ascend or descend a few segments in Lissauer's tract before synapsing in the dorsal horn lamina II and III (substantia gelatinosa of Rolandi) and lamina V (Fig. 17.15). These synapse either directly or indirectly (via interneurones) with the second-order neurones.

2 *Second-order neurones.* These are in the dorsal horn and give rise to axons, which cross to the opposite side in front of the central canal (i.e. anterior commissure) and ascend as the spino-thalamic tract through the spinal cord. The fibres

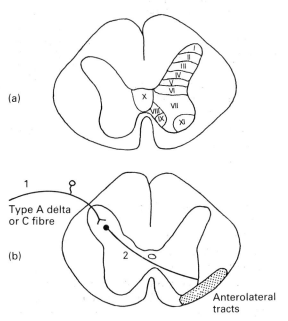

Fig. 17.15 (a) Cross-section of the spinal cord showing the anatomical laminae I to XI of the cord grey matter. (b) Cross-section of the spinal cord showing the first- and second-order neurones in the anterolateral spinothalamic tract. 1, first-order neurone; 2, second-order neurone. (Reproduced from Bray, J.J., Cragg, P.A., Macknight, A.D.C., Mills, R.G. & Taylor, D.W. (eds) (1989) *Lecture Notes in Physiology*, 2nd edn. Blackwell Scientific Publications, Oxford.)

ascend through the brain stem to end in the ventrobasal nuclear complex of the thalamus.

In the anterior commissure of the spinal cord, the temperature fibres are nearest to the central canal. Next to these are the pain fibres, and the furthest from the central canal are the touch fibres. Therefore, in *syringomyelia*, a developmental defect of the central canal, the temperature and pain fibres are damaged and the touch fibres escape. This leads to loss of temperature and pain sensation with preservation of other sensations — a phenomenon known as dissociated anaesthesia or dissociated sensory loss.

3 *Third-order neurones*. These are in the thalamus and they send axons along the sensory radiation to the somatosensory areas of the cerebral cortex.

Table 17.3 shows some major differences in the functional and anatomical characteristics between the two major sensory pathways. One of these is a difference in the spatial orientation of nerve fibres in each. Due to this difference, tumours arising outside the spinal cord (extramedullary tumours) first damage the sacral fibres of the spinothalamic pathway, whereas intramedullary tumours damage the sacral fibres last of all (i.e. sacral sparing).

Role of the thalamus and the sensory cortex in the appreciation of sensation

All sensory tracts, except the olfactory pathway, synapse in the thalamus on their way to the cerebral cortex. When impulses mediating a given sensation reach the thalamus, the subject becomes crudely aware of the sensation but he cannot perceive all of its fine details: e.g. a person will be aware of a change in temperature if he contacts a hot object but he will not be able to indicate how hot the object is. Gradations and other spatial and temporal characteristics are appreciated at the level of the sensory cortex and not at the level of the thalamus (Fig. 17.16). Pain, however, seems to be the only sensation that seems to be fully appreciated at the thalamic and probably even at the reticular formation level or even lower. Still, interpretation of the quality and localization of pain occurs at the level of the cerebral cortex.

Table 17.3 Comparison between the different characteristics of the two major sensory pathways

Dorsal column lemniscal pathway	Anterolateral pathway
Is concerned with fine type of transmission	Is concerned with a cruder type of transmission
e.g. There is a high degree of spatial orientation of nerve fibres with respect to their origin on the surface of the body	e.g. Poor degree of spatial orientation
Gradations of intensities are acute	Gradations of intensities are far less acute
Responsiveness is not greatly altered by stimuli from other areas of the nervous system	Responsiveness can be greatly altered by stimuli from other areas of the nervous system (i.e. brain and spinal cord analgesic system)
Fibres from the lower parts of the body lie towards the centre, while fibres from higher levels form successive layers laterally	Fibres from the lower parts lie laterally, while those from higher levels form successive layers towards the centre
Receives afferent sensory fibres of the dorsal root, which belong to type Aβ fibres, which ascend directly in the dorsal columns	Receives afferent sensory fibres, which are thin myelinated type Aδ fibres or unmyelinated type C fibres. These travel in Lissauer's tract and relay in the dorsal horn, to give rise to the anterolateral pathway
Has a very fast velocity of transmission (i.e. 30–110 metres/s)	Has a relatively slow velocity of transmission (8–40 metres/s)
Has great ability to transmit rapidly repetitive sensations (i.e. vibration sense)	Has poor ability to transmit rapidly repetitive sensations
Is limited to transmission of mechanoreceptive sensations only (i.e. fine touch and pressure sense, vibration sense, position sense)	Has the ability to transmit a broad spectrum of modalities (i.e. pain, thermal sensation, crude touch and pressure, itch and tickle sensations, sexual sensations)

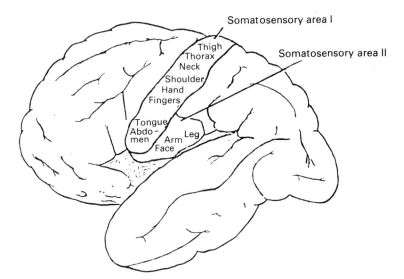

Fig. 17.16 The two somatosensory cortical areas, somatosensory areas I and II. (Reproduced with permission from Guyton, A.C. (1987) *Textbook of Medical Physiology*, 7th edn. Saunders Co, Philadelphia.)

The somatosensory cortex

The somatosensory area is that part of the cerebral cortex to which sensory signals are projected. It comprises two areas, which receive direct afferent fibres from the specific nuclei of the thalamus. Somatosensory area I is the primary sensory area and lies in the postcentral gyrus, and somatosensory area II lies in the wall of the Sylvian fissure (Fig. 17.16).

Somatosensory area I (Fig. 17.17) of one cerebral hemisphere is stimulated by impulses arriving from the contralateral side of the body, with the exception of the face, which is bilaterally represented in both hemispheres. The legs are represented on top and the head at the bottom of the gyrus. Moreover, the area of representation of a specific part of the body is related to the density of receptors in the part and not to its size. In man the lips are represented by the greatest area, followed by the face and thumb, while the entire trunk is represented by a relatively much smaller area. Recordings from somatosensory area I have also shown that it is organized functionally into columns. The cells in a given column are stimulated by afferents from a particular part of the body and all respond to the same sensory modality.

The sensory cortex is concerned with three discriminative faculties:

1 *Spatial recognition.* This includes localization of the site of the stimulus and two-point discrimination. The latter is defined as the ability of a person to perceive two touch stimuli applied simultaneously as two separate points while both eyes are closed. The ability of tactile discrimination varies according to the site. Normally, 2 mm of separation of points can be recognized as two separate stimuli on the fingertips but about 30 mm of separation of points are needed in the back region to perceive two separate stimuli. The acuity of two-point discrimination increases with increase in the number of touch receptors per unit area of skin and with the increase in width of area of representation in the sensory cortex. The ability of tactile discrimination is lost in dorsal column and parietal cortical (i.e. somatosensory cortex) lesions.

2 *Recognition of relative intensities of different stimuli.* An increase in intensity of stimuli is transmitted to the brain in the form of an increase in the number of afferent fibres stimulated and increased frequency of action potentials in these fibres. These two features are perceived as an indication of the strength of the stimulus.

3 *Stereognosis.* This is defined as the ability to recognize objects by touch without the aid of vision. Loss of this ability is called *astereognosis*, which may occur due to a dorsal column or a

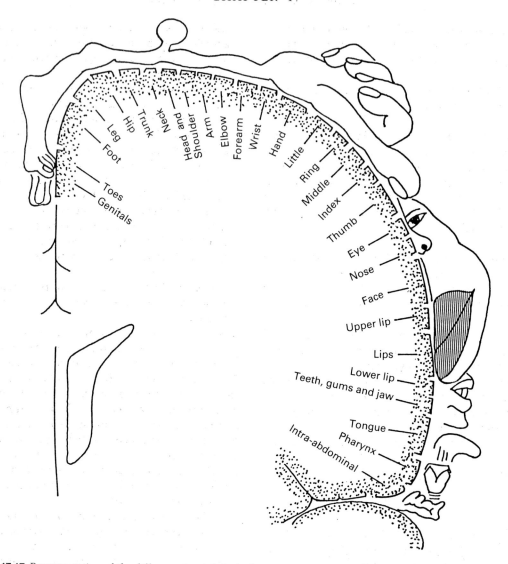

Fig. 17.17 Representation of the different areas of the body in somatosensory area I of the cortex.

parietal lobe lesion. Due to the former, other sensations subserved by the dorsal columns are also lost (i.e. position, vibration and fine pressure sense). When it is due to a parietal lobe lesion, position sense and light touch are normal but tactile discrimination is lost.

DISORDERED FUNCTION OF THE SENSORY SYSTEM
In discussing features that accompany disordered

function of the sensory system, it is helpful to recall that, at the level of the spinal cord: (i) the spinothalamic pathway is crossed; and (ii) the dorsal column pathway is uncrossed. It is also of importance to note that crossing of the dorsal column lemniscal system occurs higher up in the medulla. Due to the crossing of the two major sensory tracts, sensory information from one half of the body goes to the cerebral hemisphere of the opposite side.

We have already noted that lesions can occur at any level of a sensory tract, from the afferent peripheral nerve to the somatosensory cortex.

Localization of the site of lesion in disorders of the sensory system

Lesion of a peripheral nerve In such a case, all sensations are lost in the area supplied by the nerve. When many peripheral nerves are diffusely affected, as in polyneuritis or polyneuropathy, all forms of common sensation are impaired in the distal parts of the limbs (e.g. glove-and-stocking anaesthesia).

Lesion of the dorsal root Here, all sensations are lost in the relative dermatome, i.e. area of skin supplied by the dorsal root. The tendon reflexes mediated by fibres in the root are also lost.

Spinal cord lesions The features in spinal cord lesions depend on the location of damage. The three commonest lesions are:

1 *Brown-Séquard syndrome (hemisection of the spinal cord).* In this condition one-half of the spinal cord is damaged (Fig. 17.18). The patient will show the following:

(a) Sensory disturbances at the level of the

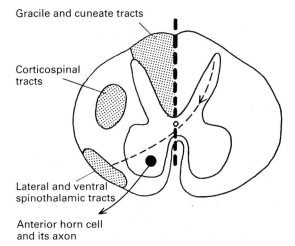

Gracile and cuneate tracts

Corticospinal tracts

Lateral and ventral spinothalamic tracts

Anterior horn cell and its axon

Fig. 17.18 Brown-Séquard syndrome. The pathways (tracts) damaged in Brown-Séquard syndrome are shown by the shaded areas. The thick broken line indicates the limits of the lesion.

lesion: loss of all sensations from the area supplied by the dorsal roots that enter the spinal cord at the damaged segments on the ipsilateral side.

(b) Sensory disturbances below the level of the lesion: (i) loss of position, vibration and tactile discrimination sense on the ipsilateral side of the lesion; (ii) loss of pain and temperature sense on the contralateral side; (iii) touch is not lost on the either side, owing to the dual pathway for touch, i.e. one crossed and the other uncrossed at spinal cord level.

(c) Motor disturbances, including: (i) lower motor neurone lesion manifestations at the level of the lesion on the same side; and (ii) an ipsilateral upper motor neurone lesion below the level of the lesion.

Figure 17.19 gives a summary of the effects of hemisection of the spinal cord (Brown-Séquard syndrome).

2 *Syringomyelia.* In this condition, damage is to the central part of the cord, where the crossing fibres of pain, temperature and touch decussate. This leads to loss of these sensations on both sides of the body at the affected segments. However, fine touch including tactile discrimination and position sense are not affected, as they are carried in the dorsal column/lemniscal pathway. Thus, the result is dissociated sensory loss.

3 *Tabes dorsalis.* In tabes dorsalis, the damage is confined to the dorsal root central to the dorsal root ganglion. This leads to:

(a) Loss of tactile discrimination, position and vibration sense. Loss of position sensation leads to sensory ataxia, which is incoordination of voluntary movements. This can be confirmed by testing for the positive Romberg's sign, in which the patient will be unable to stand steadily when closing his eyes. Ataxia also causes the patient to have a stamping gait.

(b) Loss of pain and temperature sensation.

(c) Loss of all reflexes (i.e. superficial, deep and visceral) mediated by fibres in the dorsal root affected. Certain areas are more susceptible to be affected by the disease than others (i.e. cervicothoracic and lumbosacral region).

Brain stem lesions At the level of the brain stem, the main sensory tracts have already crossed the

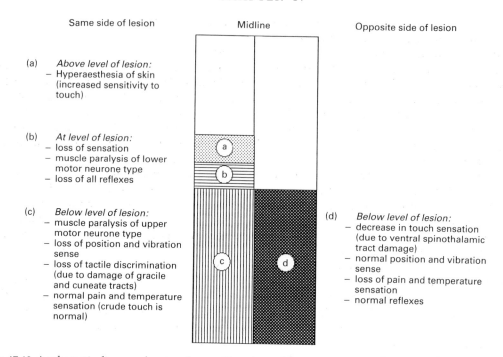

Fig. 17.19 A schematic diagram showing the manifestations of Brown-Séquard syndrome.

midline. Therefore, lesions in this area lead to loss of all sensations on the contralateral side.

Thalamic lesions Spontaneous pain of a most unpleasant quality, together with exaggerated response to painful stimuli due to facilitation of neurones in the thalamus (i.e. secondary hyperalgesia), is characteristic of this condition. Other sensations are lost on the opposite side of the body.

Cortical lesions Lesions in the sensory cortex lead to loss of *topognosis* (i.e. ability to localize the site of stimulus), loss of two-point discrimination and astereognosis.

A small area lying posterior to the main sensory area in the cerebral cortex is called the *somatic association area*. This area plays an important role in interpreting the meaning of sensations perceived by the main sensory area. Damage to this area may produce a disturbance of the body image and of spatial orientation. The patient may ignore one side of his body—a condition called

amorphosynthesis or sensory inattention.

When lesions occur at the highest cortical level, the condition arising is known as *hysterical sensory loss*. In this condition the sensory loss does not correspond to the anatomical nerve supply of the part. There are sharp lines of demarcation between normal and abnormal areas of sensation. The areas of sensory loss may change in response to suggestions made by the doctor.

The physiology of pain

PAIN SENSATION

Pain sensation is produced by stimuli that cause or threaten to cause tissue damage (i.e. noxious stimuli). Sherrington defined pain as 'the physical adjunct of an imperative protective reflex', i.e. pain is the mental accompaniment of an urgently operating reflex for protecting the body from injury.

Pain has three components, besides the distinct sensation of hurt when a lesion is inflicted on the body:

1 *Motor reactions.* These are reflexes which remove part or all the body from the painful stimulus, e.g. withdrawal reflex.

2 *Emotional reactions.* Pain has a built-in unpleasant effect in normal individuals. The reactions seen are those of anxiety, anguish, crying, depression, etc. Emotional reactions vary widely among people, although the threshold for pain is nearly the same for all individuals.

3 *Autonomic reactions.* These include tachycardia, peripheral vasoconstriction, a rise in blood-pressure, pupillodilatation and sweating. It is noteworthy that some visceral lesions, such as testicular crushing, may produce vasodilatation and a fall in blood-pressure.

The distinct sensation of pain and the other components are apparently subserved by different parts of the brain (i.e. brain stem, hypothalamus, frontal lobe, etc.). This is clearly demonstrated in patients undergoing frontal lobotomy for intractable pain. These patients still feel the pain after the operation but it does not bother them, i.e. the sensation is felt but the emotional reaction does not occur.

QUALITIES OF PAIN

Classically, the sensation of pain is subdivided into two components:

1 *Pricking pain.* This is often referred to as first pain. It is a fast acute sensation, which occurs within 0.1 s after application of a painful stimulus. It is usually well localized, the kind of sensation felt when a pin is stuck into the skin or the skin is cut with a knife. Pricking pain is usually superficial and is not felt in most of the deeper tissues. It is transmitted via type Aδ fibres.

2 *Burning or aching pain.* This is often referred to as second pain. It is a slow pain, which increases slowly over a period of many seconds or minutes. This component is the type that is difficult to endure and can occur both in the skin and in the deeper tissues. A good example is intestinal colic, toothache or a burn. Slow pain is transmitted by unmyelinated type C fibres.

The two qualities reflect not only the dual nature of the input (i.e. Aδ and C fibres), but also the two sets of connections within the nervous system.

ANATOMICAL BASIS

Receptors

Pain receptors are free nerve endings that respond to stimuli which cause tissue damage, which is the adequate stimulus. The adequate stimulus for pain receptors is therefore not as specific as for other receptors, i.e. pain receptors can be stimulated by mechanical, chemical, thermal or electrical energy.

Most pain receptors (nociceptors) are activated by a chemical mediator produced by tissue damage and, although a large number of speculations have been made concerning the nature of the pain-producing substance (PPS), no clear answer is known. However, it is evident that the PPS is a kinin similar, but not identical, to bradykinin and that it is only produced if the nociceptive nerve fibres are present. This suggests that tissue damage may cause the release of a PPS from nerve fibres and this substance, in turn, depolarizes the nerve fibres.

Peripheral nerves

Nociceptive nerve fibres are of two kinds:

1 Type Aδ fibres: fast fibres that conduct at a rate of about 20–30 m/s. These mediate pricking pain.

2 Unmyelinated C fibres: slow fibres that conduct at a rate of about 1 m/s. These mediate burning and aching pain.

Fast and slow pain fibres enter the spinal cord along the lateral division of the dorsal nerve roots.

Spinal cord

The pain fibres ascend or descend a few segments in Lissauer's tract (Fig. 17.20) to end on neurones in the dorsal horn. Type Aδ fibres end in lamina I and V and type C fibres end in lamina I and II (see Fig. 17.15) (i.e. substantia gelatinosa of Rolandi). These are considered the first-order neurones. Second-order neurones send axons, some of which end in the spinal cord and brain stem, while others cross to the opposite side and ascend as the lateral spinothalamic tract.

Brain stem

As the pain fibres pass into the brain stem, they separate into two pathways:

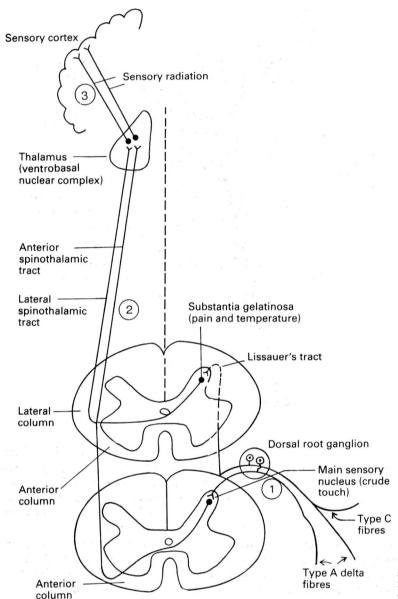

Fig. 17.20 The anterolateral pathway. 1, 2 and 3 indicate first-, second- and third-order neurones.

1 The pricking pain pathway ends in the ventrobasal complex of the thalamus. Axons of third-order neurones from the thalamus project to the somatic sensory cortex (Fig. 17.21).

2 The burning pain pathway is formed of collaterals given off by the spinothalamic tract to the reticular formation as it ascends the brain stem. These fibres end in the intralaminar nuclei of the thalamus (which are part of the reticular activating system). The reticular activating system projects to all parts of the brain, but especially to all areas of the cerebral cortex. Thus, they arouse one from sleep, create a state of excitement, create a sense of urgency and promote defence reactions to rid the person of pain.

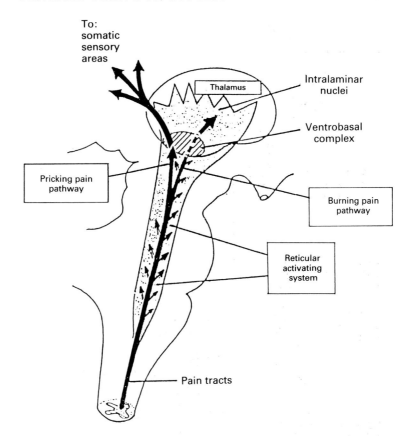

Fig. 17.21 Transmission of pain signals into the hind-brain, thalamus and cortex via the 'pricking pain' pathway and the 'burning pain' pathway. (Reproduced with permission from Guyton, A.C. (1987) *Textbook of Medical Physiology*, 7th edn. Saunders, Co, Philadelphia.)

Thalamus

When impulses reach the thalamus (and maybe lower) the subject experiences the sensation of pain.

Cerebral cortex

The cortex plays an important role in the localization of pain and in interpreting its meaning. It also mediates the emotional reaction to pain.

Cortical reaction to pain involves the parietal, frontal and temporal lobes.

TYPES OF PAIN

According to the site of stimulation, pain can be classified into cutaneous, deep somatic and visceral pain.

Cutaneous pain

Cutaneous pain is produced by stimulation of pain receptors of the skin. The pain elicited occurs in the two phases of fast pricking followed by slow burning pain. In some diseases, e.g. tabes dorsalis, fast fibres are damaged while the slow fibres are spared. Therefore, a painful stimulus in such a condition produces pain only after an abnormally long latent period.

Unlike the other two types, cutaneous pain can be accurately localized. This is due to the large number of receptors in the skin. Besides this, touch and vision aid greatly in localization.

Deep somatic pain

Deep somatic pain is produced by stimulation of pain receptors in deep structures, i.e. muscles, bones, joints and ligaments. Unlike cutaneous pain, deep somatic pain is dull, diffuse, intense and prolonged. It is usually associated with autonomic stimulation, e.g. sweating, vomiting and changes in heart rate and blood-pressure. Pain from deeper structures can also initiate reflex contractions of nearby muscles, e.g. the muscle spasm associated with bone fractures.

The adequate stimuli for deep somatic pain include:

1 Mechanical forces, e.g. severe pressure on a bone, traction of a muscle, ligament, etc.

2 Chemicals, e.g. venoms.

3 Ischaemia, e.g. muscle ischaemia.

Ischaemic muscular pain is called angina pectoris when it occurs in heart muscle and intermittent claudication when it occurs in the calf of the leg. By a series of simple experiments, Lewis concluded that the factor responsible for producing ischaemic pain must be a chemical substance (P factor) resulting from muscular contraction. This substance presumably accumulates in the muscle and causes pain when its concentration reaches a critical level.

Visceral pain

Visceral pain is that produced by stimulation of pain receptors in the viscera. Pain receptors in the viscera are sparsely distributed and therefore severe visceral pain indicates diffuse stimulation of pain receptors from a wide area of the viscus. Localized damage does not usually cause severe pain.

Pain impulses are conducted along type C fibres. Those from the thoracic viscera and lower pelvic viscera reach the CNS along parasympathetic nerves (i.e. the vagi and the pelvic nerves), while those from the rest of the viscera reach the CNS through sympathetic nerves.

The adequate stimuli for visceral pain include distension of hollow viscera. A good example is the pain we experience when the urinary bladder is full or when we overfill our stomachs. Spasm of a viscus is another potent stimulus, which gives rise to pain when the viscus becomes ischaemic, e.g. labour pain and intestinal colic. The last, but not least, stimuli are chemical irritants. Indeed, damage to the abdominal viscera resulting from leaking HCl acid from a perforated gastric ulcer gives rise to one of the most severe pains known to man.

Visceral pain is characterized by the following:

1 It is poorly localized.

2 It is often referred or radiates to other sites.

3 It is often associated with autonomic disturb-

ances, e.g. vomiting, sweating, tachycardia.

4 It can be associated with rigidity and tenderness of nearby skeletal muscles.

5 It can be associated with hypersensitivity to mild painful stimuli, i.e. hyperalgesia. Thus, even a mild painful stimulus produces a lot of pain. Hyperaesthesia might also accompany visceral pain. In such a case, even touching the area produces pain.

REFERRED PAIN

Visceral and deep somatic pain are often referred, i.e. the pain is not felt in the diseased structure itself, but at another place in the body far away from the site of its origin. Sometimes pain radiates from the diseased structure to the distant referral site.

Pain is referred according to the dermatomal rule, i.e. it is referred to the dermatome (area of skin) supplied by the dorsal nerve roots through which impulses from the diseased structure reach the CNS. But pain is projected to the area of skin instead of the viscus or deep somatic structure. Referred pain is therefore an error of projection.

The sites to which pain from different structures may be referred can be deduced from a knowledge of the segmental and peripheral innervation of the body.

Table 17.4 shows some important examples of usual sites of referral of pain from some organs.

Table 17.4 Important examples of referred pain

Organ	Site of referral
Heart	Precordium; inner aspect of left arm; epigastrium
Appendix	Umbilicus
Small intestine	Umbilicus
Central part of diaphragm	Tip of shoulder
Pleura	Abdomen
Kidney	Costovertebral angle (loin)
Ureter	Testicle
Trigone of bladder	Tip of penis
Tongue	Ear
Teeth	Head
Hip	Knee
Uterus	Low back radiating to lower abdomen

Mechanism of referred pain

Convergence and facilitation both play a part in the production of referred pain.

Convergence theory According to this theory, pain fibres from an area of skin and a diseased viscus supplied by the same spinal segment converge on the same second-order neurone in the dorsal horn (Fig. 17.22). The skin has a much richer nerve supply than any viscus, and it is more exposed to stimulation than any viscus. As a result, the somatosensory area of the cerebral cortex is more used to receiving impulses from skin than from a viscus. Thus, the brain misinterprets impulses coming along the common pathway as coming from the skin (i.e. the sensation is projected to the skin area) and not from the viscus.

Facilitation theory Pain fibres from the skin are always carrying impulses, but under normal conditions these are not enough to produce pain, i.e. the second-order neurones on which they converge are only subliminally stimulated (subthreshold stimulation). When pain receptors in a viscus from the same dermatome are stimulated, they send impulses that converge on nearby second-order neurones (which also receive afferents from the skin), thus increasing their excitability, i.e. facilitating them to reach the threshold level of stimulation (see Fig. 17.27). As a result, minor stimuli in the pain pathway from the skin, which normally would have died out in the spinal cord, pass on to the brain. Since the brain is more used to receiving impulses from the skin than from the

viscera or a deep somatic structure, the sensation is projected to the skin area. The result is pain felt in the area of skin and a lowering of its pain threshold. The pain felt in the skin is referred pain, and the lowering of its pain threshold accounts for the *hyperalgesia* and *hyperaesthesia* of the skin.

MUSCULAR RIGIDITY AND TENDERNESS ASSOCIATED WITH VISCERAL AND DEEP SOMATIC PAIN

The rigidity of nearby skeletal muscle associated with disease of a viscus is distributed regionally and not segmentally, and so its position varies according to the anatomical position of the diseased viscus. The rigidity is most marked when the parietal peritoneum or pleura becomes irritated by the diseased viscus. For example, the rigidity in the right iliac fossa that accompanies acute appendicitis is secondary to irritation of the parietal peritoneum by the inflamed appendix. However, rigidity can occur without involvement of the pleura or peritoneum. The anatomical details of the reflex pathway by which impulses from a diseased viscus initiate skeletal muscle rigidity are still not clear. The spasm protects the underlying inflamed structure from trauma. Indeed, this reflex spasm is sometimes called guarding. Yet, if it continues for long periods, the muscles become ischaemic and chemicals accumulate in them which reduce their pain thresholds. This accounts for the soreness and tenderness of the rigid muscles.

CENTRAL INHIBITION OF PAIN

Several observations have led to the belief that pain transmission and perception are subject to inhibition in the central sensory pathways by ascending and descending impulses. These observations include the following:

1 Stroking the area around the injury or rubbing it often reduces the pain of injury.

2 The degree to which each person reacts to pain is subject to wide variation.

3 Injuries caused while playing a game or in battle may be completely ignored at the time. The same injury, if inflicted, for example, in surgery without anaesthesia, would cause agonizing pain.

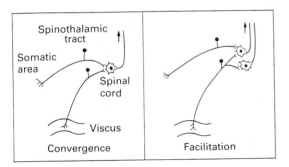

Fig. 17.22 Mechanism of referred pain. The diagrams show the convergence and facilitation theories.

4 Irritation of the skin overlying a diseased viscus, for example, with a mustard plaster, relieves pain.

5 Acupuncture has been used for years to prevent or relieve pain.

The *gate control theory* of pain inhibition (proposed by Melzack and Wall) can account for these observations. According to this theory, the dorsal horn of the spinal cord and, in particular, the neurones of the substantia gelatinosa of Rolandi form the 'gate' through which pain impulses must pass in order to reach the brain (see Fig. 17.23). The sensation of pain not only is dependent on an input from afferents transmitting pain signals, but also is significantly affected by the input from large myelinated mechanoreceptive afferents (type Aβ fibres). Impulses coming along small-diameter (pain) fibres (type C fibres) cause the release of substance P from these fibres and open the gate, while impulses coming along the large diameter type Aβ fibres close the gate, probably by the process of presynaptic inhibition. Stimu-lation of the large type Aβ fibres, therefore, can reduce or eliminate some types of pain. The common practice of rubbing the skin surface when one is hurt provides a stimulus to the large-diameter fibres. Clinically, the use of cutaneous electrodes over peripheral nerves (transcutaneous stimulation) and the implantation of dorsal column stimulators in treating certain painful conditions have proved to be very successful. Both methods electrically stimulate large-diameter fibres. Also, destruction of these fibres in certain peripheral nerve neuropathies (e.g. herpetic neuralgia) often results in a severe, prolonged, painful state in which minor stimuli can cause severe unbearable pain. Large-diameter fibres are also presumably stimulated by counter-irritants, such as mustard plaster applied to the skin to relieve pain of a diseased viscus, and also by acupuncture, although the latter appears to work mainly by the release of certain transmitters.

The 'gate' is also under the control of higher centres by means of an analgesic system of corti-cospinal and reticulospinal fibres.

The opiate system of the brain

Morphine is a powerful pain-killer (analgesic) found in opium. It acts by combining with receptors in the CNS. Because it seems unlikely that these receptors had been specifically evolved to combine with morphine, a search was made for morphine-like substances in the brain itself, and two major types of compounds with morphine-like properties (i.e. opioids) have been isolated from the brain, the encephalins and the endorphins.

The encephalins are a group of naturally occurring peptides found in high concentrations in the periaqueductal grey matter. Their effects are thought to be mediated by acting on opiate receptors found in a number of places, including the dorsal horn, periaqueductal grey matter, hypothalamus and limbic system. Like morphine, encephalins influence pain by producing two sets of effects: one is by directly inhibiting transmission of painful stimuli through the dorsal horn (i.e. closing the gate) and the other is by modifying the emotional state of the individual via their action on the limbic system, thus promoting a sense of well-being in the individual.

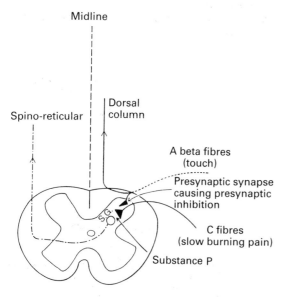

Fig. 17.23 The gate control of pain. Substance P is released from the first-order neurone in the 'slow burning' pain pathway to stimulate the substantia gelatinosa. The presynaptic synapse releases a transmitter which inhibits release of substance P and thus inhibits pain transmission. S.G., substantia gelatinosa.

Endorphins form another group of peptides, which are plentiful in the hypothalamus and pituitary. They also function via interacting with opiate receptors. Indeed, endorphins were found to prevent withdrawal symptoms in opium addicts. Their presence in high concentrations in some women during parturition suggests that these naturally occurring peptides form, in a sense, the body's own analgesic system (Fig. 17.24).

The discovery of the transmitters involved in pain sensation has explained some of the various observations we spoke about earlier. Acupuncture acts partly by causing the release of encephalins. This is indicated by the blockage of acupuncture analgesia by naloxone, a drug that competes with morphine and opioids for receptor sites. Endorphins, on the other hand, are secreted together with ACTH from the anterior pituitary in stressful situations. This could explain why injuries are often painless during playing a game or during a period of fighting.

METHODS OF TREATMENT

There are many procedures used to alleviate or treat pain. However, they differ in their mode of action and their relative merits. They include the following:

1 Treatment and cure of the cause. This is definitely the best, if possible.

2 Use of analgesics (i.e. pain-killing drugs).

3 Large-fibre stimulation. This can be done by using counterirritants, e.g. hot-water bottles, balms, cold packs and antiphlogistic plasters.

Other methods include cutaneous electrodes and the use of dorsal column stimulators.

4 Use of local anaesthetics in minor surgery, e.g. procaine, and general anaesthetics in major surgery, e.g. Pentothal, nitrous oxide, etc.

5 Use of vasodilators in ischaemic pain, e.g. use of nitrites in angina pectoris or increasing blood supply to an ischaemic part, e.g. coronary bypass in angina pectoris.

6 Use of antacids in peptic ulcers and antispasmodics in, for example, intestinal or ureteric colic.

7 Suppression of inflammation by use of steroids.

8 Acupuncture. This works in certain limited circumstances.

9 Electrical stimulation of the periaqueductal grey matter.

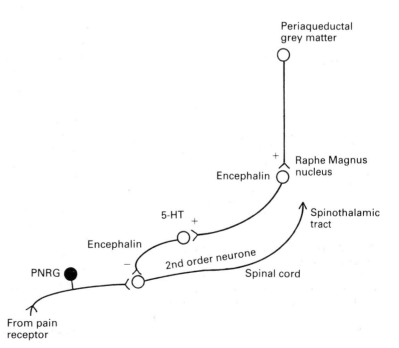

Fig. 17.24 Mechanism of the role played by descending fibres in determining the threshold of pain. +, excitation, −, inhibition, PNRG, peripheral nerve root ganglion.

10 Surgical denervation (Fig. 17.25) of diseased organs by:

(a) Neurectomy; sympathectomy.

(b) Rhizotomy (i.e. dorsal root sectioning).

(c) Myelotomy (i.e. section of spinothalamic fibres in the anterior commissure).

(d) Anterolateral cordotomy, i.e. sections of pain and temperature fibres in the antero-lateral tract of the spinal cord.

(e) Medullary tractotomy.

(f) Mesencephalic tractotomy.

(g) Thalamotomy.

(h) Gyrectomy.

11 Frontal lobotomy. By this procedure abolition of the emotional reaction to pain is achieved. This is a method of last resort for relief of intractable pain in terminal cancer.

12 Psychotherapy, hypnosis and placebos can be used in cases with pain of psychological origin.

The motor system

I Motor functions of the spinal cord and spinal reflexes

INTRODUCTION

Sensory information is integrated at all levels of the nervous system and causes different types of motor reflexes:

1 Sensory information integrated in the spinal cord leads to simple spinal reflexes.

2 Sensory information integrated in the brain stem leads to more complicated brain stem reflexes.

3 Sensory information integrated in the cerebral cortex initiates the most complicated motor responses and also leads to the perception of sensations.

THE REFLEX ARC

The basic unit of integrated neural activity is the reflex arc (Fig. 17.26). It consists of:

1 A sense organ (i.e. the receptor).

2 An afferent neurone.

3 One or more synapses (in the CNS or a sympathetic ganglion).

4 An efferent neurone.

5 An effector.

The simplest reflex arc is monosynaptic, i.e. a single synapse separates the afferent from the efferent neurone. Few of the sensory impulses that enter the spinal cord pass directly to the anterior horn cells in the anterior horn across one synapse, e.g. the stretch reflex. More complex reflex arcs contain one or more interneurones interposed between the afferent and efferent neurones. Most of the known reflex arcs mediate polysynaptic reflexes, e.g. the withdrawal reflex.

Some of the components of the reflex arc (i.e. the receptor and synapses) have been previously

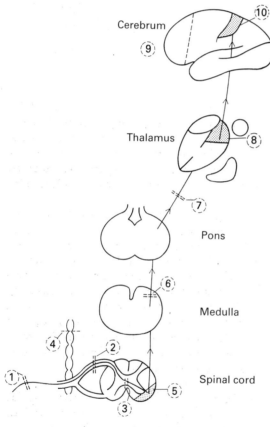

Fig. 17.25 Diagram of the various surgical procedures designed to relieve pain. 1. Neurectomy. 2. Rhizotomy. 3. Myelotomy to section spinothalamic fibres in anterior white commissure. 4. Sympathectomy (for visceral pain). 5. Anterolateral cordotomy. 6. Medullary tractotomy. 7. Mesencephalic tractomy. 8. Thalamotomy. 9. Frontal lobotomy. 10. Gyrectomy.

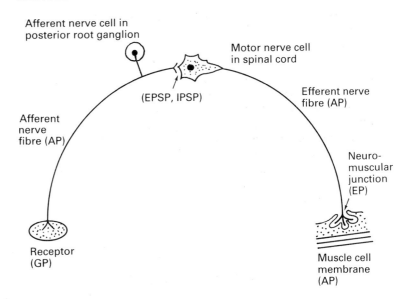

Fig. 17.26 The reflex arc. GP: generator potential (graded); AP: action potential (all or none); EPSP, IPSP; excitatory and inhibitory postsynaptic potentials (graded); EP: motor end-plate potential (graded). Note that at the receptor and at each of the junctions in the arc there is a non-propagated graded response that is proportionate to the magnitude of the stimulus. In the portions of the arc specialized for transmission (afferent and efferent nerve fibres, muscle membrane) the responses are action potentials which are not graded and obey the all-or-none law.

discussed. The rest deserve discussion in some detail.

The afferent neurone

Afferent impulses entering the spinal cord via the posterior root are conducted to two separate destinations. Some end on cells in the grey matter of the spinal cord of the same segment or nearby segments. These cells are either posterior horn cells (e.g. the substantia gelatinosa, or the main sensory nucleus), which form the origin of sensory tracts, or interneurones interposed between the afferent fibre and the efferents, or anterior horn cells, which are actually motor effector neurones. A large number of the afferent impulses, however, ascend to higher levels of the nervous system, i.e. brain stem, cerebellum, thalamus or even the cerebral cortex.

Afferent neurones carry impulses from receptors to the CNS. They can undergo divergence (Fig. 17.4), thus helping to spread a single stimulus to a wide area of the CNS, or convergence, thus helping the process of spatial summation. At the level of the spinal cord, afferent impulses will lead to facilitatory effects (i.e. raising excitability of neurones, but not enough to fire) or local excitatory effects, or to the induction of spinal cord reflexes.

Interneurones

Interneurones are small, highly excitable cells, which may be excitatory or inhibitory. In the spinal cord they lie in all areas of the cord's grey matter, making connections with one another, and many of them directly synapse with anterior motor neurones.

Interneurones also are arranged to allow convergence and divergence to occur. Their organization also allows the important process of after-discharge. In this process an impulse in an afferent neurone is not merely relayed but causes a prolonged output discharge even after the incoming impulse is over. After-discharge can last from a few milliseconds to as long as many minutes to hours. Two types of circuits formed by interneurones allow the process of after-discharge — parallel and reverberating circuits.

Parallel circuits In this type, the input impulse reaches the output neurone through various numbers of interneurones which converge on it (Fig. 17.27). As synaptic delay is about 0.3–0.5 ms at each synapse, the impulses reach the output neurone one by one after varying periods of delay. Thus, the output neurone continues to discharge for many milliseconds.

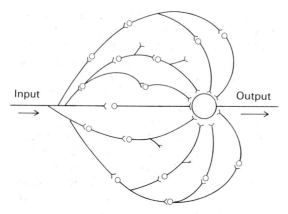

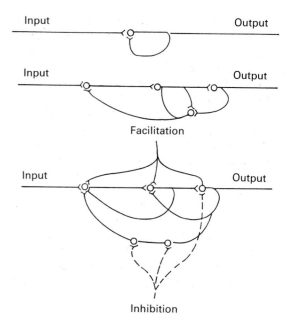

Facilitation

Inhibition

Fig. 17.27 Diagrammatic representation of the parallel after-discharge circuit. (Reproduced with permission from Guyton, A.C. (1987) *Textbook of Medical Physiology*, 7th edn. Saunders Co, Philadelphia.)

Reverberating circuits These form the most important circuits responsible for after-discharge. They are positive feedback pathways where the output of one neurone feeds back to restimulate itself (i.e. via interneurones) for a long time (Fig. 17.28). The system gets more complex as the number of interneurones increase in the circuit.

The duration of reverberating circuits varies and they only stop as a result of synaptic fatigue or by inhibitory impulses from other parts of the CNS. Examples of important reverberating circuits in the nervous system include the following:
1 During respiration, the inspiratory centre in the medulla becomes excited for about 2 s during each respiratory cycle for life by reverberating circuits. These circuits can be stopped by inhibitory impulses from other areas of the CNS, for example, pulmonary vagi stimulation.
2 One theory of wakefulness is that arousal impulses set up within the brain stem continual reverberation to keep wakefulness lasting for up to 18 hours. Sleep sets in when the synapses in the circuit fatigue.

The efferent neurone
In the spinal cord, the efferent neurones are the anterior motor neurones present in the grey matter of the anterior horn. The axons of these cells leave the spinal cord through the anterior roots to innervate skeletal muscles.

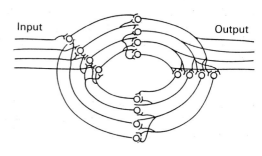

Fig. 17.28 Diagrams of reverberating circuits of increasing complexity. (Reproduced with permission from Guyton, A.C. (1987) *Textbook of Medical Physiology*, 7th edn. Saunders Co, Philadelphia.)

Anterior motor neurones (i.e. anterior horn cells or AHC) are of two types: alpha and gamma motor neurones. The alpha motor neurones are the larger cells, which give rise to large myelinated type Aα fibres (9–20 μm diameter). Each fibre excites from three to several hundred skeletal muscle fibres. The fibre and the muscle fibres it supplies are collectively called a motor unit. The size of the latter varies according to the type of movements; the more delicate the movement is, the fewer the number of muscle fibres in a single motor unit; e.g. in hand muscles, each motor unit contains around five fibres while, in the leg muscles, it contains about 100 muscle fibres. Gamma motor neurones are the smaller cells, with small axons which are type Aγ fibres (5 μm diameter). They innervate special skeletal muscle fibres, the intrafusal fibres of the *muscle spindle*.

Besides motor neurones, the anterior horn contains a large number of special interneurones, which are located in close association with motor neurones. These are known as *Renshaw cells*. They are inhibitory cells, which are excited by transmitters released from collateral branches of the motor neurone (Fig. 17.29). Renshaw inhibition, often called recurrent inhibition, plays an important role in dampening the activity of motor neurones; in particular, it appears to limit the frequency of discharge of alpha motor neurones. This, in turn, focuses and sharpens the area stimulated.

TYPES OF REFLEXES

During examination of the nervous system, three types of reflexes are usually investigated. These are deep reflexes, superficial reflexes and visceral reflexes.

DEEP REFLEXES

Deep reflexes are a group of reflexes that result from stimulation of receptors present in muscles and tendons. They include:

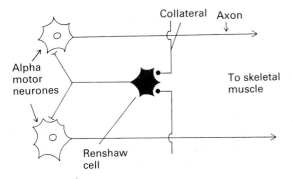

Fig. 17.29 Circuit diagram for recurrent inhibition of a motor neurone by a Renshaw cell. (Reproduced from Bray, J.J., Cragg, P.A., Macknight, A.D.C., Mills, R.G. and Taylor, D.W. (eds) (1989) *Lecture Notes in Physiology*, 2nd edn. Blackwell Scientific Publications, Oxford.)

1 The stretch reflex. This is a monosynaptic reflex elicited by applying sudden stretch to any muscle. The result is reflex contraction of the stretched muscle. The stretch reflex is the basic reflex of muscle tone and posture.

2 A group of postural spinal reflexes, the most important of which are:

(a) Rhythmic stepping reflex of a single limb. Forward flexion of a limb is followed by backward extension of the same limb. Flexion then occurs again and the cycle repeats itself.

(b) Reciprocal stepping of the opposite limb. Forward flexion of a limb is accompanied by backward extension movement of the contralateral limb.

(c) The tendon reflex (inverse stretch reflex).

The stretch reflex (the myotatic reflex)

Sherrington investigated the physiology of the stretch reflex in his classical studies on the decerebrate cat. This is an animal preparation in which a transverse section is made between the superior and inferior colliculi. Such an animal assumes a characteristic posture with limbs extended, head retracted, jaws closed and tail erect (*decerebrate rigidity*). Decerebrate rigidity results from the facilitatory action of centres in the brain stem (i.e. vestibular nuclei and facilitatory reticular formation) which are unantagonized by the inhibitory effects of the higher brain centres (cerebral cortex and basal ganglia), which lie above the level of the transection. Cutting the dorsal roots supplying the rigid limb of a decerebrate animal abolishes the rigidity of the limb. This means, therefore, that decerebrate rigidity depends on a spinal reflex, the stretch reflex. The receptors of this reflex are muscle spindles, found in the fleshy part of the muscle in between ordinary muscle fibres (i.e. extrafusal muscle fibres) but attached in parallel with them.

Functional anatomy of the muscle spindle Each spindle consists of 8–10 muscle fibres enclosed in a spindle-shaped connective tissue capsule. These are called intrafusal fibres (Fig. 17.30), which are smaller and less developed than extrafusal fibres. They lie in parallel with the muscle fibres because the ends of the spindle capsule are attached either to the muscle tendon or to the sides of the extrafusal fibres. Each intrafusal fibre consists of a central non-contractile area, which is the receptor area, and a peripheral contractile part.

There are two types of intrafusal fibres: (i) nuclear bag fibres whose central area is dilated with an aggregation of nuclei; and (ii) nuclear chain fibres, which are smaller than nuclear bag fibres and have one line of nuclei spread in a chain along the receptor area.

Innervation of the muscle spindle

1 *Afferent fibres.* These are the sensory fibres that supply the receptor area of the intrafusal muscle fibres. They are of two types:

(a) *Primary (annulospiral) endings.* These are group Ia fibres, which are thick and rapidly conducting (i.e. diameter 16–17 μm). They encircle the receptor areas of both nuclear bag and nuclear chain fibres, and synapse directly with the anterior horn motor neurones that supply the extrafusal fibres of the same muscle. Thus, the reflex pathway here is monosynaptic. Primary endings give information on both the rate and the degree of stretch of the muscle.

(b) *Secondary (flower-spray) endings.* These are terminals of smaller myelinated fibres (8 μm diameter). They are group II fibres and they encircle the receptor area of the nuclear chain fibres only. Unlike the first type, these terminate principally on interneurones of the spinal cord, although some fibres also terminate on the anterior horn cells directly. The secondary endings give information of the degree of stretch only (Fig. 17.30).

It is important to note that both primary and secondary endings are stimulated when the central receptor area of the intrafusal fibre is stretched. Sensory information from both primary and secondary endings is important. This information is used as a feedback mechanism to prevent oscillations and adjust muscle length during muscle contractions.

2 *Efferent fibres to the muscle spindle.* These are axons of the gamma motor neurones, situated in the anterior horn or brain stem. They supply the peripheral contractile parts of the intrafusal muscle fibres and are of two functional types (see Fig. 17.30):

(a) Dynamic gamma efferents, which end mainly on the nuclear bag fibres (as plate endings).

(b) Static gamma efferents, which end mainly on nuclear chain fibres (as trail endings) while some end on nuclear bag fibres.

Let us assume that the gamma motor neurones have been stimulated. Impulses will pass along the gamma efferents to cause contraction in the peripheral contractile part of the intrafusal fibres. As a result, the receptors in the midportion of the intrafusal fibres will be stretched. This, in turn, causes stimulation of the primary and secondary endings encircling the receptor area (see Fig. 17.30). Afferent impulses enter the spinal cord and stimulate alpha motor neurones of the anterior horn, which send impulses along type Aα efferent fibres to ordinary muscle fibres, causing the muscle to contract. This is the basis of the stretch reflex. (see Fig. 17.31.)

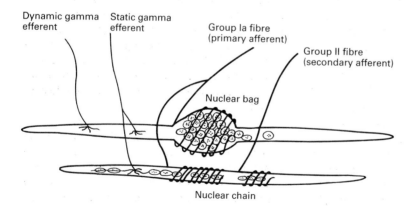

Dynamic gamma efferent   Static gamma efferent   Group Ia fibre (primary afferent)

Group II fibre (secondary afferent)

Nuclear bag

Nuclear chain

Fig. 17.30 The muscle spindle. Note the two types of intrafusal muscle fibres and also the sensory and motor innervation of the muscle spindle.

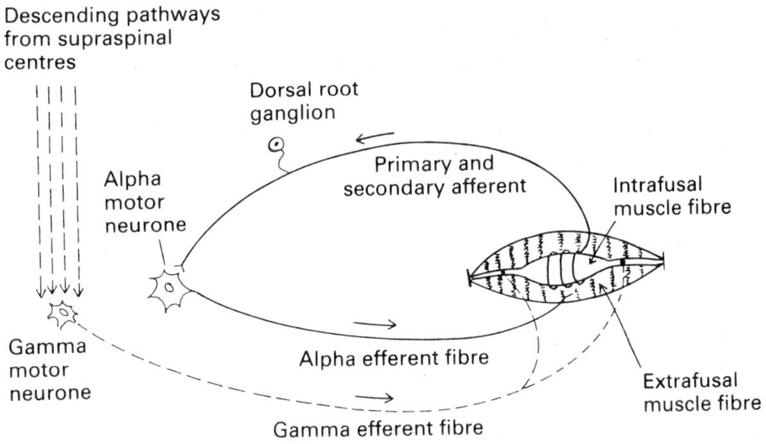

Fig. 17.31 Diagram illustrating the circuit involved in the stretch reflex and the effects of gamma motor neurone stimulation. The dashed lines represent descending pathways bringing impulses from supraspinal centres to the gamma motor neurone and its axons, which supply the peripheral edges of the intrafusal muscle fibres.

Function of muscle spindles

The function of the muscle spindle is to keep the CNS informed about muscle length and the rate of change in muscle length. The spindle can be stimulated under two conditions:

1 By stretching the whole muscle, as, for example by gravity or by tapping a muscle tendon with a hammer during examination of tendon jerks.
2 By contraction of intrafusal fibres (i.e. when stimulated by gamma efferents), even in the absence of stretch of extrafusal fibres.

Under both conditions, the central receptor area is stretched, causing stimulation of afferents, and the sequence of events explained earlier leads to contraction of the whole muscle reflexly.

Maximal stimulation of the muscle spindle occurs whenever there is increased rate of discharge of gamma efferents (so that intrafusal fibres are contracted) and the muscle is stretched. Minimal or no stimulation of the spindle occurs when the muscle actively contracts, because the spindles lie in parallel to the extrafusal fibres.

Control of gamma efferent discharge Gamma motor neurone discharge is controlled by both facilitatory and inhibitory pathways from supraspinal centres (Fig. 17.31). Facilitatory pathways increase gamma motor neurone discharge which increases the degree of the stretch of muscle spindles and increase their excitability; inhibitory pathways do the reverse. Selective simultaneous facilitation of gamma motor neurons supplying

some muscles while inhibiting those supplying other muscles leads to variation in the degree of reflex contraction of the different muscles of the body, i.e. *muscle tone*. Such variations in tone are important for the maintenance of posture. Gamma motor neurone discharge is influenced by psychic factors. The hyperreflexia of anxious patients is a good example. It is also enhanced by cutaneous stimulants.

From the above, it is obvious that a skeletal muscle contracts reflexly either directly, via alpha motor neurone discharge, or indirectly, via gamma motor neurone discharge. Normally, whenever alpha motor neurone discharge increases, gamma motor neurone discharge also increases (i.e. coactivation of alpha–gamma motor neurones). Thereby the spindles adjust alpha motor neurone discharge throughout a contraction. Coactivation of alpha–gamma motor neurones is the basis of Jendrassik's reinforcement manoeuvre, i.e. eliciting tendon jerks while the patient is strongly contracting some other voluntary muscle. The reinforcement acts by increasing gamma motor neurone discharge and thereby increasing the sensitivity of muscle spindles to stretch.

Components of the stretch reflex The stretch reflex can be divided into two separate components, the dynamic and the static stretch reflex.

The dynamic stretch reflex is elicited when a muscle is suddenly stretched. This increases the discharge in the primary endings (Fig. 17.32), and the impulses enter the spinal cord along the dorsal root, to excite anterior motor neurones. Stimulation of the anterior motor neurones brings about contraction of the muscle stretched. Soon the discharge from the primary endings declines as the muscle contracts, causing the muscle to relax rapidly. Thus, the reflex functions to try to bring the muscle to its original length. This forms the basis of the tendon jerks examined clinically.

The static stretch reflex is the second component and is elicited by maintained stretch to a muscle. As long as stretch is maintained, impulses enter the spinal cord, mainly along secondary endings, and pass to alpha motor neurones, stimulating them to cause muscle contraction. Because different spindles are subjected to different degrees of stretch, they discharge at different rates. The result is a highly asynchronous discharge bombarding alpha motor neurones, which, in turn, discharge with corresponding asynchrony to the extrafusal fibres. The net result is a sustained contraction of the muscle as long as it is stretched. This forms the basis of *muscle tone*, which is simply defined as the resistance of muscle to stretch. Since the extent of reflex muscle contraction (i.e. tone) depends on gamma efferent discharge, then a low level of gamma efferent discharge is associated with hypotonia (or flaccidity), while a high level of discharge results in hypertonia (or spasticity).

Supraspinal influence on gamma efferent discharge (on the stretch reflex)
It was already mentioned that the stretch reflex is influenced by both facilitatory and inhibitory impulses from higher levels of the central nervous system, via their effects on gamma efferent discharge (Figs 17.33 and 17.34). These central nervous system centres include:

1 *Cortical centres.* The motor cortex contains both facilitatory areas (i.e. mainly area 4) and inhibitory areas (i.e. area 4s, a narrow band between area 4 and area 6). The net effect, however, is inhibitory to the stretch reflex, due to the predominant effect of inhibitory areas.

It is important to note, however, that the pyramidal system by itself is facilitatory, and yet hypertonia is more commonly seen in lesions of the motor cortex. This is because lesions usually involve both facilitatory and inhibitory regions.

2 *Extrapyramidal centres.* Some of these are facilitatory, e.g. vestibular nuclei and the dorsolateral part of the pontine reticular formation, while others are inhibitory, e.g. basal ganglia and the ventromedial part of the medullary reticular formation. Thus, hypotonia appears in some extrapyramidal disorders (e.g. chorea), while hypertonia occurs in others (e.g. Parkinsonism).

3 *The cerebellum.* Parts of the cerebellum are facilitatory to the stretch reflex (i.e. neocerebellum), but others are inhibitory (i.e. palaeocerebellum). In humans, the net effect is facilitatory, and cerebellar lesions are characterized by hypotonia.

Table 17.5 lists some of the causes of disturbance of muscle tone.

The tendon reflex (inverse stretch reflex)
If a muscle is subjected to excessive tension due to passive stretch or muscle contraction, it relaxes

Action potential in:

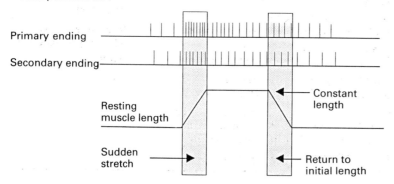

Fig. 17.32 Action potentials from muscle spindle afferents. The primary and secondary endings increase their discharge when the muscle is suddenly stretched. Note that the discharge decreases when a constant length is maintained and returns to a basal level when the muscle goes back to its initial length.

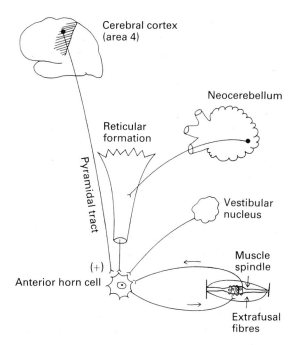

Fig. 17.33 Diagram of the supraspinal centres that facilitate gamma efferent discharge and consequently the stretch reflex: area 4 of the cerebral cortex, the dorsolateral part of the pontine reticular formation, the neocerebellum and the vestibular nucleus. +, stimulation.

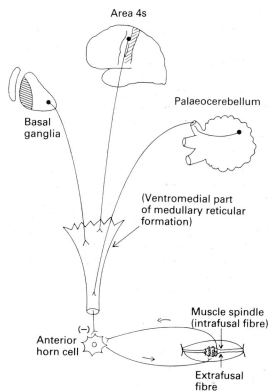

Fig. 17.34 Diagram of the supraspinal centres which inhibit gamma efferent discharge and consequently the stretch reflex: area 4s, the basal ganglia, the ventromedial part of the medullary reticular formation and the palaeocerebellum. −, inhibition.

contrary to the response in the stretch reflex. Thus the reflex is known as the inverse stretch reflex; or as the *tendon reflex* because the receptors for this reflex are in the *Golgi tendon organs*. The receptors are found in the muscle tendon and transmit their impulses via fast type Aα fibres (group 1b fibres).

It appears that different degrees of muscle stretch cause different responses. Within limits muscle stretch produces reflex contraction proportional to the degree of stretch (the receptor is the muscle spindle). But if the muscle is overstretched, muscle contraction is suddenly replaced by relaxation and the muscle lengthens (the receptor is the Golgi tendon organ). This protects muscles against rupture and/or evulsion of their tendons. Such behaviours of a muscle when subjected to different degrees of stretch resembles that of a clasp knife and is obvious in hypertonic muscle. This is described by the term

Table 17.5 Causes of disturbance of tone

Hypotonia	Hypertonia
Neural shock (cerebral or spinal)	Corticospinal lesions (UMNL) after the stage of neural shock
Lower motor neurone lesions	Some extrapyramidal lesions, e.g. Parkinsonism
Cerebellar lesions	Psychogenic
Some extrapyramidal lesions, e.g. chorea	
Interruption of the afferent limb of the stretch reflex (e.g. tabes dorsalis)	
Myopathies	

'clasp knife spasticity' or *'the lengthening re-action'* (Fig 17.35).

Pathway of the reflex When these receptors are stimulated, they produce excitation of inhibitory interneurones, which end on alpha motor neurones supplying the muscle containing the receptors (see Fig. 17.36). As a result, motor neurones are inhibited and the muscle relaxes.

Tendon jerks: clinical significance
Tendon jerks are examples of the stretch reflex. It should be clear that elicitation of normal tendon reflexes reflects, besides other factors, a normal rate of gamma efferent discharge. Exaggerated tendon reflexes are seen when the rate of gamma efferent discharge is high, as in upper motor neurone lesion, anxiety states, tetanus, thyrotoxicosis and strychnine poisoning. Except during neural shock, absence of a tendon reflex almost always means that the reflex arc subserving it has been interrupted, as in poliomyelitis, tabes dorsalis and peripheral neuropathy.

Investigating tendon reflexes is very useful in medical diagnosis. It enables deduction of the side of the lesion when reflexes are abnormal on one side of the body. In addition, the localization of the segmental level of the lesion becomes possible, by marking the level along the longi-

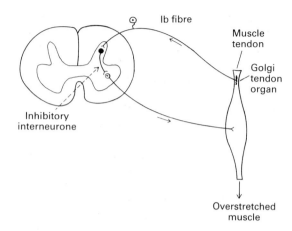

Fig. 17.36 Diagram illustrating the circuit of the tendon reflex.

tudinal axis of the body at which reflexes are abnormal. Tendon jerks also change characteristically in certain diseases, such as cerebellar disease, in which they become pendular. In myxoedema, the contraction and relaxation phases of tendon jerks become prolonged.

Clinical examples The most clinically examined tendon jerks are the knee and ankle jerks in the lower limbs and the biceps and triceps jerks in the upper limbs. Details of how to elicit these reflexes are important in clinical practice and may be studied in the practical physiology course.

SUPERFICIAL REFLEXES
Superficial reflexes result from stimulation of receptors present in the skin. They are clinically important because they aid in localizing the sites of lesions.

Superficial abdominal reflexes
When the skin on one side of the abdomen is stroked with a blunt object, the underlying muscles contract. The centres for these reflexes lie in thoracic segments T7 to T12.

Plantar reflex
In healthy adults stimulation of the sole of the foot produces plantar flexion of the toes. The centre for this reflex lies in lumbar segment L5 and sacral segment S1.

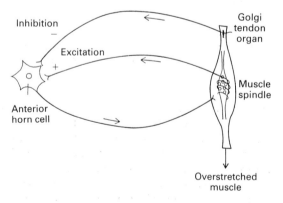

Fig. 17.35 The lengthening reaction. Moderate stretch causes reflex contraction. The receptor is the muscle spindle. Greater stretch increases the tension, the contraction stops and the muscle relaxes. The receptor is the Golgi tendon organ.

Cremasteric reflex

When the skin of the upper and inner aspect of the thigh is stroked, the cremasteric muscle contracts, causing elevation of the testicle. The centre of the reflex lies in lumbar segments L1 and 2.

The flexor reflex (withdrawal reflex)

The flexor reflex is elicited by an injurious and usually painful stimulus to a limb, and the limb is withdrawn away from the injurious stimulus. The response is reflex contraction of the flexor muscles. Figure 17.37 shows the pathway.

Stimulation of pain receptors in the hand region causes a volley of impulses to travel to the spinal cord via type Aδ or C fibres. These impulses pass via several interneurones before reaching the anterior horn cells: hence the term polysynaptic reflex. Figure 17.37 also shows that excitation of flexors is accompanied by inhibition of extensors, a phenomenon occurring through inhibitory interneurones synapsing with extensor motor neurones. This is known as reciprocal innervation (or reciprocal inhibition). In addition to flexion and withdrawal of the stimulated limb, the figure also shows extension of the opposite limb. This usually occurs when the strength of the stimulus becomes severe enough and is called the crossed extensor response (or reflex), considered by many physiologists to be a part of the withdrawal reflex. The process of reciprocal innervation also occurs in the *crossed extensor reflex*, i.e. flexors are inhibited while extensors are excited. This pushes the entire body away from the injurious agent and supports the body-weight.

Scratch reflex

The scratch reflex is initiated by the itch and tickle sensation, which could, for example, be caused by irritation of the skin by a moving insect. It results in a to-and-fro scratching movement. The pathway is also endowed with reciprocal innervation and can occur in decerebrate animals.

Positive supporting reaction

If deep pressure is applied to the sole of the foot, the reaction will be reflex contraction of both flexors and extensors of the limb so as to change it into a rigid column to support body-weight. The location of the pressure on the sole of the foot determines the position to which the limb will extend. Thus, if pressure is applied to the side, front or back of the foot, the foot will move in the direction of the applied pressure. This is called the magnet reaction.

VISCERAL REFLEXES

The clinically important visceral reflexes are those concerned with micturition, defecation and erection.

Micturition reflex

The micturition reflex is basically a spinal reflex, which becomes subject to higher voluntary control after the first 2—3 years of life, via cortical centres located in the paracentral lobule.

Nervous control of the bladder is exercised through the autonomic nervous system. Parasympathetic nerves (S 2, 3 and 4) cause contraction of the bladder wall and relaxation of the internal sphincter. The sympathetic nerves (L1 and 2), when stimulated, relax the bladder wall and con-

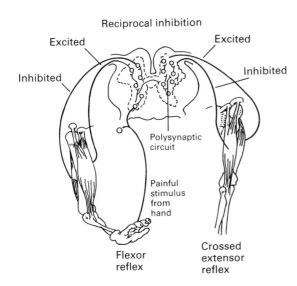

Fig. 17.37 The flexor reflex, the crossed extensor reflex and reciprocal innervation (reciprocal inhibition). (Reproduced with permission from Guyton, A.C. (1987) *Textbook of Medical Physiology*, 7th edn. Saunders Co, Philadelphia.)

tract the internal sphincter, thus stopping emptying of the bladder. Somatic supply, in the form of the pudendal nerve (S 2, 3 and 4), controls voluntary voiding of urine by contracting the external sphincter of the urethra.

It is obvious from the above account that various neural lesions may interfere with bladder function. Table 17.6 shows the various lesions and states which disturb bladder function.

Defaecation reflex

Distension of the rectum results in reflex contraction of the rectum and relaxation of the internal anal sphincter, leading to emptying. The reflex is also under voluntary control and the external anal sphincter is supplied by the pudendal nerve. The nerves involved are the same ones that are involved in the micturition reflex. Neural lesions which disturb bladder function usually also disturb rectal function in much the same way, but consequences are less severe.

Erection reflex

This reflex is initiated by psychic stimuli from the brain and also by physical stimuli from the genetalia. The response is the function of the parasympathetic nervous system. Parasympathetic impulses pass from the sacral spinal cord to the penis causing dilatation of the penile arteries. Blood accumulates in the corpora cavernosa (erectile tissue of the penis) under great pressure causing swelling and stiffness of the penis called erection.

GENERAL PROPERTIES OF REFLEXES

Adequate stimulus

Each reflex has a precise stimulus called the adequate stimulus. For example, a painful stimulus to the sole of the foot produces the flexor withdrawal reflex, while deep pressure applied to the same area gives rise to the positive supporting reaction.

Local sign

The pattern of response in a reflex depends on the afferent nerve fibre stimulated. This is called 'local sign'. A painful stimulus, if applied to the lateral side of the arm, produces flexion and adduction, but, if the same stimulus is applied to the medial aspect, it results in flexion and abduction of the limb.

Irradiation

The extent of the response in a reflex depends on the intensity of the stimulus. The more intense the stimulus, the greater is the spread of activity in the spinal cord (i.e. by divergence) to involve more and more motor neurones (i.e. irradiation of the stimulus). For example when the sole of the foot is stimulated by a weak painful stimulus, the big toe only is dorsiflexed. A stronger stimulus will cause reflex flexion of the big toe plus the ankle. The strongest stimulus will cause withdrawal of the whole leg by causing reflex flexion of the big toe, ankle, knee, and hip. Impulses will also cross to the other side of the spinal cord to cause extension of the other leg (see the crossed extensor reflex).

Summation

In spatial summation, two or more stimuli, through different sensory routes, which, by themselves, are insufficient to elicit a reflex contraction, may elicit a contraction when given together (see Fig. 17.6).

In temporal summation, two or more stimuli

Table 17.6 Effects of various lesions on bladder function

Lesion	Defect in bladder function
Damaged parasympathetic nerves (nervi erigentes)	Initially retention, with overflow; later bladder becomes decentralized and functions as an autonomous bladder
Damaged dorsal roots or columns	Patient does not feel bladder sensation but can empty bladder at will
Spinal cord transection above sacral segments	Retention, with overflow initially; later bladder function as automatic bladder through sacral reflexes
Injury to bladder area in cortical centres	Inability to start or stop micturition at will
Emotional stress	Frequency of micturition due to loss of bladder's ability for adaptation

given at different times, which, by themselves, are insufficient to elicit a reflex contraction, may elicit a contraction if given within a short time of each other (see Fig. 17.6). Spatial summation seems to be more important than temporal summation in reflex action.

Occlusion

The tension obtained in a certain reflex by strong stimulation of two adjacent afferents simultaneously (i.e. which give rise to the same reflex) is less than the sum of tensions developed by stimulation of each of them separately. This is due to overlap of some of the neurones excited by the two afferents.

In Fig. 17.38, the upper part shows that strong stimulation of afferent 'a' alone causes four motor neurones into action and stimulation of afferent 'b' alone also causes four motor neurones to discharge. Simultaneous stimulation of the two afferents, because of their overlap, brings only six, not eight, motor neurones into action.

Summation of subliminal fringes

The tension obtained in a certain reflex by weak (but threshold) stimulation of two adjacent affer-

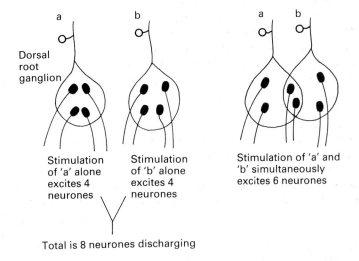

Total is 8 neurones discharging

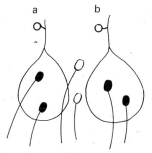

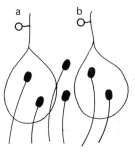

Fig. 17.38 Upper diagram shows occlusion; lower diagram shows summation of subliminal fringes.

ents simultaneously is more than the sum of tension developed when each is stimulated separately. Stimulation of each afferent separately causes discharge of some neurones but only facilitates some nearby neurones (i.e. excites them but not enough to discharge). Those facilitated neurones are said to lie in the *subliminal fringe* zone of those discharging. Simultaneous stimulation of the two afferents, however, causes overlapping of the subliminal fringe zones and raises the excitability of the neurones (within the subliminal zones) to discharge.

In the lower part of Fig. 17.38, stimulation of 'a' alone causes two motor neurones to discharge and two to be facilitated. Also, stimulation of 'b' alone causes two motor neurones to discharge and two to be facilitated. In total, stimulation of the two afferents separately results in four neurones discharging. However, stimulation of a and b simultaneously causes six neurones to discharge, due to overlap of the subliminal fringe.

Recruitment and after-discharge

If a repetitive stimulus is maintained, the strength of the reflex contraction slowly increases to a final level. The slow build-up is due to gradual activation of more motor neurones (i.e. recruitment) (Fig. 17.39).

The reflex response may continue some time after cessation of the stimulus, due to after-discharge, the main type being reverberating circuit after-discharge.

Reciprocal innervation (or reciprocal inhibition)

Reflex stimulation of an agonist muscle also causes inhibition of the antagonist (see Fig. 17.40). Inhibition occurs via the presence of an inter-

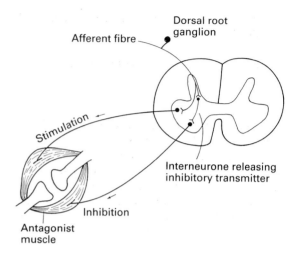

Fig. 17.40 The circuit in reciprocal inhibition. Note the presence of an interneurone releasing an inhibitory transmitter at its junction with the anterior horn cell supplying the antagonistic muscle.

neurone which releases an inhibitory transmitter at its junction with the anterior horn cell supplying the antagonistic muscle.

Final common path

Many sensory inputs converge on the motor neurone. Each motor neurone is thus the common efferent path for several reflexes.

Response time

Response time is an indicator of the number of synapses in the reflex arc. Very short response times indicate a smaller number of synapses in the reflex path. The knee jerk, which has the shortest response time, is a monosynaptic reflex.

Rebound phenomena

This is the exaggeration of a reflex after a temporary period of inhibition. For example, a flexor withdrawal reflex in one limb involves stimulation of flexors and inhibition of extensors. If this is followed by another reflex which involves stimulation of extensors of the same limb (such as the crossed extensor reflex), the extensor response will be greatly exaggerated.

Rebound is one of the important mechanisms for co-ordinating the rhythmic to-and-fro move-

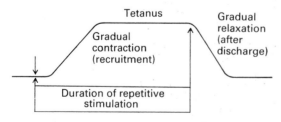

Fig. 17.39 A diagram showing recruitment and after-discharge in reflex action.

ment required in locomotion (i.e. walking and running.)

II Motor functions of the brain stem and cerebral cortex

THE BRAIN STEM

The brain stem is a complex extension of the spinal cord, which performs sensory, motor and reflex functions. Besides containing centres that regulate cardiovascular, respiratory and gastro-intestinal functions, the brain stem plays a major role in the control of eye movements, in the support of the body against gravity and also in the control of other complex reflex movements.

The reticular formation of the brain stem

The reticular formation is a large structure occupying the core of the brain stem from the caudal medulla to the rostral midbrain (Fig. 17.41). It consists of areas of diffuse neurones of two types: sensory neurones, which are greater in number and which make multiple connections within the reticular formation itself, and motor neurones, which are larger in size and receive

impulses from sensory neurones. Motor neurones give rise to axons, which divide into ascending branches and descending branches. The former pass to the non-specific thalamic nuclei, to the basal ganglia and to the cerebral cortex via the thalamus. This is known as the *reticular activating system*, which plays a major role in the control of brain activity and related phenomena, such as consciousness and alertness. The descending branches, however, pass to the spinal cord to supply the anterior motor neurones. These are the lateral and ventral reticulospinal tracts.

Motor functions

Functionally the reticular formation can be divided into:

1 The pontine reticular system, often called the facilitatory reticular formation, is located in the dorsolateral part of the pons. It has spontaneous intrinsic activity which is enhanced by impulses from motor area 4 of the cerebral cortex, the vestibular nucleus, the neocerebellum, and also from the classical sensory pathways. This area transmits facilitatory impulses to the gamma motor neurones supplying extensor anti-gravity

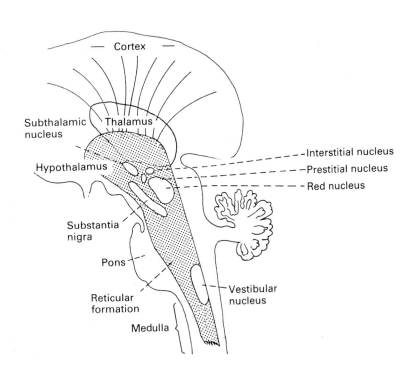

Fig. 17.41 The reticular formation of the brain stem. (Reproduced with permission from Guyton, A.C. (1987) *Textbook of Medical Physiology*, 7th edn. Saunders Co, Philadelphia.)

muscles via the ventral reticulospinal tracts; which increase the sensitivity of muscle spindles. **2** The medullary reticular system (or inhibitory reticular formation) occupies the ventromedial part of the medulla. Unlike the pontine reticular system, it does not discharge spontaneously, but it is driven by impulses from the suppressor area of the motor cortex, the basal ganglia, and the palaeocerebellum. It transmits inhibitory impulses via the lateral reticulospinal tracts to the gamma motor neurones supplying the same antigravity muscles. This counterbalances the excitatory signals from the pontine reticular system so as to maintain normal muscle tone.

The positive supporting reflex is an example of increased facilitatory activity of the reticular formation; the antigravity muscles stiffen when the foot comes in contact with the ground and the weight of the body is supported against gravity.

EQUILIBRIUM AND THE VESTIBULAR APPARATUS

Equilibrium is maintained by variation in the degree of contraction of the individual antigravity muscles; i.e. if a person begins to fall to one side, the extensor muscles on that particular side contract while those on the opposite side relax. This variation in the degree of contraction of individual antigravity muscles is controlled by signals, mainly from the vestibular apparatus.

The vestibular apparatus

The vestibular apparatus is the sensory organ for equilibrium (Fig. 17.42). It is composed of the bony labyrinth, which consists of a group of bony cavities and canals connected with each other in the temporal bone. The bony labyrinth is formed of the vestibule, three bony semicircular canals and a bony cochlea. This encloses the membranous labyrinth, composed mainly of the cochlear duct (membranous cochlea), three membranous semicircular canals and two chambers: the utricle and saccule (Fig. 17.43). Between the bony and membranous labyrinth is a space filled with fluid, the *perilymph*. Inside the membranous labyrinth is another fluid, called the *endolymph*. The cochlear duct is the sensory organ for hearing and is not concerned with equilibrium, while the utricle, semicircular canals and possibly, the saccule contain the receptors concerned with equilibratory mechanisms.

The utricle and saccule These are two chambers each containing a small sensory area called the macula. Each macula is constructed of a ridge of columnar epithelial cells, between which are intermingled hair cells, which project cilia (Fig.

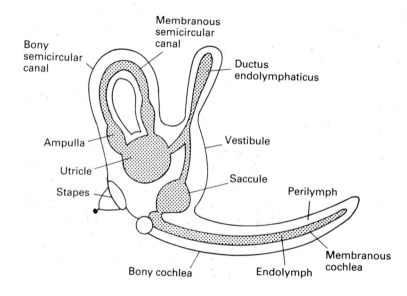

Fig. 17.42 The vestibular apparatus (bony and membranous labyrinth).

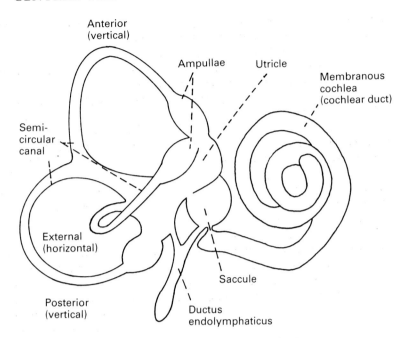

Fig. 17.43 The membranous labyrinth. (Reproduced with permission from Guyton, A.C. (1987) *Textbook of Medical Physiology* 7th edn. Saunders Co, Philadelphia.)

17.44). A single hair cell projects on average about 50 small cilia, called *stereocilia*, plus one very large cilium, called the *kinocilium*. Stimulation of the hair cell occurs when the stereocilia bend towards the kinocilium, and inhibition occurs when they bend away from it (Fig. 17.45). The hair cells project into a gelatinous layer, which surrounds each macula and in which are embedded small calcium carbonate crystals, called *statoconia* (or otoliths) (Fig. 17.44). Each macula has groups of hair cells oriented on its surface in all directions, so that different groups are stimulated with different positions of the head. In the

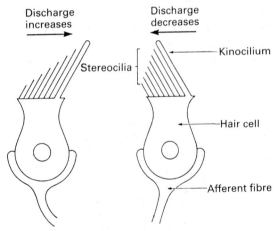

Fig. 17.45 Diagram showing two hair cells of the membranous labyrinth and their mode of stimulation. A tonic level of discharge is recorded from the nerve fibre leading from the hair cell. This level of discharge, however, increases when stereocilia bend towards the kinocilium, and decreases when stereocilia bend away from the kinocilium. (Reproduced from Bray, J.J., Cragg, P.A., Macknight, A.D.C., Mills, R.G. & Taylor, D.W. (eds) *Lecture Notes in Physiology*, 2nd edn. Blackwell Scientific Publications, Oxford.)

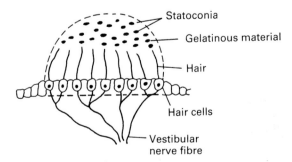

Fig. 17.44 The macula of the saccule and utricle (stimulated by the pull of gravity).

upright position with the head vertical, the macula of the utricle lies in the horizontal plane (Fig. 17.46); the processes of the hair cells pointing upwards. This position allows different groups of hair cells to be stimulated whenever the head bends forwards, backwards or laterally creating a different pattern of excitation each time. It is these changing patterns which make it possible for the brain to determine the orientation of the head in space and in relation to the pull of gravity or the forces of acceleration.

In the saccule, the macula lies in a vertical plane with processes of the hair cells pointing laterally (Fig. 17.46). This allows them to operate only when the head is not in the vertical position, e.g. when one is lying down.

Function of the utricle and saccule The macula in the utricle (and to a lesser extent in the saccule) plays two important functions:

1 *Maintenance of static equilibrium.* It was mentioned earlier that different hair cells are oriented in different directions in the macula of the utricle, so that different patterns of stimulation of the hair cells indicate a certain position of the head in space. In the vertical position of the head, the impulses from the right and left utricles balance each other and there is no sensation of malequilibrium. However, if malequilibrium occurs, e.g. the head is tilted to one side, calcium carbonate crystals in the gelatinous material fall by their weight to that side. In doing so, they pull on the cilia projecting from the hair cells. The stereocilia on that side bend towards the kinocilium on the same side. Consequently, the rate of discharge of impulses increases on the first side but decreases on the other side. Unbalanced discharge from the right and left utricle is carried from the bases of hair cells by axons which form the vestibular division of the 8th cranial nerve. This gives a sensation of tilt towards one side, i.e. a sensation of malequilibrium is created. In response, the equilibratory centres in the brain stem and cerebellum, stimulate the appropriate muscles to restore equilibrium.

2 *Detection of linear acceleration.* Statoconia are also displaced when the body accelerates. Forward acceleration as when a person stands in a bus which suddenly accelerates, causes the statoconia to fall backwards on the hair cell's cilia. This creates a sense of malequilibrium (i.e. the person feels he is falling backwards) which is corrected by leaning forwards so that the statoconia shift anteriorly. Backward acceleration on the other hand, as when the bus suddenly decelerates, causes the statoconia to fall forwards and the sense of malequilibrium (i.e. the person feels he is falling forwards) is corrected by leaning backwards so as to shift the statoconia posteriorly.

In conclusion, maculae are as important for detection of linear acceleration (not linear velocity) as for maintenance of static equilibrium.

The semicircular canals There are three semicircular canals on each side, known as external (horizontal) and anterior and posterior (both are vertical), which lie in three planes perpendicular to each other (Fig. 17.47). Each semicircular canal has a dilated end, called the ampulla, containing a sensory organ, the *crista ampullaris*. Similarly to the macula, the crista is composed of hair cells intermingled with supporting cells. The hair cells' cilia are embedded in a gelatinous mass, similar to that in the utricle, known as the *cupula* (Fig. 17.48). Again, cilia are of two types, stereocilia and a large kinocilium, located always on one and the same side of the cell with respect to its orientation in the ampulla. The hair cells are also connected to nerve fibres, which pass into the

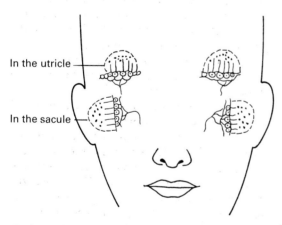

In the utricle

In the sacule

Fig. 17.46 Orientation of the maculae in the utricle and saccule.

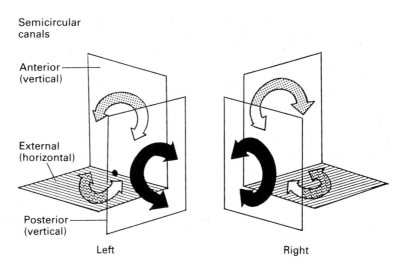

Fig. 17.47 Planes of the semicircular canals showing directions of rotation.

vestibular nerve. As the semicircular canals are filled with endolymph, its flow in the canals excites the sensory organ in the ampulla, i.e. hair cells of the crista.

Function of the semicircular canals The semicircular canals are the sensory organs that detect head rotation in any direction. This is called angular acceleration.

Under resting conditions, the semicircular canals on both sides of the head transmit from

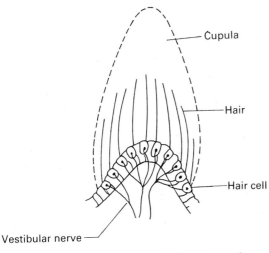

Fig. 17.48 The crista ampullaris of the semicircular canals (stimulated by rotation).

their cristae a continuous series of impulses averaging about 200/s. But when the head suddenly begins to rotate in any direction, suppose from left to right in the horizontal plane (Fig. 17.49), the horizontal semicircular canal moves in the same plane. The endolymph inside it, however, tends to remain stationary for a while because of its inertia. Thus, it is as though the endolymph is flowing in the canal in a direction opposite to the rotation of the head, i.e. from right to left (Fig. 17.49a). The cupula is bent by the endolymph and so are the projecting cilia. Since the kinocilium is located on one side of the hair cell with respect to its orientation in the ampulla, the frequency of impulses from the ampulla of the right canal is increased, as the cilia are bent towards the kinocilium, while that from the ampulla of the left canal is decreased, because the cilia are bent away from it. This unbalanced discharge from the two sides will give the sensation of rotation to the right.

With continued rotation, the endolymph will soon rotate (in the same direction of rotation, i.e. from left to right) as rapidly as the semicircular canal itself due to friction inside the canal, and the cupula, by virtue of its elasticity, returns to the resting position (Fig. 17.49b). This causes the discharge from the two sides to decline back to tonic level (i.e. resting level). The equal and balanced discharge from the two sides causes the person not to perceive the sensation of rotation.

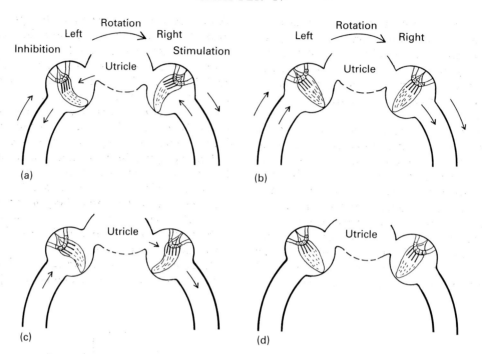

Fig. 17.49 Mode of action of semicircular canals during rotation in the horizontal plane. The right and left horizontal canal ampullae are shown. The arrows drawn inside the canal indicate movement of the endolymph and the arrows outside indicate direction of rotation of the canal. Note the direction of bending of the hairs at the base of the ampulla. See text for explanation of stages (a), (b), (c) and (d).

Figure 17.50 shows the response of one hair cell when the semicircular canal is stimulated at the beginning of rotation and then the decline in the number of impulses to tonic level as rotation continues.

If the rotation suddenly stops, the opposite sequence of events occurs. The canal stops rotating but the endolymph continues to flow for some time due to its momentum in the same direction of rotation (from left to right), bending the cupula — this time in the opposite direction to when rotation was first started (Fig. 17.49c). Thus, discharge from the left canal ampulla increases and that from the right decreases. Imbalance in the discharge between the two sides gives the person a false sensation of rotation, this time from the right to the left side. This is known as vertigo. In a few seconds, then, the endolymph stops moving and the cupulae recoil to the resting position. Discharge will consequently return to tonic level (Figs 17.49d and 17.50) and the false sensation of rotation (vertigo) disappears.

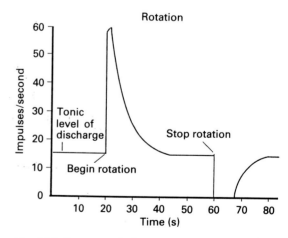

Fig. 17.50 Response of a hair cell when a semicircular canal is stimulated first by rotation and then by stopping rotation. (Reproduced with permission from Guyton, A.C. (1987) *Textbook of Medical Physiology*, 7th edn. Saunders Co, Philadelphia.)

Thus, it seems that the semicircular canal in the particular plane of rotation transmits a positive signal when the head begins to rotate and a negative signal when it stops rotating. It is important to remember that each semicircular canal is stimulated by movements taking place in its own plane.

In conclusion, the function of the semicircular canals is detection of head rotation or stoppage of head rotation in one direction or another. This forms an important mechanism whereby they detect early that a person is turning. Consequently, they send impulses to the nervous system, which applies its central equilibratory corrective measures (i.e. relative degree of contraction of individual antigravity muscles) ahead of time so that a person does not start to fall before he begins to correct the situation.

Neural connections of the vestibular apparatus
Figure 17.51 shows that fibres from the maculae and cristae ampullaris pass as the vestibular nerve to the medulla to synapse in the ipsilateral vestibular nucleus. From the vestibular nucleus, nerve fibres pass to four main sites:

1 *Cerebellum.* Fibres pass to the flocculonodular lobe and dentate nucleus. Some of the fibres pass directly from the receptors to the cerebellum without synapsing in the vestibular nucleus.

2 *Motor cranial nuclei of cranial nerves 3, 4 and 6.* These supply the extraocular muscles. These fibres are called the medial longitudinal fasciculus. They serve to correct eye movements and so help fixing objects seen in the visual fields when the head is rotating.

3 *Reticular formation.* This in turn, sends fibres to the spinal cord (i.e. reticulospinal tracts).

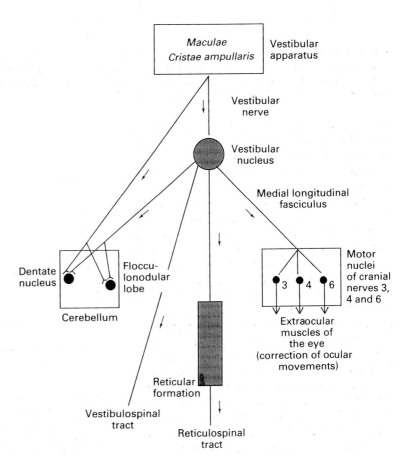

Fig. 17.51 Neural connections of the vestibular apparatus.

4 *Spinal cord.* The fibres form the vestibulospinal tract.

The impulses from the reticular formation and vestibular nuclei maintain equilibrium by sending impulses that facilitate or inhibit the stretch reflex and thus regulate the tone of various muscles.

Effects of stimulation of the semicircular canals
The semicircular canals can be stimulated in the laboratory or clinic using various methods, including the rotational method, in which a subject is rotated in a rotatory chair at a high speed with the head fixed at certain angles to stimulate a particular semicircular canal. Other methods include the caloric method and the electrical method, which one will find more details about in a physiology practical schedule.

Stimulation of the semicircular canals results in a number of changes, including vertigo and nystagmus.

Vertigo This is the false sensation of rotation that a person feels when rotation is stopped. The sensation is always in the opposite direction to that of the original rotation. The pathway involves impulses that pass from the stimulated semicircular canals to the cerebellum and reticular formation. These then pass to the superior temporal gyrus of the opposite side, via the thalamus.

Nystagmus This is a series of jerky movements of the eye, observed at the beginning and end of rotation, and is another sign of stimulation of the semicircular canals. This is a reflex directed towards fixing objects in the visual field during rotation. It has two components:
1 A slow component in the opposite direction to that of rotation (vestibulo-ocular reflex). Its pathway involves impulses that travel from the semicircular canals via the medial longitudinal fasciculus to stimulate the 3rd, 4th and 6th cranial nerves, supplying the extraocular muscles.
2 A fast component in the same direction of rotation. The centre for the latter is in the midbrain and it sends impulses also via the medial longitudinal fasciculus. By convention, the direction of eye movement in nystagmus is identified by the direction of the fast component. Thus, during rotation the fast component is in the

same direction as that of rotation, and, when rotation is stopped, it is in the opposite direction to that of rotation.

Stimulation of the semicircular canals may also cause nausea, vomiting, bradycardia, hypotension and sweating. These are autonomic reactions that occur due to stimulation of the autonomic centres in the brain stem by impulses from the vestibular apparatus. It also leads to an increase in muscle tone on the ipsilateral side and a decrease on the opposite side. This, obviously, is a reflex mechanism to support the body on the side of stimulation. The effects of stimulation of semi-circular canals include what is known as the *postrotatory reactions*. These are cortical voluntary movements of the limbs and body to correct the false sensation of vertigo. The aim of these compensatory movements is to prevent falling of the body in the direction of false sensation. The result is falling of the body towards the opposite side, i.e. in a direction opposite to the false sensation.

POSTURAL REFLEXES
The upright posture characteristic of man is produced, maintained and restored, when upset, by a series of co-ordinated reflexes called the postural reflexes.

Posture depends on the degree and distribution of muscle tone, and muscle tone depends principally on the stretch reflex. Therefore, the stretch reflex is the basic postural reflex.

The role of the stretch reflex in the maintenance of posture is reinforced and modified by afferent impulses from proprioceptors of the muscles of the neck, trunk and limbs, from the eyes, from the vestibular apparatus and from exteroceptors in the skin. Impulses from these sources, which stream into the central nervous system, are integrated by the co-ordinated activity of the spinal cord, brain stem, cerebellum, basal ganglia and cerebral cortex. Postural reflexes are divisible into two groups: (i) static reflexes; and (ii) phasic reflexes (*statokinetic reflexes*).

Static reflexes
Static reflexes maintain posture when the body is at rest. They are of two types, stance and rightening reflexes.

Stance reflexes These are static reflexes that keep the body the right way up. They also give attitude. There are three subtypes:

1 Local static reflexes, which are confined to the stimulated limb. By means of these reflexes, a limb which is freely movable at a given moment can be converted at the next moment to a rigid pillar for bearing weight, i.e. *positive supporting reaction or magnet reaction.*

2 Segmental static reflexes are mediated by one segment of the spinal cord for adjusting the position of one limb in relation to movements of the opposite, as in the *crossed extensor reflex.*

3 General static reflexes produce reflex alterations of tone and posture of the trunk and limbs in accordance with the position of the head in space. They are initiated by impulses arising from the neck muscles, when they are known as neck *statotonic reflexes*, and/or from the otolith organs, when they are called *labyrinthine statotonic reflexes.*

Neck statotonic reflexes are studied in decerebrate animals after destroying the labyrinth. They depend on stimulation of the abundant muscle spindles in the neck muscles by changes in the head position. The result is movement involving limbs. For example, turning the head to the right relative to the body causes the right limb to extend and the left limb to flex. Turning to the left produces the opposite effect. Dorsiflexion of the head will extend the forelimbs and flex the hind limbs, while ventroflexion has the opposite effect.

Labyrinthine statotonic reflexes are also studied in the decerebrate animal after elimination of the neck proprioceptors. The receptors are the otolith organs (receptors in equilibratory vestibular apparatus) and the response consists of extension of four limbs on dorsiflexion of the head and flexion on ventroflexion.

Rightening reflexes These are reflexes which come into operation when the upright position is disturbed, as in falling down. The afferent impulses come from the eyes, which mediate the optical rightening reflexes, i.e. an animal can appreciate the position of its body with respect to the surroundings by the visual images from the periphery. If the position is disturbed, it is corrected. Afferent impulses also come from the labyrinth, which mediates the labyrinthine rightening reflexes. These correct the position of the head if the body is not in the proper position, e.g. an animal held in the air from its pelvis by a sling keeps its head level. The receptors are the otolith organs. The body rightening reflexes are mediated by afferent impulses from the trunk muscles which arise as a result of pressure on one side of the body (e.g. when an animal is laid on its side). The response is reflex correction of the position of the head and body.

Although all the previously mentioned reflexes are functional in man, the optical rightening reflexes are particularly important. They are the reason why a person with sensory ataxia compensates quite well for his/her deficit with his eyes open but tends to fall when they are closed (positive *Romberg's sign*).

Phasic or statokinetic reflexes
These operate to maintain posture when the body is in motion. The afferent impulses originate in the vestibular apparatus, semicircular canals being stimulated during angular acceleration and otoliths (maculae) during linear acceleration. Some examples of phasic reflexes that come into play when the body is displaced in the horizontal plane are the hopping and placing reactions. The *hopping reaction* consists of the hopping movements that keep the limbs in position so as to support the body-weight when a standing animal is pushed laterally. Its sensory input comes from the labyrinths. The *placing reaction* can best be elicited by suspending a blindfolded animal in the air. When the animal is moved towards a supporting surface, the feet are placed firmly on the supporting surface as soon as they touch it. Its sensory input comes from touch receptors in the soles of the feet. Both these reflexes are integrated in the cerebral cortex.

THE MOTOR CORTEX

The upper and lower motor neurones
The lower motor neurones consist of the anterior horn cells, the homologous cells in the brain stem. Their efferent fibres pass via the anterior spinal roots and peripheral nerves to skeletal muscles. Thus the lower motor neurones form

the final common path via which various motor influences are delivered to skeletal muscle fibres.

Several inputs converge on the lower motor neurone. Some arise from the spinal cord and brain stem and result in various reflex movements, but others arise from the motor cortical areas and result in the initiation of voluntary movement. It is important, however, to note that most cortical motor function also involves patterns of functions arising in the lower brain areas (i.e. brain stem, basal ganglia and cerebellum) and the spinal cord.

From the above, it is apparent that, for the performance of normal voluntary movement, the integrity of two sets of neurones is important:

1 *Upper motor neurones.* These consist of the motor neurones originating in the cerebral cortex and the brain stem, which synapse directly or indirectly with the anterior horn cells or with the motor neurones of the cranial nerves. Upper motor neurones are grouped into pyramidal and extrapyramidal systems.

2 *Lower motor neurones.* These include the motor cranial nuclei and their axons, i.e. motor fibres of the cranial nerves (3rd, 4th, 5th, 6th, 7th, 9th, 10th, 11th and 12th). In the spinal cord they include the anterior horn motor neurones and their axons, i.e. the motor nerves to skeletal muscles (Fig. 17.52).

The primary motor and premotor areas

The motor cortex constitutes three separate areas, which produce movement on stimulation: the primary motor area, the supplemental motor area and the premotor area. Motor responses are also obtained on stimulating somatosensory areas I and II. It is possible that all these areas contribute fibres to the pyramidal and extrapyramidal systems.

The primary motor area, also known as motor area 4, occupies the precentral gyrus (see Fig. 17.53) and is highly excitable, due to the presence of the highly excitable *Betz cells*. This area forms the main region of origin of the pyramidal tracts. The primary cortex of one side controls skeletal muscles of the opposite side of the body.

Representation of the body is unilateral, except for the upper part of the face (i.e. forehead and eyes), which is represented in both cerebral hemispheres. The various parts of the body are represented in an inverted manner, i.e. the feet at the top and the face at the bottom and medially. The area of representation of any part is proportional to the degree of complexity of skilled movement of that part rather than to its size. Thus, the areas representing muscles of speech (i.e. lips, tongue and vocal cords) and that representing muscles that control hand movement are large. Cortical representation of trunk muscles is noticeably small (Fig. 17.54). It is also noted that there is a high degree of control over discrete muscular movement in areas controlling complex, fine or skilled movements. In other areas, which control gross movement, there is control over groups of muscles rather than separate muscles. For instance, threshold stimuli applied to the fingers and thumb regions of motor area 4 cause contraction of single muscles or even small fasciculi of muscles, while stimulation of the trunk region causes as many as 30–50 small trunk muscles to contract together.

The supplemental motor area is a small area located on the lateral side of the brain in front of

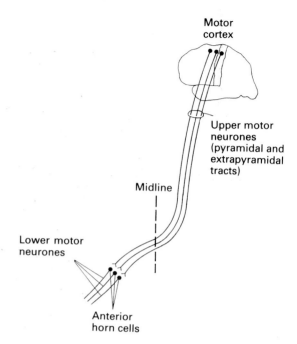

Fig. 17.52 The upper and lower motor neurones.

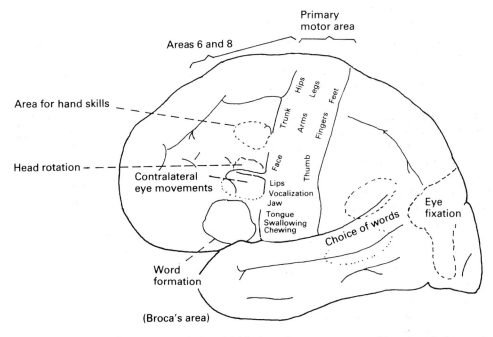

Fig. 17.53 Representation of the different muscles of the body in the motor cortex and location of other cortical areas responsible for certain types of motor movements.

area 4 and above the premotor area. It extends on the medial side of the hemisphere down to the cingulate sulcus. This area projects mainly to the motor cortex and appears to be concerned with programming motor sequences. Indeed, in humans, blood flow which matches neuronal activity, increases in this area whenever the movements performed are complex.

The premotor area constitutes areas 6 and 8, which lie in front of the primary motor area (Fig. 17.53). It is also known as the motor association area. It contains no Betz cells, which makes it less excitable than area 4. Body representation in this area is also in an inverted manner, but the area contains very few neurones that project nerve fibres directly to the spinal cord. Stimulation of the premotor area produces complex co-ordinated movement. This is achieved by sending signals to area 4 to excite multiple groups of muscles.

Figure 17.53 shows a premotor area that lies immediately above the Sylvian tissue in front of the primary motor area. This is known as *Broca's* area or the word formation centre, and its stimu-

lation leads to vocalization. Damage in Broca's area, however, does not prevent a person from vocalizing, but it does make his vocabulary limited to very few words, a condition known as motor aphasia.

The voluntary eye movement field is another area that lies immediately above Broca's area. It controls conjugate eye movement and is connected to the occipital visual centres. Damage to this area prevents movement of the eyes towards different objects.

Above the voluntary eye movement area lies the head rotation area. As the name implies, its stimulation leads to head rotation and the area has connections with the eye movement area so as to direct the head towards different objects.

The area for hand skills is an important region in the premotor area which lies immediately anterior to the primary motor area for the hands and fingers. A lesion destroying this area leads to *motor apraxia*, i.e. hand movement becomes uncoordinated and useless.

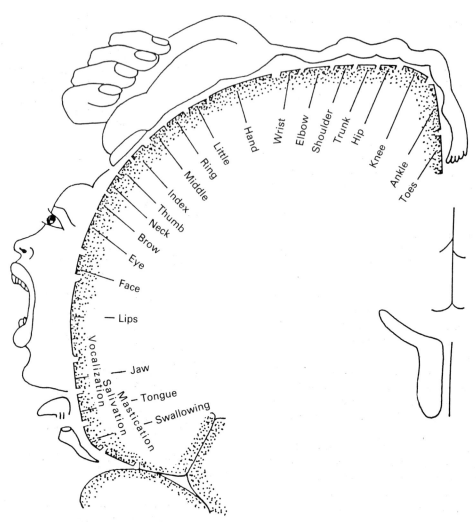

Fig. 17.54 Degree of representation of the different muscles of the body in the motor cortex.

CONNECTIONS OF THE MOTOR CORTEX

Afferent fibres

The motor cortex is excited by signals from numerous and different sources including:

1 Subcortical fibres from nearby areas, especially the somatic sensory areas.

2 Subcortical fibres from the corresponding areas of the motor cortex of the other side.

3 Somatic sensory fibres, arriving via the ventrobasal complex of the thalamus.

4 Fibres from the ventroanterior and ventro-

lateral nuclei of the thalamus, which receive signals from the basal ganglia and cerebellum. These cause coordination of function between cortex, basal ganglia and cerebellum.

5 Fibres from the non-specific nuclei of the thalamus. These control the level of excitability of the cortex generally.

Efferent fibres

The motor areas of the cerebral cortex send their impulses through the descending tracts to the spinal cord anterior horn cells and motor cranial nerve nuclei (i.e. the pyramidal and extrapyramidal tracts).

The pyramidal system

The pyramidal system can be subdivided into three tracts according to the site of muscles they supply:

1 *The corticospinal tracts.* These consist of approximately one million fibres. Of these, 30% originate from the primary motor area, 30% from the premotor areas and 40% from the somatic sensory areas (areas 3, 1, 2) posterior to the central sulcus. About 3% of the fibres are large myelinated fibres, derived from the large pyramidal Betz cells of the motor cortex.

From the cortex the fibres descend into the corona radiata (see Fig. 17.55), through the genu and posterior limb of the internal capsule, the cerebral peduncle of the midbrain and the pons, and enter the medulla. There, the fibres collect to form the pyramids, after which 85% of the fibres cross to the opposite side (i.e. pyramidal or motor decussation) and descend through the spinal cord as the crossed or lateral corticospinal tract, to synapse principally with small interneurones in the intermediate region of the cord grey matter. These interneurones, in turn, synapse on anterior horn cells.

The remaining 15% of fibres descend to the spinal cord on the same side, as the direct or anterior corticospinal tract. Even these fibres, however, cross to the opposite side in either the neck or the upper thoracic region.

2 *The corticobulbar tract.* This originates mainly from the lower part of the primary motor and premotor areas. The fibres follow the same route as the corticospinal tract until they reach the brain stem (see Fig. 17.56a). There, they cross to the opposite side and synapse with motor nuclei of cranial nerves 5, 7, 9, 10, 11 and 12 in the pons and medulla. The axons from cells of these nuclei supply the muscles of the head.

3 *The corticonuclear tract.* Figure 17.56b shows that this originates mainly from area 8, and the fibres also follow the corticospinal tract route. At the level of the brain stem, the fibres cross to synapse on the motor nuclei of cranial nerves 3, 4 and 6 on both sides. The fibres from these nuclei supply the extraocular muscles. This part of the

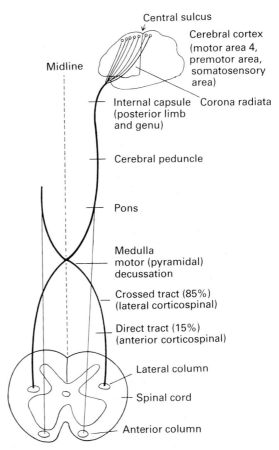

Fig. 17.55 The corticospinal (pyramidal) tract to the whole body below the head.

pyramidal system controls conjugate deviation of both eyes to the opposite side.

It is important to note that uncrossed fibres from the ipsilateral motor cortex may exist. This serves the purpose of providing bilateral innervation of muscles of both sides which move simultaneously, e.g. muscles of the upper face and abdominal and respiratory muscles.

Function of the pyramidal systems It is concerned with the execution of complex, fine, skilled, voluntary movements, especially of the fingers, toes and face. It is facilitatory to muscle tone and deep reflexes. This is evidenced by the fact that sectioning of the pyramids (which contain only pyramidal fibres) in monkeys results in hypotonia.

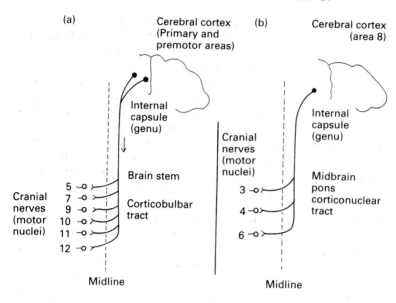

Fig. 17.56 (a) The corticobulbar (pyramidal) tract to the head. (b) The corticonuclear (pyramidal) tract to the eye muscles.

The extrapyramidal system

The extrapyramidal system is made up of all those parts in the central nervous system that are concerned with motor control, other than the pyramidal system and cerebellum. These consist of:

1 Cortical motor areas, especially the premotor area.

2 The basal ganglia.

3 The reticular formation, the red nuclei, the tectum of the brain and the vestibular nuclei.

Axons from the cortical areas descend in the corona radiata and internal capsule, intermingled with the pyramidal fibres. In the basal ganglia, these fibres synapse with neurones of the caudata, putamen and globus pallidus. The globus pallidus, being the efferent centre of the basal ganglia, sends impulses to the thalamus, superior and inferior colliculi (the tectum), red nuclei, substantia nigra, reticular formation, vestibular nucleus and inferior olivary nuclei. From these areas, fibres descend to the spinal cord, to supply the anterior horn cells in the following extrapyramidal tracts:

1 Rubrospinal tracts.

2 Reticulospinal tract.

3 Tectospinal tracts.

4 Vestibulospinal tracts.

5 Olivospinal tracts.

Although extrapyramidal tracts pass to the spinal cord mixed to some extent with the pyramidal fibres, the pyramids in the medulla contain purely pyramidal and no extrapyramidal fibres.

Functions of the extrapyramidal system This system is concerned with the planning, programming and initiation of movement (operating together with the motor cortex and cerebellum as one unit). It is also responsible for the subconscious gross movements occurring in groups of muscle, such as the associated movements which underlie voluntary actions, e.g. swinging of arms while walking. The extrapyramidal system also provides the necessary postural background for performance of skilled movements by the pyramidal system. Some tracts in this system are facilitatory while others are inhibitory to muscle tone; however, fibres from the cortical areas 6 and 8 have a strong inhibitory effect over lower motor neurones, especially gamma motor neurones. Therefore, damage of these areas leads to rigidity of muscles. This is an example of facilitation of alpha neurones of the anterior horn of the spinal cord, due to inactivity of the inhibitory reticular formation, which is excited by these extrapyramidal fibres.

Disordered function of the pyramidal and extrapyramidal systems

Pyramidal lesions (upper motor neurone lesions)
In man, pure pyramidal tract lesions do not occur. However, pyramidal fibres are most commonly involved in cerebrovascular accidents, such as strokes due to haemorrhage, thrombosis or embolism, as they pass through the internal capsule. In such lesions, many extrapyramidal fibres are also damaged. Thus, clinicians talk of upper motor neurone (UMN) lesions and not of pyramidal lesions.

An upper motor neurone lesion is characterized by the following:
1 *Paralysis*. This occurs on the opposite side of the body (i.e. half of face, upper limb and lower limb, known as hemiplegia) if the lesion is at or above the pyramidal decussation. This type of paralysis is usually widespread and is characterized by poor recovery.
2 *Hypertonia*. The paralysed muscles show increased tone of the spastic type. There is stiff resistance to passive movement, which suddenly gives way (clasp-knife spasticity). Hypertonia is due to loss of the inhibitory effect of the cortical extrapyramidal area on the gamma motor neurones of the spinal cord. The facilitatory centres lower down (i.e. facilitatory reticular formation and vestibular nucleus) become hyperactive, stimulating the gamma motor neurones.
3 *Exaggerated tendon reflexes*. These are seen on the affected side, exemplified by the knee and ankle jerks, and are due to the release of the stretch reflex from cerebral inhibition.
4 *Clonus*. This is the occurrence of rhythmic contractions of muscles when they are subjected to sudden sustained stretch, e.g. ankle clonus. The precise cause of clonus is not known. This phenomenon is associated with increased gamma efferent discharge, occurring as a result of the release of the stretch reflex from inhibition.
5 *Loss of superficial reflexes*. This occurs on the affected side, due to loss of supraspinal facilitation. The abdominal and cremasteric reflexes are absent but the plantar reflex becomes extensor, known as a positive *Babinski's sign* (i.e. scratching the outer aspect of the sole by a blunt object results in dorsiflexion of the big toe and fanning

of the other four toes). The abnormal response is thought to be a primitive reflex that reappeares following injury of the pyramidal fibres.

The Babinski's sign is considered physiological during the first year of life, due to immaturity of the pyramidal tract, and in adults during sleep, deep anaesthesia or coma, due to the depressed activity of the motor cortex.
6 *Absence of significant muscle-wasting*. This is because, although the muscles are not used voluntarily, they still contract reflexly and the spasticity saves them from disuse atrophy.

In conclusion, it seems that in upper motor neurone lesions, the paralysis of movements represents loss of function, while the hypertonia (spasticity) represents a release of function. This refers to release of the stretch reflex, the basis behind muscle tone, from the effects of the inhibitory pathways damaged by the lesion, as explained above.

Extrapyramidal lesions Extrapyramidal disease is characterized by three features:
1 *Disturbance of voluntary movement*. There is difficulty in initiating voluntary movement (i.e. *akinesia* or *hypokinesia* of Parkinsonism). Associated movements are lost (poverty of movements).
2 *Alteration of tone*. Tone may be increased (as in Parkinsonism) or decreased (as in chorea). The hypertonia of extrapyramidal disease is called *rigidity*, to differentiate it from the clasp-knife spasticity of pyramidal disease, e.g. lead-pipe and cog-wheel rigidity of Parkinsonism.
3 *Involuntary movements*. These are four types: tremors, chorea, athetosis and ballismus.

DISORDERED FUNCTION OF THE LOWER MOTOR NEURONES
As discussed earlier, normal motor function is also affected by the integrity of the lower motor neurone which connects skeletal muscles with the central nervous system. Lower motor neurone cells are situated in the anterior horn of the spinal cord and in the brain stem.

Lower motor neurone lesion (LMN lesion)
A lesion of the lower motor neurones or their axons is characterized by the following features:

1 *Paralysis.* This occurs in the muscles supplied by the affected segments only (e.g. muscles of a limb only) on the same side of the lesion.

2 *Hypotonia or atonia.* The paralysed muscles show decrease or loss of tone; thus it is often referred to as flaccid paralysis. This occurs because of interruption of the stretch reflex pathway.

3 *Absent deep reflexes.* This occurs in the muscles supplied by the affected segments or motor nerves.

4 *Absent superficial reflexes.* This is also seen in the affected segments only.

5 *Marked muscle-wasting (atrophy).* The affected muscles show trophic change due to disuse. These changes are due to the inability of the muscles to contract reflexly or voluntarily. Intact motor innervation seems to be essential for the well-being of muscles. A denervated muscle shows wasting even if it is kept contracting by suitable electrical stimulation.

6 *Presence of fasciculations.* These are visible spontaneous contractions of bundles of fibres in the affected muscles (i.e. motor units), due to pathological discharge of spinal motor neurones. Fasciculation continues during the degenerative process and stops once the nerves have atrophied. Persistent fasciculation in muscle indicates neuropathy.

7 *Changes in the electrical excitability of the affected muscles.* These include appearance of fibrillation potentials, e.g. fine irregular contractions of individual muscle fibres, which occur due to hypersensitivity to acetylcholine. Other changes include an increase in chronaxie, i.e. the time required to elicit a response by a current twice the rheobasic strength. The rheobase is the minimum strength of current required to elicit a response.

INTERNAL CAPSULE AND SPINAL CORD LESIONS

Having discussed the sensory system fully and the relevant part of the motor system, it is convenient to introduce the reader to some of the disturbances of function that result from damage of the two systems simultaneously.

The internal capsule

The internal capsule is a collection of axons lying between the basal ganglia and the thalamus (Fig. 17.57). When sectioned horizontally, the internal capsule appears V-shaped, with the point of the V looking medially. This point is the genu of the internal capsule, with the anterior limb lying anteriorly and the posterior limb lying posteriorly.

Pyramidal tract fibres occupy the genu and the anterior two-thirds of the posterior limb in an orderly fashion, i.e. the most anterior fibres are concerned with control of the head muscles, followed by those of the arm, trunk and leg muscles. Extrapyramidal fibres are intermingled with the pyramidal fibres.

Posteriorly and immediately behind the pyramidal fibres lies the sensory radiation, followed by the optic radiation and, most posteriorly, the auditory radiation.

Effects of a unilateral lesion in the posterior limbs of the internal capsule The posterior limb is a common site for lesions due to haemorrhage or thrombosis in the lenticulostriate artery. Such a lesion is characterized by widespread sensory and motor disturbances. There is paralysis, of the upper motor neurone lesion type, on the opposite side of the body (i.e. *hemiplegia*). Hemianaesthesia, that is, loss of sensation, occurs also on the opposite side of the body, due to damage of the thalamocortical fibres. Special senses fibres are also affected. The patient may show homonymous hemianopia, which is loss of vision in two corresponding halves of both eyes, due to injury of the optic radiation. For example, injury to the left optic radiation causes inability to see objects in the right halves of the visual fields of the two eyes (Fig. 17.58). Hearing may also be slightly diminished, but deafness does not occur, because each cerebral hemisphere receives impulses from the two cochleae.

Spinal cord lesions

Spinal cord transection usually occurs as a result of road accidents that result in fracture of the vertebrae. It can also occur as a result of primary tumours or metastasis from malignant tumours. The extent of the lesion determines the clinical manifestations.

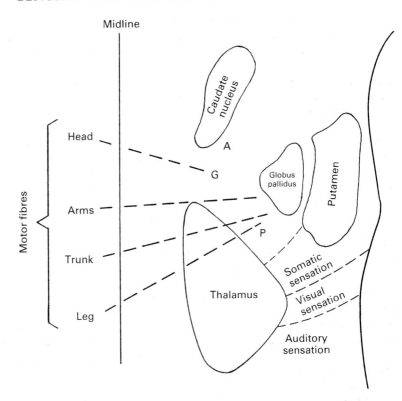

Midline

Caudate nucleus

A

G

Globus pallidus

Putamen

P

Head

Motor fibres

Arms

Trunk

Leg

Thalamus

Somatic sensation

Visual sensation

Auditory sensation

Fig. 17.57 The internal capsule. A, anterior limb; G, genu; P, posterior limb. (Redrawn from (1982) *Samson Wrights Applied Physiology*, 13th edn, by permission of the Oxford University Press, Oxford.)

Complete transection of the spinal cord Obviously, the higher the level of the section, the more serious are the consequences. If the transection is in the upper cervical region, immediate death follows, due to paralysis of all respiratory muscles; however, if it occurs in the lower cervical region, diaphragmatic respiration is still possible, but the patient suffers complete paralysis of all four limbs (i.e. *quadriplegia*). Transection lower down in the thoracic region allows normal respiration but the patient ends up with paralysis of both lower limbs (i.e. *paraplegia*).

When the cord is completely sectioned, the patient passes through various stages. In the immediate period following transection, there is complete loss of spinal reflex activity below the level of transection. This is known as the *stage of spinal shock* and is thought to occur due to the sudden withdrawal of facilitatory supraspinal influence (i.e. corticospinal, reticulospinal and vestibulospinal tracts). The stage of spinal shock is characterized by the following features, all of which occur below the level of the lesion:

1 Loss of all sensation and voluntary movement, due to section of all sensory and motor tracts respectively.

2 Loss of all reflexes (superficial, deep and visceral). Absent deep reflexes are manifested in the absence of tendon jerks and the loss of muscle tone, which results in flaccidity. The flaccid muscles lead to a decrease of the efficiency of the muscle pump in the affected part and so venous return is reduced, causing the part to become cold and blue.

Absent visceral reflexes include the micturition, defecation and erection reflexes. The walls of the urinary bladder and rectum are paralysed but the tone in the internal vesical sphincter returns very rapidly. The result is retention of urine, which accumulates in the bladder until the pressure in the bladder overcomes the resistance offered by the tone of the internal sphincter, and dribbling occurs. This is known to clinicians as 'retention with overflow'.

3 Loss of vasomotor tone occurs, due to interruption of fibres that connect the vasomotor

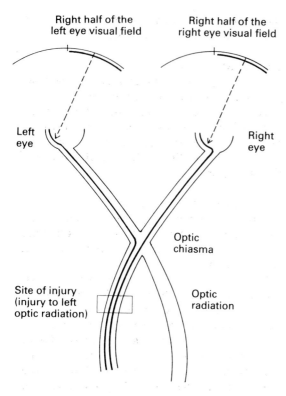

Right half of the
left eye visual field

Right half of the
right eye visual field

Left
eye

Right
eye

Optic
chiasma

Site of injury
(injury to left
optic radiation)

Optic
radiation

Fig. 17.58 Diagram showing site of lesion in the left optic radiation resulting in loss of vision in the right half of the visual field of both eyes (homonymous hemianopia).

centres with the lateral horn cells of the spinal cord. The relative vasodilatation causes a fall in blood-pressure. The higher the level of the section the lower the blood pressure.

4 Bedsores are seen where the body-weight hinders circulation of blood to the skin. This occurs particularly over bony prominences, i.e. back, heel and gluteal region. As a result, the skin sloughs off at these points, forming ulcers that heal with difficulty, known as decubitus ulcers.

This stage varies in duration but usually occupies a period of 2–6 weeks, after which some reflex activity is recovered. This is known as the *stage of recovery* of reflex activity and it occurs because of an increase in the natural degree of excitability of the spinal cord neurones, presumably to make up for the loss of supraspinal facilitatory influences.

Recovery is gradual and includes the following features:

1 Gradual rise of arterial blood-pressure towards normal, due to regain of activity in the lateral horn cells of the spinal cord, which form the spinal vasomotor centre. As a result, excitatory impulses are sent to the muscle walls of arterioles and venules. The blood-pressure returns to normal levels, but vasomotor control from the medulla is absent. The patient develops a sudden fall in blood-pressure on being made to sit up or stand.

2 Spinal reflexes also return. Flexor reflexes return earlier than extensor ones; however, the plantar reflex, when tested, shows a positive Babinski sign. Deep reflexes also recover earlier in flexors. As a result, flexor tone causes the lower limbs to take a position of slight flexion, a state referred to as *paraplegia in flexion*. The muscle tone is generally increased, due to facilitation of the stretch reflex by suprasegmental facilitation. The return of spinal reflexes and tone in arterioles and venules improves the circulation through the limbs.

3 Visceral reflexes return, as evidenced by return of automatic function of both bladder and rectum. For example, distension of the urinary bladder causes impulses to travel to the spinal centres, leading to reflex contraction of the bladder wall and relaxation of the internal urethral sphincter. Consequently, the bladder is emptied (automatic evacuation) but voluntary control over micturition and the sensation of bladder fullness are permanently lost.

4 Mass reflex, a state that reflects the heightened excitability of the spinal cord neurones, appears in this stage. A minor painful stimulus to the skin of the lower limbs will not only cause withdrawal of that limb but will excite many other reflexes by spreading to many levels of the cord. So the bladder and rectum will also empty, the skin will sweat, the blood-pressure will rise, etc.

5 Sexual reflexes, consisting of erection or even ejaculation on genital manipulation, also recover.

Since effective regeneration never occurs in the human central nervous system, patients with complete transection never recover fully. Voluntary movements and sensations are permanently lost; however, patients can be rehabilitated by proper management and can then enter into a more advanced stage of recovery. During the latter, the tone in extensor muscles gradually returns and becomes greater than in the flexors, so that the lower limbs become extended — a state known as *paraplegia in extension*. Features of upper motor neurone lesion become very evident below the level of transection. The positive supporting reflex becomes well developed and the patient can stand without support.

Hemisection of the spinal cord (Brown-Séquard syndrome) The manifestations of the Brown-Séquard syndrome depend on the level of the lesion.

At the level of the lesion, all manifestations occur on the same side. These consist of:

1 Paralysis of the lower motor neurone type, involving only the muscle supplied by the damaged segments.

2 Vasodilatation of the blood-vessels that receive vasoconstrictor fibres from the damaged segments.

3 Loss of all sensations in the areas supplied by the afferent fibres that enter the spinal cord in the damaged segments.

Below the level of the lesion (see Fig. 17.19), the following manifestations occur:

1 Widespread paralysis of the upper motor neurone lesion type occurs on the same side, due to the interruption of pyramidal and extrapyramidal tracts.

2 Vasodilatation also occurs, due to interruption of the pathway of fibres from the medullary vasomotor centres.

3 Sensory loss occurs, due to interruption of the ascending pathways. Crude touch, pain and temperature are lost on the opposite side of the lesion, due to damage of ventral and lateral spinothalamic tracts respectively, while fine touch, position and vibration sense are lost on the same side, as a result of damage of the gracile and cuneate tracts.

III Motor functions of the basal ganglia and cerebellum

The power and precision of muscle activity is the product of normal function not only of the cerebral cortical areas, but also of two other essential brain structures, namely the basal ganglia and the cerebellum. These do not function by themselves, but work in close association with the cerebral cortex and corticospinal tracts.

THE BASAL GANGLIA

Anatomically the motor portions of the basal ganglia are composed of the caudate nucleus and the putamen, collectively known as the *neostriatum*, and of the *globus pallidus*. The substantia nigra, the subthalamus, part of the thalamus and the reticular formation are functionally related to the basal ganglia.

Connections of the basal ganglia
Functionally important pathways connect the basal ganglia with each other and with other areas of the central nervous system. The main connections are:

1 Pathways from the cerebral cortex to the basal ganglia and back to the cortex. These consist of two circuits, one passing through the putamen and the other through the caudate nucleus.

(a) The putamen circuit (see Fig. 17.59a). Nerve fibres of this circuit originate mainly in the premotor and supplemental motor areas of the motor cortex and also in the primary somatic sensory area of the sensory cortex. These are all areas adjacent to the primary motor area, and yet no fibres arise from there. Next, the fibres pass to the putamen, then to the internal part of the globus pallidus and then to the ventro-anterior and ventrolateral nuclei of the thalamus, from where the fibres finally return to the cerebral cortex, but now mainly to the primary motor area. A few fibres go to some portions of the premotor and supplemental motor areas. Thus, the putamen has its input mainly from accessory motor areas, concerned with patterns of movement, and sends its output back mainly to the primary motor cortex, responsible for execution of this pattern.

(b) The caudate circuit (see Fig. 17.59b). A large proportion of the nerve fibres in this circuit arises from the cortical association areas (i.e. motor association area, somatic sensory association area, visual association area and auditory association area). These areas process and integrate different types of motor and sensory information into usable patterns of thought. From these areas, the fibres pass to the

(a)

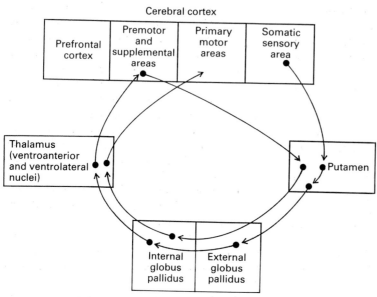

(b)

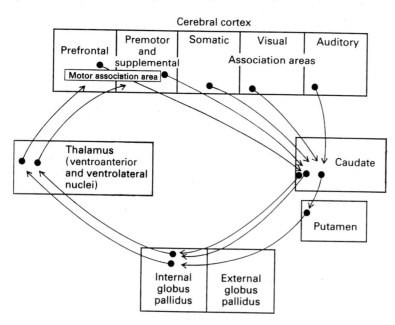

Fig. 17.59 (a) The putamen circuit. Inputs to the putamen arise from the premotor, supplemental motor and somatosensory areas (none from the primary motor cortex). Returning signals pass to the primary motor cortex mainly and to portions of the premotor and supplemental motor cortex. (b) The caudate circuit. Input signals to the caudate arise from the cortical association areas. Returning signals pass directly to the prefrontal, premotor and supplemental motor areas of the cerebral cortex with no fibres passing to the primary motor area.

caudate nucleus, next to the internal globus pallidus and then to the ventroanterior and ventrolateral nuclei of the thalamus. From the thalamus, the fibres pass back to the prefrontal, premotor and supplemental areas of the cerebral cortex, which are areas concerned with patterns of movement. None of the fibres, however, pass directly to the primary motor area of the cerebral cortex.

The fibres from the neostriatum (caudate and putamen) to the globus pallidus secrete gamma-aminobutyric acid (GABA), which is an inhibitory transmitter. Thus, the circular pathways between the cerebral cortex and basal ganglia operate as a negative feedback system.

2 Pathways between the neostriatum and the substantia nigra (see Fig. 17.60). Axons from neurones in the neostriatum (i.e. caudate and putamen) pass to the substantia nigra and secrete GABA. Another group of axons from neurones in the substantia nigra pass to the neostriatum. These secrete dopamine, which is also inhibitory in function. Lesions in one or more of these pathways lead to a number of different clinical abnormalities.

3 Other pathways include nerve tracts passing from the cerebral cortex to the neostriatum. These fibres secrete *acetylcholine*, which is excitatory. Figure 17.60 also shows nerve fibres that originate in the brain stem raphe nuclei and pass to the neostriatum, secreting *serotonin*, which is inhibitory. This pathway is thought by some authors to function in relation to the sleep-producing effect of the raphe nucleus. However, this is still a matter of debate. It is important at this point to note that the normal function of the basal ganglia is brought about by a balance between the various excitatory and inhibitory influences of the various transmitters.

4 The main efferent pathway from the basal ganglia. The globus pallidus is the chief efferent pathway of the basal ganglia (Fig. 17.61). It sends impulses to the subthalamic nuclei and substantia nigra. From these areas the impulses descend into the reticular formation, which sends impulses via the extrapyramidal tracts to the motor neurones of the spinal cord. Some fibres pass directly from the globus pallidus to the reticular formation.

Note that the descending extrapyramidal tracts include the reticulospinal, vestibulospinal and rubrospinal tracts.

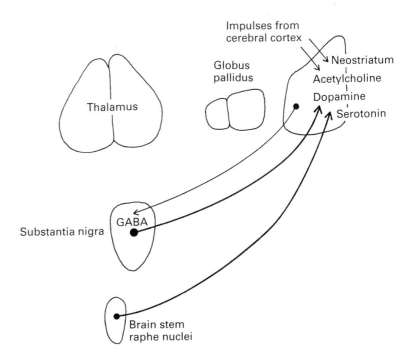

Fig. 17.60 Afferent and efferent pathways of the neostriatum showing several transmitter substances: acetylcholine in corticostriatal endings, GABA in the substantia nigra and globus pallidus, and dopamine and serotonin in the neostriatum.

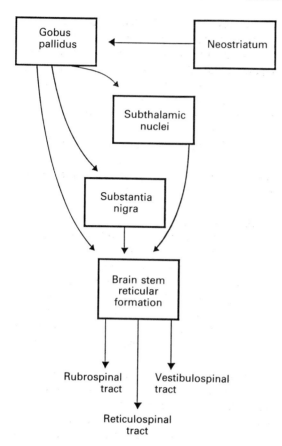

Fig. 17.61 Diagram showing the main efferent pathways from the basal ganglia to the brain stem reticular formation.

Functions of the basal ganglia

Although they play a very important role in motor control, the precise functions of the basal ganglia were not clear for a long time, due to their inaccessible position inside the brain. The most important functions include the following:

1 They help the corticospinal system in executing subconscious learned patterns of movement, such as writing letters of the alphabet, cutting with scissors, some aspects of vocalization, etc. This is thought to be executed by the *putamen circuit*. Damage to the basal ganglia causes gross abnormality of these patterns, even in the presence of a normally functioning cortical motor system.

2 The *caudate nucleus*, in association with the cerebral cortex, helps in planning sequences of patterns of movements, which the mind must integrate together to achieve a complex goal. This process begins in the areas of thought in the brain and results in an overall sequence of actions appropriate to arising situations. A good example is when one is subjected to an approaching danger; he/she responds immediately by turning away from it and then begins to run and may even try to seek appropriate shelter. The caudate circuit subserves this function. Impulses arrive from the cortical association areas, which are centres for integration of sensory and motor experiences into thoughts, reaching the caudate nucleus. The latter sends output signals (via the thalamus) to the motor areas concerned with patterns of movement (see Fig. 17.59b).

The caudate circuit is also responsible for modifying the timing of these patterns of movement, i.e. they can occur rapidly or slowly, and also modifying their spatial dimensions, e.g. one can write very small or very large.

3 The basal ganglia are inhibitory to muscle tone throughout the body. This is probably brought about by inhibitory impulses from the basal ganglia to the motor cortex and/or to the reticular formation. Therefore, widespread damage of the basal ganglia produces rigidity.

4 The basal ganglia are also responsible for initiation and regulation of the gross intentional movements of the body (e.g. swinging of the arms while walking, facial expressions, and crude walking).

5 The globus pallidus is thought to be responsible for the posture taken by the body to perform a particular voluntary movement (e.g. one assumes a certain posture of trunk and limbs to be able to write).

Clinical syndromes associated with damage to the basal ganglia

Chorea This is a syndrome which occurs due to degeneration of the *caudate nucleus* mainly but the putamen may be involved too. It can occur as a hereditary disorder, when it is known as Huntington's chorea, or as a complication of rheumatic fever in children. It is characterized by spontaneous, uncontrolled, flicking movements, which occur one after the other at rest and are increased by muscular activity and emotions. As a result, the normal progression of voluntary movement cannot occur and, instead, the person forms one pattern of movement for a few seconds, followed suddenly by another pattern, as in dancing. Thus, these movements are called dancing movements. The abnormal movements disappear during sleep and are associated with hypotonia and muscle weakness. Examination of the knee jerk shows it to be pendular.

The mechanism behind the abnormal movement in chorea or all the other syndromes is not very clear. However, it is proposed that, with damage of the neostriatum, the GABA-producing neurones are also damaged, leading to removal of inhibition from the substantia nigra. The latter will consequently become very active and secrete dopamine excessively into the neostriatum, inhibiting it. This interrupts the negative feedback loop from cortex to basal ganglia, leading to the abnormal movements. Degeneration of acetylcholine (i.e. excitatory) neurones in the neostriatum adds further to its inhibition and contributes to the abnormality.

Athetosis This occurs due to damage of the lateral portion of the *globus pallidus*. The damage sometimes involves the neostriatum. It is characterized by involuntary, slow, writhing movements of the hand, neck, face and tongue or some other part of the body, which occur continuously. The contracting muscles show a high degree of spasm and the movements are increased by emotions or by excessive signals from sensory organs. As a result, voluntary movements in the affected part become impaired or even impossible.

The mechanism behind the succession of abnormal movements is proposed to be due to the interruption of the feedback circuit between the basal ganglia, thalamus and cerebral cortex which occurs with damage to the globus pallidus. This causes impulses to take unusual routes, leading to abnormal movements.

Ballism This is a condition which occurs due to lesions in or near the *subthalamic nuclei*. It is characterized by continuous gross abrupt contractions of axial and proximal muscles of the

extremities, resulting in flailing movements. In most cases, this disorder affects one side of the body (hemiballism). This may be associated with hypotonia and chorea. When the movement occurs in the legs, it is likely to cause the person to fall. In these patients, attempts at performing voluntary movements often cause ballistic movements instead.

Hemiballismus occurs due to the inability of the damaged subthalamic nucleus to perform its function of integrating smooth, progressive movements of the body parts.

Parkinson's disease (paralysis agitans) This is a syndrome that occurs due to widespread destruction of the *substantia nigra*, which secretes dopamine, but is often associated with lesions of the globus pallidus. The causes for this damage are not yet clearly identified but cerebral atherosclerosis, prolonged treatment with phenothiazines (a tranquillizer that blocks dopamine receptors) and repeated trauma to the head are all proposed.

The syndrome is characterized by three principal features: (i) *rigidity* of muscles, either in a widespread area or in isolated areas; (ii) *tremor*; and (iii) *akinesia*. The mechanism behind these is not known but is definitely related to the loss of dopamine secretion. Rigidity is different from that seen in decerebration, i.e. spasticity, as here increased motor neurone discharge occurs to both agonist and antagonist muscles. However, flexors are more affected than extensors and proximal more than distal parts, which causes the body to be bent forwards. On examining the patient, passive movement (e.g. flexion of a limb) is met with a plastic, dead-feeling resistance (similar to bending a lead pipe). This lead-pipe rigidity is sometimes interrupted by a series of catches during movement; it is then known as cog-wheel rigidity. Rigidity results from excess impulses transmitted in the corticospinal system, which activate the alpha as well as the gamma efferent neurones in the anterior horn of the spinal cord.

Tremors are involuntary regular contractions of muscle, which occur at rest and, unlike cerebellar tremor, disappear during movement and are thus called *static tremors*. They consist of flexion and extension or pronation and supination, usually affecting the distal joints. Pill-rolling movement describes the movement that occurs due to tremors of the thumb and index finger. It is thought that, with lack of the inhibitory effects of dopamine, there is enhancement of activity of the feedback mechanism between cerebral cortex and basal ganglia through the thalamus. This produces oscillations, which cause the muscle tremors. Stoppage of tremors during movement is probably because motor signals from the cerebral cortex or cerebellum override the abnormal signals from the basal ganglia.

Akinesia is the most serious abnormality that patients experience. It is the inability to initiate voluntary movements due to excessive rigidity. This affects speech and gait. The speech becomes slow and monotonous and the gait of the patient consists of short steps without lifting the feet off the ground. A noticeable abnormality is the loss or decrease in associated movements, such as the normal unconscious swinging of the arms during walking and the facial expressions that are related to emotions or content of speech. Due to the lack of facial expression and rigidity of facial muscles, the face becomes like a mask; hence the term mask face.

Generally speaking, the basal ganglia are stimulated by acetylcholine and inhibited by dopamine. Thus, treatment of Parkinsonism consists of administration of L-dopa, which is converted in the body to dopamine. This substitutes for the dopamine no longer secreted by the basal ganglia. Some patients, however, respond to the administration of anticholinergic drugs, which also leads to inhibition of the basal ganglia. Neurosurgeons often use electrocoagulation of the ventrolateral nucleus of the thalamus. This reduces the enhanced activity of the circular pathway between the cerebral cortex and the basal ganglia due to loss of dopamine secretion. It is particularly effective in relieving the tremors.

THE CEREBELLUM

The cerebellum occupies a prominent position beside the main sensory and motor systems in the brain stem (Fig. 17.62). It is connected to the brain stem by the three cerebellar peduncles: superior, middle and inferior. Various fibres enter and leave the cerebellum via these peduncles.

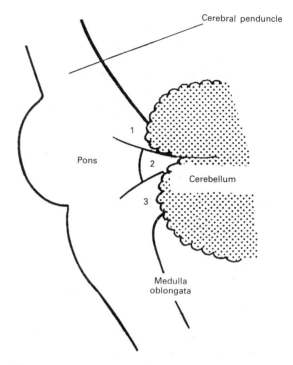

Fig. 17.62 Diagram of the brain stem to show the relationship to the pons and to the superior (1), middle (2) and inferior (3) cerebellar peduncles.

Divisions of the cerebellum

Anatomically, the cerebellum is divided into three areas (Fig. 17.63):

1 *Archicerebellum*. This is represented by the flocculonodular lobe and it is concerned with equilibrium.

2 *Palaeocerebellum*. This is represented by the entire midline area (vermis) and is 2–3 cm wide in both the anterior and posterior lobes. It plays a role in the regulation of muscle tone and posture.

3 *Neocerebellum*. This is represented by the lateral parts of the posterior lobe, which form the cerebellar hemispheres. Its main function is the co-ordination of rapid muscular activities and it has a role in the regulation of muscle tone.

Functionally, the anterior and posterior lobes are organized along the longitudinal axis (see Fig. 17.64):

1 The vermis controls muscle movements of the axial body, neck, shoulders and hips.

2 The intermediate zones (on either side of the

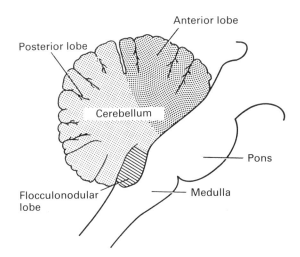

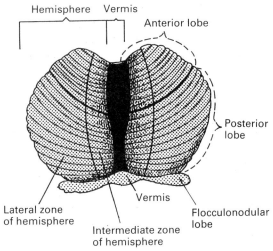

Fig. 17.63 Side-view of cerebellar lobes. (Reproduced with permission from Guyton, A.C. (1987) *Textbook of Medical Physiology*, 7th edn. Saunders Co, Philadelphia.)

Fig. 17.64 Posteroinferior view of the cerebellum. (Reproduced with permission from Guyton, A.C. (1987) *Textbook of Medical Physiology*, 7th edn. Saunders Co, Philadelphia.)

vermis control muscular contractions in the distal portions of both the upper and the lower limbs (especially the hands, fingers, feet and toes).

3 The lateral zones (on either side of the intermediate zones). These areas seem to help in the planning of sequential motor movements.

Structure and connections of the cerebellum

The cerebellum has a cortex of grey matter

covering a mass of white matter. All afferent fibres to the cerebellum (climbing and mossy fibres) end in the cortex, which consists of three layers (Fig. 17.65):

1 A superficial molecular layer.

2 A middle layer of Purkinje cells.

3 A deep granular layer.

All afferent fibres coming to the cerebellum synapse directly or indirectly with the Purkinje

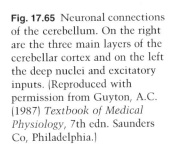

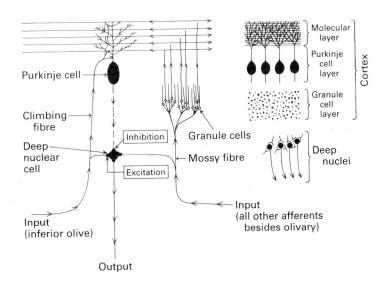

Fig. 17.65 Neuronal connections of the cerebellum. On the right are the three main layers of the cerebellar cortex and on the left the deep nuclei and excitatory inputs. (Reproduced with permission from Guyton, A.C. (1987) *Textbook of Medical Physiology*, 7th edn. Saunders Co, Philadelphia.)

cells. The axons of the Purkinje cells carry efferent impulses to the cerebellar nuclei, i.e. dentate, interpositus and fastigial nuclei. The axons arising from the cerebellar nuclei transmit impulses to various areas of the brain (i.e. brain stem and thalamus).

Table 17.7 gives a summary of the important afferent and efferent connections of the cerebellum. See also Figs 17.66 and 17.67.

Note that the dentatorubrospinal tract is a double crossing tract. It arises from the dentate nucleus of each cerebellar hemisphere and passes to the red nucleus of the opposite side. From the red nucleus of each side arises the rubrospinal tract, which crosses to the opposite side and passes down the brain stem to the spinal cord. Because of this double crossing, each cerebellar hemisphere acts on the ipsilateral muscles, i.e. muscles on the same side as itself.

Topographic representation of the body in the cerebellum

There are two separate representations. In the anterior lobe the body is represented upside down, while it is erect in the posterior lobe. The axial portions of the body lie in the vermis (Fig. 17.68) whereas the limb and facial regions lie in the intermediate zones. Obviously, the lateral zones would have no topographic representations of the body as these areas have different functions, i.e. planning and co-ordination of sequential patterns of muscular activity.

Functions of the cerebellum

The cerebellum is called the silent area, because its stimulation does not give rise to any sensation and causes almost no motor movement. However, it is vitally important in the precise execution of rapid muscular movement. Damage to the cerebellum can cause almost total incoordination of muscular movement, although the muscles are not paralysed.

The cerebellum functions to co-ordinate motor activity initiated elsewhere in the nervous system. These activities may originate in motor areas of the cerebral cortex, where voluntary movements are controlled, or in the basal ganglia, reticular formation or even the spinal cord, where postural and equilibratory movements are controlled.

Function of the cerebellum in voluntary movements The motor areas of the cerebral cortex on one side are connected to the intermediate zone of the cerebellar hemisphere of the opposite side by a closed feedback circuit (corticoponto-interpositus-thalamocortical) (Fig. 17.69).

When the motor cortex transmits a signal to a group of muscles (via pyramidal and extrapyramidal tracts) to execute a particular movement, it transmits the same information simultaneously to the intermediate zone of the cerebellum through the corticopontocerebellar tracts to inform the cerebellum about the orders given from cortex to muscles. As the muscles respond to cortical signals by contraction, receptors such as muscle spindles, Golgi tendon organs, joints and other peripheral receptors are stimulated and, in turn, respond by transmitting signals upwards along the spinocerebellar tracts to the cerebellum. The latter, now being informed about the performance of movements, is able to compare that performance with the orders received from the cortex. If not appropriate, the cerebellum initiates corrective signals, which are sent to the motor cortex (via cerebellointerpositus-thalamocortical

Table 17.7 Afferent and efferent connections of the cerebellum

Peduncle	Afferent	Efferent
Superior cerebellar peduncle	Ventral spinocerebellar tract	1 Dentatothalamocortical tract 2 Dentatorubrospinal tract
Middle cerebellar peduncle	Corticopontocerebellar tract	Fibres to reticular formation of pons
Inferior cerebellar peduncle	1 Dorsal spinocerebellar tract 2 Vestibulocerebellar tract 3 Olivocerebellar tract	Fibres to reticular formation of medulla

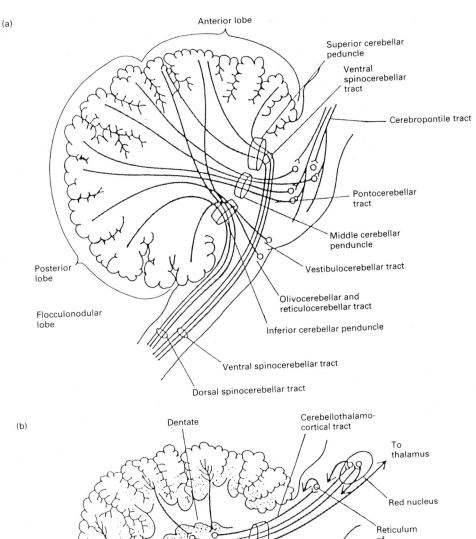

Fig. 17.66 (a) The main afferent tracts to the cerebellum. (b) The main efferent tracts from the cerebellum. (Reproduced with permission from Guyton, A.C. (1987) *Textbook of Medical Physiology*, 7th edn. Saunders Co, Philadelphia.)

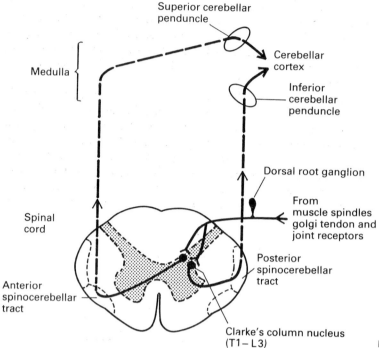

Fig. 17.67 The spinocerebellar tracts.

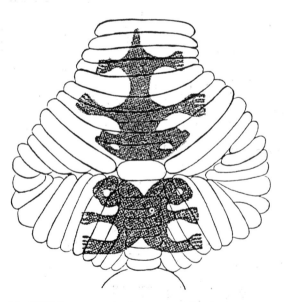

Fig. 17.68 Sensory projection areas of the cerebellar cortex. (Reproduced with permission from Snider, R.S., (1958) *The cerebellum.* © 1958 by Scientific American, Inc. All rights reserved.)

pathways), where the stimulus first originated.

There are various theories as to how voluntary motor function is adjusted, some of which are discussed below:

1 *The braking effect of the cerebellum.* The cerebellum seems to compare the intentions of the planning areas of the motor cortex with the actual performance of the part concerned. In everyday life, and especially during rapid movements, the motor cortex transmits far more impulses than are needed to perform each intended movement. If the precise intended points is not to be surpassed, the cerebellum must act to inhibit the motor cortex at the appropriate time after the muscles have begun to move. The cerebellum is able to perform in this manner by assessing the rate of movement, calculating the length of time needed to reach the intended point and then transmitting inhibitory impulses to the motor cortex to inhibit the agonist and excite the antagonist muscles. Thus, brakes are applied to stop the movement at the precise intended point.

2 *The damping function of the cerebellum.* Essentially, all voluntary movements being performed develop momentum which would cause overshooting of the movements beyond the intended point (i.e. pendular movement). To overcome this, the cerebellum sends appropriate signals which stop the movement at the required point and prevent overshooting.

If the cerebellum is damaged, overshooting occurs. The cerebral cortex recognizes the overshoot and initiates a movement in the opposite direction to bring the moving part, for example an arm, to the intended point. But, due to its momentum, the arm again overshoots and cor-

recting signals are again sent from the cortex. Thus, the arm oscillates beyond the point of intention several times before it settles on the intended point. This forms the basis of the kinetic or intention tremors of the neocerebellar syndrome.

3 *The control of ballistic movements by the cerebellum.* Ballistic movements are rapid movements that are planned to travel to a specific distance and settle precisely at the point of intention. An example of such rapid precise movements is the movements of fingers in typing or piano playing or the movements of eyes while reading. This special function is possible due to the

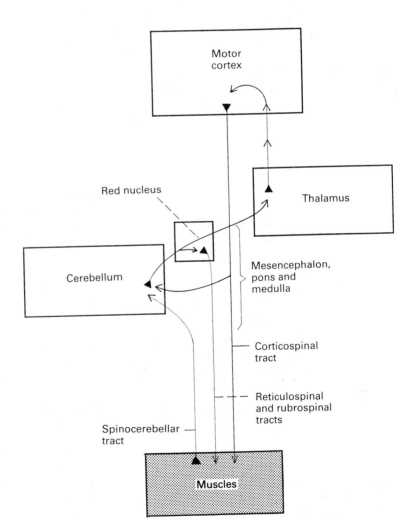

Fig. 17.69 Cerebellar control of voluntary muscular movement. (Reproduced with permission from Guyton, A.C. (1987) *Textbook of Medical Physiology*, 7th edn. Saunders Co, Philadelphia.)

beautifully organized circuitry of the cerebellum, which allows, first, excitatory function, which sends strong excitatory signals to the motor area initiating the motor movement so as to reinforce the onset of ballistic movements. The cerebellum then follows this by inhibitory signals, which stop the ballistic movement by the braking effect.

This function allows the smooth progression of movements required for the performance of rapid movements.

4 *The planning and timing function of the cerebellum.* Recent studies have shown that electrical activity is recorded in the cerebellum before the beginning of voluntary movements, particularly rapid ballistic movements. The lateral zones of the cerebellum receive afferent impulses from the cortical association areas, which are believed to be the site of origin of commands for voluntary movement. It seems that these impulses are transmitted into the cerebellum, which, with other areas of the central nervous system, plans movements. The cerebellum then sends efferent signals to the motor areas of the cerebral cortex (via the thalamus) and movement is initiated via corticospinal tracts and other pathways. Thus, the lateral zones of the cerebellum act as planners for voluntary movements.

Another important function of the lateral cerebellar hemisphere is to provide appropriate timing for each movement. Thus, lesions to these zones lead to inability to judge the distance moved by a particular part in a given time. This, in turn, makes a person unable to control the beginning of the next movement, so that successive movements become either too separate or pile up on each other. Incoordination of rapid movements, in particular, will be very evident. This is known as failure of smooth progression of movement.

Other functions of the cerebellum include:

Function of the cerebellum in involuntary movements The cerebellum co-ordinates involuntary postural movements initiated by the extrapyramidal system, by acting as a comparator (in the same way as in voluntary movement) and correcting errors, so that movements do not overshoot.

Function of the cerebellum in equilibrium The flocculonodular lobe in the cerebellum is the site which plays an important role in maintenance of equilibrium. It is connected by a feedback circuit with the vestibular apparatus. Inputs from the latter are used by the cerebellum for immediately adjusting postural motor signals so that balance is maintained. Damage to the flocculonodular lobe leads to evident disturbance of equilibrium.

Function of the cerebellum in muscle tone The cerebellum receives information from muscle spindles and sends signals to excite the brain stem, which stimulates the gamma efferent neurones. The neocerebellum adds additional support to the stretch reflex, thus facilitating muscle tone, but, since its feedback time is considerably longer than the simple cord stretch reflex, it prolongs the effect. This helps further in the maintenance of posture. On the other hand, the palaeocerebellum is inhibitory to muscle tone, through the inhibitory reticular formation.

Defects produced by cerebellar lesions in humans (the neocerebellar syndrome)
This is due to damage of the deep cerebellar nuclei as well as the cerebellar cortex. The manifestations occur on the same side of the lesion, i.e. a lesion of the left cerebellar hemisphere produces its effects on the left side of the body, etc.

The manifestations include:
1 *Dysmetria and ataxia.* Dysmetria is the inability of the person to judge properly ahead of time the range of his movements. Therefore, the movements usually overshoot their intended mark, i.e. past pointing. Dysmetria results in incoordination of movements, which is called ataxia.
2 *Failure of progression of voluntary movements.* This results in:
  (a) *Dysdiadochokinesia* — the inability to perform rapid successive alternating movements, e.g. repeated supination and pronation of the forearm.
  (b) *Dysarthria* — defective speech. Articulation is jerky and explosive. Syllables of words are separated from one another (scanning speech).
3 *Kinetic or intention tremors.* These are coarse,

rhythmic, involuntary movements, which occur during voluntary movements but disappear at rest and during sleep. They occur due to the absence of damping function of the cerebellum. These intention tremors disappear at rest or during sleep.

4 *Nystagmus.* This is a tremor of the eyeball, which occurs when the patient attempts to fix his gaze on an object to the side of his head (horizontal nystagmus). It is also due to absence of the damping function.

5 *Rebound phenomenon.* This is elicited by asking the patient to flex his forearm against resistance. When the resistance is suddenly removed, the limb overshoots the normal range (i.e. the forearm will fly quickly to strike the patient's face). This is due to the loss of the braking effect of the cerebellum.

6 *Hypotonia.* This is due to the loss of the facilitatory effect of the cerebellum on the stretch reflex. When the knee jerk is elicited, it is reduced and pendular. Sometimes hypotonia is the first sign of cerebellar disease.

7 *Decomposition of movements.* This is the inability to perform successive components of a complex motor act. It is tested by the heel–knee test.

8 *Rough movements.* Delicate movements, such as threading a needle or buttoning of a shirt, become very difficult.

9 *Disturbance of posture and gait.* The head is tilted to the side of the lesion. The gait is unsteady and the patient walks with a wide base in a drunken fashion and tends to fall to the side of the lesion.

CONTROL OF VOLUNTARY MOVEMENT

At this point in the discussion, it seems useful to try to describe the functional organization of the motor system and the control of voluntary movement. Signals initiating voluntary movement start in the cortical association areas (see Fig. 17.70). According to the command, movements are then planned in various areas of the brain, including the motor cortex and also the basal ganglia and the lateral parts of the cerebellar hemispheres, both of which deliver their information to the motor cortex via the thalamus. The planned movements from the motor cortex are

coded in terms of movements rather than individual muscle contractions. Movement is then initiated via the corticospinal tract and other pathways to the spinal and cranial motor neurones, which activate the motor nerves.

When muscle movement occurs, it stimulates the receptors in muscles, tendons, joints and the skin, which send feedback information (via sensory tracts) directly to the motor cortex and to the intermediate portions of the cerebellum. The output from the latter goes both to the motor cortex, to adjust and smooth the movement, and to motor neurones (via brain stem pathways), to adjust posture continuously during movement.

Table 17.8 shows disordered motor function, which may result from a lesion at one of seven different levels.

The thalamus

The thalamus presents the gate for essentially all sensory signals on their way to the cerebral cortex, with the single exception of those from the olfactory system. It is thus the highest subcortical sensory centre.

The thalamus is made up of a number of nuclei, which can be functionally divided into:

1 Nuclei which project to the whole cerebral cortex (i.e. non-specific projection nuclei).

2 Nuclei which project to specific areas of the cerebral cortex and limbic system (i.e. specific projection nuclei).

Table 17.8 Various levels of lesions which can result in disturbed motor function

Level of lesion	Examples of disorders
Muscular level	Familial periodic paralysis
Neuromuscular junction	Myasthenia gravis
Lower motor neurone	Poliomyelitis
Cerebellum	Medulloblastoma
Extrapyramidal system	Parkinsonism
Pyramidal system (upper motor neurone)	Cerebrovascular accident in the internal capsule
Cerebral cortex	Apraxia (i.e. inability to carry out a movement although the patient understands its purpose and his muscles are not paralysed)

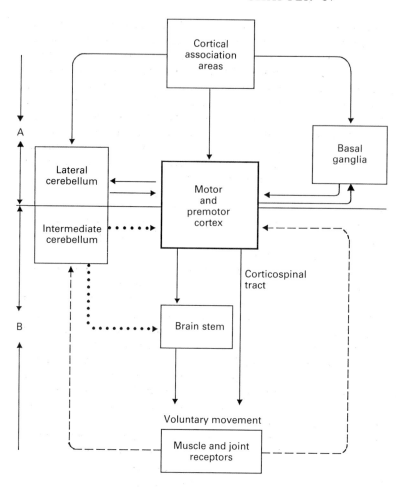

Fig. 17.70 Planning and execution of voluntary movement. A: represents parts concerned with planning voluntary movement; B: represents parts concerned with execution of voluntary movement. Solid lines indicate tracts concerned with motor information, the broken lines indicate sensory information and the dotted lines indicate tracts carrying corrective signals.

NON-SPECIFIC THALAMIC NUCLEI

The non-specific projection nuclei constitute the midline and intralaminar nuclei. They receive signals from the reticular activating system and project to all areas of the cerebral cortex.

SPECIFIC THALAMIC NUCLEI

These specific nuclei receive somatosensory, visual and auditory information from spinal cord neurones and brain stem nuclei. These specific sensory modalities are projected to specific sensory areas of the cerebral cortex. The postero-ventral (or ventrobasal) nuclei relay somatosensory inputs while the medial and lateral geniculate bodies are concerned with auditory and visual signals respectively.

In addition to the above other nuclei receive information from the hypothalamus, the cerebellum and the basal ganglia. The hypothalamic inputs are relayed to the limbic areas of the cortex while those from the basal ganglia and cerebellum are relayed to the motor cortex.

FUNCTIONS OF THE THALAMUS

The thalamus is a relay station for all types of sensations—fine and crude somatic sensations, pain, temperature, visual and auditory sensations—on their way to their respective centres in the cerebral cortex. Besides relaying, the thalamus is a major centre in perceiving pain, and its intralaminar nuclei may be one of several important sites at which the analgesic system of the brain can inhibit pain transmission. As far as motor function is concerned, the thalamus also relays signals from the basal ganglia and cerebellum to the motor areas of the cerebral cortex. Moreover, it relays autonomic signals and others related to emotional reactions, due to its connections with the hypothalamus and limbic system respectively. The thalamus may also play an important role in coding, storing and recalling memories.

THE THALAMIC SYNDROME

The thalamic syndrome arises due to thrombosis of the arterial supply of the thalamus, with consequent damage to the posterior thalamic nuclei, leaving the medial and anterior group of nuclei intact.

The patient presents with loss of almost all sensations on one side of the body and *ataxia* due to damage of the relay nuclei. The pain threshold is raised, but when reached the pain produced is prolonged, severe and very unpleasant. This is an example of secondary hyperalgesia which arises due to the increased sensitivity of the medial thalamic nuclei to pain impulses arising along the spinorecticular pathway. The patient also suffers emotional disturbances. Partial recovery of sensations might occur after some time.

The hypothalamus

The hypothalamus is situated at the base of the brain, just rostral of the brain stem. It is an important centre for the integration of visceral reflexes, so as to maintain the constancy of the internal environment.

CONTROL CENTRES OF THE HYPOTHALAMUS

Figure 17.71 shows division of the hypothalamus into various centres, each of which constitutes a number of nuclei or areas. The anterior centres include the supraoptic and paraventricular nuclei, the medial and posterior preoptic area and the anterior hypothalamic area. Medial centres include the dorsomedial and ventromedial nuclei. Lateral centres constitute the lateral hypothalamic area and the posterior centres include the posterior hypothalamus, the perifornical nucleus, the mammillary body and the arcuate and periventricular zone.

For the hypothalamic centres to be able to perform their function of homoeostasis, numerous to-and-fro connections between it and various areas of the brain must exist. The hypothalamus receives afferents from the preolfactory area, the limbic lobe, the amygdala, the hippocampus, the non-specific thalamic nuclei and the globus pallidus, and collaterals from the sensory ascending tracts. It sends efferents to the neurohypophysis, the anterior thalamic nuclei and the brain stem reticular formation.

FUNCTIONS OF THE HYPOTHALAMUS

Many visceral, behavioural and emotional reactions are reflexly regulated by the hypothalamus:

1 *Autonomic functions.* Almost all autonomic functions are modified by the hypothalamus. Previously, the hypothalamus was called 'the head ganglion of the autonomic nervous system', which may not be strictly true as far as in the control of autonomic functions is concerned.

Stimulation of the anterior hypothalamic centres produces parasympathetic effects, while stimulation of the posterior dorsomedial and lateral hypothalamic centres produces sympathetic effects.

2 *Temperature regulation.* Homothermic animals subjected to transection just below the hypothalamus become poikilothermic. The reflex responses activated by cold (e.g. cutaneous vasoconstriction, shivering) are controlled by the posterior hypothalamus. Heat-losing mechanisms (e.g. cutaneous vasodilatation, sweating) are controlled by the anterior hypothalamus.

3 *Body water regulation.* Electrical stimulation of particular areas — the drinking centre — in the hypothalamus of an animal makes it drink, whereas its destruction abolishes the sensation of thirst. Injection of hypertonic saline into the hypothalamus causes polydipsia. The thirst centre is located in the supraoptic region of the hypothalamus.

4 *Food intake regulation.* This is regulated by a balance between two centres. The *feeding centre* is located in the lateral hypothalamus and, when stimulated, it leads to extreme hunger, a voracious appetite and an intense desire to search for food. This eventually leads to obesity. However, dam-

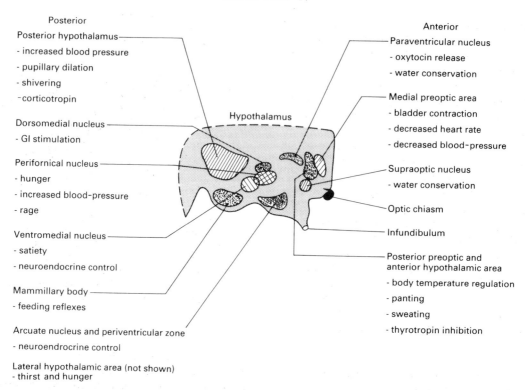

Posterior

Posterior hypothalamus
- increased blood pressure
- pupillary dilation
- shivering
- corticotropin

Dorsomedial nucleus
- GI stimulation

Perifornical nucleus
- hunger
- increased blood-pressure
- rage

Ventromedial nucleus
- satiety
- neuroendocrine control

Mammillary body
- feeding reflexes

Arcuate nucleus and periventricular zone
- neuroendrocrine control

Lateral hypothalamic area (not shown)
- thirst and hunger

Hypothalamus

Anterior

Paraventricular nucleus
- oxytocin release
- water conservation

Medial preoptic area
- bladder contraction
- decreased heart rate
- decreased blood-pressure

Supraoptic nucleus
- water conservation

Optic chiasm

Infundibulum

Posterior preoptic and anterior hypothalamic area
- body temperature regulation
- panting
- sweating
- thyrotropin inhibition

Fig. 17.71 Hypothalamic control centres and functions. (Reproduced with permission from Guyton, A.C. (1987) *Textbook of Medical Physiology*, 7th edn. Saunders Co, Philadelphia.)

age to this area produces severe loss of appetite, known as *anorexia*. Another centre, called the *satiety centre*, is located in the ventromedial nucleus and, when stimulated, anorexia occurs. Damage to this area causes the hunger centre to become overactive, leading to hyperphagia and obesity, known as hypothalamic obesity.

5 *Control of sexual functions.* The hypothalamus, through its influence on the anterior pituitary, regulates sexual function. Oestrus, maternal behaviour and care of the young in lower animals are under direct hypothalamic control.

6 *Defence reactions.* As a component of the limbic system, the hypothalamus is part of the effector mechanism of emotional expression. It is involved in the expression of rage and fear.

7 *Wakefulness and sleep.* The hypothalamus is part of the reticular activating system, which is indispensable to the initiation and maintenance of the alert wakefulness of the healthy animal. Also, stimulation of the mediorostral suprachiasmal portion of the anterior hypothalamus is found to promote sleep, and lesions in the same area can sometimes cause such intense wakefulness that the animal actually dies of exhaustion.

8 *Relation of cyclic phenomena.* Lesions of the hypothalamus disrupt the circadian rhythm in the secretion of ACTH and melatonin. It therefore appears that the hypothalamus adjusts bodily rhythms to the 24-hour light–dark cycle.

9 *Behavioural functions.* The hypothalamus is part of the limbic system, which is concerned with the affective nature of sensations. Stimulation of certain regions of the hypothalamus pleases the animal whereas stimulation of other parts causes terror, pain, fear and other types of aversion. However, the prefrontal lobe overrides the hypothalamus and limbic system in determining behaviour, including sexual, feeding and emotional reactions.

10 *Neuroendocrine functions.* The hypothalamus controls the posterior pituitary by a neural mechanism and the anterior pituitary by a humoral mechanism.

Antidiuretic hormone and oxytocin are produced in the supraoptic and paraventricular nuclei and transported to the posterior pituitary along the axons of the hypothalamohypophyseal tracts. The hypothalamus also produces at least nine releasing hormones, which are carried to the anterior pituitary by the hypothalamohypophyseal portal system. Most of these release anterior pituitary hormones, e.g. corticotrophin-releasing hormone (CRH) and growth hormone-releasing hormone (GHRH), while others inhibit the release of anterior pituitary hormones, e.g. prolactin inhibitory hormone (PIH) and somatostatin.

Arousal mechanisms, sleep, wakefulness and the electrical activity of the brain

THE RETICULAR ACTIVATING SYSTEM (RAS)
The RAS is a multisynaptic pathway located within the reticular formation of the brain stem (Fig. 17.72). It receives collaterals from the classical sensory pathways as they ascend to their respective relay nuclei in the thalamus; and sends ascending branches to all areas of the cerebral cortex. Some of their branches pass directly but most of them synapse in the non-specific nuclei of the thalamus (i.e. intralaminar and midline nuclei) before they pass to the cortex.

The RAS is activated to an equal degree by impulses from all the classical sensory pathways. This led to the use of the term 'non-specific or diffuse system' to distinguish it from the classifical sensory pathways which are specific in the sense that they are activated by only one sensory modality. While impulses arriving along sensory pathways determine activity in the RAS, the RAS determines the overall activity of the cerebral cortex.

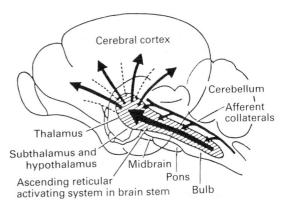

Fig. 17.72 Diagrammatic representation of ascending reticular system projected to the cerebral cortex in the cat brain. (Reproduced from Starzl, T.E., Taylor, C.W. & Magoun, H.M. (1951) Collateral afferent excitation of the reticular formation of the brain stem, *J. Neurophysiol.*, 14, 479, with permission from the American Physiological Society, Bethesda.)

EVOKED CORTICAL RESPONSES
Electrical activity produced in the cortex in response to stimulation of a receptor can be monitored by electrodes placed on the surface of the cerebral cortex of an anaethetized animal. The response consists of two positive waves separated by a negative wave (Fig. 17.73). The first positive wave is called the *primary evoked potential (EP)*. It has a short latency of 5–12 ms and is only recorded from a specific localized area of the cortex (i.e. if the hand is stimulated, the wave is observed over the 'hand area' of the sensory cortex). The primary evoked potential is followed by a small negative wave due to hyperpolarization.

The second positive wave is called the *diffuse secondary response*. It is a large prolonged positive wave. It has a longer latent period of 20–80 ms. It is wide spread over the cortex and other brain structures. This wave is probably evoked by stimulation of the non-specific thalamic nuclei. Its longer latency is due to the slower passage of impulses through multiple synapses in the RAS and also to the appearance of wide spread reverberations in cortical pathways.

In clinical practice evoked responses can be recorded over the brain or the spinal cord while

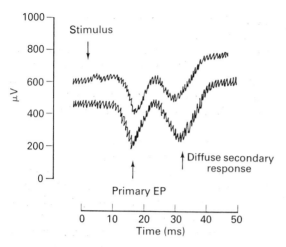

Fig. 17.73 Stimulation (arrow) of median nerve giving evoked responses in contralateral sensory cortex. Upward wave is surface-negative. EP, evoked potential.

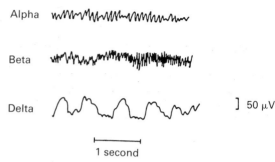

Fig. 17.74 Normal electroencephalogram. (Reproduced with permission from Guyton, A.C. (1987) *Textbook of Medical Physiology*, 7th edn. Saunders Co, Philadelphia.)

applying repeated stimuli (100–1000 times) to a specific receptor area. The recording is made possible by a computerized averaging technique which summates the EPs and cancels interference. When somatosensory, auditory or visual stimuli are applied the evoked potentials are known as (SSEP), (AEP) or (VEP) respectively. This technique, especially VEP, has proved to be useful in the detection, localization and follow-up of lesions in the sensory pathways being investigated.

THE ELECTROENCEPHALOGRAM (EEG)

The EEG is a record of the electrical activity of the brain. It can be recorded by applying electrodes on the scalp in patients or by placing them on the surface of the brain in experimental animals. The electrical activity appears on a multichannel recorder as waves of variable intensity (0–200 µV) and frequency (1–50 Hz).

Types of normal EEG waves

There are four types of normal EEG waves (Fig. 17.74):

1 *Alpha waves.* These are rhythmic waves occurring at a frequency of 8–13/s (8–13 Hz) with an amplitude of about 50 µV. These are most marked in the parieto-occipital region but they are sometimes observed in other locations. Alpha waves are recorded in a mentally and physically relaxed person who is awake but with the eyes closed.

2 *Beta waves.* These are seen in the frontal regions. They are of lower voltage than alpha waves but are faster, i.e. frequency of about 18–30 Hz. They occur during intense activation of the central nervous system or during tension.

3 *Theta waves.* These are large 100 µV waves, having a frequency of 4–7 Hz. These occur in the parietal and temporal regions of normal children and in adults under emotional stress.

4 *Delta waves.* These are the slowest $(1-3\frac{1}{2}$ Hz, of high voltage). They occur in infancy and in deep sleep in both children and adults.

When theta and delta waves occur in an adult during the waking state, it always indicates abnormality. Overbreathing or hyperventilation induce theta and delta waves bilaterally.

Desynchronization or alpha block

The intensity (i.e. voltage) of brain waves is determined mainly by the number of neurones that discharge synchronously. Random nonsynchronous firing of whatever number of neurones, however strongly they may discharge, will cancel each other and produce waves of weak voltage. Frequency of brain waves, on the other hand, is determined by the total level of cerebral activity. This can be clearly seen in Fig. 17.75.

Eyes open   Eyes closed

Fig. 17.75 Alpha rhythm replaced by asynchronous discharge on opening of the eyes. (Reproduced with permission from Guyton, A.C. (1987) *Textbook of Medical Physiology*, 7th edn. Saunders Co, Philadelphia.)

While the eyes are closed, synchronous discharge of many cerebral neurones produces alpha waves, but when the eyes are opened, faster low voltage irregular beta waves are recorded. The higher frequency is due to the increased brain activity (enhanced by visual stimuli), and the lower voltage is due to desynchronization of the cerebral neural discharge which causes the disappearance of the alpha waves. This is known as 'alpha block'.

FACTORS AFFECTING THE ACTIVITY OF THE RAS

The activity of the RAS is increased by impulses along collaterals from the specific sensory pathways. Proprioceptive and painful stimuli are particularly effective and can arouse a person from sleep. Descending impulses from the temporal lobe and certain areas of the frontal lobe of the cerebral cortex have a strong excitatory effect on RAS. Such impulses may be responsible for the alerting responses due to emotions and related psychic phenomena. Another important region in the cerebral cortex that activates the RAS is the motor cortex. This is probably the cause why voluntary movements help in keeping a person awake. Epinephrine and norepinephrine secreted from the adrenal medulla produce alerting responses. The effect is probably mediated by blood-pressure, because any increase in blood-pressure increases the excitability of the reticular formation.

Activity in the RAS can be diminished (or inhibited) by many factors, which include:
1 Impulses from the sleep-producing centres of the reticular formation.
2 Various lesions that damage cells of the brain stem, e.g. vascular lesions, poisons, brain tumours, prolonged hypoxia and infectious diseases.

These disorders can actually lead to coma. Coma differs from sleep in that a person cannot be aroused from coma. In some comatosed patients, other areas in the brain are inactivated too. In such cases, the electrical activity of the brain stops and the brain waves are said to be 'flat'. This is the condition known as brain death.

CLINICAL USES OF THE EEG
The EEG can be used clinically in:
1 *Localizing brain tumours.* If the tumour is large enough to block electrical activity from a given portion of the cerebral cortex, the voltage of the brain waves in the region of the tumour will be considerably reduced. On the other hand, the presence of a tumour can lead to very high-voltage waves in the EEG. The latter is more frequent and results from abnormal electrical excitation of the compressed areas surrounding the tumour.
2 *Diagnosis of different types of epilepsy.* Epilepsy is a condition characterized by uncontrolled excessive activity of either part or all of the nervous system. It can be divided into two major types:
(a) Grand mal *epilepsy.* In this type there is characteristically increased neuronal discharge from all areas of the brain and the lower reticular activating system. It is characterized by an aura, i.e. sensory hallucinations, followed by generalized tonic convulsions of the entire body, leading to clonic jerks. During the attack the person may bite his tongue, has difficulty in breathing and may also urinate or defecate as the extreme signals reach the viscera. *Grand mal* seizures last from a few seconds to as long as 3−4 minutes and are followed by depression of the whole nervous system, during which the person remains in a state of stupor. The EEG shows characteristic high-voltage spikes, each followed by a slow wave. This pattern is seen all over the brain (Fig. 17.76).
(b) Petit mal *epilepsy.* This type is characterized by 3−30 s of unconsciousness, during which the patient may show several twitch-like contractions of muscles, usually in the head region, i.e. muscles of the corner of the eye, blinking. This is followed by return of consciousness or resumption of previous ac-

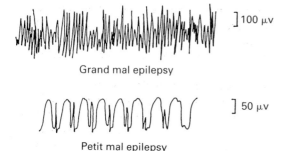

Grand mal epilepsy   ] 100 μv

] 50 μv

Petit mal epilepsy

Fig. 17.76 Electroencephalograms in epilepsy.
(Reproduced with permission from Guyton, A.C. (1987)
Textbook of Medical Physiology, 7th edn. Saunders
Co, Philadelphia.)

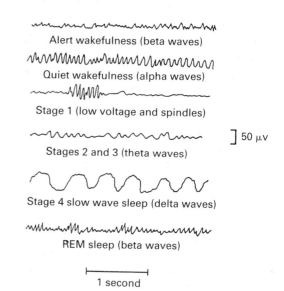

Alert wakefulness (beta waves)

Quiet wakefulness (alpha waves)

Stage 1 (low voltage and spindles)

] 50 μv

Stages 2 and 3 (theta waves)

Stage 4 slow wave sleep (delta waves)

REM sleep (beta waves)

|————————|
1 second

Fig. 17.77 Electroencephalogram during the different
stages of sleep. (Reproduced with permission from
Guyton, A.C. (1987) *Textbook of Medical Physiology*,
7th edn. Saunders Co, Philadelphia.)

tivity. *Petit mal* attacks may initiate *grand mal*
attacks and they too almost certainly involve
the reticular activating system. The EEG during
the attack shows a characteristic spike and
dome pattern all over the brain (3 Hz).
3 *Diagnosis of psychopathic disturbances.* This
needs an experienced electroencephalographer
and is useful in only certain types of psychopathic
disturbances.

SLEEP

Sleep is a state of loss of consciousness from
which a subject can be aroused by appropriate
stimuli. As a person falls asleep, different stages
can be identified from an EEG recording (Fig.
17.77):
1 Alert wakefulness, characterized by high-
frequency beta waves.
2 Quiet wakefulness, in which the person is
relaxed and quiet with the eyes closed and which
is associated with alpha waves.
3 Light sleep (stage 1), which is characterized by
low-voltage waves, broken by sleep spindles, i.e.
bursts of alpha waves (12−14 Hz).
4 Slow-wave sleep (stages 2, 3 and 4), which
is characterized by very low-frequency waves
(2−3 Hz), i.e. delta waves.

Types of sleep

During each night a person goes through stages
of two different types of sleep, which alternate
with each other. These are called:

1 *Slow-wave sleep.* This type is exceedingly restful and is associated with a decrease in peripheral
vascular tone, a 10−30% decrease in blood-
pressure and a decrease in respiratory rate and
basal metabolic rate. The EEG pattern is typically
theta and delta waves. Although slow-wave sleep
is often called dreamless sleep, dreams do occur.
But, unlike the other type of sleep, such dreams
are not remembered, because they are not consolidated in memory.
2 *Rapid eye movement (REM) sleep.* This type
occurs in episodes of 5−30 min, which recur
about every 90 min. Tiredness shortens the duration of each episode, but, as the person becomes
restful through the night, the duration greatly
increases.

REM sleep is associated with active dreaming,
which is remembered later. During this type of
sleep, muscle tone is extremely depressed, due to
excitation of the reticular inhibitory centres, the
heart and respiratory rates are irregular, there are
rapid movements of the eye and the person is
more difficult to awaken than during the slow-
wave sleep. The EEG pattern in that period shows

beta waves indicating a high level of activity in the brain during this period of sleep. Thus, this type is often known as paradoxical sleep, because it is a paradox that a person can still be asleep despite the marked activity in the brain.

Mechanism of sleep

Sleep is most probably caused by an active process, in which centres located below the midpons and probably the hypothalamus actively cause sleep by inhibiting other parts in the brain. These centres include:

1 *The raphe nuclei.* The neurones of the raphe nuclei secrete serotonin at their endings in the reticular formation, spinal cord and other areas. Serotonin was assumed to be the major transmitter associated with the production of slow-wave sleep for a long time. This was based on animal experiments, in which drugs causing selective depletion of brain serotonin produced prolonged wakefulness. However, wakefulness was not produced in humans but, instead, serotoninergic neurones were found to discharge rapidly in the awake state, more slowly, with bursts, during sleep and not at all during REM sleep. Moreover, in humans serotonin agonists appear to suppress sleep, while its antagonist ritansiren was found to increase slow-wave sleep. It thus appears that the relation of serotonin to sleep is still debatable.

New evidence suggests that adenosine is a causal factor of sleep and that methylxanthines, such as caffeine, result in wakefulness by blocking adenosine receptors. Recently, the presence of a sleep peptide has been described; however, its chemistry and possible physiological role are still uncertain.

2 *The rostral part of the hypothalamus and an area in the diffuse nuclei of the thalamus.* These are areas which also promote sleep when stimulated. Indeed, lesions of the anterior hypothalamus can sometimes cause such intense wakefulness that the animal may actually die of exhaustion.

3 *The locus ceruleus on each side of the brain stem.* This area has neurones that secrete norepinephrine at their nerve endings when stimulated. It is believed that this causes excess activity in certain regions of the brain during REM sleep.

The reason why such signals do not cause wakefulness is that they do not follow the appropriate channels for that to occur. Lesions in the locus ceruleus can reduce REM sleep or even prevent it.

Behaviour and the limbic system

Behaviour is a broad term, involving a number of different functions. However, the limbic system is responsible for those special types of behaviour associated with emotions, subconscious motor and sensory drives and the feelings of punishment, pleasure or reward.

Anatomically, the limbic system consists of a ring of cortex on the medial side of each hemisphere which surrounds the corpus callosum, known as the allocortex, and another ring, the juxtacortex, lying between the allocortex and the non-limbic, well-developed, cortical tissue, the neocortex. Subcortical nuclei of the limbic system include the septum, the paraolfactory area, the epithalamus, the anterior nuclei of the thalamus, portions of the basal ganglia, the hippocampus and the amygdala (Fig. 17.78). From a physiological point of view, the hypothalamus is one of the central elements of the limbic system. The reticular formation and allied regions of the brain stem also function in close association.

There are important routes of communication between the components of the limbic system and also between the limbic system and the neocortex. The limbic system circuits are characterized by prolonged after-discharge. This probably explains the fact that emotional responses are generally prolonged and continue long after the stimuli that initiate them.

FUNCTIONS OF THE LIMBIC SYSTEM

Olfaction

The limbic system was earlier called the rhinencephalon, due to its role in olfaction.

Autonomic responses and feeding behaviour

The limbic system produces the autonomic responses that accompany emotional states and behaviours. It is also associated with activities related to feeding such as movements of the mouth. Lesions of the amygdala cause overeating that can involve all kinds of food.

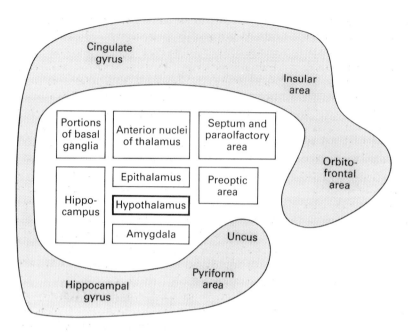

Fig. 17.78 Diagrammatic representation of the different components of the limbic system. (Reproduced with permission from Gayton, A.C. (1987) *Textbook of Medical Physiology*, 7th edn. Saunders Co, Philadelphia.)

Sexual behaviour

Sexual intercourse, or coitus, is basically reflexive in nature, but the component of behaviours that lead to it are regulated by the limbic system and hypothalamus. In animals, sexual behaviour depends to a great extent on the gonads and their hormonal output. Human sexual behaviour, however, is more complex and is far from being purely instinctual or hormonal. The complexity is due to the modification of sexual functions and behaviours by higher cerebral functions such as learning and by social and psychic factors. Damage of the limbic system would be expected to affect sexual behaviour.

Fear and rage

Fear and rage are two related phenomena which are instinctive and protective against threats of the environment. Fear, often called the fleeing or avoidance reaction, occurs when an animal is threatened, and rage, often called the fighting or attack reaction, occurs when the animal is threatened and cornered.

Besides the mental component, fear has external manifestations, such as sweating, pupillary dilatation, tachycardia, dryness of the mouth and escape reactions. The fear reaction can be pro-duced by stimulation of the hypothalamus and amygdaloid nuclei and can disappear in lesions of the amygdala.

Normality in animals and humans is attained by a balance between *rage* and its opposite *placidity*. However, some lesions may tip the balance. When this occurs, trivial stimuli provoke violent episodes of rage, or the reverse, i.e. the most provoking stimuli do not disturb the animal's placidity. Either state would obviously depend on the site of the lesion. The rage reaction can be produced by removal of the neocortex or after lesions involving the ventromedial hypothalamic nuclei and septal nuclei while the neocortex is left intact. Placidity, on the other hand, can be produced after bilateral destruction of the amygdaloid nuclei.

Motivation

The limbic system also contains some areas whose stimulation gives a *rewarding* sensation and other areas whose stimulation gives the sense of *punishment* or avoidance. The former include points located in a medial band of tissue passing from the frontal cortex through the hypothalamus to the midbrain tegmentum. The greatest sense of reward is obtained from stimulation of points in the medial forebrain bundle, the tegmentum and the septal nuclei. The points where stimulation gives a sensation of punishment include the lateral portion of the posterior hypothalamus, the dorsal midbrain and the entorhinal cortex.

The sensations of reward and punishment are important in motivating our behaviour, i.e. we continue to do things that are rewarding and avoid those that result in punishment. They also play an equally important role in the learning process, e.g. a sensory experience causing neither reward nor punishment is not remembered very well because the animal or human being gets habituated to the stimulus. However, if the stimulus causes either reward or punishment, the cortical response becomes more intense (i.e. reinforced).

Administration of some tranquillizers, such as chlorpromazine, inhibits both the reward and punishment centres, whereby the affective reactivity of the subject is greatly reduced. It is thus presumed that tranquillizers function in psychotic states by suppressing many of the behavioural areas in the central nervous system.

Higher functions of the nervous system

MEMORY
Memory is concerned with the storage and recall of information at the conscious or unconscious level. Formation and recall of a memory involve at least four stages:
1 Reception of the information.
2 Formation of a memory trace.
3 Consolidation of the memory trace.
4 Recall of the memory trace.

Information is received and interpreted in the sensory areas of the cerebral cortex, but the location of the memory trace in the nervous system is not very clear. However, it is very clear from experimental and clinical evidence that conscious memory involves the cerebral cortex. Penfield reported that stimulation of portions of the temporal lobes in humans evokes detailed memories of events that occurred in the remote past. Also, brain concussion or electroshock therapy is frequently accompanied by loss of memory for events immediately preceding the event.

Consolidation of the memory trace is the pro-

cess that encodes the memory and makes it remarkably resistant to erasing. Before this process and during it, the memory trace is vulnerable, and, following it, there is a stable, resistant, memory engram. Experiments in which rats were taught to run a maze showed that, if given a large shock soon after running the maze, they forgot what they had learnt. If the shock was given a few hours later, the memory was not affected. The process of consolidation involves the hippocampus and its connections. Damage in this region causes defects in recent memory and humans with such lesions have intact remote memory. Also, recent memory loss in alcoholic patients correlates with the presence of pathological changes in the mammillary bodies, which have afferent and efferent connections to the hippocampus.

Loss of memory, known as amnesia, can occur following injuries. The inability to remember events which happened before the injury is called *retrograde amnesia*, while the inability to store in the memory events that happened after it is called *antegrade amnesia*.

Types of memory
There are three types of memory:
1 *Sensory memory*. This is the ability to retain sensory signals in the sensory areas of the brain for an extremely short period of time (several hundred milliseconds) following the actual sensory experience. This can then be used for further processing, i.e. it is scanned to pick out important points. Sensory memory forms the initial stage of the memory process.
2 *Primary memory*. This is the memory of a few facts, words, numbers, letters or other information for a few seconds to a few minutes. This type of memory is limited to about seven bits of information (e.g. telephone numbers, cars, recall of a visual scene, etc.). It is characterized by the fact that information in this memory store is instantaneously available, i.e. a person does not have to search through his mind for it. It is also clear that old information in the primary store is easily displaced when new bits of information are placed in it, e.g. a newly learnt telephone number is forgotten if another number is learnt.
3 *Secondary memory*. This is the storage in the

brain of information that can be recalled at a later time – hours, days, months or even years later. This type is also known as long-term, fixed or permanent memory.

Information in this memory store is not instantaneously available, except when the memory is deeply ingrained. Memory can be ingrained deeply by the sudden impact of important events (e.g. death of a close relative) or by repeated learning (e.g. days of the week, number and times of prayers).

Mechanisms of memory

Mechanisms of primary memory Primary memory requires a neuronal arrangement that can hold specific information for a short time (a few seconds to a few minutes). Several possible mechanisms have been proposed.

1 *Reverberating circuit theory.* According to this theory, it is proposed that sensory signals, when they arrive at the cerebral cortex, can set up reverberating oscillations for some time. Reverberating circuits form the most important mechanism of after-discharge (i.e. a signal causes a prolonged discharge even when the incoming signal is over). These are positive feedback pathways, where the output of one neurone feeds back via interneurones for a long time. These circuits are thought to be the basis for primary memory.

As the reverberating circuit fatigues or as new signals interfere with the reverberations, the primary memory fades away.

2 *Post-tetanic facilitation.* It is known that tetanic stimulation of a synapse for a few seconds causes a short period of synaptic fatigue, followed by a period (a few seconds to a few hours) of increased excitability of the same synapse. If during this time the synapse is stimulated again, the neurone responds much more vigorously than normal. This is known as post-tetanic facilitation and is obviously a type of memory that depends on change of excitability of synapses.

Information in the primary memory store is stored as a continuous electrical activity. Thus, this type can easily be lost by events that inactivate such electrical activity, e.g. accidents, electroconvulsive therapy.

Mechanisms of secondary memory This type of memory is resistant to electroshock, general anaesthesia, etc., which suggests that memory might be stored as a biochemical change in the neurones. There is evidence from animal studies that protein synthesis is involved in some way in the process responsible for memory. Changes in protein could result in the release of more transmitter with each impulse or, alternatively, it could make more postsynaptic receptors available for a given transmitter. However, these speculations do not allow us to draw an understandable conclusion on the relationship between protein synthesis and the formation of secondary memory.

LEARNING

Learning is defined as the alteration of behaviour or modification of innate responses by experience and training. It involves many levels; the most advanced types of learning are largely cortical phenomena but the brain stem and even the spinal cord are also involved in these processes.

Conditioned reflexes are an important type of learning. A conditioned reflex is acquired by repeated association of two stimuli: one which produces an inborn reflex and one which does not. After some time the latter stimulus alone (the conditioned stimulus) will produce the inborn response. In the classical experiments of Pavlov, placing meat in the dog's mouth (unconditioned stimulus) produced salivation (unconditioned response). But, when the unconditioned stimulus was immediately preceded by the ringing of a bell (conditioned stimulus) on several different occasions, the animal would eventually salivate when the bell was rung, without providing any meat. It should be noted that a vast number of somatic, visceral and other neural changes can be made to occur as conditioned reflex responses. The essential feature of a conditioned reflex is what appears to be the formation of a new functional connection in the nervous system (e.g. in Pavlov's experiment, a connection was made between the bell ringing and salivation; in other words, between the auditory pathway and the autonomic centres responsible for salivation).

The limbic system is very much involved in

the process of learning. Events with a strong emotional content (i.e. pleasant or unpleasant) are most easily remembered, and damage to the hippocampus — one important component of the limbic system — seriously impairs the learning ability.

In operant conditioning, an animal learns to do something in order to get a reward or avoid punishment. The conditioned stimulus is a signal that alerts the animal to perform the task, and the unconditioned stimulus is the rewarding event. For example, a bell rings and the animal presses a lever and is rewarded by getting food. With repetition, this will condition the animal to perform the task whenever the stimulus (e.g. bell ringing) is presented. Alternatively, the stimulus may precede the punishing event, e.g. a bell rings and the animal gets an electric shock. In time, the animal is conditioned to take an avoiding reaction, such as jumping away, whenever the stimulus (e.g. bell ringing) is presented.

SPEECH AND LANGUAGE

Speech is the highest function of the nervous system. Basically, it involves the understanding of spoken and printed words, besides the ability to express ideas in speech and writing.

Various areas in the neocortex subserve the complex function of speech. The primary visual cortex in the occipital lobe and the primary auditory cortex in the temporal lobe enable a person to see visual images (e.g. written words) and to hear spoken words, respectively. Further analysis and interpretation of the incoming visual and auditory sensations are the function of the sensory association areas. Thus, damage of the visual association area results in a learning disability called word blindness or dyslexia, in which there is an impaired ability to learn to read, due to failure to recognize the meaning of written words. Alternatively, lesions in the auditory association area result in word deafness. In such a condition, the person fails to understand sounds or spoken words, even though they are heard.

A region in the posterior part of the superior temporal lobe, called Wernicke's area, is an area of confluence where the somatic, visual and auditory association areas meet. This is also known as the general interpretative area, suggest-ing its global importance in the higher levels of brain function, i.e. cerebration. Wernicke's area is usually much more highly developed in one cerebral hemisphere than the other. There it plays the greatest role in the interpretation of auditory and visual information to form a thought that is expressed. Lesions in Wernicke's area allow a person to hear different words perfectly but make him unable to arrange these words into a coherent thought. Likewise, the person is capable of reading but cannot understand the ideas conveyed by the read words.

Immediately behind Wernicke's area is the angular gyrus, which fuses posteriorly into the visual cortex. This region seems to be concerned with the interpretation of information obtained from words that are read or obtained from other visual scenes. Damage to this area interrupts the flow of visual experience into Wernicke's area from the visual cortex and results in dyslexia.

The dominant hemisphere

It is well established that language function depends more on one cerebral hemisphere than the other — a fact which led scientists to call the hemisphere that is concerned with categorization and verbal symbolization the dominant hemisphere. This concept of dominance has now been replaced by the concept of hemispheric specialization, i.e. each of the two hemispheres specializes in a particular line of function. That hemisphere for language functions and analytic process is called instead the categorical hemisphere, while the other hemisphere, which is specialized in spatiotemporal relations (i.e. recognition of faces, identification of objects by form, understanding, and interpretation of music) is called the representational hemisphere.

In 95% of people, the categorical hemisphere is the left hemisphere. Hemispheric specialization is also related to handedness. Right-handed individuals constitute 91% of the population and in these the categorical hemisphere is the left hemisphere. In 15% of left-handed individuals, the right hemisphere is the categorical hemisphere and in another 15% lateralization is not obvious. In the remaining 70% of left-handed individuals, the categorical hemisphere is the left hemisphere, as in right-handed persons.

Language disorders

It is important to emphasize again the importance of Wernicke's area as the centre in the 'mind' for interpreting sensory experiences, for formation of thought in response to that interpretation and for the choice of words to express thoughts. Wernicke's area projects via the arcuate fasciculus to Broca's area (area 44) of the categorical hemisphere (if speech is involved) or to the area of hand skills in premotor area 6 (if writing is involved). Broca's area processes the information received from Wernicke's area into a co-ordinated pattern of vocalization and then projects that pattern to the motor cortex. The latter initiates movement of the muscles of speech in the tongue, lips and larynx, via the corticobulbar fibres of the pyramidal tract. If writing is concerned, the information received from Wernicke's area is processed in the area of hand skills. The result is a co-ordinated pattern of muscle movements projected to the arm and hand region of the motor cortex, which initiates the necessary muscle movements in the hand and arm required for writing a particular word.

Aphasias are abnormalities of language functions due to injury of language centres in the cerebral cortex. Impairment may affect the comprehension or expression of words. This usually follows thrombosis or embolism of cerebral vessels. Two major and many minor forms of aphasia have been described:

1 *Motor or Broca's aphasia.* In this type, the lesion is in Broca's area. The patient understands spoken and written words but has trouble in speech and finds writing difficult or impossible. The speech is poorly articulated, produced slowly and with great effort, and abnormal in rhythm. In severe cases, the patient may be limited to two to three words with which to express a whole range of emotions. For these reasons it is sometimes termed a non-fluent aphasia.

2 *Sensory or Wernicke's aphasia.* The lesion in this type is in Wernicke's area or in the arcuate fasciculus joining Wernicke's area with Broca's area. This type is different from the motor type in that the patient's comprehension is impaired. He tends to lose almost all intellectual functions

Table 17.9 Some characteristic forms of naming errors given by patients with different types of aphasia when the stimulus is a picture of a chair*

Naming error	Type of aphasia	Site of lesion
'Tssair'	Motor	Broca's area
'Stool' or 'chossl'	Sensory	Wernicke's area
'I know what that is... I have some'	Anomic	Angular gyrus
'Flair...no, swair...tair'	Conduction	Arcuate fasciculus

* Based on Goodglass, H. (1980) Disorders of naming following brain injury. *American Scientist,* **68**, 647.

associated with language or verbal symbolism. Thus, the patient has great difficulty in finding the correct words to express his thoughts (see Table 17.9) and fails to interpret the meaning of written or spoken words. This type is often called 'fluent aphasia' referring to the meaningless and excessive talk characteristic of these patients in severe cases.

A minor type of aphasia occurs when the lesion affects the nerve fibres of the arcuate fasciculus, and leads to a condition known as *conduction* aphasia. The patient understands the speech of others but is unable to repeat it. As in Wernicke's aphasia, their own speech is fluent but full of words that make no sense. Another type of aphasia is called *anomic* aphasia, in which the lesion damages the angular gyrus in the categorical hemisphere, leaving Wernicke's and Broca's areas intact. In these cases, there are no difficulties of speech and auditory comprehension, but visual comprehension is abnormal (see table 17.9). This arises due to the fact that visual information is not processed and transmitted to Wernicke's area.

Further reading

1 Carpenter, R.H.S. (1990) *Neurophysiology*, 2nd edn. Edward Arnold, Sevenoaks, Kent.
2 Guyton, A.C. (1991) Organisation of NS, sensory receptors and somatic sensations. In *Textbook of Medical Physiology*, 8th edn, pp. 477–532. Saunders, Philadelphia.

18: The Special Senses

Smell and taste

INTRODUCTION

Smell (olfaction) and taste (gustation) are specialized chemical senses. They are closely related, as evidenced by the common observation that the common cold not only depresses the sense of smell but also alters the sense of taste. The two senses are related to food intake: if the smell and

---

Objectives

On completion of the study of this chapter, the student should be able to:

1 Understand the mechanisms involved in the chemical sensations of taste and smell.

2 Comprehend the visual pathways so as to recognize the manifestations of various lesions.

3 Explain the optical aspects of image formation on the retina.

4 Understand the nature and function of visual pigments.

5 Explain the mechanisms of colour vision.

6 Appreciate the visual explanation of perception of distance and three-dimensional characteristics of objects.

7 Understand the properties of sound so as to explain the mechanisms of auditory receptors.

8 Understand the mechanisms of hearing so as to explain the perception of pitch and loudness.

9 Develop an awareness of some clinical applications of the knowledge on the special senses.

10 Become aware of some commonly used clinical tests of vision and hearing.

---

taste of a food are agreeable, the food is ingested; but it is rejected if either or both its smell and taste are unacceptable. The combined effect of the smell and taste of a food is included under the term flavour. There are, however, differences between the two senses. While smell receptors are telereceptors (distance receivers), i.e. the sensation is coming from a distance outside the body and is projected to the environment, taste is entirely confined to the mouth. Smell pathways do not relay in the thalamus and do not reach the sensory cortex. Instead, they end in the part of the brain referred to previously as the rhinencephalon but now considered as part of the limbic system. Taste pathways, on the other hand, relay in the thalamus and are finally projected to the sensory cortex.

SMELL

The sense of smell is developed in some animals, such as the dog and rabbit, which are referred to as macrosmatic, but is greatly reduced in primates, including man, which are referred to as microsmatic.

The olfactory mucosa in man is located in the roof of the nasal cavity near the septum; it has a yellowish-brown colour. Its total area on both sides is about 5 cm^2. As air enters the nose it is warmed by the lining epithelium, and convection or eddy currents arise from the airstream passing through the lower part of the nose to reach the olfactory receptors at the roof. Sniffing may be necessary in order to push up a larger sample of air and may occur voluntarily, but sometimes it occurs semireflexly when a strange smell is sensed in the environment.

The olfactory receptors are bipolar neurones (Fig. 18.1), which are about 10–20 million in number. Their dendritic zone is expanded into olfactory rods, which end in cilia. Between the receptor neurones there are supporting cells; these end in microvilli and secrete mucus, which overlies the mucosa. The axons of the receptor neurones are unmyelinated and collected in bundles called fila olfactoria, which go through the holes in the cribriform plate of the ethmoid bone to enter the olfactory bulb (Fig. 18.2). The receptor axons synapse with mitral and tufted cells in the glomeruli of the bulbs. A great deal of spatial summation occurs at this step, with thousands of receptors synapsing with tens of

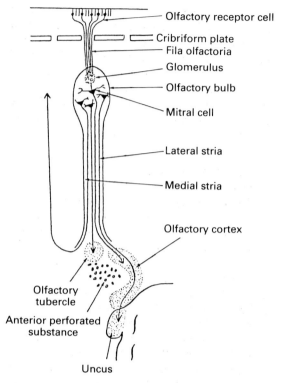

Fig. 18.2 The olfactory pathway. (From Keele, C.A., Neil, E. & Joels, N. (eds) (1982) *Samson Wright's Applied Physiology*, 13th edn, kindly supplied by Professor E.W. Walls. Reproduced by permission of the Oxford University Press, Oxford.)

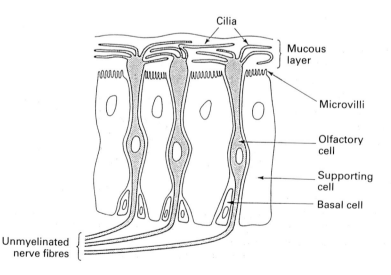

Fig. 18.1 Olfactory receptor and support cells. (Reproduced from Bray, J.J., Cragg, P.A., Macknight, A.D.C., Mills, R.G. & Taylor, D.W. (1989) *Lecture Notes in Physiology*, 2nd edn. Blackwell Scientific Publications, Oxford.)

mitral and tufted cells. From the mitral cells arise the lateral and intermediate olfactory stria, which end respectively in the ipsilateral olfactory cortex and uncus and the olfactory tubercle. The medial olfactory stria arise from the tufted cells and cross the midline in the anterior commissure to end on granular cells in the opposite bulb. These commissural fibres are concerned with the transfer of olfactory memories from one side to the other.

PHYSIOLOGY OF OLFACTION

Molecules of substances to be smelled dissolve in the mucus layer overlying the olfactory epithelium and combine with receptors on the cilia of the olfactory rods. The receptor cell is stimulated by means of a specific G protein (Golf). Adenylate cyclase is activated, thus increasing intracellular cyclic adenosine monophosphate (cAMP). The latter causes opening of Na^+ channels and an influx of Na^+. This is followed by a receptor potential. The receptor potential depolarizes the first segment of the axon which leads to action potentials in the olfactory pathways.

There is no generally accepted classification of the basic types of smells recognized by man. The sense of smell can be very sensitive, sensing minute concentrations of some substances in air. There is also no consistent correlation between strong smell and chemical structure, but substances with strong smells seem to be either highly water-soluble or lipid-soluble. Discrimination within a certain smell is poor, requiring a change in concentration of at least 30% before a difference can be detected.

Man can distinguish between 2000–4000 different odours. There is considerable individual variation in the acuity of the sense of smell and, in general, women have a more keen sense of smell than men. Adaptation can occur to pleasant as well as nasty smells, due to changes in both the receptors and the central connections.

CLINICAL CONSIDERATIONS

Loss of the sense of smell is called anosmia and its alteration is referred to as parosmia or dysosmia. Hyperosomia refers to an increased and hyposmia to a decreased acuity in the sense of smell.

Damage to the olfactory epithelium or pathways by trauma or disease may lead to anosmia or parosmia. Patients with adrenal insufficiency develop a more acute sense of taste and especially of smell. Hyposmia occurs in vitamin A deficiency and hypogonadism.

TASTE

Gustatory receptors are found in the taste-buds. These number about 10 000 in man and are found on the tongue, epiglottis, soft palate and pharynx. On the tongue taste-buds are found in the fungiform, foliate and vallate papillae. They are absent in the small filiform papillae in the mid-dorsum of the tongue; consequently, the mid-dorsum of the tongue is insensitive to taste.

In the taste-buds (Fig. 18.3) the receptor cells end in cilia, which project through the taste pore. Supporting cells are found between the receptors. Nerve fibres probably start as minute filaments within the receptors and form a plexus beneath the basement membrane before they emerge. Taste fibres from the anterior two-thirds of the tongue at first run with the fibres subserving touch, pain and temperature in the lingual nerve. They then separate to join the chorda tympani nerve, which enters the medulla as part of the facial nerve (Fig. 18.4). Fibres from the posterior one-third of the tongue run in the glossopharyngeal nerve and those from the epiglottis, palate and pharynx travel in the vagus nerve. The

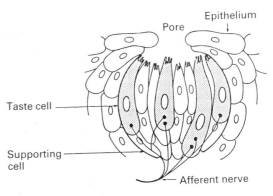

Fig. 18.3 Structure of the taste-buds. (Reproduced from Bray, J.J., Cragg, P.A., Macknight, A.D.C., Mills, R.G. & Taylor, D.W. (1989) *Lecture Notes in Physiology*, 2nd edn. Blackwell Scientific Publications, Oxford.)

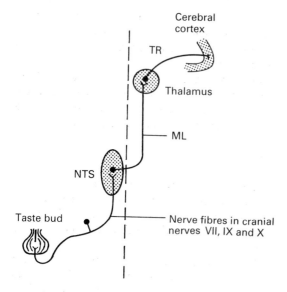

Fig. 18.4 The taste pathway. NTS: nucleus of tractus solitarius; ML: medial lemniscus; TR: thalamic radiation.

PHYSIOLOGY OF TASTE

In order to be tasted, substances must be dissolved in saliva. Their molecules attach to receptors on the cilia of gustatory receptor cells, leading to a generator potential and action potentials in the gustatory pathways. The combination with receptors must be weak, since taste can be abolished quite easily by washing the mouth with water.

In the case of taste, the primary modalities (types) have been identified. They are sour, salt, sweet and bitter. All other tastes are produced by various combinations of these four primary modalities.

The mechanisms of stimulation of specific taste receptors are as follows. Sour is detected when

taste fibres in the three cranial nerves form the tractus solitarius, whose nucleus lies in the medulla. After synapse in the nucleus of the tractus solitarius, second-order neurones arise and cross the midline to ascend in the medial lemniscus to the thalamus. Third-order neurones arise after the synapse in the posteroventral nucleus of the thalamus and, finally, project to the lower part of the postcentral gyrus along with other afferents from the face.

acids containing H^+ block K^+ channels. Salt probably acts by depolarization of salt receptors due to influx of Na^+. Bitter stimuli act by a G protein (Go), which activates phospholipase C, leading to an increase of intracellular inositol triphosphate (IP_3), leading to Ca^{2+} release from the endoplasmic reticulum. Sweet stimuli work through a G protein (Gs), which activates adenylate cyclase, thus increasing cAMP, which leads to decreased K^+ conductance.

It is interesting that, in human beings and rats, distilled water is tasteless and actually inhibits electrical discharge in taste nerve fibres; but, in cats, dogs, pigs and rhesus monkeys, distilled water has a taste, since it produces electrical responses in the gustatory nerves. Each of the primary taste modalities is sensed maximally in a specific area of the tongue; thus, sweet is best appreciated at the tip, bitter at the back, sour along the edges and salt on the dorsum anteriorly near the edge. In addition, sour and bitter substances are tasted on the palate.

Taste is generally less sensitive than smell. There is also individual variation, and acuity of taste declines in old age. Discrimination is as poor as in smell, similarly requiring a 30% change. Adaptation occurs with taste and, in man, is due entirely to the receptors. To savour a pleasant taste, the sapid (taste-producing) substance is moved around the mouth in order to avoid adaptation.

RELATIONSHIP OF CHEMICAL STRUCTURE TO TASTE

Some sapid substances combine taste modalities but many substances have a taste belonging to one of the primary types.

A sour taste is given by acids, whether organic or inorganic. For any acid, sourness is directly proportional to the hydrogen ion concentration, thereby providing a definite link between chemical structure and the taste sensation. The reference acid is HCl, with a normal threshold of pH 3.5.

The salt taste is typically given by NaCl, which is used as the reference substance, with a threshold of 0.02 M. In the case of NaCl, the taste is due to the sodium ion but, in salts with other elements or metals, both the cation and the anion contribute and other tastes may be combined with the salty taste.

Many substances, organic and inorganic, are bitter. The reference substance is quinine sulphate, at a threshold of 0.000008 M. Urea, caffeine, nicotine, strychnine hydrochloride and morphine are bitter. The inorganic salts of magnesium, ammonium and calcium are also bitter and no definite chemical structure can be assigned to the bitter taste.

The most pleasant taste modality, sweet, is given by a variety of substances. The reference substance is sucrose, at a threshold of 0.01 M. Other sugars are sweet, including lactose, glucose and fructose, but polysaccharides, glycerol, some alcohols (not ethanol), aldehydes and ketones are also sweet. Chloroform is sweet. The amides of aspartic acid and some recently discovered proteins are sweet, as are the inorganic salts of lead and beryllium. Artificial sweetening substances are used by diabetics and in reducing diets; they include saccharine, dulcin, cyclamates and aspartame.

CLINICAL CONSIDERATIONS

Complete taste blindness is termed ageusia and disturbed taste sensation is dysgeusia. Some diseases cause hypogeusia. Hypergeusia occurs in patients with adrenal insufficiency. Administration of drugs that have sulphydryl groups in their structure, such as penicillamine, may cause temporary ageusia.

Inability to taste phenylthiocarbamide (PTC) in dilute threshold concentrations is a defect which is inherited as an autosomal recessive factor and may be used in human genetic studies.

Vision

ANATOMICAL CONSIDERATIONS

The eyeball has a diameter of about 2.5 cm. It is basically composed of three layers, which are, from without inwards, the sclera, the choroid and the retina (Fig. 18.5).

The sclera consists of connective tissue, which is opaque, except anteriorly where it is modified by becoming transparent and more curved to form the cornea. The choroid is a pigmented layer with numerous blood-vessels. It is modified anteriorly to give the ciliary body and the iris. The ciliary body is a circular structure consisting of the ciliary glands and smooth muscle. The ciliary muscle consists of circular fibres and longitudinal fibres that are inserted near the corneoscleral junction. Suspensory ligaments from the ciliary muscle are joined to the zonule or lens ligament.

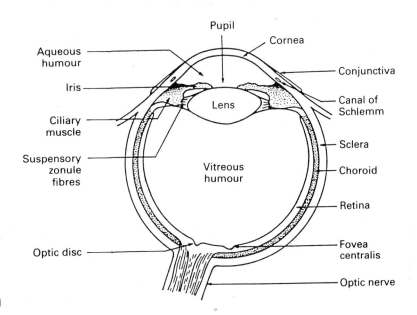

Fig. 18.5 Horizontal section of the eye. (Reproduced from Bray, J.J., Cragg, P.A., Macknight, A.D.C., Mills, R.G. & Taylor, D.W. (1989) *Lecture Notes in Physiology*, 2nd edn. Blackwell Scientific Publications, Oxford.)

The lens is crystalline and biconvex, consisting of concentric laminae of elongated epithelial cells and surrounded by a capsule. The iris is the coloured part of the eye, with the aperture or pupil in its centre. It has circular smooth muscle fibres, which constrict the pupil when they contract, and radial or longitudinal fibres, which dilate the pupil, in this way varying the amount of light that enters the eye. The circular fibres are supplied by parasympathetic cholinergic nerves, while the longitudinal fibres are innervated by sympathetic adrenergic nerves. Clinically, the pupil is usually dilated by instilling drops of atropine-like drugs into the eye, which paralyse the parasympathetic nerves. The retina lines the posterior two-thirds of the choroid and contains the receptors for light, the rods and cones. Medial or nasal to the anteroposterior axis of the eyeball is the spot where the optic nerve fibres leave the eye, called the optic disc, which measures about 1.5 mm in diameter. About 3 mm lateral or temporal to the optic disc and very close to the anteroposterior axis is the macula lutea (yellow spot), at the centre of which is a depression called the fovea centralis.

The space between the retina and the lens is filled with a transparent gelatinous material, called the vitreous humour or body. The space between the cornea anteriorly and the iris and lens posteriorly is called the anterior chamber. The posterior chamber is the narrow circular space bounded anteriorly by the iris and posteriorly by the ciliary body and lens. Both the anterior and posterior chambers are filled with aqueous humour. This is a fluid similar in composition to the plasma without the proteins, but it has lower glucose and urea concentrations than in the plasma. The lens is avascular and metabolizes glucose taken from the aqueous humour. Aqueous humour is produced in the ciliary glands by diffusion and active transport. It moves forward in the posterior chamber, going between the lens and the iris into the anterior chamber, and moves on to the iridocorneal angle, between the cornea and iris and through vacuoles in the lining endothelial cells, to drain into the sinus venosus sclerae or *canal of Schlemm*. This continuous production and drainage into venous blood of aqueous humour maintains intraocular pressure at 10–25 mmHg. A rise of pressure above 25 mmHg results in glaucoma, one cause of which is blockage of the canal of Schlemm.

External protection of the eye

The eye is protected by the bony orbit and by the eyelids, which have lashes on their edges and are lined on the inside by conjunctiva. Tear fluid is produced by lacrimal glands, which are located in the upper and outer part of the orbit. This fluid normally drains through the nasolacrimal duct. It is an isotonic solution of Na^+, Cl^- and HCO_3^- with a low protein content and a bactericidal enzyme called lysozyme. There is normally a film of tear fluid lining the conjunctiva and cornea. Blinking, occurring spontaneously at the rate of about 20 times per minute, is responsible for renewing this film. Blinking also occurs when the front of the eye is touched, when bright light is shone into the eye and in the case of a suddenly approaching object. In the case of irritation and when a foreign body enters the conjunctiva, the rate of tear secretion is increased, through a nervous reflex involving efferent parasympathetic fibres, in order to wash out the foreign body or irritant. Drainage capacity may be exceeded, in which case tear fluid spills over and runs on the cheeks; this is termed lacrimation (weeping). Lacrimation may also occur under emotional circumstances.

The retina

The retina has 10 layers (Fig. 18.6). The arrangement of these layers can be appreciated when it is realized that the light receptors, the rods and cones, face the pigment layer towards the outer side of the eyeball, and that the rods and cones synapse with bipolar neurones, which synapse in turn with ganglion cells, the axons of which constitute the optic nerve fibres. The 10th layer is an internal limiting membrane. Horizontal cells make synaptic connections with and between receptors, while amacrine cells, having no axons but numerous processes, make horizontal connections between ganglion cells.

Each eye has about 120 million rods and 6 million cones. Each rod and cone has an outer segment, consisting of discs or saccules of membrane that contain a photosensitive pigment, an

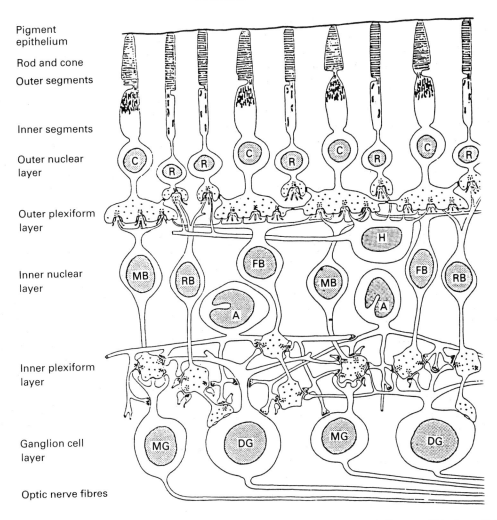

Pigment epithelium

Rod and cone

Outer segments

Inner segments

Outer nuclear layer

Outer plexiform layer

Inner nuclear layer

Inner plexiform layer

Ganglion cell layer

Optic nerve fibres

Fig. 18.6 The neural structure of the retina. R: rods; C: cones; MB: midget bipolar cell; RB: rod bipolar cell; FB: flat bipolar cell; H: horizontal cell; A: amacrine cell; MG: midget ganglion cell; DG: diffuse ganglion cell. Note that the axon of the horizontal cell makes no synapses with the cone. These axons in mammals only synapse with the rods. (Reproduced with permission from Dowling, J.E. & Boycott, B.B. (1966) *Proceedings of the Royal Society (Series B)* 166, 80.)

inner segment, which is rich in mitochondria and includes the nucleus, and a synaptic terminal. The numerous mitochondria signify high metabolic activity, which is manifested by the presence of Na^+, K^+ ATPase in the membrane of the inner segments. Rods are extremely sensitive to light and operate under dim light conditions. Cones have a higher threshold and operate under bright light conditions.

There are no rods or cones on the optic disc, which is consequently blind (the blind spot). The fovea centralis contains only cones. The density of cones falls sharply in the periphery of the retina. The density of rods increases from outside the fovea towards the periphery of the retina. The fovea centralis is the spot with highest acuity of vision: details of objects are distinguished, colour is appreciated. To see an object clearly, the head and eyes may be turned to face the object so that light reflected from it falls on the fovea centralis.

The ophthalmoscope is basically an instrument that illuminates the interior of the eye and the examiner, in effect, uses the eye lens of the subject as a magnifying glass to see the fundus. The optic disc is easily recognized with retinal blood-vessels emerging from it; about two disc diameters laterally is the slightly darkish macula lutea, which is not crossed by blood-vessels. Examination of the fundus may reveal conditions like optic atrophy, papilloedema and retinal detachment. In addition, viewing of the retinal blood-vessels constitutes the only opportunity to directly examine blood-vessels clinically. Examination of the fundus is therefore important in all diseases that may affect blood-vessels, particularly hypertension and diabetes mellitus. The retinal vessels supply the bipolar and ganglion cells of the retina. However, the receptor cells are nourished mainly from the capillary plexuses present in the choroid. Therefore, retinal detachment is damaging to the receptor cells.

There are 1.2 million fibres in the optic nerve. With 126 million receptors, one would expect converging connections of rods and cones with optic nerve fibres. Actually, one foveal cone synapses with one bipolar neurone and one ganglion cell, but about 10 cones in the periphery of the retina synapse with one bipolar neurone and one ganglion cell. In the case of rods, about

300 of them synapse with one bipolar cell and one ganglion neurone.

THE VISUAL PATHWAYS (Fig. 18.7)

Since light travels in straight lines, the right half of the retina views the left half of the field of vision and the upper half of the retina looks at the lower half of the visual field, and vice versa. It is noteworthy that, in the study of vision, the terms temporal and nasal are preferred to lateral and medial, respectively. The optic nerves emerge from the eyes to meet in the optic chiasma, where decussation of fibres occurs in a regular pattern: fibres from the temporal half of the retina on each side proceed on the same side, while fibres from each of the nasal halves of the retinae cross to the opposite side. Fibres from the macula on each side behave in exactly the same way. The optic tracts carry fibres from the retinal and macular halves of the same side, i.e. the right optic tract carries fibres from the right halves of both retinae and maculae, and fibres from the left halves of the retinae and maculae are carried in the left optic tract. For the field of vision, the directions are reversed, e.g. the left optic tract carries impulses coming from the right half of each field of vision. Optic tract fibres synapse in the lateral geniculate body of the thalamus. The grey matter of the lateral geniculate body has six

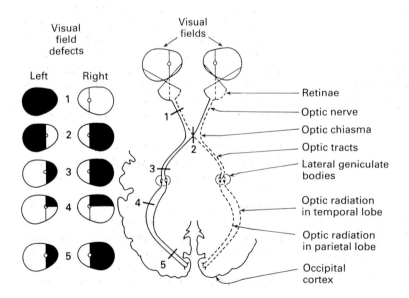

Fig. 18.7 The visual pathway and the effect of various lesions, 1–5. (Reproduced from Macleod, J. (ed) (1974) *Davidson's Principles and Practice of Medicine*, 11th edn. With permission from Churchill Livingstone, Edinburgh.)

layers; contralateral fibres end in layers 1, 4 and 6, while fibres from equivalent spots on the ipsilateral side of the retina end in layers 2, 3 and 5. Fibres arising from the lateral geniculate body ascend in the optic radiation to the occipital cortex. The primary cortical visual area for the same-side halves of the field of vision are projected to the inner side of the opposite hemisphere above and below the calcarine fissure (Brodman's area 17; see Fig. 18.8). The macula is represented by a much larger area than the periphery of the retina, which is represented by small forward areas above and below the calcarine fissure. Near by, areas 18 and 19 are visual association areas, where recognition of objects and other cognitive functions associated with vision are performed.

Some fibres branch off from the optic tract just before entering the lateral geniculate body; they go to the pretectal area of the superior colliculus of the midbrain, where a synapse occurs and new fibres arise to go to the oculomotor (3rd cranial nerve) nucleus on both sides. This is the pathway for the *light reflex*: when light is shone on an eye, the pupil constricts; the opposite pupil also constricts — the *consensual light reflex*. The consensual reaction is due to the fact that illumination of the retina in one eye sends impulses in both optic tracts and that the fibres from the superior colliculus go bilaterally to the oculomotor nucleus. The pathway for the pupillary constriction that accompanies accommodation is somewhat different, as will be explained below.

LESIONS OF THE VISUAL PATHWAYS (Fig. 18.7) A lesion interrupting the whole of the optic nerve results in complete blindness in the corresponding eye. Beyond the optic nerve, lesions may result in blindness in half the field on each side; this is termed *hemianopia*. When the same sides are affected, the hemianopia is homonymous; when opposite sides are affected it is heteronymous. The terms temporal and nasal, as well as right and left, are also used.

If a lesion affects the central part of the optic chiasma, such as by pressure from a pituitary tumour, fibres from the nasal halves of both retinae are interrupted, leading to bitemporal hemianopia, which is heteronymous. A lesion pressing on the lateral sides of the chiasma would interrupt the fibres from the temporal halves of both retinae and would lead to binasal hemianopia, which is also heteronymous. Thus, chiasmal lesions lead to *heteronymous hemianopia*.

Lesions of the optic tracts, optic radiation or one complete side of the occipital cortex, i.e. retrochiasmal, lead to *homonymous hemianopia*. There are, however, two special characteristics of cortical lesions. First, since the macula is represented by a much larger area than the periphery of the retina, a lesion may not destroy all macular representation, resulting in macular sparing in the field of vision. Secondly, since fibres from the upper and lower quarters of both sides of the retinae and maculae are represented on the occipital cortex respectively above and below the calcarine fissure, a lesion acting entirely above or below the fissure may result in quadrantic visual field defects.

An important clue as to the location of the lesion in the visual pathways may be obtained by trying to elicit the light reflex. If light is shone on the blind half of the retina and the pupil constricts, it means that the pathway for the light reflex is intact; therefore, the lesion must be

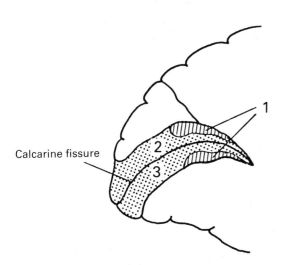

Fig. 18.8 Medial view of occipital lobe showing location of the visual areas on the calcarine fissure of the occipital lobe. 1: projection area for peripheral retina; 2: area for upper macula; 3: area for lower macula. Note that the projection of the macula occupies the largest part of the visual cortex.

Calcarine fissure

beyond the point where fibres take off from the optic tract, just before the lateral geniculate body. If the light reflex is absent, the lesion must be in the optic nerve, chiasma or tracts before the fibres branch off.

OPTICS OF THE EYE

Light rays coming from a distance greater than 6 metres are considered parallel; those coming from less than 6 metres away are considered diverging. When light passes from one medium to another medium of different density, it is refracted unless it strikes the interface between the two media perpendicularly. A biconvex lens converges parallel rays to a principal focus behind it, while a biconcave lens diverges parallel light rays, which may be extrapolated to give a principal focus in front (see Fig. 18.9). The more curved the lens, the more it will bend the light and the shorter will be the focal length. The power of a lens (positive in case of convex and negative in case of concave) is measured in dioptres (D). The dioptric power of a lens is the reciprocal of the focal length is metres. Thus, a lens with a focal length of 25 cm has a power of 1/0.25 cm or 4 dioptres, while a lens with a focal length of 10 cm has a power of 10 dioptres.

The optically normal eye is termed *emmetropic*. In such an eye, parallel rays of light are refracted first by the cornea and then by the lens to be focused on the retina (Fig. 18.10A). When a normal individual is looking at the horizon, light is coming in parallel rays, the ciliary muscle is relaxed and the lens is under tension, so that it is in its most flattened shape. The power of this resting eye is about 60 D, with the cornea accounting for about 40–45 D and the lens 15–20 D. While more refraction occurs at the cornea, the advantage of the lens is, as will be explained, its ability to become more convex and therefore increase its power.

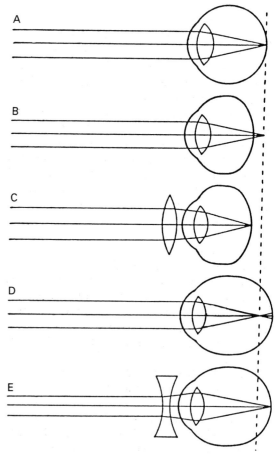

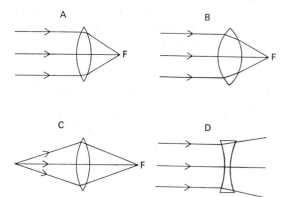

Fig. 18.9 Light refraction by lenses. A: Biconvex lens; B: biconvex lens of greater strength than A; C: same lens as A showing the refraction of light rays coming from a near point; D: biconcave lens. The line in the middle is the principal axis and F is the principal focus.

Fig. 18.10 Correction of common defects of refraction. (Reproduced from Emslie-Smith, D., Paterson, C.R., Scratcherd, T. & Read, N.W. (eds) (1988) *BDS Textbook of Physiology*, 11th edn. With permission from Churchill Livingstone, Edinburgh.)

When the gaze is transferred from a far object to a near object (from which light rays would be diverging), certain changes occur in the eye in order to focus the image of the near object on the retina. These changes are:

1 Increasing the convexity or power of the lens, a process referred to as *accommodation*.

2 Pupillary constriction, so that light is concentrated on the more powerful centre of the crystalline lens.

3 Convergence of the visual axes, such that they meet at the near object.

The above changes are sometimes included under the term the *near response*.

Accommodation

At rest, the ciliary muscle is relaxed and the suspensory ligament is under tension, pulling the elastic lens into a flattened shape. In accommodation, both the circular and longitudinal ciliary muscle fibres are contracted and the whole of the ciliary body moves inward and forward, thereby relaxing tension on the suspensory ligament and allowing the elastic lens to fall into a more convex shape. The increased convexity occurs mainly in the anterior surface of the lens. A number of dioptres is added to the power of the lens, allowing light rays to be focused on the retina (up to 12 dioptres in children).

Neural pathways for the near response

The ciliary muscle, the circular muscle in the iris and the two medial rectus muscles (responsible for convergence of visual axes) are all supplied by the oculomotor nerve, i.e. it is the final common pathway for the near response. Afferent fibres travel in the optic nerves through the chiasma to the optic tracts and most probably synapse in the lateral geniculate body and continue to the occipital cortex. From there the fibres probably go to the frontal lobes and then descend through the internal capsule to reach the oculomotor nucleus in the midbrain. It should be noted that fibres to and from the cerebral cortex are exclusive to the neural pathways for the *near response*. The fibres branching from the optic tract to the superior colliculus and proceeding to the oculo-motor nucleus are exclusive to the pathway of the *pupillary light reflex*. A lesion in the pretectal region may occur in neurosyphilis, interrupting the light reflex but leaving the pathway for the near response intact, and resulting in the *Argyll Robertson pupil*, i.e. absent or sluggish response to light but constricting when looking at a near object.

Amplitude of accommodation and presbyopia

If an object is brought gradually nearer and nearer to the eye, a point is reached when the object cannot be focused and appears blurred. The nearest point to the eye where an object can be clearly seen is called the near point. At the near point the dioptric power of the eye is maximal, being equal to the sum of the power of the eye at rest plus the number of dioptres added by increasing the convexity of the lens. These additional dioptres are the amplitude of accommodation, which can be obtained by subtracting the power of the eye at rest from the total power at the near point. If the resting dioptric power of the eye is taken as zero, the amplitude of accommodation in dioptres is given by the reciprocal of the distance of the near point from the eye in metres, e.g. if the near point is 10 cm away, the amplitude of accommodation would be 10 dioptres.

With advancing age, the lens gradually loses its elasticity and therefore its capacity to increase its convexity. Consequently, the near point recedes and the amplitude of accommodation decreases with advancing age—a process called *presbyopia*. This is clearly illustrated in Table 18.1. By the age of 40–45 years, the near point starts to recede at a faster rate and close work becomes increas-

Table 18.1 Variation of near point and accommodation with age

Age (years)	Near point (cm)	Amplitude of accommodation (dioptres)
10	9	11
20	10	10
30	12.5	8
40	18	5.5
50	50	2
60	83	1.2
70	100	1

ingly difficult. Many people require glasses with convex lenses for close work beyond the age of 40 years.

The retinal image

The refractive index of each of the cornea, aqueous humour and vitreous humour is 1.33, i.e. about the same as water, while the lens has a refractive index of 1.42. Normally, light is refracted towards the normal by the cornea and normal by the anterior surface of the lens, and away from the normal by the posterior surface of the lens, to be focused on the retina. This process may be simplified by assuming that all refraction is occurring at the cornea, and by defining the optic centre or nodal point of the eye, as in the schematic or reduced eye of Listing (Fig. 18.11). Any light ray passing through the nodal point is not refracted, but rays passing through any other point are refracted to be focused on the retina. In man, the distance of the nodal point from the retina is about 15 mm. For an object AB the retinal image is represented by ab in Fig. 18.11. Thus, the retinal image is inverted and this inversion continues to the occipital cortex. From infancy, the brain is genetically programmed to reverse the inversion so that we see the world right way up.

Triangles AnB and anb in Fig. 18.11 are similar and therefore:

$$\frac{AB}{ab} = \frac{Bn}{bn}$$

If the height of an object and its distance from the eye are known, and given that bn is equal to 15 mm, the size of the retinal image can be calculated.

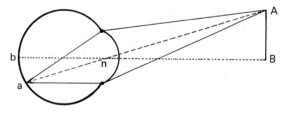

Fig. 18.11 Reduced or schematic eye showing the nodal point n. In this diagram it is assumed that all refraction takes place at the cornea. The refracted dotted lines converge on the retina to give an inverted image.

Angle AnB is called the visual angle. It is the angle subtended by an object at the nodal point of the eye. Assessment of visual acuity, i.e. how well we can distinguish details in the field of vision, depends upon this angle. As long as two lines are separated by an arc of at least 1 minute and thus subtend a minimum visual angle of 1 minute at the nodal point, the distance between the two images on the retina will be 2 μm. In this case, two separate cones are stimulated by each point on the lines, leaving one unstimulated cone in between (the diameter of a cone is 1.5 μm). This means that the two lines are perceived as separate. Separation by an arc of less than 1 minute will stimulate two adjacent cones and the two lines will be perceived as a single line. Letters or symbols on charts which are used for assessing visual acuity are written so that the thickness of a line subtends a visual angle of 1 minute and the whole letter or symbol subtends an angle of 5 minutes at the nodal point. A subject stands at 6 metres away from the chart and his/her visual acuity is given by the ratio d/D, where d is the distance from the chart (6 metres) and D the distance indicated on the chart for the row that the subject has been able to read or identify correctly. Thus, normal visual acuity is 6/6, while a visual acuity of 6/12 is subnormal and indicates that the subject sees at 6 metres what a normal subject can see at 12 metres away.

Errors of refraction

Abnormalities of the length of the eyeball or its refractive power result in either *hypermetropia* (long-sightedness) or *myopia* (short-sightedness) (see Fig. 18.10). These abnormalities may be referred to as ametropia.

In hypermetropia the eyeball is too short and parallel rays of light are focused behind the retina. An affected individual has to employ some accommodation even for distant objects, leading to headaches and hypertrophy of the ciliary muscles from excessive use. The near point is further than the position in accordance with age so that glasses are required for close work at an earlier age. Glasses with convex lenses are used to correct the defect.

In myopia the eyeball is too long and parallel

rays of light are brought to a focus in front of the retina. The far point is finite, i.e. the subject cannot see far, while the near point is too near for age and consequently the amplitude of accommodation is reduced. A myopic individual would not require glasses for close work, even in his old age. The defect is corrected by biconcave lenses.

Astigmatism This is a condition in which the curvature of the cornea is not uniform; rarely, the lens may be affected. Light refracted by one meridian is focused differently from light refracted by another unequal meridian. Typically, an individual affected by astigmatism cannot focus vertical and horizontal lines on graph paper at the same time. The defect is corrected by cylindrical lenses, which focus light to a line, thus correcting the curvature in a certain meridian and ensuring the overall uniformity of curvature.

THE DUPLICITY THEORY

The visible spectrum extends from the 397 nm to the 723 nm wavelength of light. Within this range, the eye can function under two conditions of illumination: dim light and bright light. Vision in dim light is termed *scotopic vision* and in this type only the outline of objects in the field of vision can be distinguished; no details or colour can be appreciated. *Photopic vision* is employed in bright light and in it the details of objects, as well as their colour, can be distinguished. The presence of scotopic and photopic vision and the consequent double input from the eye to the central nervous system are the elements of the duplicity theory.

The scotopic visibility curve peaks at the 505 nm wavelength while the photopic curve peaks at the 550 nm wavelength of light (Fig. 18.12). Scotopic vision is served by the rods, while photopic vision is due to the cones. This has been proved by the finding that the light absorbance of the rod pigment, rhodopsin, coincides almost exactly with the scotopic visibility curve, peaking at the 505 nm wavelength of light. The cones have three pigments, each most sensitive to one of the primary colours (see below). The light absorbance curve of a mixture of the three pigments also coincides with the photopic visibility curve, peaking at the 550 nm wave-

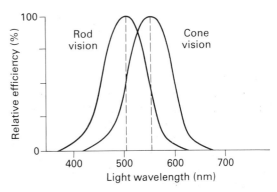

Fig. 18.12 Visibility curves for photopic (cones) and scotopic (rods) vision showing the wavelength range for each. The relative sensitivity has been equalized in this diagram. In reality, rod sensitivity is far greater.

length. It should be noted that the Y axis in Fig. 18.12 is relative sensitivity, i.e. maximal sensitivity in both cases is taken as 100% and lesser sensitivities are expressed as a percentage. If absolute sensitivity values were employed, the curves would be of the same shape but the scotopic curve would be much higher than the photopic curve.

Figure 18.13 shows the range of luminance within which the eye may function. Note the great sensitivity of rod or scotopic vision and that there is a transition zone between rod vision and cone vision.

PHOTOSENSITIVE PIGMENTS

The pigments in the outer segments of rods and cones have two components: *retinene*, which is the aldehyde of vitamin A, and a protein called *opsin*. Retinene is the same in all the pigments; differences between the pigments are in the amino acid sequence of the opsin.

The opsin in rhodopsin is termed *scotopsin*. The amino acid sequence of scotopsin is determined by a gene on chromosome 3. Light bleaches rhodopsin or separates it into retinene and scotopsin. Recent work has shown that light actually acts only on retinene, changing its isomerism from the 11-*cis* form to the all-*trans* form. The shape of retinene is changed, giving rise to *prelumirhodopsin*, which subsequently changes to some intermediates, including

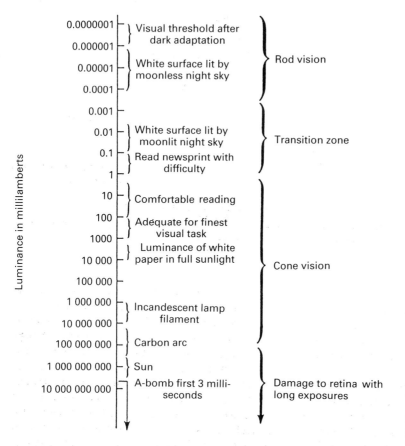

Fig. 18.13 The range of luminance to which the human eye responds. The receptive mechanisms are indicated in the column on the right. (Reproduced from Emslie-Smith, D., Paterson, C.C., Scratcherd, T. & Read, N.W. (eds) (1989) *BDS Textbook of Physiology*, 11th edn. With permission from Churchill Livingstone, Edinburgh.)

metarhodopsin II, before being hydrolysed to retinene and scotopsin. Retinene and scotopsin may recombine to reconstitute rhodopsin. More slowly, retinene may be changed back to vitamin A, under the influence of the enzyme alcohol dehydrogenase, and vitamin A with scotopsin may regenerate rhodopsin. These changes are summarized in Fig. 18.14.

The action of light on the cone pigments is probably similar. The action of light on the photosensitive pigments in the outer segments of rods and cones results in the generation of receptor potentials, which lead to action potentials in the optic nerve fibres, as will be explained below.

The role played by vitamin A in rod function explains the fact that deficiency of this vitamin leads to impairment of the ability to see in dim light and may result in night-blindness or *nyctalopia*. With persistent vitamin A deficiency, cone function may also be affected.

THE CONE PIGMENTS AND COLOUR VISION

The primary colours are red, green and blue. When red light (wavelength 723–647 nm), green light (575–429 nm) and blue light (492–450 nm) are mixed, white light or any spectral colour may be obtained. The *Young–Helmholtz* theory explaining colour vision in man is now widely accepted. According to this theory, there are three types of cones, each containing a pigment most sensitive to one of the primary colours. Light entering the eye stimulates the cones in proportion to its spectral components and the differential discharge is coded and conveyed to the occipital cortex. Colour is sensed by cones in the fovea centralis and appreciated within photopic vision.

Human cone pigments have been named as follows:

erythrolabe — red-catching
chlorolabe — green-catching
cyanolabe — blue-catching

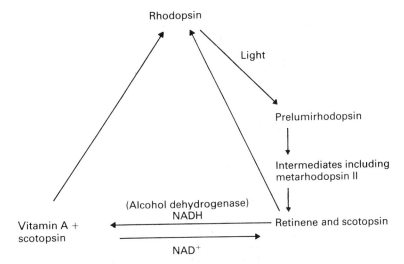

Fig. 18.14 The rhodopsin cycle in man.

Each of these pigments contains retinene and an opsin which differs between pigments in some of its amino acid sequence. Erythrolabe was first isolated from the retina of birds and has been called iodopsin, with its opsin termed photopsin. The amino acid sequences of the opsins of erythrolabe and chlorolabe are very similar (96% homology), as they are both determined by adjacent genes on the X chromosome. The amino acid sequence of the opsin of cyanolabe is determined by a gene on autosomal chromosome 7 and has only 43% homology with each of the opsins of erythrolabe and chlorolabe.

Colour blindness
Colour blindness may be due to weakness in detecting one of the primary colours, or to blindness to one or even two of the primary colours. Human beings may be divided into:
1 *Trichromats.* These include people who are normal in all respects in relation to colour perception, plus those with weakness in detecting red, green or blue, i.e. they need more of the affected primary colour to perceive colour normally.
2 *Dichromats.* These are completely blind to red or green or blue and they get their colour sensation by mixing only two of the primary colours.
3 *Monochromats.* These have only one cone pigment; to them the world is black and white and shades of grey.

Colour-blindness commonly affects red and green and in this case is usually inherited as a recessive factor on the X chromosome. As such, it affects men more than women and, although heterozygous women do not show the defect, they transmit it to half of their sons. Abnormalities in relation to blue are rare and are not sex-linked when inherited. Monochromats are extremely rare and they would have other neurological defects. One of the convenient methods to test for colour blindness is to use polychromatic plates (*Ishihara charts*). These indicate numbers or lines made of spots of certain colours against a background of confusing colours; a key gives the expected responses by normal people and by people affected by colour blindness.

DARK ADAPTATION
If a person is in brightly lit surroundings, he would be using his cones. If the light is suddenly switched off, at first nothing can be seen, but gradually the outline of objects in the field of vision starts to appear and the situation improves with time, reaching a maximum in about 20 minutes (Fig. 18.15). This is called dark adaptation, during which only the gross features of objects can be distinguished—not their details or colour.

It has been shown that, during the first 5 minutes, the threshold for cones decreases (or their sensitivity increases). From 5–20 minutes

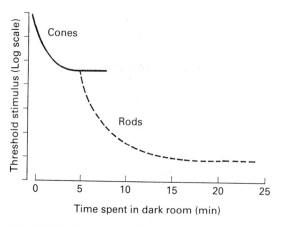

Fig. 18.15 Dark adaptation for rods and cones. The threshold decreases as the subject stays for a longer time in the dark.

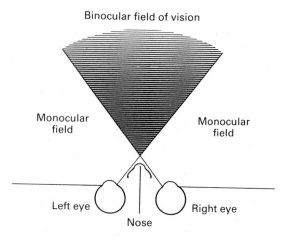

Fig. 18.16 Monocular and binocular fields of vision. The shaded area enclosed by the two fields is seen by both eyes (binocular vision).

in the dark, a great increase occurs in the sensitivity of the rods, which thus account for the greater share of dark adaptation. The main part of the time required for dark adaptation to reach its maximum is needed for regeneration of rhodopsin so that rods can function optimally. The changes that occur in the cones during the first 5 minutes are not fully understood. If light is switched on again, the rods are knocked out of action and the cones start to function, adjusting to the level of brightness within 5 minutes; this process may be called light adaptation.

BINOCULAR VISION

The field of vision for each eye may be mapped by using an instrument called the perimeter, the process being called perimetry. Basically, the subject looks with one eye at the point of fixation, while a target is moved along an arc at a certain meridian. The points where the target just enters into view for each meridian are marked on a chart. The field of vision for each eye is mapped on a separate chart. When in real life we use both eyes, the areas in the centre of the field of vision for the two eyes overlap and any object in this area will be seen by both eyes, i.e. the vision is binocular. At the temporal part of each field, vision is monocular (Fig. 18.16).

In binocular vision there would be two retinal images. The eyes are actually moved by the extra-

ocular muscles in such a way as to ensure that the two retinal images fall on corresponding points on the retinae, the connections of these being organized so that the two images completely fuse at the cortical level. If one of the extraocular muscles is paralysed, e.g. the lateral rectus muscle, the affected eye is deviated inward as a result of the unopposed action of the medial rectus, and double vision or diplopia results, since the retinal image in the affected eye is not on the corresponding point.

Depth perception

Appreciation of depth in the field of vision is essentially monocular but is improved when using the two eyes. Two clues from the field of vision requiring only one eye are used to gauge depth: first, the relative sizes of objects — the further the object, the smaller it appears; secondly, movement parallax, by which near moving objects move right across the field of vision but distant moving objects appear to hardly move. Binocular vision contributes to depth perception through the process of *stereopsis*: a near object has its retinal image on the temporal part of the two retinae while the image of a far object falls on the nasal part. This creates a kind of parallax in relation to location on the retina, which is, in turn, taken to the visual cortex and analysed.

Critical fusion frequency (CFF)

If a light source is interrupted regularly, as by a rotating notched disc, the light flickers. As the rate of interruption is increased, a frequency is reached when the flicker disappears and the light appears continuous. The frequency at which complete fusion of the successive flickering images just occurs to give an illusion of continuity is called the critical fusion frequency (CFF). To ensure complete fusion, the light stimulus has to be delivered at a rate just exceeding the CFF. In motion pictures, frames are usually delivered at a rate of about 20 per second, which is more than the CFF. If the projector slows down and the frequency of the frames drops to below the CFF, the picture on the screen starts to flicker.

ELECTROPHYSIOLOGY OF VISION

When light strikes the outer segments of rods and cones, it causes changes in the photosensitive pigments. These changes lead to receptor potentials, which are converted to action potentials in the optic nerve fibres. The changes in rods have been studied in more detail than those in cones but the process is basically similar in the two receptors.

There is Na^+, K^+ ATPase in the inner segment of rods. In the dark, Na^+ is pumped to the outside but it re-enters through sodium channels in the membranes of the outer segments and the synaptic zone, creating a continuous sodium current. When light strikes the outer segments, it changes the isomerism of retinene and produces several intermediates before complete bleaching of rhodopsin. One of the intermediates is *metarhodopsin* II, which leads to activation of transducin (a Gt protein). This causes activation of phosphodiesterase, which decreases intracellular cyclic guanosine monophosphate (cGMP), and, as a result, closure of some of the sodium channels takes place in the membrane of the outer segments. This leads to less sodium entering and creates a negative potential. Thus, rods respond to light by *hyperpolarization* (Fig. 18.17). Cones also respond by hyperpolarization.

A number of synaptic transmitters have been found in the retina, including acetylcholine, dopamine, serotonin, GABA, substance P and other peptides. It seems that in the dark these

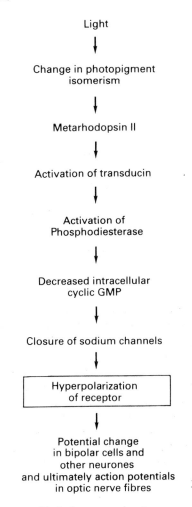

Fig. 18.17 Possible linkage sequence between photopigment changes and electrical events in the retina.

transmitters are continuously released. By causing hyperpolarization in the receptors, light decreases the release of the transmitters, leading to electric potentials in horizontal cells, bipolar neurones and amacrine cells. The depolarizing potentials in amacrine cells have been identified as the generator potentials leading to action potential spikes in the ganglion neurones and optic nerve fibres.

Ganglion cells in the retina respond to light in the form of circles. One-half of the ganglion neurones discharge when the centre of the circular field is illuminated but are inhibited by illumination of the periphery of the field (on-centre cells). The other half discharges to illumination of the periphery but is inhibited by stimulation of the centre (off-centre cells). Cells in the lateral geniculate body and some neurones in cortical area 17 respond in a similar fashion. This type of response may be responsible for demarcating the edges of objects in the field of vision.

Two types of cortical neurones have been identified: simple cells, in area 17, and complex cells, mainly in the association areas 18 and 19 but also in area 17. Both respond to light in the form of lines. Simple cells respond according to the orientation of the line stimulus and neurones responding to the same orientation are arranged in vertical columns, called orientation columns, in the occipital cortex. Complex cells respond to the orientation of the light stimulus but discharge more when the line stimulus is moved and may thus be responsible for detecting form and movement. Most neurones concerned with vision respond either to one eye or to the other. Neurones responding to the ipsilateral eye are arranged in vertical columns, alternating with columns of neurones responding to the contralateral eye; these are called ocular dominance columns. About half of the complex cells respond to stimuli from both eyes and may be concerned with stereopsis, which requires binocular vision.

EYE MOVEMENTS

The actions of the six extraocular muscles should be ascertained from an anatomy text. It is, however, important to understand what each type of movement involves in terms of a structure like the eyeball. Abduction and adduction refer to rotation of the eyeball around the vertical axis, elevation and depression around the transverse horizontal axis and intorsion and extorsion around the horizontal anteroposterior axis. The nerve supply of the extraocular muscle may be conveniently remembered by referring to the following nonsensical formula: $LR_6 (SO_4)_3$, i.e. the lateral rectus is supplied by the 6th or abducent nerve, the superior oblique is supplied by the 4th or trochlear nerve, and the rest are innervated by the 3rd or oculomotor nerve. The actions of extraocular muscles are highly co-ordinated in order to

ensure that retinal images fall on corresponding points; otherwise, diplopia results.

There are four types of eye fixation movements which are related to visual function:

1 *Saccadic movements.* These are very rapid conjugate movements of the eyes, occurring when inspecting an object, when the gaze shifts from one object to another or during reading. Since they are present most of the time, they may be responsible for prevention of adaptation to the retinal image, which, if allowed, might lead to disappearance of objects under constant view! Sometimes eye movements responsible for preventing visual adaptation are referred to as physiological nystagmus.

2 *Smooth pursuit movements.* As the name implies, by these movements the eyes follow or track moving objects.

3 *Vergence movements.* Convergence occurs when the gaze is changed from a far to a near object or when fixing on an approaching object. Divergence of the visual axes occurs when the gaze is transferred from a near to a far object or when watching an object moving away.

4 *Vestibular movements.* These occur in order to maintain visual fixation when the head moves. During rotation the eyes move in the opposite direction in order to maintain fixation. When the limit is reached, the eyes snap back, and the process is repeated, leading to nystagmus.

All these movements are initiated in neural pathways related to their function, e.g. vestibular movements are initiated by stimuli in the semicircular canals and travel in the vestibular pathways, but the final common pathway is constituted by the nuclei of the 3rd, 4th and 6th cranial nerves, which innervate the extraocular muscles.

Hearing

ANATOMICAL CONSIDERATIONS

The ear consists of the outer, the middle and the inner ear. The outer ear consists of the pinna and the external auditory canal, directed medially and forwards to the ear-drum or tympanic membrane. The tympanic membrane is shaped like a shallow funnel, with the tip of the funnel or umbo pointing inwards. It is lined by skin on the outside and mucous membrane on the inside. The middle ear is a cavity in the temporal bone filled with air and containing the three ossicles: the malleus, the incus and the stapes. The handle of the malleus (manubrium) is attached to the back of the tympanic membrane with its tip at the umbo; the head of the malleus articulates with the incus, the process of which articulates with the head of the stapes. The footplate of the stapes is attached to the wall of the oval window of the cochlea by an annular ligament, which is fixed firmly at a posterior point but is slightly loose otherwise, to allow for movement of the footplate.

The air in the cavity of the middle ear communicates with the nasopharynx through the Eustachian or auditory tube and then with the outside air. The opening of the auditory tube in the nasopharynx is normally closed but it opens during chewing, swallowing and yawning, to allow equalization of pressure between the middle ear and atmospheric pressure, i.e. on both sides of the tympanic membrane. A severe head cold may cause inflammation which closes off the auditory tube. The air trapped in the middle ear is partly absorbed, leading to atmospheric pressure pressing on the tympanic membrane and causing discomfort and reduced acuity of hearing.

There are two small striated muscles in the middle ear: (i) the tensor tympani, attached to the neck of the malleus; when it contracts, it pulls the drum medially; and (ii) the stapedius, attached to the neck of the stapes; when it contracts, it tends to pull the footplate of the stapes out of the oval window. Simultaneous contraction of the two muscles reduces conduction of sound to the cochlea. Very loud sounds lead to reflex contraction of the tensor tympani and the stapedius, which thus reduces transmission of these loud sounds to the auditory receptors. This is termed the tympanic reflex, which is protective; however, it has a latent period of 40–160 ms, which makes it ineffective against brief sounds of high intensity.

The inner ear or labyrinth consists of the bony and membranous labyrinth. The bony part is a series of channels in the petrous part of the temporal bone, and membranes run inside the bony channels to form the membranous labyrinth. Between the bone and membranes is a fluid called perilymph, while inside the membranous labyrinth the fluid is endolymph; the two fluids do not mix. In the inner ear the semicircular canals, the utricle and the saccule are concerned with balance and are part of the vestibular system, which has been described in Chapter 17. The cochlea is concerned with hearing. In man the cochlea is about 35 mm long; it is coiled spirally around a central bony pillar, called the modiolus, making two and three-quarter turns. Throughout its length it is divided into three chambers by two membranes: Reissner's membrane and the basilar membrane (Fig. 18.18). The upper chamber is called the scala vestibuli, which ends laterally at the oval window. The lower chamber is called the scala tympani, which ends laterally at the round window, which is closed by a secondary tympanic

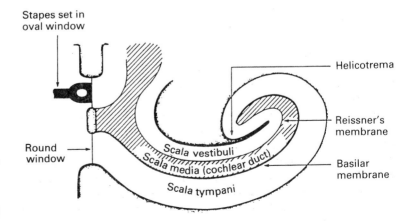

Fig. 18.18 Diagram of the inner ear. (Reproduced from Bray, J.J., Cragg, P.A., Macknight, A.D.C., Mills, R.G. & Taylor, D.W. (eds) (1989) *Lecture Notes in Physiology*, 2nd edn. Blackwell Scientific Publications, Oxford.)

Stapes set in oval window

Round window

Helicotrema

Reissner's membrane

Basilar membrane

Scala vestibuli

Scala media (cochlear duct)

Scala tympani

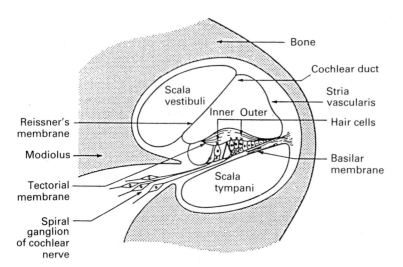

Fig. 18.19 Cross-section through the canals of the cochlea showing the organ of Corti (containing the hair cells) sitting on the basilar membrane. The inner and outer hair cells are separated by the tunnel of Corti. (Reproduced from Bray, J.J., Cragg, P.A., Macknight, A.D.C., Mills, R.G. & Taylor, D.W. (eds) (1989) *Lecture Notes in Physiology*, 2nd edn. Blackwell Scientific Publications, Oxford.)

membrane. Between the two membranes is the scala media, containing endolymph. Both the scala vestibuli and the scala tympani contain perilymph and communicate with each other at the apex of the cochlea (the helicotrema).

The receptors for hearing are hair cells, which are found in the organ of Corti (Fig. 18.19). The organ of Corti sits on the basilar membrane and extends from the base to the apex of the cochlea. Between the rods of Corti is the tunnel of Corti, which contains perilymph. There is a single row of inner hair cells medial to the tunnel of Corti and numbering about 3500. Lateral to the tunnel of Corti are three rows of outer hair cells, numbering about 20 000. The hair cells have processes that pierce the tough reticular lamina. The processes of the outer, but not inner, hair cells are embedded in the elastic ribbon-like tectorial membrane. Nerve fibres, with cell bodies in the spiral ganglion, branch extensively around the bases of the hair cells. There are 28 000 fibres in the auditory nerve. About 90–95% innervate the inner hair cells, while only 5–10% supply the more numerous outer hair cells.

THE AUDITORY PATHWAYS
The axons of the neurones that innervate the hair cells constitute the auditory division of the 8th nerve and synapse in the dorsal and ventral cochlear nuclei in the medulla (Fig. 18.20).

Second-order neurones arise from these and go to synapse in the olive and trapezoid body of the same and opposite side. Fibres then ascend in the lateral lemniscus to synapse bilaterally in the inferior colliculi and then in the medial geniculate bodies of the thalamus. Some collaterals from the lateral lemniscus go directly to the medial geniculate body, while other collaterals enter the reticular formation. From the medial geniculate body, fibres are finally projected bilaterally to the auditory cortex, Brodman's area 41, in the superior portion of the temporal lobe. There are auditory association areas adjacent to the primary auditory cortex. It should be noted that, once fibres leave the cochlear nuclei, they proceed bilaterally up to the cortex.

The auditory nerve also contains efferent fibres arising from the ipsilateral and contralateral superior olivary complex and ending around the bases of the outer hair cells in the organ of Corti; these fibres are referred to as the olivocochlear bundle.

SOUND
Sound, unlike light, requires a medium to travel in. It travels in air at the speed of 344 m/s and at a faster speed in water. Sound travels in the form of waves, which, in their simplest form, can be represented as sine waves (Fig. 18.21). There are two basic properties of sound waves:

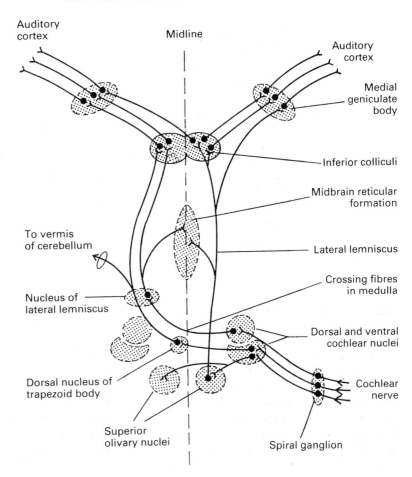

Fig. 18.20 Diagram of the main auditory pathways.

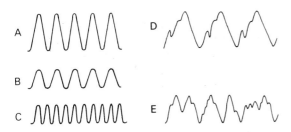

Fig. 18.21 Amplitude and frequency of sound waves. A: Record of a pure tone; B: same frequency (pitch) but lower amplitude (loudness) than A; C: same amplitude but greater frequency than B; D: shows a regularly repeated wave pattern, usually perceived as a musical sound; E: shows no regular wave pattern—this is perceived as noise.

(i) amplitude, which determines the *intensity* of sound or its *loudness*—the higher the amplitude, the greater the intensity of sound; and (ii) frequency or the number of cycles per second (Hz), which determines the pitch or tone of sound—the higher the frequency, the higher the pitch of sound. Loudness is not the same as intensity of sound. Intensity refers to the strength of the sound in physical terms, while loudness is a psychological term denoting appreciation of the sound intensity by the hearing apparatus. Loudness is affected by the frequency of the sound, and this is best illustrated by the use of dog whistles; these emit sounds of frequency higher than 20 000 Hz, which the dog can hear and respond to, but humans cannot hear sounds of such high frequencies and suffer no sensation of loudness, even though the intensity may be so high as to make the dog put its paws over its ears. Complex sounds with regularly repeating patterns are perceived as music, while sound waves of irregular frequency give a sensation of noise.

Sound intensity may be measured in terms of energy, in microwatts or ergs per second, or pressure, in dynes/cm^2. The range in intensity between the threshold of sound and the loudest sound that can be tolerated is huge. It would therefore be inconvenient to employ a linear absolute scale to measure sound intensity in everyday life. A comparative logarithmic scale is employed, in which the basic unit is the bel (after A.G. Bell):

$$\text{bel} = \log \frac{\text{intensity of sound}}{\text{intensity of standard sound}}$$

Since intensity is directly proportional to the square of the sound pressure:

$$\text{bel} = 2 \log \frac{\text{pressure of sound}}{\text{pressure of standard sound}}$$

The bel is too large for everyday use and the decibel is used instead:

decibel(dB) = 0.1 bel

$$\text{Intensity of sound in decibels} = 20 \log \frac{\text{pressure of sound}}{\text{pressure of standard sound}}$$

The standard sound adopted as reference is the auditory threshold in healthy young people; in physical terms it is 0.000 204 dynes/cm^2. The decibel scale is illustrated in Fig. 18.22. It should be noted that it is a logarithmic scale, in which 0 dB refers to the auditory threshold, while a sound of 120 dB indicates a sound intensity of 10^{12} times that of the intensity of the auditory threshold.

AUDIBILITY

The human audibility curve shows that humans hear sounds in the range of 20–20 000 Hz. Within this range, the threshold is lowest in the range 1000–4000 Hz and greater at lower and higher frequencies (Fig. 18.23). Pitch discrimination is best in the range 1000–3000 Hz. Human beings can distinguish up to 2000 pitches and training can even improve on this, as musicians do. In

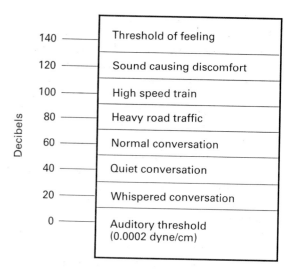

Fig. 18.22 Scale (in decibels) for common sounds in everyday life.

ordinary conversation the average male voice has a frequency of about 120 Hz while that of the female voice is about 250 Hz. Very loud sounds are felt as well as heard and are also partly conducted through the bones of the skull.

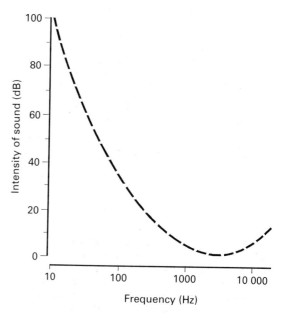

Fig. 18.23 Part of the human audibility curve showing the threshold of hearing at different frequencies.

THE MECHANISM OF HEARING

When sound waves strike the tympanic membrane, it vibrates with them but the vibration stops as soon as the sound ceases. Movement of the tympanic membrane is conducted through the ossicles to the oval window of the cochlea (Fig. 18.24). The ossicles constitute a system of levers which increases the force acting on the tympanic membrane by 1.3 times. Since the area of the tympanic membrane (about 50 mm^2) is approximately 17 times that of the oval window of the cochlea (3 mm^2), the pressure created by movement of the tympanic membrane is multiplied 17 times when it arrives at the oval window. This means that the force of sound vibration is multiplied 1.3 × 17 times or 22 times when it arrives at the oval window. Indeed, it has been calculated that about 60% of the sound energy is transmitted to the cochlea, which is a high efficiency when considering the inevitable loss due to resistance.

Basically, sound may be conducted to the cochlea either by air or through bone. Air conduction constitutes the normal situation, when sound waves travelling in air cause vibration of the tympanic membrane, which is transmitted by the ossicles to the oval window of the cochlea. In bone conduction the sound causes vibration of the bones of the skull, directly transmitting the sound vibrations to the cochlea. Under normal circumstances, air conduction is more efficient than conduction through bone.

As indicated in Fig. 18.24, the footplate of the stapes rocks at the oval window like a hinged door fixed at a posterior point. This movement causes a pressure wave, which travels through the perilymph in the scala vestibuli and, through vibration of Reissner's membrane, to the endolymph of the scala media; this, in turn, conveys the wave to the perilymph of the scala tympani, through the vibration of the basilar membrane, and the wave eventually dies away at the round window of the cochlea. Normally the basilar membrane is not under tension. Stimulation of the hair cells is effected through vibration of the basilar membrane, on which the organ of Corti sits.

Like the arterial pulse wave, travelling waves in the cochlea reach a peak and then decline, to dissipate at the round window. High-frequency sounds peak at the base of the cochlea, while low-frequency sounds peak at the apex of the cochlea, with displacement of the basilar membrane closely following this pattern (Fig. 18.25). Nerve fibres from each part of the basilar membrane are distinct and signal the place at which the wave has peaked to the auditory cortex. This is the main mechanism for discrimination of pitch or tone of sound and has been referred to as the *place theory*. It has been found that, in addition to this, sound frequencies below 2000 Hz are also signalled to the cortex by having one burst of action potential synchronized with each sound wave, i.e. a volley effect. The place theory plus the volley effect has recently been referred to as the *duplex theory* of hearing.

COCHLEAR MICROPHONIC AND THE ENDOCOCHLEAR POTENTIAL

Since the hair cells are sitting on the basilar membrane, its displacement moves the bodies of the hair cells, thus bending the processes of the

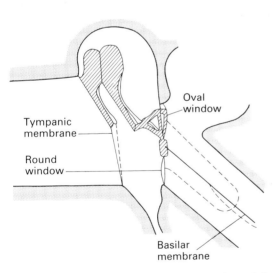

Fig. 18.24 The movements of the bony ossicles and the oval window are converted into waves in the fluids of the inner ear. The waves die out at the round window. The tympanic membrane, bony ossicles, oval window and basilar membrane are indicated. (Reproduced from Passmore, R. & Robson, J.S. (eds) (1976) *Companion to Medical Studies*, 2nd edn, Vol. 1. Blackwell Scientific Publications, Oxford.)

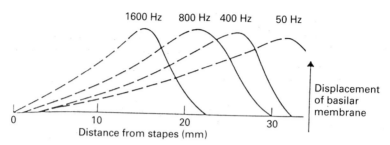

Fig. 18.25 Displacement of various parts of the basilar membrane produced by waves in the cochlear fluids. The frequencies of the waves are shown at the top of each wave. It is seen that each part of the membrane shows maximal displacement at a different frequency, depending on the distance from the stapes. This gave rise to the place theory of the discrimination of pitch (frequency). (Reproduced from Bray, J.J., Cragg, P.A., Macknight, A.D.C., Mills, R.G. & Taylor, D.W. (eds) (1989) *Lecture Notes in Physiology*, 2nd edn. Blackwell Scientific Publications, Oxford.)

hair cells, which are fixed by piercing the reticular lamina. Bending of the processes of the hair cells generates the cochlear microphonic, by which mechanical energy is directly transformed into electrical energy. The cochlear microphonic is directly proportional to the displacement of the basilar membrane and faithfully reproduces the characteristics of the sound. It is considered as the main component of the generator potential, leading to the action potential in auditory nerve fibres.

Perilymph, which is formed mainly from plasma, has a composition similar to that of the extracellular fluid, with high Na^+ and low K^+ concentrations, while the composition of endolymph is nearer to that of the intracellular fluid, with low Na^+ and high K^+ concentrations. Endolymph is formed by the stria vascularis which has a highly active Na^+, K^+-ATPase and an electrogenic K^+ pump. Studies with microelectrodes have shown that endolymph in the scala media has a positive potential (+ 80 mV) in relation to perilymph in either the scala vestibuli or the scala tympani; it is referred to as the *endocochlear potential*. Since the basilar membrane is relatively permeable to perilymph in the scala tympani, the tunnel of Corti and consequently the bases of the hair cells are bathed in perilymph; the processes of the hair cells are bathed in endolymph.

FUNCTION OF THE HAIR CELLS
It has recently been recognized that the inner

and outer hair cells have somewhat different functions. The processes of the inner hair cells are not embedded in the tectorial membrane and are actually bent by fluid movement under the tectorial membrane. It is now clear that the inner hair cells are the primary receptors for sound, transducing the fluid movement in the cochlea into action potentials in the auditory nerve. The outer hair cells are contractile and are supplied by efferent cholinergic fibres from the superior olivary complex, travelling in the olivocochlear bundle. Discharge in the olivocochlear fibres causes the outer hair cells to contract, which exerts tension on the basilar membrane. It seems that the outer hair cells are the means by which the basilar membrane is 'tuned', in a similar way to that used by musicians to tune string instruments.

ACTION POTENTIALS AND BRAIN MECHANISMS
Action potentials in the auditory nerve signal to the brain the intensity and frequency of sound. At low-threshold sound intensity, an auditory nerve fibre discharges to one frequency, depending upon which part of the basilar membrane it is coming from. With increasing intensity, the range of frequency carried widens; especially to frequencies below that at which threshold stimulation occurs and at the same time the rate of action potential spikes increases. This pattern is also exhibited by neurones in the cochlear nuclei in the medulla and thus reaches the auditory cortex. Neurones

in the auditory cortex respond to the onset, duration and repitition rate of sound, and especially to the direction it is coming from.

The basilar membrane is represented on the auditory cortex with low tones anterolaterally and high tones posteromedially. In addition to the perception of tone, the cortex is also essential for sound localization and for cognitive functions associated with hearing.

SOUND LOCALIZATION

The pinna transforms the front of the sound wave differently, according to the direction it is coming from. This constitutes a clue, which can be employed by a single ear. The main mechanism for sound localization, however, involves using the two ears. Differences in the time of arrival of the sound wave at the two ears, as well as the differences in loudness, by which the nearer ear hears louder than the other, are employed as clues for localization.

MASKING

The presence of a background sound affects the ability to hear another sound in the environment, i.e. background noise can mask the sound wholly or partly. Masking is due to the fact that the receptors are already stimulated and therefore refractory when the sound desired to be heard arrives. Masking is more effective when two sounds have similar rather than widely differing frequencies. Masking is present all the time in natural environments and is automatically taken into consideration, as when a speaker raises his voice in order to be heard.

EFFECT OF NOISE ON HEARING

Modern industrial societies are noisy and noise pollution is another environmental hazard. Exposure to sound intensities above 80 dB may result in damage to the outer hair cells depending upon the extent of the intensity and the duration of exposure. The damage usually starts at about 4000 Hz, and the subject is usually unaware of any deficit. The damage then affects lower frequencies and, when it reaches frequencies below 1000 Hz, the ability to hear normal conversation is impaired, inducing the victim to seek medical advice.

DEAFNESS

Deafness may be classified, according to its cause, into conduction deafness and nerve or perceptive deafness. *Conduction deafness* is due to impairment of sound transmission through the external ear and middle ear. It may result from too much wax or a foreign body in the external auditory canal, repeated middle ear infections, a perforated drum or otosclerosis. Otosclerosis is a disease characterized by pathological fixation of the stapes in the oval window. Since the masking effect of background noise is greatly reduced, sounds are heard better by bone rather than by air conduction and sound frequencies are uniformly affected. Nerve deafness may be congenital or may result from damage to the cochlea or auditory pathways. The cause may be toxic, e.g. administration of streptomycin, or due to inflammation, e.g. cerebrospinal meningitis. Vascular lesions in the medulla or tumours, e.g. acoustic neuroma, may interrupt the auditory pathways and lead to some degree of deafness. In this connection, it should be remembered that the auditory pathways are characterized by bilateral representation and that, in contrast to vision, cortical lesions hardly ever result in deafness. In nerve deafness, hearing by both air and bone conduction is depressed.

TESTS OF HEARING AND AUDIOMETRY

The audiometer is an instrument which is used to determine the auditory threshold for sound frequencies which may range from 50 Hz to 16 000 Hz. Pure tones are delivered through earphones (air conduction) or directly through a vibrating piece on the mastoid bone (bone conduction). The test is conducted in a soundproof, or at least very quiet, room. The responses are plotted on a chart (audiogram) where 0 dB is the normal threshold and any hearing loss in dB is indicated for each frequency. An audiogram gives a measure of the hearing loss, shows the frequencies affected and, by comparing air conduction with bone conduction, gives an important clue as to whether the patient is suffering from conduction deafness or nerve deafness.

Before the audiometer became available, some simple tests using only a tuning fork were devised to differentiate between conduction and nerve deafness. These tests are still used by doctors as part of their examination of the ear.

Weber's test

The base of a vibrating tuning fork is placed on the vertex of the skull. Normally, the sound should be heard equally well on both sides. In the case of unilateral conduction deafness, the sound is heard better (or lateralized) to the diseased side because of the absence of the masking effect of environmental noise on that side. In nerve deafness affecting one ear, the sound is lateralized to the normal side.

Rinne's test

The base of the vibrating tuning fork is placed on the mastoid process until the subject cannot hear by bone conduction; then the prongs of the fork are held in the air near the ear. Normally, since air conduction is better than bone conduction, the subject will hear the vibration in the air after bone conduction is over, and this is referred to as a positive Rinne's test. If the subject does not hear the vibration in the air, the test is repeated but reversing the sequence, i.e. starting with the tuning fork near the ear and, after hearing the vibration in the air is over, the base of the tuning fork is placed on the mastoid process. Normally, the vibration should not be heard through bone after hearing by air is over. If it is, this means bone conduction is better than air conduction and Rinne's test is said to be negative in this case. A negative Rinne's test is thus indicative of conduction deafness. A positive Rinne's test with hearing loss is indicative of nerve deafness.

Further reading

1 Ganong, W.F. (1991) Vision. In *Review of Medical Physiology*, 15th edn, pp. 136–157. Appleton & Lange, San Mateo, California.
2 Ganong, W.F. (1991) Hearing and equilibrium. In *Review of Medical Physiology*, 15th edn, pp. 158–170. Appleton & Lange, San Mateo, California.
3 Ganong, W.F. (1991) Smell and taste. In *Review of Medical Physiology*, 15th edn, pp. 171–177. Appleton & Lange, San Mateo, California.
4 Keele, C.A., Neile, E. & Joels, N. (eds) (1982) *Samson Wright's Applied Physiology*, Part VI: *The Special Senses*, 13th edn, pp. 370–394. Oxford University Press, Oxford.

19: An Introduction to Statistics

Objectives

On completion of the study of this section, the student should be able to:

1 Define commonly used statistical terms in order to understand data presented in health statistics reports and research publications.

2 Describe the various numerical and graphical methods so as to use them for presentation of quantitative and qualitative data. For this purpose the student should be able to recognize and draw the following:
   (a) Histogram.
   (b) Scatter diagram.
   (c) Bar diagram.

3 Become familiar with commonly used tests of significance so as to use them in the appropriate situations. For this purpose the student should:
   (a) Appreciate the meaning of statistical significance.
   (b) Know the meaning of a P-value.
   (c) Calculate a t-value.
   (d) Calculate a chi-square value.

Introduction

Statistics is an important tool in medicine and biology. Knowledge of statistics helps students and doctors to understand medical literature. It is also essential for the design of research projects and the analysis and interpretation of observations. All counts and measurements done on human beings are very variable. Individuals differ in sex, weight, pulse rate, haemoglobin levels and many other variables. The function of statistics is to help in the analysis and presentation of data. As an example, one may imagine a doctor who has measured the height of a large number of children. If he records the results as they are and tries to show them to others, it will be very difficult for him to draw conclusions from these and difficult for others to understand his observations. By applying statistical techniques to his data, he will find it easier to draw conclusions from them and others will find it more interesting and simpler to understand his data.

In this chapter, some basic statistical methods are described. However, it would be useful first to review briefly the types of data that are encountered in medical sciences.

Types of data

Data in medicine are of two main types:

1 *Quantitative data.* These are data of observations or things that can be measured and/or counted. Examples of measurements are height, weight, blood-pressure, skin-fold thickness, haemoglobin, electrolytes and liver function tests. Examples of counts are total white blood cell count, total red cell count, pulse rate, respiratory rate and parasite ova count.

2 *Qualitative data.* These are only counted, never measured. In describing the incidence of a disease, one can count the number of persons with that disease, e.g. 500 cases of malaria, 100 cases of diabetes, etc. It is also applicable for persons sharing a common characteristic other than a disease state, e.g. 150 African men, 200 Europeans or 100 males and 90 females.

Samples

Data are usually collected from a sample of the population. It is important to bear in mind that the word 'population' in statistics does not refer to people only. It may be used for insects or animals, e.g. the population of mosquitoes and rats, or for things, e.g. a population of automobiles or a population of hospital beds.

The purpose of collecting data from a sample of the population rather than observing the whole population is to reduce the cost and effort in data-gathering while achieving relatively accurate results. There are several types of samples:

1 *The simple random sample.* In this type, each unit is chosen in a random position. Statistical reference books give tables of random numbers which are used for selection of the sample. This ensures that each unit in the population from which the sample was drawn has a fair chance of being selected. In this way, the sample will be fairly representative of the population. The simple random sample is useful in cases of populations with relatively small numbers.

2 *The stratified random sample.* In this type, the population is first divided into homogeneous non-overlapping strata or subgroups. A simple random sample is drawn from each stratum and all simple random samples are added together to give one stratified random sample.

Example: A stratified random sample of workers in a city can be selected as follows. The workers are classified according to occupation. From each occupational subgroup a simple random sample is chosen. All such samples are added up to give a stratified random sample.

3 *Other types of samples.* These include the multistage sample, the systematic sample, the quota sample and the purposive sample.

Presentation of data

GRAPHICAL PRESENTATION OF QUANTITATIVE DATA

After collection, quantitative data are presented in the form of a frequency distribution in a table or a diagram. The following is an example to illustrate this:

Table (19.1) gives results of measurements of

Table 19.1 Systolic blood-pressure measurement for 198 people

105	150	110	135	130	110	110	110	95	130	105	135
130	150	135	130	130	135	140	110	125	110	130	120
120	120	125	135	130	135	125	120	120	125	130	120
130	95	145	135	125	135	120	135	130	120	125	125
120	120	120	110	125	120	130	155	125	140	100	140
130	120	110	120	130	130	125	125	140	130	100	120
110	135	100	125	125	125	114	130	120	120	115	120
120	120	130	120	120	120	115	110	125	125	120	115
135	130	120	115	140	110	105	125	120	110	125	125
125	115	110	100	125	115	105	110	120	105	115	120
130	135	120	130	110	125	110	130	115	115	110	125
120	110	120	114	105	125	120	125	125	120	110	125
95	120	120	115	110	110	145	125	120	135	115	95
130	120	120	110	110	125	125	120	140	95	120	115
120	110	130	120	125	120	110	140	115	110	130	100
125	115	110	130	120	120	110	135	110	115	125	120
125	110	115	120	125	110						

systolic blood-pressure (SBP) in 198 people:

These measurements of SBP can be better presented in a frequency distribution table (Table 19.2). The word 'frequency' means the number of times the same measurement (or range of measurements) is observed. For example, the SBP in the interval between 110 and 119 mmHg is found in 49 persons.

The same data can be presented in a diagram, known as a histogram, in which the intervals of measurement are plotted on the x-axis and the frequencies (or number of occurrences) are plotted on the y-axis.

The columns forming the frequencies are drawn

Table 19.2 Frequency distribution of systolic blood-pressure

SBP in mmHg	Frequency
90–99	5
100–109	11
110–119	49
120–129	81
130–139	40
140–149	9
150+	3
Total	198

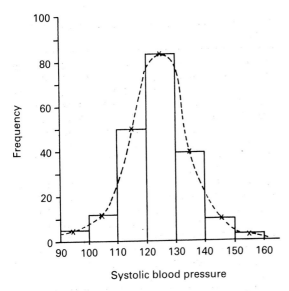

Fig. 19.1 A histogram (and a polygon) showing the frequency of systolic blood-pressure values in 198 subjects.

side by side with no gaps between them (Fig. 19.1). If the midpoints of the tops of the frequency columns in a histogram are joined by a continuous line (shown by dots in Fig. 19.1), then a frequency diagram (or polygon) will be formed.

MATHEMATICAL PRESENTATION OF QUANTITATIVE DATA

Mathematical presentation of quantitative data utilizes the measures of central tendency and measures of scatter. As we have seen above, the frequency distribution is a complete summary of data. Measures of central tendency and scatter form a further step in summarizing the data. We will consider them separately.

Measures of central tendency

In any number of measurements, there is a central tendency. Imagine a number of horses in a race: one or two horses will be in front, one or two at the back and the rest crowded in the middle. In all kinds of observations, the majority of measurements crowd in the middle and this is what is known as the central tendency. The central tendency (or measurement of the middle) is measured by three values: mean, median and mode. The

mean (also known as the arithmetic mean) is the average value. It is calculated as the sum of observations divided by their number. It is usually denoted by an x with a dash on top of it, i.e. $\bar{x}$. It is calculated as follows:

$$\bar{x} = \frac{\Sigma x}{n}$$

where x represents individual measurement, Σ is the sum of, and n is the number of observations.

In the example of systolic blood-pressures, the sum of the observations was 23 963 and there were 198 readings of SBP. The mean SBP will be:

$$\bar{x} = \frac{\Sigma x}{n} = \frac{23\,963}{198} = 121.03 \approx 121 \text{ mmHg}$$

The median is defined as the value that divides a set of observations into an upper half and a lower one.

The mode is defined as the most frequently occurring value in a set of observations.

The median and mode are illustrated in the following example of SBP readings:

105, 120, 120, *125*, 130, 135, 140

The median is 125, with three observations above and three observations below it. The mode is 120 which occured twice in the above sample values. If the number of observations (n) is even, then we obtain the median by averaging the two middle values, as in the following example:

100, 105, 120, *120*, *125*, 130, 135, 140

The median in this case is 122.5 or 123. Note that, in order to calculate the median of a relatively small series of measurements, they are arranged in an ascending or descending order. Calculation of the median for a very large sample will require a different approach, which is beyond the scope of this text.

Measures of dispersion (scatter)

A measure of dispersion illustrates how far the observations deviate from the mean. After obtaining the mean of a group of measurements, one would still want to know how representative that mean is of the whole group, i.e. how far that mean can give a summary of these measurements.

It is generally accepted that, if the degree of variation between all the measurements and their mean can be calculated, it will give a measure (or index) of how far the mean is representative of the whole sample. The smaller the index of variation, the more representative the mean is. There are three such indices or measures of variation: the range, the variance and the standard deviation.

To illustrate these measures, let us consider the following sample of SBP measurements:

105, 120, 120, 125, 125, 125, 125

In this example, the range will be the lowest and highest value, i.e. 105–125 mmHg. The range is not a good measure of dispersion because it does not refer to the mean or tell us where the other observations lie between the lowest and highest ones.

The variance and standard deviation are indices that give information on the degree to which the mean represents the individual measurements. The variance is the square of the standard deviation. The variance is calculated as the sum of the squares of the difference of each observation from the mean, divided by the number of observations. This applies to the above example as follows:

Mean = 120.7 or 121 mmHg

$$\text{Variance} = (105 - 121)^2 + (120 - 121)^2 + \\ (120 - 121)^2 + (125 - 121)^2 + \\ (125 - 121)^2 + (125 - 121)^2 + \\ (125 - 121)^2 = \frac{322}{6} = 53.7$$

The standard deviation (SD) is the square root of the variance, i.e.:

$$SD = \sqrt{\frac{322}{6}} = \sqrt{53.7} = 7.33$$

Now, we can express these values in the form of:

$$\text{Variance} = \frac{\Sigma(x - \bar{x})^2}{n - 1} \tag{1}$$

$$\text{Standard deviation} = \sqrt{\frac{\Sigma(x - \bar{x})^2}{n - 1}} \tag{2}$$

where in both cases Σ is the sum of, x is each individual observation, $\bar{x}$ is the mean of a set of observations and n is the number of observations.

The formula can be rewritten in the following way:

$$\text{Variance} = \frac{fx^2 - \dfrac{\Sigma(fx)^2}{n}}{n - 1} \tag{3}$$

$$\text{Standard deviation} = \sqrt{\frac{fx^2 - \dfrac{\Sigma(fx)^2}{n}}{n - 1}} \tag{4}$$

The reason why we divide by $n-1$ instead of n is that, in this way, we obtain a better estimate of the sample standard deviation and variance. Formula (2) is used in the case of a small number of observations while formula (4) is used to calculate the standard deviation from a large number of observations, usually presented in the form of a frequency distribution table.

The variance and the standard deviation are used as part of the data summary to give more information about the mean. Moreover, they have an important role in more advanced statistical analysis techniques.

STANDARD ERROR OF THE MEAN

The standard error of the mean is a value which measures the degree to which the mean of a sample is representative of the population. It is calculated as:

$$\text{Standard error of the mean} = \frac{\text{standard deviation}}{\sqrt{n}} \tag{5}$$

where n is the size of a sample (also equal to the number of observations in a sample).

If the sample size is large, the value of the standard error becomes smaller indicating that the value of a sample mean is nearer to the population mean.

Normal distribution

Any biological quantitative data, when plotted as a frequency histogram, assume what is called the normal distribution or appear similar to it. It is called normal because it is the distribution normally found in biological measurements, not because it represents normal (or reference) values

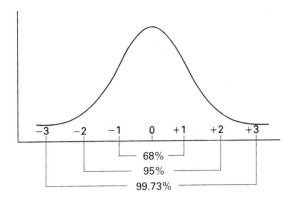

Fig. 19.2 The shape of the normal curve.

(see Fig. 19.2). Its shape is similar to a bell and has the following properties:

1 It has two symmetrical halves if a line is drawn vertically at its middle.

2 The mean, median and mode are equal in a normal distribution and they all coincide at the midpoint of the curve.

Most biological measurements conform to the properties of the normal distribution. If they do not, they can be made to do so by mathematical methods known as transformations. All normal distributions in biology can be related or approximated in a standard normal distribution. The standard normal distribution has a mean of zero and a standard deviation equal to 1 (see Fig. 19.2). The standard normal distribution of a large number of samples can be presented in a table known as the *t*-table, which is available in all textbooks of statistics. Now, if one considers the areas under the standard normal curve, the following facts apply (see Fig. 19.3a, b and c):

1 That the mean ±1 standard deviation includes about 68% of the area under the curve.

2 That the mean ±2 standard deviations includes about 95% of the area under the curve.

3 That the mean ±3 standard deviations includes 99.73% of the area under the curve.

To illustrate the usefulness of these facts, consider the following example. The results of measurements of body temperature in a sample of normal young adults were given as follows: mean = 36.7 °C (with a standard deviation ±0.2 °C). If the body temperature of a young adult is found to be 37.4 °C (i.e. more than 3 standard

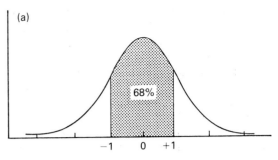

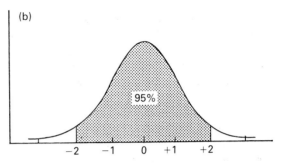

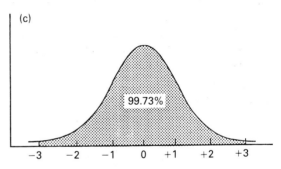

Fig. 19.3 The normal curve: area under the curve (a) within 1 standard deviation, (b) within 2 standard deviations and (c) within 3 standard deviations.

deviations above the mean), there is less than one chance in a hundred that this young adult belongs to a population of adults with normal body temperature. It is quite reasonable, in fact, to assume that this subject has an abnormal temperature, but this assumption would be wrong at least once in every hundred times it was made.

Probability and statistical significance

In mathematics, probability (*P*) is a measure of likelihood of an event occurring. It has a scale of 0–1 (i.e. none to 100%). Therefore, it is always a fraction within this scale, e.g. $P = 0.5$, $P = 0.05$, $P = 0.001$, etc. In statistics, probability is used to measure the likelihood of occurrence of an event by chance. This is the basis of what is known in statistics as the null hypothesis. The null hypothesis means that you assume that any event has occurred due to chance only. Then, by calculating a P value, you will be able to measure the likelihood of chance. If the value of calculated P is small, then the likelihood of chance is remote and hence the result is referred to as significant. The conventional value of P at which the likelihood of chance is considered as remote and hence the result is considered as significant is P equal to or less than 0.05. The smaller the P value than 0.05, the more significant the result is. Hence, a P value of 0.001 is more significant than a P value of 0.01 which, in turn, is more significant than a P value of 0.05.

As far as quantitative data are concerned, we shall consider the applications of probability and statistical significance in a situation where there are two samples, each of which has a mean and a standard deviation. The question is: do the two means differ significantly or not?

Example: In two samples A and B of systolic blood-pressure measurements, sample A is composed of 28 readings with a mean systolic BP of 120 mmHg and a standard deviation of 12 mmHg. Sample B is composed of 28 measurements with a mean systolic BP of 119 mmHg and a standard deviation of 10 mmHg. Do the two means differ significantly from each other?

The answer to this question is to perform a test of significance. According to this test, a P value is obtained which will confirm whether the difference between the two means is significant or not. For quantitative data and for this situation mentioned in the above example, the t-test is used.

THE t-TEST
The t-test depends on calculation of a t-value for which there is a corresponding P value read from t-distribution tables. The t-value is the ratio of the difference between two means to the standard error of the difference between means:

The difference between the two means $= \bar{x}_1 - \bar{x}_2$
Standard error of the difference between the two

$$\text{means} = \sqrt{\frac{SD_1^2}{n_1} + \frac{SD_2^2}{n_2}}$$

$$t = \frac{\bar{x}_1 - \bar{x}_2}{\sqrt{\dfrac{SD^2}{n_1} + \dfrac{SD^2}{n_2}}} \tag{6}$$

Regarding our example above:

mean of the first sample $\bar{x}_1 = 120$ mmHg
standard deviation of the first sample $SD_1 = 12$ mmHg
size of the first sample $n_1 = 28$
mean of the second sample $\bar{x}_2 = 115$ mmHg
standard deviation of the second sample $SD_2 = 10$ mmHg
size of the second sample $n_2 = 28$

$$t = \frac{\bar{x}_1 - \bar{x}_2}{\sqrt{\dfrac{SD_1^2}{n_1} + \dfrac{SD_2^2}{n_2}}} = \frac{120 - 115}{\sqrt{\dfrac{12 \times 12}{28} + \dfrac{10 \times 10}{28}}}$$

$$= 1.6938$$

The degree of freedom in the above example
$= (n_1 - 1) + (n_2 - 1)$
$= (28 - 1) + (28 - 1)$
$= 54$

So, in a t-distribution table, the t-value of 1.6938 is read against the degree of freedom (which, in this case, is 54). It is found to correspond to a P value greater than 0.05, which is not *significant*. Hence, according to the t-test, the difference between the two means is not significant.

CORRELATION
Correlation is the study of a relationship between two variables, i.e. to study whether variable y increases as x increases or y decreases as x increases. Examples of this are height and weight, weight and blood-pressure, etc. Note that x is usually the independant variable and y the dependent variable.

Correlation is illustrated by a graph known as a scatter diagram (see Figs 19.4a, b and c). In a scatter diagram, points which represent x and y are plotted. A straight line is drawn which passes through most of the points. The types of correlation obtained usually fall approximately into one of the following types:

1 *Positive correlation*: when the graph demon-

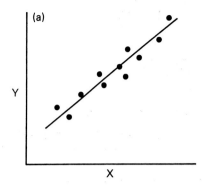

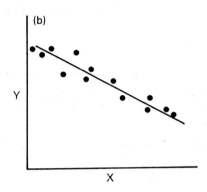

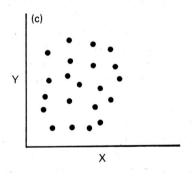

Fig. 19.4 Graphs showing (a) positive correlation, (b) negative correlation and (c) no correlation.

strates a direct relationship, i.e. *y* increases as *x* increases (Fig. 19.4a).

2 *Negative correlation*: when the graph demonstrates an inverse relationship, i.e. *y* decreases as *x* increases (Fig. 19.4b).

3 *No correlation at all*: when the points in a scatter graph cannot be joined by a straight line (Fig. 19.4c).

The correlation coefficient (r)

Correlation can be expressed mathematically as a value known as the correlation coefficient, denoted by the letter *r*. The formula for calculation of *r* is as follows:

$$r = \frac{(x - \bar{x})(y - \bar{y})}{(x - \bar{x})^2(y - \bar{y})^2}$$

where *x* is any individual value of the first quantitative variable, *y* is any individual value of the second quantitative variable, $\bar{x}$ is the mean of all values of *x* and $\bar{y}$ is the mean of all values of *y*.

Values of *r* lie between 0 (i.e. no correlation) and 1.0 (perfect correlation). When *r* is 0.5 or more the correlation is significant. The exact level of significance can be obtained from statistical tables.

Correlation is a study of relationships but does not necessarily give proof for a cause-and-effect relationship; e.g. one may obtain a correlation between watching television and cancer of the lung, but such correlation does not mean that cancer is caused by watching television!

Qualitative data

Qualitative data are those that are counted but not measured. The following are illustrative examples:

1 Blood groups: numbers of persons are counted as having blood group A, B, AB, or O.

2 Persons with a disease under treatment are described as cured or not cured.

3 Persons may be counted according to sex, i.e. males or females, or according to occupations, e.g. workers, soldiers, doctors, businessmen, clerks, etc.

PRESENTATION OF QUALITATIVE DATA

Qualitative data can be presented as follows:

1 *Numerically*. In the form of numbers, percentages or proportions.

2 *Graphically*. The following are two commonly used types of graphs:

   (a) The bar diagram (Fig. 19.5).

   (b) The pie diagram (Fig. 19.6).

Example: In a group of medical students of 200, the blood groups are distributed as follows: 70 students were group A, 80 were group B, 30 were

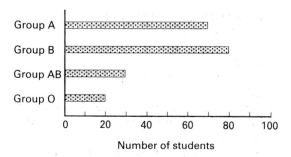

Fig. 19.5 Bar diagram showing blood group distribution in 200 medical students.

group AB and 20 were group O. This example is shown in the form of a bar diagram (Fig. 19.5) and as a pie diagram (Fig. 19.6).

STATISTICAL SIGNIFICANCE APPLIED TO QUALITATIVE DATA: THE CHI-SQUARE TEST

A common situation regarding qualitative data is to compare the results of two types of treatment or drugs on two groups of persons. The aim is to check whether one treatment is superior to the other. A test commonly used for this purpose is the chi-square test.

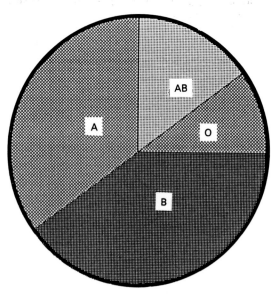

Fig. 19.6 Pie diagram showing blood group distribution in 200 medical students.

Example: In a trial of two drugs A and B for treatment of a certain disease, a total of 20 patients were divided into two groups. The first group consisted of 11 patients, who were given drug A. The second group consisted of 9 patients, who were given drug B. Five patients receiving drug A were cured and only 2 receiving drug B were cured. Which drug is superior in treating the disease?

The apparent answer is drug A. Let us do a test of significance before making a judgement. The chi-square (χ^2) test principle is to compare the observed values with those that are expected, i.e.:

$$\chi^2 = \Sigma \frac{(O - E)^2}{E}$$

where χ is the Greek letter chi, O is the observed values, E is the expected values and Σ is the sum of.

The χ^2 value is used to obtain a P value from χ^2 tables with the use of degrees of freedom as in the t-test. The application of the chi-square test on the above example is as follows.

The observed and expected values are given in Table 19.2. The horizontal lines in Table 19.3 are known as rows and the vertical lines as columns. The observed values are the values given in the example. The expected values are calculated as follows. An expected value corresponding to 5 cured patients treated with drug A is calculated by multiplying the row total of drug A by the column total of all patients cured (given drug A

Table 19.3 Observed and expected values

	Cured	Not cured	Total
Observed values			
Drug A	5	6	11
Drug B	2	7	9
Total	7	13	20
Expected values			
Drug A	4	7	11
Drug B	3	6	9
Total	7	13	20

and B) and dividing the result by the grand total as follows:

$$\frac{11 \times 7}{20} \approx 4$$

The expected value corresponding to 6 patients not cured by drug A in the observed values is calculated similarly as:

$$\frac{11 \times 13}{20} \approx 7$$

$$\text{Expected} = \frac{\text{Row total} \times \text{column total}}{\text{Grand total}}$$

The chi-square value is calculated by deriving the difference of each observed and expected value, squaring it, dividing by the expected value and adding them all together. In the above example this will be as follows:

$$\chi^2 = \frac{(5-4)^2}{4} + \frac{(6-7)^2}{7} + \frac{(2-3)^2}{3} + \frac{(7-6)^2}{6} \approx 0.9$$

The degrees of freedom = (number of rows − 1) × (number of columns − 1) = (2 − 1) × (2 − 1) = 1.

Hence, from a χ^2 table, a χ^2 value of 0.9 read against a degree of freedom of 1 corresponds to a P value greater than 0.1, which is not significant.

Therefore, drug A is not superior to drug B in treating the disease.

Summary

1 Data are classified as quantitative or qualitative.

2 Quantitative data are presented by frequency of occurrence, e.g. distribution tables, histograms or polygons.

3 The relationship between two quantitative sets of data is studied by looking for correlation. A correlation coefficient (r) is calculated. If r is greater than 0.5, it is significant.

4 The t-test is used to analyse quantitative data. The test compares two means for significance of difference. Calculated t-values are looked up in statistical tables to obtain P values. When P is less than 0.05, the difference is significant.

5 Qualitative data are presented as whole numbers or counts, percentages or proportions. They can be graphically presented as bar diagrams or pie diagrams.

6 The chi-square test is used to analyse qualitative data. Calculated values of χ^2 are looked up in statistical tables to obtain P values.

Further reading

1 Bradford Hill, A. (1971) *Principles of Medical Statistics*. The Lancet, London.

Appendix 1

Ranges of reference values in human whole blood (B), plasma (P) or serum (S)

| Determination | Reference value (varies with procedure used) | |
	Traditional units	SI units
Acetoacetate (S)	<1.0 mg/dl	<0.1 mmol/l
Adrenocorticotrophin (ACTH) (P)	15−70 pg/ml	3.3−15.4 pmol/l
Aldosterone (P)	3−10 ng/dl	83−277 pmol/l
Alpha-amino nitrogen (P)	3.0−5.5 mg/dl	2.1−3.9 mmol/l
Ammonia (B)	12−55 μmol/l	12−55 μmol/l
Amylase (S)	4−25 units/ml	
Ascorbic acid (B)	0.4−1.5 mg/dl (fasting)	23−85 μmol/l
Bilirubin (S)	Conjugated (direct): up to <0.2 mg/dl	<3.4 μmol/l
	Total (conjugated plus free): up to 1.0 mg/dl	up to 17 μmol/l
Blood volume (B)	8.5−9.0% of body weight in kg	80−85 ml/kg
Calcitonin (S)	Male: 0−14 pg/ml	0−4.1 pmol/l
	Female: 0−28 pg/ml	0−8.2 pmol/l
Calcium (total) (S)	8.5−10.5 mg/dl	2.1−2.6 mmol/l
Carbon dioxide content (S)	24−30 meq/l	24−30 mmol/l
Carotenoids (S)	0.8−4.0 μg/ml	1.5−7.4 μmol/l
Ceruloplasmin (S)	27−37 mg/dl	1.8−2.5 μmol/l
Chloride (S)	100−106 meq/l	100−106 mmol/l
Cholesterol (S)	120−220 mg/dl	3.1−5.7 mmol/l
Cholesteryl esters (S)	60−75% of total cholesterol	
Copper (total) (S)	100−200 μg/dl	16−31 μmol/l
Cortisol (a.m.) (S)	5−25 μg/dl	140−690 nmol/l
Cortisol (p.m.) (S)	3−16 μg/dl	83−441 nmol/l
Creatinine (S)	0.6−1.5 μg/dl	53−133 μmol/l
Follicle-stimulating hormone (FSH) (S)	Male: 3−18 mU/ml	3−18 arb. unit
	Female: 4.6−22.4 mU/ml	4.6−22.4 arb. unit
Glucose (true) (B)	70−110 mg/dl	3.9−5.6 mmol/l
Growth hormone (S)	Male: <5 ng/ml	<233 pmol/l
	Female: 0−30 ng/ml	0−1395 pmol/l
Insulin (fasting) (S)	6−26 μunits/ml	43−187 pmol/l
Iodine, protein-bound (S)	3.5−8.0 μg/dl	0.28−0.63 μmol/l
Iron (S)	50−150 μg/dl	9.0−26.9 μmol/l
Lactic acid (B)	5.4−17.1 mg/dl	0.6−1.9 mmol/l
Lipase (S)	Up to 2 units/ml	4.5−10 g/l
Lipids, total (S)	450−1000 mg/dl	0.8−1.3 mmol/l
Luteinizing hormone (LH) (S)	Male: 3−18 mU/ml	3−18 arb. unit
	Female: 2.4−34.5 mU/ml	2.4−34.5 arb. unit
Magnesium (S)	1.5−2.0 meq/l	0.8−1.3 mmol/l
Non-protein nitrogen (S)	15−35 mg/dl	10.7−25.0 mmol/l
Osmolality (S)	280−296 mosm/kg water	280−296 mmol/kg

P_{CO_2} (arterial) (B)	35−45 mmHg	4.7−6.0 kPa
Pepsinogen (P)	120−140 ng/ml	120−140 μg/l
pH (B)	7.35−7.45	
Phenylalanine (S)	0−2 mg/dl	0−120 μmol/l
Phosphatase, acid, total (S)	Male: O.13−0.63 sigma units/ml	36−175 nmol.s^{-1}/l
	Female: 0.01−0.56 sigma units/ml	3−156 nmol.s^{-1}/l
Phosphatase, alkaline (S)	13−39 IU/l (adults)	0.22−0.65 μmol.s^{-1}/l
Phospholipids (S)	9−16 mg/dl as lipid phosphorus	2.9−5.2 mmol/l
Phosphorus, inorganic (S)	3.0−4.5 mg/dl (infants in first year, up to 6.0 mg/dl)	1−1.5 mmol/l
P_{O_2} (arterial) (B)	75−100 mmHg	10.0−13.3 kPa
Potassium	3.5−5.0 meq/L	3.5−5.0 mmol/l
Prolactin (S)	2−15 ng/ml	0.08−6.0 nmol/l
Progesterone (S)	Male: <1.0 ng/ml	<3.2 nmol/l
	Female: 0.2−0.6 ng/ml (follicular phase)	0.6−1.9 nmol/l
	0.3−3.5 ng/ml (midcycle peak)	0.95−11 nmol/l
	6.5−32.2 ng/ml (postovulatory)	21−102 nmol/l
Protein:		
Total (S)	6.0−8.4 g/dl	60−84 g/l
Albumin (S)	3.5−5.0 g/dl	35−50 g/l
Globulin (S)	2.3−3.5 g/dl	23−35 g/l
Pyruvic acid (P)	0−0.11 meq/l	0−110 μmol/l
Sodium (S)	135−145 meq/l	135−145 mmol/l
Sulphate (S)	2.9−3.5 mg/dl	0.3−0.36 mmol/l
Testosterone (P)	Male: 300−1100 ng/ml	10.4−38.1 nmol/l
	Female: 25−90 ng/ml	0.87−3.12 nmol/l
Thyroid-stimulating hormone (TSH) (S)	0.5−5.0 μU/ml	0.5−5.0 arb. unit
Total tri-iodothyronine (T_3) (S)	75−195 ng/dl	1.16−3.00 nmol/l
Total thyroxine (T_4) (S)	4−12 μg/dl	52−154 nmol/l
Transaminase (SGOT) (S)	7−27 U/l	0.12−0.45 μmol.s^{-1}/l
Triacylglycerol (fasting) (S)	40−150 mg/dl	0.4−1.5 mmol/l
Urea nitrogen (BUN) (B)	8−25 mg/dl	2.9−8.9 mmol/l
Uric acid (S)	3.0−7.0 mg/dl	0.18−0.42 mmol/l
Vitamin A (S)	0.15−0.6 μg/ml	0.5−2.1 μmol/l

Mainly drawn from Scully, R.E., McNeely, B.U. & Mark, E.J. (1986) Normal reference laboratory values. *N. Engl. J. Med.* **314**, 39.

Appendix 2

The SI units

SI stands for the Système International d'Unités (International System of Units) used in scientific reporting. In accordance with the decision of several scientific societies to employ a universal system of metric nomenclature, the SI system of units has been introduced.

The fractions and multiples used in the SI system (and their abbreviations) are given below.

FRACTIONS AND MULTIPLES

deci-	(d)	$= 10^{-1}$	kilo-	(k)	$= 10^3$
centi-	(c)	$= 10^{-2}$	mega-	(M)	$= 10^6$
milli-	(m)	$= 10^{-3}$	giga-	(G)	$= 10^9$
micro-	(µ)	$= 10^{-6}$	tera-	(T)	$= 10^{12}$
nano-	(n)	$= 10^{-9}$			
pico-	(p)	$= 10^{-12}$			

UNITS OF CONCENTRATION

Physical and chemical properties and consequently the biological activity of substances depend on the number of molecules in a solution. In SI units molar concentrations, which are based on the number of molecules per litre of fluid, are used instead of concentrations based on weight per unit volume. Thus:

$$1 \text{ Mole of a substance} = \frac{\text{Mass (g)}}{\text{Molecular weight}}$$

Moles depend on the number of atoms. The larger the molecular weight, the lower the molar concentration. In the traditional system, blood glucose, urea and cholesterol concentrations were expressed as 100 mg/100 ml, 30 mg/100 ml and 200 mg/100 ml respectively. When converted into SI units (i.e. mmol/litre) they all become 5 mmol/litre.

UNITS OF VOLUME

Litre	(l)	Centilitre	(cl)
Decilitre	(dl)	Millilitre	(ml)

UNITS OF MASS

Kilogram	(kg)
Gram	(g)
Milligram	(mg)

UNITS OF PRESSURE

The basic unit of pressure as expressed in SI units is the newton per square metre (Nm^{-2}). This unit is also known as the pascal (Pa), and is usually expressed in 1000 pascals or kilopascals. In clinical practice, however, blood-pressure is usually measured using a mercury manometer and expressed in mm of mercury. As the pascal is too small for clinical use, the kilopascal is used. The relationship between the kilopascal and mmHg is as follows: 1 kPa = 7.5 mmHg (see Table A2.1) and 0.133 kPa = 1.0 mmHg.

UNITS OF ENERGY

The traditional unit of heat and energy is the calorie and kilocalorie (kcal). In SI units, the joule (J) is used to represent the unit of both heat and energy. Like the pascal, calories are too small for clinical use, and kilocalories are used. In the SI system they are converted to the kilojoule (kJ), corresponding to 1000 joules, and the megajoule (MJ), corresponding to 1 000 000 joules. In SI units: 1 kcal = 4.2 kJ.

Table A2.1 Relationship between the kilopascal and millimetres of mercury

kPa	mmHg	kPa	mmHg
1	7.5	11	82.5
2	15	12	90
3	22.5	13	97.5
4	30	14	105
5	37.5	15	112.5
6	45	16	120
7	52.5	17	17.5
8	60	18	135
9	67.5	19	142.5
10	75	20	150

Bibliography

Åstrand, O. & Rodahl, K. (eds) (1986) *Textbook of Work Physiology*, 3rd edn. McGraw-Hill, New York.

Bray, J.J., Cragg, P.A., Macknight, A.D.C., Mills, R.C. & Taylor, D.W. (1989) *Lecture Notes in Physiology*, 2nd edn. Blackwell Scientific Publications, Oxford.

Carpenter, R. (1990) *Neurophysiology*, 2nd edn. Edward Arnold, Sevenoaks, Kent.

Edwards, C.R.W. & Bouchier, I.A.D. (eds) (1991) *Davidson's Principles and Practice of Medicine*, 16th edn. Churchill Livingstone, Edinburgh.

Emslie-Smith, D., Paterson, C.R., Scratcherd, T. & Read, N.W. (eds) (1988) *BDS Textbook of Physiology*, 11th edn. Churchill Livingstone, Edinburgh.

Firkin, F., Chesterman, C., Pennington, D. & Rush, B. (eds) (1989) *de Gruchy's Clinical Haematology in Clinical Practice*, 5th edn. Blackwell Scientific Publications, Oxford.

Ganong, W.F. (1991) *Review of Medical Physiology*, 15th edn. Appleton & Lange, San Mateo, California.

Guyton, A.C. (1991) *Textbook of Medical Physiology*, 8th edn. Saunders Co, Philadelphia.

Newsholme, E.A. & Leech, A.R. (1988) *Biochemistry for Medical Sciences*, 2nd edn. John Wiley & Sons, Chichester.

Passmore, R. & Eastwood, M.A. (1986) *Human Nutrition and Dietetics*, 8th edn. Churchill Livingstone, Edinburgh.

Patton, H.D., Fuchs, A.F., Hille, B., Scher, A.M. & Steiner, R. (1989) *Textbook of Physiology*, 21st edn. Saunders, Philadelphia.

Thomas, J.H. & Gillham, B. (1989) *Will's Biochemical Basis of Medicine*. Wright, London.

Weatheral, D.J., Ledingham, J.G.G. & Warrell, D.A. (eds) (1987) *Oxford Textbook of Medicine*, 2nd edn. Oxford University Press, Oxford.

Whitby, L.G. & Beckett, G.J. (1988) *Lecture Notes on Clinical Chemistry*, 4th edn. Blackwell Scientific Publications, Oxford.

Index